AF767865

The Philosophy and Practice of Medicine and Bioethics

INTERNATIONAL LIBRARY OF ETHICS, LAW, AND THE NEW MEDICINE

Founding Editors

DAVID C. THOMASMA†
DAVID N. WEISSTUB, *Université de Montréal, Canada*
THOMASINE KIMBROUGH KUSHNER, *University of California, Berkeley, U.S.A.*

Editor

DAVID N. WEISSTUB, *Université de Montréal, Canada*

Editorial Editor

TERRY CARNEY, *University of Sydney, Australia*
MARCUS DÜWELL, *Utrecht University, Utrecht, the Netherlands*
SØREN HOLM, *University of Cardiff, Wales, United Kingdom*
GERRIT K. KIMSMA, *Vrije Universiteit, Amsterdam, the Netherlands*
DAVID NOVAK, *University of Toronto, Canada*
EDMUND D. PELLEGRINO, *Georgetown University, Washington D.C., U.S.A.*
DOM RENZO PEGORARO, *Fondazione Lanza and University of Padua, Italy*
DANIEL P. SULMASY, *Saint Vincent Catholic Medical Centers, New York, U.S.A.*
LAWRENCE TANCREDI, *New York University, New York, U.S.A.*

VOLUME 47

For other titles published in this series, go to
http://www.springer.com/series/6224

Barbara Maier · Warren A. Shibles†

The Philosophy and Practice of Medicine and Bioethics

A Naturalistic-Humanistic Approach

Springer

Dr. Barbara Maier
Department of Gynecology and Obstetrics
Paracelsus Medical University
SALK
Müllner Hauptstrasse 48
5020 Salzburg
Austria
b.maier@salk.at

Prof. Warren A. Shibles†
University of Wisconsin–Whitewater
53190 Whitewater Wisconsin
USA

ISSN 1567-8008
ISBN 978-90-481-8866-6 e-ISBN 978-90-481-8867-3
DOI 10.1007/978-90-481-8867-3
Springer Dordrecht Heidelberg London New York

Library of Congress Control Number: 2010935875

© Springer Science+Business Media B.V. 2011
No part of this work may be reproduced, stored in a retrieval system, or transmitted in any form or by any means, electronic, mechanical, photocopying, microfilming, recording or otherwise, without written permission from the Publisher, with the exception of any material supplied specifically for the purpose of being entered and executed on a computer system, for exclusive use by the purchaser of the work.

Printed on acid-free paper

Springer is part of Springer Science+Business Media (www.springer.com)

For Anna and Edith,
for Colin and Ernst,
for the physicians, midwives and nurses of the University Clinic of Gynecology and Obstetrics,
for the students of the Paracelsus Medical School Salzburg,
for all who provided so much appreciated encouragement and support.
They will find themselves honored in this book. . . .

Contents

1 Metaphor in Medicine: The Metaphorical Method 1
1.1 Introduction 1
1.2 Types of Metaphor 4
1.2.1 Substitution 4
1.2.2 Juxtaposition 4
1.2.3 Analogy, Simile, or Comparison 4
1.2.4 Symbolism 4
1.2.5 Metonymy 4
1.2.6 Synecdoche 5
1.2.7 Synesthesia 5
1.2.8 Reversal 5
1.2.9 Personification 5
1.2.10 Oxymora or Combination of Opposites 6
1.2.11 Deviation 6
1.2.12 Metaphor-to-Myth Fallacy 6
1.3 Metaphorical Methods Should be Considered for Analysis of and Writing Research Papers 7
1.4 Case Example: A Healthcare Worker (H) – Patient (P) Metaphoric: H/P Modeling in Medicine 11
1.5 H/P Models 12
1.5.1 H = P 12
1.5.2 H versus P 12
1.5.3 Not H and not P 12
1.5.4 H or P 13
1.5.5 H and P 13
1.5.6 H not P 13
1.5.7 P not H 13
1.5.8 H (verb) P 13
1.5.9 P (verb) H 13
1.5.10 H ≠ P 13
1.5.11 H ? P 13
References 14

2 Definition 15
2.1 Where Does It Come from that We Think We Need to Define? 15
2.2 Distinction Between Types of Definition 18
References 21

3 Decision-Making: Fallacies and Other Mistakes 23
3.1 Conditions of Decision-Making 23
3.2 Frequent Causes of Irrational Medical Thinking and Decision-Making 24
3.3 Five Levels of Decision-Making in Medicine 28
3.4 Fallacies in Decision-Making 29
3.5 Mistakes 34
3.5.1 What are Mistakes? 34
3.5.2 What are Indications of Errors? 35
3.5.3 Indications that Mistakes are Often Preventable Ones 36
3.5.4 What are the Reasons for the Mistakes? 37
3.5.5 Case Example: Misleading Diagnosis 41
3.5.6 Personal Experiences: Mistakes 42
References 43

4 Analysis of Causation in Medicine 45
4.1 Decision-Making and Cause 46
4.2 Synonyms of Cause 48
4.3 Antonyms for Cause 49
4.4 Metaphorical Models for Cause 49
4.5 Substitutions for Cause 49
4.6 Temporal Factors in Causality 49
4.7 Types of Causality 50
4.8 Summary 62
References 63

5 Ethics and Non-ethics 65
5.1 Introduction 65
5.2 A Naturalistic Theory of Ethics 69
5.3 What Is Ethics in Actual Usage? 72
5.4 Ethics and Morals: An Unethical Society 80
5.5 Value Contradictions 83
5.6 Examples of Contradictions 84
5.7 On Being Non-ethical and Anti-Inquiry 87
5.8 Brief Conclusion 89
5.9 An Ethics Text for British Medical Schools 90
5.10 Case Example: Medicine and Dysfunctional Culture (Made Available by Dr Wolf Michael Luetje, Head of the Women's Hospital Viersee in Germany) 91

5.11 Case Example: Military Medical Service as Contradictory to Medical Practice 92
5.12 Insensitivity to Killing: The Failure to be Embarrassed 95
5.13 Case Example: On Sensitivity 97
5.14 Case Example: Tsunami Disaster and Cultural Irresponsibility 98
5.15 Case Example: Culture and Family as Anti-Medicine: Female Circumcision 100
References 103

6 Medicotheology and Biotheology 111
6.1 Introduction: How Many People Have Religious Beliefs? . . . 111
6.2 The Influence of Religion on Bioethics and Medicine 112
6.3 Church Opposition to Medicine 115
6.4 Should Medicine Be Based on Supernaturalism? 117
6.5 Science and Metaphysical Causes? 117
6.6 Case Example: Religion and Autonomy 118
6.7 Religion Versus Medicine: A Common Ground? 119
6.8 Religion as Ethics 120
6.9 Ethics Committees 121
6.10 Humanism Versus Religion 122
6.11 Absolute Religious Ethics Versus Consequentialism 123
6.12 Case Example: Deprogramming Religion in Medicine 123
6.13 Case Example: A Real Woman 124
6.14 The Person as a Soul 125
6.15 Sanctity-of-Life (Human) 125
6.16 General Observations Regarding the Value of Human Life . . 126
6.17 Contradictions Regarding the Sanctity-of-Life Doctrine 127
6.18 Selected Arguments from the Philosophy of Religion 128
6.19 Prayer as Medical Treatment 129
References 132

7 Emotion In Medicine 137
7.1 Introduction 137
7.2 Case Example: Non-mental Associations Provide Complexity to Cognitions 140
7.3 Emotion Is Not an Internal State 140
7.4 Emotions Can Be Changed 141
7.5 The Happy Stoics: Passionate Rational Emotion 142
7.6 Virtually All Judgments Involve Emotion 142
7.7 Emotion Can Change with Bodily Feeling 143
7.8 Emotion Is Not Passive 144
7.9 Emotions Are Unique 145
7.10 Rejection of the Release Theory of Emotions 145
7.11 Case Example: Emotion Requires Assessment 146
7.12 Negative Emotions Are Philosophical Language Fallacies . . . 147

7.13 Some Traditional Examples of Philosophy of Language Fallacies 148
7.14 Pity 149
7.15 Hope and Humor 150
7.16 Case Example: Patients' Negative Emotions 150
7.17 Can Emotions Be Reduced to Physiology? 151
7.18 How Are Diseases and Emotions Classified? 152
7.19 Case Example: Legal Recognition of Emotional Harm 153
7.20 Brief Summary of the Cognitive Theory of Emotion 154
7.21 Case Example: The Cognitions Involved in the Emotion of Interest 155
References 156

8 Enlightened Versus Normative Management: Ethics versus Morals 161
8.1 Introduction 161
8.2 Requirements for Good Management 163
8.3 Special Section on Overwork: A Failed Metaphor of the Medical System 182
8.3.1 Introduction 182
8.3.2 How Many Hours Do Physicians, Nurses and Healthcare Workers Work? 183
8.3.3 Attempts to Limit the Number of Work Hours 183
8.3.4 Do Physicians and Nurses Also Cause the Problems of Overwork? 185
8.3.5 What Is the Effect of Overwork? 187
8.3.6 Overwork Harms Health of Staff 188
8.3.7 Overwork Increases Sick Leave 188
8.3.8 Overwork Causes Stress and Burnout and Addictions 189
8.3.9 Overwork and Suicide 190
8.3.10 Overwork Causes Loss of Quality of Life 190
8.3.11 Overwork Is a Cause of Negative Emotions 190
8.3.12 Overwork Causes Loss of Interest in Medical Practice 191
8.3.13 What Is the Legal Result of Overwork? 191
8.3.14 Denial that Healthcare Workers Overwork and/or that It Is Harmful 192
8.3.15 Is There Evidence for the Harm of Overwork? 193
References 196

9 Care: A Critique of the Ethics and Emotion of Care 201
9.1 Introduction 201
9.2 Care Theories 202
9.3 The Word-Field Meanings of Caring 206
9.3.1 Introduction 206

9.3.2 The Synonyms and Word-Field of the Term "Caring." 207
9.4 Irrational Forms of Caring (See Also Empathy and Sympathy) 208
9.5 The Cognitive-Emotive Theory of Caring 211
9.5.1 Feeling 212
9.5.2 Caring Is a Value Cognition Causing Feeling [Caring = (Cognition > Feeling)] 212
9.5.3 Caring Is Based on Positive Cognitions 213
9.5.4 Emotions Can Be Changed 213
9.5.5 We Cause Our Own Emotions: Caring Is Caused by Ourselves 214
9.5.6 The Passionate Stoics: Rational Emotion, Rational Caring 215
9.5.7 Negative Emotion Changes with Feeling 215
9.5.8 Negative Emotion Is Not Passive 215
9.5.9 Each Emotion of Caring Is Unique 216
9.5.10 Rejection of the Release Theory of Caring 216
9.5.11 Judgments Generally Involve Emotion 216
9.5.12 Metaemotion 217
9.6 Caring and Negative Emotions 217
9.7 Mutuality of Caring 218
9.8 The Patient's Hippocratic Oath 218
9.9 Empathy and Caring 219
9.10 Summary 221
References 223

10 Egoism and Altruism in Medicine 227
10.1 Introduction 227
10.2 Common Definition of Altruism and Egoism 227
10.3 Definitions of Altruism 228
10.4 Definitions of Egoism 228
10.5 An Analysis of the Word-Fields of Altruism and Egoism 228
10.5.1 The Word-Field of Altruism 229
10.5.2 The Word-Field of Egoism 229
10.6 Altruism Versus Egoism 230
10.7 The Problem of the Self 231
10.8 The Ethical Basis of Altruism and Egoism 232
10.8.1 General Remarks 232
10.8.2 Utilitarian Altruism 233
10.8.3 Ayn Rand's Objectivist Egoism 233
10.8.4 The Ordinary Language Basis of Altruism and Egoism 235
10.9 Altruism and Egoism as Emotions 235
10.9.1 Altruistic and Egoistic Emotions Are Not Mere Bodily Feelings 236

10.9.2 Altruism and Egoism Are Cognitions Causing Bodily Feelings [26] 236
10.9.3 The Emotions of Altruism and Egoism Can Be Changed 236
10.9.4 Altruism and Egoism Are Based on Value Cognitions 237
10.10 Sympathy 238
10.11 Selfishness 239
10.12 Rational Altruism and Egoism 241
10.13 Summary 242
References 242

11 Letting Die 245
11.1 Introduction 245
11.2 Misuse of Ethical Terms 249
11.3 Criteria for Preferential Treatment: Non-contradiction 250
11.4 Case Example: Oregon Healthcare Prioritizing 250
11.5 What About Self-Caused Illness and How to Determine? . . . 251
11.6 The Hippocratic Oath: Pacifism in Medicine? 254
11.7 Should We Kill X to Save Y? The Numbers Game 257
11.8 Allowing Death = Killing = Murder 262
11.9 Letting-Die and the Samaritan 264
11.10 Albert Schweitzer on *Reverence for Life* 267
11.11 Negative Emotions Kill and Let Die 270
11.12 Lack of Organs for Transplantation as a Form of Letting Die 270
11.13 Suicide and Euthanasia 270
11.14 Conclusion 273
References 274

12 A Critique of Autonomy and Patient Responsibility 281
12.1 Introduction 281
12.2 Criticisms of the Principle of Autonomy 282
12.3 Patient Responsibility and a Patient Code of Ethics 291
12.4 Patients Duties 293
12.5 Case Report: Patient and Legal Irresponsibility 295
References 295

13 Philosophy and Ethics of the Body 299
13.1 Introduction 299
13.2 Definition of Philosophy and Body 300
13.3 The Scientific Method: Medicine as a Science 301
13.4 A Naturalistic Ethics of the Body 302
13.5 The Value of Life in Terms of the Body 303
13.6 The Mind 305
13.7 The *Self* as a Language Construct 305

13.8 The Un-philosophical Body 307
13.9 Outward Physical Appearances: Beauty 308
13.10 The Face 309
13.11 The Body as a Whole and Body Parts: Organs and Transplantation Medicine 310
13.12 Reproduction of Bodies? 311
13.13 *Leib*: Living to the Full 311
13.14 The Philosophical Body: The Body as an Aesthetic Whole 312
13.15 Summary of the Ethics of the Body 313
References 313

14 Organ Donation: Mandatory Organ Donation Declaration 317
14.1 How Many People Need Organs? 317
14.2 Death Requirement 319
14.3 Opposition to Organ Donation 320
14.4 Support of Organ Donation 321
14.5 Presumed Organ Donation 322
14.6 Family Approval as an Adverse Policy 324
14.7 Recommendations for Obtaining Organs for Transplantation 325
14.7.1 Lottery 325
14.7.2 The Economic Incentive Approach 326
14.7.3 Irresponsible Lifestyles and Organ Preference 326
14.7.4 Mandatory Organ Donation Declaration 329
References 331

15 Embryonic Stem Cell Research: A Question of Beliefs? 335
15.1 Introduction 335
15.2 Definitions and Clarifications of Morals and Ethics 336
15.3 Facts and Beliefs About Stem Cells 337
15.3.1 What We Already Know About Stem Cells 337
15.3.2 The Promise of Stem Cell Research in General . . . 339
15.4 The Controversy About What an Embryo Is 340
15.4.1 Definitions 340
15.4.2 Embryonic Development in Its Context 341
15.4.3 The Moral Status Ascribed to an Embryo 341
15.4.4 Life Is Not Just Life: When Is a Human a Human? 342
15.5 Ethical Issues in Stem Cell Research 343
15.5.1 How to Perform an Ethical Examination? 343
15.5.2 Inquiry into Language 344
15.5.3 The Abortion Argument All Over Again? 345
15.5.4 Adult Stem Cell Research-an Alternative to Embryonic Stem Cell Research? Other Alternatives? 346

15.5.5 IVF "Left Over" Embryos Versus "Created for Research" Embryos . . . 346
15.5.6 Public Funding? . . . 347
15.5.7 The Ethical Challenge of Research . . . 349
15.6 Conclusions: Humaine Medicine – Medicine for Suffering People . . . 350
References . . . 350

16 The Philosophy of Prevention . . . 355
16.1 Introduction . . . 355
16.2 An Analysis of Prevention . . . 356
16.3 Unethical Behavior and Irresponsible Lifestyles . . . 360
16.4 Lifestyle as Philosophical and Critical Thinking . . . 361
16.5 Areas of Prevention . . . 362
16.5.1 Education: The Greatest Preventative of Disorder . . . 362
16.5.2 Cancer Prevention . . . 362
16.5.3 AIDS . . . 365
16.5.4 Alzheimer's Disease (AD) . . . 365
16.5.5 Lack of Exercise: The Obvious Escapes Us . . . 366
16.5.6 Sexually Transmitted Disease (STD) . . . 366
16.5.7 Longevity . . . 367
16.5.8 Death and Disease . . . 368
16.5.9 Hand-Washing: The Obvious Escapes Us Again . . . 369
16.5.10 Drugs and Toxins . . . 369
16.6 Hidden Prevention Possibilities . . . 371
16.7 Summary . . . 378
References . . . 378

17 Ethics Counseling: Philosophy of Medicine Counseling Instead of Medical Ethics Counseling . . . 383
17.1 Introduction . . . 383
17.2 What is Ethics Counseling (EC)? . . . 386
17.2.1 Task Force on Standards of Bioethics Consultation (USA) . . . 386
17.2.2 Basic Ethical Principles in European Bioethics and Biolaw . . . 388
17.3 Criticism of Bioethical Principlism . . . 396
17.4 Case Method of Clinical Ethics . . . 401
17.5 Holistic Philosophy of Medicine . . . 405
17.6 Brief Description of Dewey's Pragmatism and Naturalistic Ethics . . . 408
17.7 Humanism Contains Many of the Elements of Contemporary Definitions of Philosophical Practice (PP) . . . 409
17.8 The Present Definitions and Methods of Philosophical Counseling (PC) are Too Restrictive . . . 410

17.9 Philosophical Counseling or Philosophical Practice (PP) . . . 412
17.10 A Proposal to Change the Title of Philosophical Practice to Philosophy Education 414
17.11 The Philosophy Practitioner and Emotion 415
17.12 Summary 417
References 417

18 Medical Language: The Ordinary Language Approach 427
18.1 Introduction 427
18.2 Formal Logic as a Pseudo-Logical Failure 431
18.3 Formal Logic is Irrelevant to Thought, Reason and Emotion 433
18.4 Formal Logic as Irrelevant to Ethics or Bioethics 435
18.5 Formal Logic as Formal Fallacy 435
18.6 Formal Logic as a Fallacy of Abstractionism 435
18.7 The Arrogance of Logicians 436
18.8 Formal Logic Reduces Language to Mathematics 436
18.9 Formal Logic as a Faulty View of Meaning 437
18.10 "Propositions": A Pseudo-Logical Term 438
18.11 Formal Logic as Dogmatism and Misuse of Symbols 439
18.12 Formal Logic Misuses the Term "Truth" 440
18.13 The Useless Syllogism 441
18.14 Formal Logic is Not Philosophy 441
18.15 The Primacy of Ordinary Language and Pragmatism 442
18.16 Formal Logic Excludes Metaphor and Creative Language 443
References 445

19 A Critique of Evidence-Based Medicine (EBM): Evidence-Based Medicine and Philosophy-Based Medicine 453
19.1 Does EBM Really Meet the Challenge of Modern Medicine? 453
19.2 What Is the View of Evidence in EBM – Is It Left Undefined? 456
19.3 EBM as Statistics 457
19.4 EBM Often Investigates the Obvious and Trivial 458
19.5 EBM Bears the Risk of Overgeneralization 459
19.6 EBM Is Often Unintelligibly Complex 459
19.7 EBM Is Often Too Abstractionistic 460
19.8 EBM as Appeal to Authority Fallacy 460
19.9 EBM and the Individual Case and Context 461
19.10 Uncritical Use of EBM and Clinical Experience 462
19.11 EBM Often Excludes Relevant Causes and Variables 463
19.12 EBM Has Limited Self-Criticism 465
19.13 EBM and Psychiatry 465
19.14 EBM and Human Emotions 466

19.15 EBM and Ethics 466
19.16 EBM Depersonalizes 466
19.17 EBM Text Reviews 467
19.17.1 Evidence-Based Spirituality 467
19.17.2 EBM and Practical Medicine 467
19.17.3 Evidence-Based Nursing 467
19.17.4 EBM and Logic 468
19.17.5 EBM and Gender Medicine 468
19.18 EBM and Rational Medicine 469
19.19 EBM, Psychosomatics and Philosophy 470
19.20 EBM and the Problem of the Placebo 471
19.21 "Philosophy of Medicine" – Based Medicine Instead of Only "Evidence" – Based Medicine 479
References 481

20 Lying in Medicine 487
20.1 Introduction 487
20.2 Definitions of Lying 488
20.3 A New Theory of Lying 488
20.4 Self-Lie 489
20.5 Consequences of Lying 490
20.6 Logic of Flattery: Beneficial Lying 493
20.7 Hypocrisy 493
References 494

21 The Rhetoric of Death and Dying 497
21.1 Definitions of Death 497
21.1.1 General Definitions 498
21.1.2 Medical-Psychological Definitions of Death 498
21.1.3 Death and Abortion 498
21.2 Death: The Literature 499
21.2.1 The Poet's View 499
21.2.2 Wittgenstein on Death 502
21.3 Dying 510
21.3.1 The Cognitive-Emotive Theory 510
21.3.2 The Cognitive-Emotive Theory of Grief and Bereavement 512
21.4 Philosophy of Religion 514
21.4.1 Views in Theology 514
21.4.2 Old Testament 514
21.4.3 New Testament 515
21.5 Humanism 516
21.6 The Rhetoric of Death Using the Metaphorical Method 517
21.6.1 Introduction to the Metaphorical Method 517
21.6.2 Rhetorical Techniques for the Exploration of the Concept 519

21.7 Death and Medical Profession 530
21.8 Final Personal Remarks 531
References 531

Index 537

The Rationale Behind the Book

This is not a standard or normative textbook collection giving a few of the usual arguments to imply standards for the profession, but rather a deep and challenging analysis, which is more in line with the tradition of honest, open philosophical inquiry. The usual bioethics texts are put into question.

This is a book on the philosophy of medicine. By philosophy of medicine is meant a critique of the concepts and methods used in medicine. This is a more encompassing and philosophical discipline than bioethics. Bioethics as commonly practiced is rather seen as a pseudo-discipline, not biology, not medicine, not ethics, and not philosophy. Bioethics is a misnomer. Bioethics suffers from trying to invent a new subject instead of integrating medical issues into the already existing philosophy of medicine and philosophy generally. The new invention of the subject of bioethics is based on a defense of cultural institutions (law, church and religion, culture, custom, etc.) to control physicians and healthcare workers. But the areas of ethics and philosophy are already well established so that no new area, bioethics, needs to be created. Bioethics does not replace ethics or the philosophy of medicine. In this sense, the philosophy of medicine includes a critique of bioethics.

Regarding the present book:

1. No specific or absolute recommendations are given regarding medical treatment, moral approaches, or legal advice. Given rather is discussion about each issue involved and the strongest arguments indicated. Each argument is subject to further critical analysis. This is the same position as with any philosophical, medical or scientific view.
2. The argument that decision-making in medicine is inadequate unless grounded on a philosophy of medicine is not meant to include all of philosophy and every philosopher. On the contrary, it includes only sound, practical and humanistic philosophy and philosophers who are creative and critical thinkers and who have concerned themselves with the topics relevant to medicine. These would be those philosophers who engage in practical philosophy, such as the pragmatists, humanists, naturalists, and ordinary-language philosophers.

Such passionate, critical thinkers are also able to provide in-depth analyses of the uses and misuses of ordinary language. They are aware of and try to avoid the informal logical fallacies, e.g., circularity, ad hominem fallacy, teleological fallacy, abstractionist fallacy, appeal to majority fallacy (consensus), etc. which fallacies are prevalent in bioethics as well as in medical practice and literature. Thus, a special critical ability to clarify language and definitions is essential to such philosophies and philosophers. Language must in the first instance be extensively critiqued and not uncritically taken for granted as is almost always presently done. Thus, for example, in this book many of the most relevant basic and misused terms in the philosophy of medicine will be extensively clarified and critiqued, for example, autonomy, caring, case method, cause, death, emotion, energy, ethics, evidence-based medicine, health, medicine, mental, moral, patient, person, placebo, psychological, quality of life, statistics, etc. The main goals of medicine and the philosophy of medicine can be the same: rationality, effectiveness, humanism, caring, and bringing about optimally desired health and quality of life in all of its relevant aspects.

One cannot be a good physician, healthcare manager, or patient without knowledge of critical thinking and philosophy, including the philosophy of medicine, ethics, and emotion. It is quite usual to graduate from college and medical school with virtually no exposure to these subjects at all. These are typically not available, much less required. Few healthcare workers can tell the difference between scientific statements and moral statements. As long as medical schools and academic research centers do not make the philosophy of medicine part of their education and culture, medicine, also as a science, will be undermined.

What is also significant and tragic is that the physical lifestyle of the typical healthcare worker (physician, nurse, therapist, etc.) is often unhealthful, and in addition, their psychological lifestyle as well. Like most people, healthcare workers are usually culturally indoctrinated and far from critical and philosophical thinking, or learning about ethics or emotions even when these subjects directly concern medical practice. The result is discussion illiteracy, emotion illiteracy, and ethical illiteracy in professional as well as private lives. Medical conferences consist typically of simplistic data presentations with little or no clarification of the concepts, which are used in their largely statistical presentations. There is the experimental method, but conceptual confusion. (*Wittgenstein*)

Healthcare at present exists in a theoretical vacuum without an overall well-grounded philosophy and therefore is at the whim of politics, law, economics, religion, popular opinion and culture. A humanistic and holistic philosophy of medicine as presented in this book can provide an evaluation of goals and ethical directions. It is also one of the tasks of philosophical counseling.

The philosophy of medicine is dealt with in this book as a matter of life and death, also a matter of how to live a meaningful life. Theories, beliefs, and decisions have life and death consequences as much as and often more than individual physical events or medical treatment. There are beliefs, policies and practices today, which are causing millions of people to die. There are many ways in which we promote and cause our own as well as others' disease and death, at least, let other people die. In our discussion of medicine we wish to address the issues so as to promote long, healthy and qualitative life and show explicitly ways in which this is not now being done. We may call these examinations the philosophy of preventative medicine.

Uncritical culture, custom, and common normative morality take the place of critical humanistic ethics. But the appeal to the majority is not the common good. It is regarded as a fallacy in philosophy. Each society enculturates and indoctrinates. Medicine becomes enculturation instead of critical evaluation of our lives. Culture and business often dominate and enslave medicine. Also, law takes the place of ethics. Bioethics becomes biolaw. Cultural and prevailing medical practices are subject to philosophical critique in this book.

Ethical theories are often reduced to quantitative, formal, or arbitrary systems or fixed principles far removed from relevance to the lives of human beings. Utilitarianism is an empty quantitative formula (must presuppose an ethical system), deontology is blind obedience, universalization is a mere abstract formal principle, egalitarianism is not an ethical system but equalization for its own sake, intuitionism is self-righteousness, etc. Ethical theories are often absolutistic rather than consequentialistic. In this book a naturalistic, practical, pragmatic, consequentialistic, and humanistic theory of ethics is presented which stresses reason and humanism.

There is narrow and split decision-making rather than holistic decision-making involving the most comprehensive philosophical thinking of which humans are capable.

Abstract theories and formal quantitative systems prevail, which obliterate the human and humanity. Principlism is formal, general, fixed principles, which substitute for and take the place of contextual human reason. Formal logic removes from language: meaning, emotion, reason, style, ethics, understanding, practical problem solving, creativity and clarity. It purges from language that which we are most interested in. It dehumanizes us and violates our humanity. It especially dehumanizes our language.

From the above a new definition of our own philosophy of life emerges and it is necessary to have one. Good lifestyle no longer means just abstaining from cigarettes, alcohol and getting exercise. It also means living a holistic life, which includes all of one's thinking, personality and actions. To treat merely one aspect of a person to the exclusion of the rest is narrowing and splitting off at the expense of all involved. To have a holistic lifestyle one must know about ethics, emotion, prevention of disease, and be a critical thinker, a rational humanist, and positive altruist. One could also say that medical establishments should follow holistic decision-making. These qualities and characteristics must be put into practice and continuously reexamined and improved. Medicine need not be merely a backup for unnecessary and unhealthful lifestyles, but should aim at helping people be the best they can. This requires a philosophy of medicine.

This book also includes new ways of thinking. In this regard the "Metaphorical Method" is explained, used, and exemplified in depth, for example in the chapters on care, egoism and altruism, letting die, etc.

In accordance with the above analysis the healthcare worker has a chance not to just blindly serve often also anti-medical practice and tradition, but to instead take a leadership role in moving medical care to a higher level based on ethical and philosophical thinking and practice as exemplified by the philosophy of medicine. The patient must also be a leader in the sense of cooperatively sharing responsibility for his/her own treatment and prevention of disease by adopting a healthful, holistic lifestyle.

About the Authors

This book emerged from my clinical experience as well as from the need of philosophical clarification of what it means being a physician and especially from the discussions and sharing ideas between Warren and me, a man and a woman, a philosopher and a physician/philosopher, an American and a European. Warren Shibles died July 17th 2007. He was a senior philosophy professor at the University of Wisconsin at Whitewater and also taught courses at Tübingen, Germany. Unconventionally, he gave his lectures, involving students in a kind of Socratic dialogue. With his focus on the rich possibilities of language he explored metaphor, humor (his humor book is available on the internet as a free I-book), aesthetics, and ethics examining values as open context terms. He also wrote poetry. He has published 27 books, and over 180 professional journal articles. He also was a researcher in phonetics. Main topics of his research were philosophy of language, emotions, love, time, humanism, and philosophical counseling.

I am a senior physician, gynecologist and obstetrician at the University Hospital in Salzburg, Austria and head of the Department of Gynecological Endocrinology and Assisted Reproduction. I have a PhD in Ethics in Medicine, and MD from the University of Vienna and have been teaching ethics in medicine at the Institute for Ethics and Law at the University of Vienna and for Warren's Department *Ethics in Science* at the University of Wisconsin 2004/2005/2007. I am currently teaching at the Paracelsus Medical School in Salzburg as well as at the Medical University of Vienna. My main concern has been to combine theory with practice.

I presently am a member of the Bioethics Committee for the Austrian Chancellor.

I have translated from English to German a critical philosophy book: *Lying: a Critical Analysis* by Warren Shibles.

Salzburg, Austria May 20th, 2010

Chapter 1
Metaphor in Medicine: The Metaphorical Method

Abstract What is to be shown in this chapter is that and how metaphor may be used as a scientific method of analysis and how it functions in medical statements. The metaphorical method is used to gain insights into the philosophy of medicine and bioethics. Philosophy of medicine is metaphors about medicine. The meanings of medicine are generated by a constant stream of metaphors. Types of metaphors are presented and examples are given how to work with them (A healthcare worker (H) – patient (P) metaphoric: H/P modeling in medicine). Metaphorical methods are useful for analysis of and writing research papers (a guideline how to do that is presented). The Metaphorical Method is used throughout this book to critically examine medicine and bioethics, practice and theory and establish a philosophy of medicine relevant to its practical tasks.

Keywords Metaphorical Method · philosophy of medicine · types of metaphor · scientific method · medical language · narrative · self · therapeutic metaphor · insights · healthcare worker – patient relationship

1.1 Introduction

According to Robert Frost, *All thinking ... is metaphorical* [1]. So also is philosophy and science. What is to be shown here is that and how metaphor may be used as a scientific method of analysis and how it functions in medical statements. The style, narrative, models and language of medicine basically consist of metaphors, which need clarification. Narrative is one of the old and recently re-discovered techniques of gaining medical knowledge. As is argued in this book, the usual view that the scientific method usually mentioned in science and medicine is falsely based on naïve empiricism (sensation and observation) or abstractionistic notions of truth (formal logic and deduction). Observation and sensation are linguistic terms in need of clarification. There is, for example, the philosophy of perception by which will be argued that the scientific method rather rests on and presupposes language. Thus, any method of science, including statistics and mathematics, needs to use the techniques available in language. These are mainly rhetorical devices, the most fundamental one being metaphor and its various types. The metaphorical method

B. Maier, W.A. Shibles, *The Philosophy and Practice of Medicine and Bioethics*,
International Library of Ethics, Law, and the New Medicine 47,
DOI 10.1007/978-90-481-8867-3_1, © Springer Science+Business Media B.V. 2011

is used in this book to gain insights into the philosophy of medicine and bioethics. Philosophy of medicine is metaphors about medicine. The meanings of medicine are generated by a constant stream of metaphors. Metaphors in medicine interact and break on one another.

The first annotated metaphor bibliography contained much of the previous writing on metaphor [2]. The literature on metaphor has exploded in the last 30 years including web-based material. Metaphor involves combinations of unlike terms (oxymora), reversals, neologisms, juxtapositions, puns (especially popular with Deconstructionists), analogy, imagery, category-mistakes, tension metaphors, humor, irony, taking terms literally, being captivated by a paradigm or picture, etc. Researchers often take their models literally, for example "evidence-based medicine," or the medical model, which treats all disorders as physical ones. Metaphor involves especially deviation, such as from the normal, expected, traditional, rules, values, etc. Metaphor is basically to relate unlike things. The techniques and types of metaphor are held to be fundamental to understanding and methodology in science.

Because it cannot be literally true, the "x is y" form cannot be reduced to the literal simile form "x is like y." Metaphor is open-context. It does not tell us how "x is y," how "the world is matter," how "the body is physical matter to be medically treated." Some wish to reduce cause to statistics or to matter by means of literal simile, others are content to regard cause as reasoning in a non-literal, metaphorical way. Reasons have been presented to show that metaphor has meaning, which cannot be reduced to literal language [3, 4]. Every theory creates a new world. Metaphor has meaning of its own which cannot be reduced to literal language. Metaphors in medicine also have meaning of their own which cannot be reduced to literal language. Style is not irrelevant, but rather determines what is said. A paraphrased Hippocrates is not Hippocrates, religious humanism is not Dewey's humanism. We may therefore ask what each term in medicine means. The philosophy of medicine involves the intensive and extensive clarification of medical language.

To create a metaphor is to create a category-mistake, or produce type-crossing. Two different universes of discourse are brought together, such as "thought is chemical," or "cause is statistical." The second metaphor is used in evidence-based medicine. The unlike is related to the unlike. Therefore, if the metaphorical statement is to make sense we must find unity in difference. The metaphor appears as a contradiction, enigma, mystery, or riddle waiting to be solved. If we diagnose that someone has a disease we need specific clinical experience to determine what it really involves. Metaphor is a context-deviation. Terms are used in other than their normal or usual context or language-game e.g. in scientific research for problem solving. The result of this is surprise and apparent contradiction, which upon resolution produces the satisfaction of solution. Research departs from what has been understood and ends in wonder. The physician is like a detective or experimental researcher trying to find a workable method of treating a disease. The impossible becomes, after all, possible.

This may suggest that if apparently contradictory metaphors can be resolved, then perhaps the perverse and extensive enculturated contradictions of our lives can be resolved as well. This as we shall see is what happens with black humor in medicine [5]. The terms and methods of one universe of discourse are used to give insight into another. We speak of medical causes and description in terms of atoms, mathematics, statistics, quarks, language, physics, emotion, pictures, diagrams, etc. Metaphor becomes, then, a tool of discovery and a scientific method.

What metaphor often comes down to is breaking rules – deviation. The tool of the scientist, like of the good physician is to deviate to solve complex problems. To do so is business as usual. Ramsey pointed out *What is not verbally odd is devoid of disclosure power* [6]. It is to de-contextualize and disengage the subject so as to admit new perspectives of appreciation. More specifically, there are deviations from the usual, grammar, context, behavior, the familiar, beliefs, the proper (e.g., sinking = relating high value to low value), the practical, the logical, the obvious, the literal, the real, usual cause and effect, usual perception. We find these techniques used in science as well as in philosophy of medicine. Each theory, test, discipline, map, diagram, hypothesis, statement may be regarded as a metaphor or model which is then expanded [7]. Kuhn in *The Structure of Scientific Revolutions* [8] argued that paradigms are the basis of every theory. Statistics is not a true science, but merely a metaphor, which we may find useful. The same is true of ever growing medical theories, which we now no longer find so useful. Kuhn showed how scientists, and by extension also medical researchers, are captivated by their paradigms such that they are not open to alternative ideas. He even argues that prevailing paradigms can hold us captive and turn scientific thinking into fashionable models and dogma. When metaphors are taken literally it turns metaphor into myth, delusion, and dogma. Perspectival thinking is lost. If we unknowingly take a metaphor literally, it is a fallacy. If we deliberately take it as a way to create insight, it can be a significant tool for inquiry. As will be seen in the Chapter 2, to define is to take a model or metaphor. By thinking of definitions as metaphors, it helps us not to take them literally. It was Wittgenstein [9] who pounded one of the last nails into the coffin of fixed definitions. They no longer exist. We are left with disciplines, which are useful fictions, as-ifs. Medicine, among other disciplines, is a collection of metaphors, which define our medical experience. Even perception is perspectival, not the basis of the scientific method.

We may distinguish between cognitive metaphor and perceptual metaphor. Perceptual metaphor may be clarified in terms of the widely used concept of *seeing-as*. It is held that we never merely see or sense directly. That would be naïve empiricism. Virtually all seeing is seeing-as, seeing or hearing in terms of our thinking [4, 10]. We never have mere pure sensation. There is no innocent eye or ear. We do not have sensation neat. It is partly "cognitive" which involves language. Seeing an object as being larger than normal is the result of faulty perceptive cues due to a confusion of contexts, for example, the moon illusion whereby the moon looks larger when on the horizon. We see our illnesses and risks as larger or smaller than they are. Seeing does not work like a camera. There is no mere copying. *An image is not a picture* [9]. Images combine language and sensation inextricably [4, 10]. "According to the scientific evidence..." and "It has been scientifically shown that..." are value expressions, attempts to persuade, but lack reasons or evidence. We cannot say that, for example, evidence-based medicine is based on science. Which science and what is to be counted as evidence? (See Chapter 19).

1.2 Types of Metaphor

An analysis of some of the types of metaphor may give insight. The use of metaphor for analysis is called the "Metaphorical Method" [4, 10, 11]. A few examples of this are:

1.2.1 Substitution

Substitution is used to show a semantic connection that can be liberal, metaphorical, strange, or provocative. "An un examined war is not worth fighting," (Cf. Socrates, an un examined life is not worth living).

1.2.2 Juxtaposition

Juxtaposition combines two words and creates another (surplus) meaning. e.g. in German: lange Weile ->Langeweile

1.2.3 Analogy, Simile, or Comparison

Analogy transfers information from a particular subject (the so called source) to another particular subject (the so called target).

A simile is a figure of speech comparing two subjects by using words "like" or "as". It is often used for subjects we have not get words for describing them in our everyday language. The source then is rather familiar, the target rather strange. By analogy we try to understand.

1.2.4 Symbolism

Symbolism works with vehicles, symbols to represent ideas or concepts. e.g. God is a glass of water in the middle of the desert.

1.2.5 Metonymy

Metonymy is the substitution of attributes or associations of an object with the object itself. For example, left-handed people supposedly do not live as long as right handed people. By their stress on associations, we can see how the medical language can express both cognition and emotion. Metonymy, or non-causal or remote metaphorical associations are sometimes used and taken literally in medicine. The Life Extension Foundation maintained that "Researchers concluded that after adjusting for other risk factors, the presence of a unilateral earlobe crease was associated with a 33% increase in the risk of a heart attack; the risk increased to 77% when the earlobe crease appeared bilaterally." Kuon, on the other hand, concluded that the ear-lobe crease is associated with age and overweight (causal), but does not predict a hemodynamically relevant coronary heart disease (non-causal) [12]. Statistical myths may be thus sometimes created. (See Chapter 19).

1.2.6 Synecdoche

This is the substitution of part for whole or whole for part. Qualities merely or even remotely associated with the stimulus become capable of setting off the same response as the original stimulus.

1.2.7 Synesthesia

We can have visual emotion, kinesthetic emotion, etc. With synesthesia these become combined. We do not use only one sense at a time.

1.2.8 Reversal

A: B becomes B: A. Chiasmus is reversing the order of elements. Cause may be exchanged with effect. Reflexivity and reciprocity also apply. Withholding treatment in medicine may nevertheless be regarded as treatment. Sometimes there is no diagnosis or known cure to give. Self-reflexivity may be exemplified by iatrogenic medicine. Medicine may be reversed, for example, medicine reduced to religious or economic principles and protocols. Reciprocal metaphor is where, for example, "Medical therapy is religious practice," where also, "Religion is medical practice." Metaphors just come to us from the examination of the situation. We proceed from experience to metaphor and from metaphor to experience. We know about authors because of their metaphors, as well as about metaphors because of the authors. Meta-metaphor or metaphor about metaphor is another form of reflexivity. If metaphor renders emotion, this becomes emotion about emotion. There is double bind. We must accept a disease we cannot easily accept or it will make it worse. It is a placebo-like double bind. Anti-inflammatory medicines just cover up symptoms while they can burn a hole in your stomach, kidneys and liver. Anti-inflammation medicine can relieve pain, but increase arthritis [13]. Palliative care and pain reduction may shorten one's life – which often is a myth anyway.

1.2.9 Personification

We personify embryos, fetuses, the dead, the purposes of organs, animals, nature, medicine, etc. The theory of empathy involves anthropomorphism becoming one with one's medical practice. The distinction between the self and object disappears. The emotion is personified and anthropomorphized in the object by empathy. (German: *Einfühlen*, lit. "feel oneself into" the object) Medicine may be humanized or dehumanized. Mice are used to test treatments to be used on people, which is a form of personification. It is a literalism to conclude directly from mice behavior to human behavior, this is personification.

1.2.10 Oxymora or Combination of Opposites

Opposites yield paradox and mystery. When opposites are combined the result is contradiction on the first denotative level of meaning. This creates tension. The second or connotative level attempts to resolve the paradox. Because of the contrast of the apparent contradiction of metaphor and the abstractness of the connotative level, metaphor often remains somewhat paradoxical and open for further interpretation and appreciation.

Some oxymora are the following:

Truth is falsity (Nietzsche)
Benevolent neutrality (Chief Justice Berger)
Kill for peace.
Use force to end force.
We believe in nothing so firmly as what we least know. (Montaigne)
Rational Love [14].
The surgeon must always expect the unexpected.
Identity in difference
Enemies are friends
Futile treatment
Treat by not treating.
Letting-die is the same as killing.
Objective is the subjective. Subjective is the objective.
Patients as adversaries

Verbal and perceptual oxymora may be rendered in every aspect of medicine. Combinations of opposites may be divided into (a) combinations of the near opposite, (b) analytic contradiction (contradiction in definition), (c) synthetic contradiction (contradiction in experience or knowledge), (d) incongruity. Other forms of incongruity may be added such as the metaphorical devices of hyperbole or exaggeration, extravaganza, sinking (reducing valued to trivial), dialectic, finding unity in difference. Conceit is a far-fetched metaphor having great deviant contrast. Freud's work is basically far-fetched metaphor, or conceit.

1.2.11 Deviation

Alternative medicine expresses the very notion of deviation. When one deviates from rules one needs an experimental license to do so. This is suggested by such terms as: clinical experience, expertise, insight, probative, experimental, hypothesis, analogy, likeness, etc.

1.2.12 Metaphor-to-Myth Fallacy

The metaphor-to-myth fallacy is committed when one takes one's model literally or is captivated by it. The "medical model" is a metaphor taken literally. All is reduced to the physical and only physical treatment is allowed. Emotions are analyzed only as hormones and nerve impulses. The cognitive is excluded. It is physical medicine. Thoughts are nerve impulses or brain images. With evidence-based medicine, the nowadays, leading model, medicine is largely reduced to statistics. A holistic view of medicine would produce a different metaphor. Medicine would be regarded from the viewpoint of the philosophy of medicine. Philosophical synonyms include ethics, caring, reason, overall consequences, critical thinking, inquiry, etc. It also has been tried to reduce medicine to theology or religion as will be shown.

Examples of metaphor-to-myth fallacy are also to be found in the example of feminism and Women's Studies literature, which interprets medicine in terms of the notion of "patriarchy". However, the notion of "patriarchy" besides being an all-fallacy has been recently exposed as a nonscientific myth [15].

One may view medicine from any point of view as long as the case can be made out. However, to avoid the literalist fallacy, one needs to make out a plausible case and not take one's own interpretation literally.

Other grammatical and rhetorical possibilities are too numerous to mention. If we are conscious that these are only metaphors, we may use them to gain insight. If we treat them essentialistically as literal or true, we commit the metaphor to myth fallacy.

We are inconsistent in our ethics. We can by metaphor explore subjects not culturally or usually related, bring them together to expose the contradictions. It is a form of insight metaphor. Is it contradictory to use life-saving medicine to support life-taking war? Is letting-die a form of killing? If people do not help the over a billion starving people in need of medical care is it a form of letting-die? If the fertilized egg is a potential life is the sperm and unfertilized egg also? Is food a drug and habit forming? Withdrawal of treatment can itself be a treatment. To not use metaphor is itself a metaphor.

1.3 Metaphorical Methods Should be Considered for Analysis of and Writing Research Papers

The metaphorical method is given here as a method to provide a creative and more adequate way in which concepts (language use) can be analyzed. It is an exploration of the depth of what we can do with language, and the limits of our language, the limits of our thinking and scientific models (See Chapter 19).

The major error is to use vague and abstract terms without defining them. This failure invalidates nearly every paper published or presented at conferences.

Another error is the assumption that the scientific method is epistemologically based on observation, and on real facts rather on language use.

Another error is the mentalistic fallacy of thinking that there are such pseudo-psychological entities as concepts, ideas, thoughts, mind, imagination, memory, and emotions as such.

Faulty conclusions are given. Journal articles often err by concluding the following:

The terms investigated are indefinable.
X may cause y. (It also may not.)

We can never know the resolution or solve the problem under investigation.
The results are tentative and further research needs to be done.
We can never know everything, or all is relative, or that all positions can be argued for equally.

The following are some methods used to clarify medical language. Medicine may be thought of as a narrative. To critique metaphorical practice one needs also to know about style and the philosophy of medicine. One could follow the suggestions mentioned below.

1.3.1 (a) Give and analyze the synonyms of the terms involved in the concept to be analyzed (e.g., Emotion = feeling, mood, affect, attitude, etc.) (b) Give the antonyms. This helps clarify the terms to be analyzed. Example: "Mental process" is not like "cooking process," or "digestion process."
1.3.2 List the major metaphors people usually use regarding the concept. Listen to the narratives of patients and healthcare workers. Also list the major statements made in your research sources. Include the critique of the views read. Read the critiques of evidence-based medicine trials and literature.
1.3.3 Analyze statements to be examined for possible mistakes, confusions or misuses of terms.
1.3.4 Show naming fallacies, which are false assumptions that words such as energy, force, meaning, idea, etc. name substances or entities. Reduce such terms to concrete examples, operational definitions, or show that they cannot be so reduced and are then meaningless. E.g., "All calories are the same (e.g., from lard or vegetables)."
1.3.5 Show category-mistakes (e.g., an embryo is not a person.) "I am just a medical student". (Falsely takes "am" (is) as identity). Terms of one situation are used to apply to another.
1.3.6 The metaphor of combining identical things shows circularities or question begging. (E.g., "X is wrong because it is immoral." "Whatever happens at all happens as it should" [16]. "Each person should get what they deserve."
1.3.7 State faulty assumptions made, e.g., "To be treated equally is to be treated fairly regardless of one's condition." "All problems are quantifiable." (cf., Symbolic logic, structuralism)
1.3.8 Identify the basic definition, model or metaphors being used by you and identify those you think possible regarding the subject (e.g., Caring may be thought as love or treatment.)
1.3.9 Expand these models and arguments to attempt to clarify them. Autonomy presupposes knowledge and responsibility, both of which are often absent. In regard to medical theology it may be shown: because counterexample is irrelevant to religion, even *support* is irrelevant to religion.
1.3.10 Expand these models (and arguments) in an attempt to reduce them to absurdity. Does the model account for itself without other prior unnecessary assumptions? Descartes' cogito, ergo sum (I think, therefore I am),

presupposes language. Scientific observation presupposes language. That is, knowledge may rest on language, not on thought or naïve empiricism.

1.3.11 Ask about the major statements: What do the terms mean? What does disease, cure, certainty, idea, etc. mean? If they mean nothing, to ask if they exist, is not an intelligible question.

What is the concept (language expression) like? Give examples or illustrations of abstract definitions and critique them.

Question whether the concept or definition has any significant relevance or practical use.

1.3.12 Fallacies are whatever deviate from the arguments themselves or are mistakes. These should be identified as will be done through this book.

1.3.13 Use insight humor to clarify a concept (language use). We see that many of the things, which we usually think are true, are actually false or mistakes. They are jokes we don't realize as jokes. Humor is a genuine method of reasoning. Humor is caused by the assessment that there is a mistake or deviation, which is, however, accepted as being okay and not harmful. If not it produces ridicule or anger. The types of metaphors (as deviation) may be seen to be a basis for the types of humor, e.g., analogies, associations, juxtapositions, paradox, simile, synecdoche, etc. As satire or criticism, it shows contradictions, ambiguities, circularities, context deviations, defense mechanisms, deviations, hypocrisy, informal logical fallacies, exaggeration, impossibility, irony, personification, etc. That is, each type of humor can either produce new synthetic or constructive knowledge (insight humor), or serve to analyze or criticize present knowledge (satire, hypocrisy). Blatant vice humor concerning patient autonomy may be rendered as, "I am autonomous in my decision, and the doctor is responsible for the consequences from it."

1.3.14 State relevant epistemological methods used and discuss whether or not they are acceptable: (e.g., intuition, reason, belief, obviousness, faith, etc.) Some approaches support intuition rather than reason.

1.3.15 Give the opposite of the prevalent statements to see if they are equally true or false (e.g., change "Emotions are irrational," to "Emotions are rational." This is the metaphor of oxymoron or antithesis.

1.3.16 Reversal: People cause much of their own diseases. (See Chapter 16) By your vote against hospital funding, you cause yourself to be untreated. There is no self as such. There is no mind, memory or imagination as such.

1.3.17 Check terms and statements for personifications or anthropomorphisms or depersonalizations. "Human beings never understand how anthropomorphic they are." (Goethe) "The fertilized egg is a person." Researchers generalize from results from mice experiments to humans. The *pathetic fallacy,* is giving animals the feelings and thoughts of humans.

1.3.18 Identify which questions asked are obscure, meaningless pseudo-questions.

1.3.19 One of the most prevalent errors in medicine is claiming certainty unjustifiably. Assess the degree of certainty of the various views presented. E.g., absolute certainty (dogma) is not to be held. Is there a warranted hypothesis, which can be changed in view of future evidence? There is merely 50% possibility, but does not state fully the context and limitations of the statistics. Is it a belief or view, which will not be changed even in the light of future evidence? (See Chapter 19)

1.3.20 See which terms, models or theories are taken literally, thus committing the metaphor-to-myth fallacy. (e.g., "The self is mind and body.") Is it? E.g., statistics is not literally true, but an expanded metaphor. Show that what is thought to be literal is metaphorical.

1.3.21 Discuss the topic and its related problems with someone knowledgeable to allow new perspectives to arise. This involves team discussion and evidence from clinical practice.

1.3.22 Construct your own models, definitions or metaphors to try to answer and avoid all of the objections and criticisms above.

1.3.23 Expand your models or arguments to see if they can account for all relevant phenomena and counterarguments. E.g., if clock time is change, can it account for psychological time? Expansion of a metaphor can elucidate and serve as hypothesis. Every word in language can be expanded into a whole philosophy.

1.3.24 Determine if you omitted arguments (or diagnoses) and if the arguments are comprehensive and adequate.

1.3.25 Test your model and arguments by substituting your new definition for all the main occurrences of the word you redefined. See if and where it makes sense and whether or not it can account for what we usually say. If a disease is said to be due to old age, change "old age" to changes.

1.3.26 Deviate from what we usually say about the topic in attempt to give insight.

1.3.27 Some exercises using metaphor for insight:

Analogy. Failed conversation with management: "I threw the ball, but he did not throw it back." Analogies: We are inconsistent in our morals. Thus we can explore subjects not culturally or usually related, bring them together to expose the contradictions. It is a form of insight metaphor.

1.3.28 Using the same question, vary the situation in which it is used. Five physicians may give five different diagnoses. E.g., The *I-Ching* gives a set number of answers to any question asked. Read instead all possible answers for any question asked and note the metaphorical insights.

1.3.29 Deviate from culture, habit, the expected, familiar belief, from what is considered proper, from the practical, the logical, the self-evident, from normal cause and effect relations. As examples may serve: Gentleness is power. Every step is an arrival.

1.3.30 Make the familiar seem strange, e.g., showing that perception is like a miracle, or asking questions of which we falsely think are obvious, such as, "What is an idea?" or "How do we move our arms?"

1.3.31 Use various types of similes or comparisons with and without the use of "like" and "as." Example: If people willingly vote for war, how credible can their anti-abortion views be?

1.3.32 Juxtapose, e.g., on a global scale we give billions for defense, but not one cent for prevention of attack.

1.3.33 Use therapeutic metaphor. Metaphor may be used to avoid the literal, to escape from narrow or oppressive categories, avoid taboo or unacceptable language, provide avoidance, and give indirect ways of saying something. *Euphemism* is substitution of an agreeable word for a word we wish to avoid uttering, e.g., "pass away" for "die," "slumberroom" for "room for the dead." Schizophrenics and others are able to speak metaphorically about things they cannot face directly or more literally. By means of metaphor we are able to distance ourselves from an object, person, or situation. Therapy also involves showing that what one takes literally is really metaphorical [5].

1.3.34 Use elucidating metaphor. Metaphor is often used to clarify, to describe phenomena, which cannot otherwise be described.

1.3.35 Reduce the abstract to the concrete. Take an abstract term and give clarity to it by reducing it to concrete exemplification and illustration.

These are some methods of analysis, which are used in the medical literature and throughout this book. (See also the Chapter 3).

1.4 Case Example: A Healthcare Worker (H) – Patient (P) Metaphoric: H/P Modeling in Medicine

The various possible relationships between the healthcare worker and patient may be given by means of the metaphorical method. The following is, then, a metaphorical exploration of the physician-patient or nurse-patient relationship. We may first note that there is a problem of the self here. Whatsoever description of the self is to be used in the H/P relationship? Whichever definitions of the self of the healthcare worker and patient are given, they create a different relationship. The patient may be regarded merely as a body to be only physically treated, or as a person to be holistically treated and all gradations between. The healthcare worker (H) may be regarded only as a part of medical institution, as impersonal, functional, one to be always there.

H and P often refer to undefined or vaguely defined abstractions. Whether defined or not, individuals may only partially participate in the definition of the class. The distinction between evaluative, stipulative, descriptive and other types of definition must also be considered. The diversity of meaning invites equivocation, but also interesting comparison.

What is meant by the basic terms, healthcare worker and patient? (See Chapter 9) The philosophy and beliefs of each H and P generate numerous different relationships. The following are some traits often ascribed to each role: The healthcare worker is rational, logical, caring, professional, responsible, dependable, efficient, knowledgeable, supportive, communicative, patient, altruistic, concerned, obligated, etc. The patient is not required or expected to have such qualities. The patient can be aggressive, egoistic, irrational, non-communicative, ungrateful, coarse, uninformed, impersonal, irresponsible, uncaring, non-participatory, untruthful, superstitious, etc. These stereotypes confine to certain role-plays, often bad ones.

However, individual differences between healthcare workers may be as different from one another as patients are from one another. We may transcend physician-patient roles entirely. Looking for general psychological physician-patient differences may be counterproductive and does not help the individual person. The philosophy and beliefs of each H and P generate numerous different relationships. Ideally, both H and P would be equally emotionally positive, rational, informed, honest, humanistic, and willing to fully participate in the treatment required. Humanism transcends enculturated healthcare and patient roles.

1.5 H/P Models

H and P form the major variables, which are then related by diverse metaphorical connectives. The rules of transformation create a kind of calculus of the H/P relationship by means of which to generate arguments and theories. The metaphorical method is exploratory and avoids both absolute definition as well as dogmatic essentialism. Heuristic models may be created to explore diverse possible combinations of the relations between the healthcare worker (H) and the patient (P).

H-R-P (R = a relationship such as: treats, cares for, charges, diagnoses, is legally bound by, etc.)

1.5.1 H = P

The patient participates in his or her own healthcare equally. There may be physician-patient decision-making. This can involve equality, flexibility of roles, rejection of differences, identity, a dialectical exchange, total cooperation, and humanism. The oxymoron, "the patient becomes his/her own physician," may give the insight that a patient's autonomy may override the expertise of the physician. "The physician is the patient" may suggest that the physician has excessive concern for the patient and suffers with the patient, or that the therapist has problems as well as the patient. Physicians need not take the role of suffering patients, but be separate fully functional individuals.

1.5.2 H versus P

H versus P may make for an adversarial relationship. The patient may be seen as an opponent or challenge.

1.5.3 Not H and not P

Here the teachings of religion or the law trump the decisions of both H and P.

1.5.4 *H or P*

This is an either-or fallacy. It ignores gradations between H and P. "H or P" is like black or white. There is an element of this in both autonomy and paternalism.

1.5.5 *H and P*

This considers both individuals. It suggests cooperation, open communication, shared judgments, humanism.

1.5.6 *H not P*

The healthcare worker is made totally responsible for care of patient with no responsibility on the part of the patient. In a negative form this may involve paternalism. Power and authority are given to the physician, but not the patient.

1.5.7 *P not H*

The patient may have total autonomy not considering any consequences.

1.5.8 *H (verb) P*

E.g., physician *supports/overpowers* patient.

1.5.9 *P (verb) H*

E.g., patient *supports/overpowers* healthcare worker.

1.5.10 *H ≠ P*

Learned or unlearned differences between males and females. Often stereotypes. Treatment for women may not be the same as treatment for men.

1.5.11 *H ? P*

This suggests that the relationship between physician and patient is unclear or unknown. Certainly the relationship changes over time.

Grammatical and rhetorical possibilities are too numerous to mention them all, but the above schematic can show some of the many possible physician patient relationships. In the above models, time, context, and quantity must be specified. The above H/P chart should accordingly be changed to, e.g., (quantity or degree) H = (quantity or degree) P, at time t in context (specified) x. "=" must be clarified as well as H and P. That this is seldom done only indicates that the models H = P, etc., are used as root metaphors to be expanded in many directions. If we are conscious that these are only metaphors, we may use them to gain insight. If we treat them essentialistically as literal or positively true, we commit the metaphor-to-myth fallacy.

The Metaphorical Method is used throughout this book to critically examine medicine and bioethics, practice and theory and establish a philosophy of medicine relevant to its practical tasks.

References

1. Frost, R. 1949. Education by metaphor. In *Selected prose of Robert Frost*, eds. Cox, H., and Lathem, E. New York: Holt, Rinehart and Winston.
2. Shibles, W. 1971. *Metaphor: An annotated bibliography and history*. Whitewater, Wisconsin: The Language Press.
3. Shibles, W. 1971. *An analysis of metaphor*. The Hague: Mouton.
4. Shibles, W. 1995. *Emotion in aesthetics*. Dordrecht: Kluwer Academic Publishers.
5. Shibles, W. 2002. *Humor reference guide: A comprehensive classification and analysis*. eBook on humor at http://facstaff.uww.edu/shiblesw/ or search: Shibles, Humor.
6. Ramsey, I. 1964. *Models and mystery*. New York: Oxford University Press.
7. Burke, K. 1954. *Permanence and change: An anatomy of purpose*.Indianapolis: Bobbs-Merrill.
8. Kuhn, T. 1996. *The structure of scientific revolutions*. Chicago: University of Chicago Press.
9. Wittgenstein, L. 1958. *Philosophical investigations*, 3rd edn. New York: Macmillan.
10. Shibles, W. 1974. *Wittgenstein, language and philosophy*, Rev 3rd edn. Dubuque, Iowa: Kendall-Hunt.
11. Shibles, W. 1973. Die metaphorische Methode. Merz K tr, *Deutsche Vierteljahrsschrift für Literaturwissenschaft und Geistesgeschichte* 48:1–9.
12. Kuon, E., Pfahlbusch, K., and Lang, E. 1995. The diagonal ear lobe crease for evaluating coronary risk. *Zeitschrift für Cardiologie* 84:512–519.
13. Hyman, M., and Liponis, M. 2003. *Ultraprevention*. New York: Scribner.
14. Shibles, W. 1978. *Rational love*. Whitewater, WI: The Language Press.
15. Denfeld, R. 1995. The new Victorians: A young woman's challenge to the old feminist order. New York: Warner Books; Patai, D., and Koertge, N. 1994. *Professing feminism*. New York: Basic Books; Shibles, W. 1989. Radical feminism, humanism and women's studies. *Innovative Higher Education* 14:35–47; Shibles, W. 1991. Feminism and the cognitive theory of emotion: Anger, blame and humor. *Women and Health* 17:57–69; Shibles, W. 1991. The myth of patriarchy. *Journal of Value Inquiry* 25:305–318; Sommers, C. 1994. *Who stole feminism*? New York: Simon and Schuster.
16. Marcus, Aurelius 2006. *Meditations*. Hammond M tr. London: Penguin Classics.

Chapter 2
Definition

Definitions are only for use, not absolute or final [1].

Abstract In the philosophy of medicine, as with philosophy in general, we may assume that words are meaningless and statements false until defined and defended. With an uncritical language we do not know what we are talking about. We typically fail to understand how definition works. Because there are many types of definition we often knowingly or unknowingly argue a case by equivocating between different definitions or by giving a false, biased, or persuasive definition. We cannot have absolutely true or literal definitions for anything. Definitions may be rather regarded as perspectival seeings-as. To define non-circularly is to relate different things. To define is to take a model or metaphor. Distinction is made between several types of definition. To define and critically examine given definitions will be a main task throughout this book.

Keywords Definition · types of definition · misuse of language · circular statement · equivocation · universe of discourse · word field · literal definition · metaphorical definition · use in context definition

2.1 Where Does It Come from that We Think We Need to Define?

What does it mean to create or to have a definition? Each word may be regarded as vague and abstract (Platonic), e.g., cure, disease, autonomy, self, etc. Defining or describing can be to make definite, to de-fine, and to reduce the abstract to the concrete. Philosophical arguments in bioethics and elsewhere can seem to go along pretty well as long as we do not question our words. Then they often fall apart. This little problem of definition has become quite an inconvenience for the scientific world. All one has to do is to ask a speaker or writer what his or her terms mean and the arguments tend to dissolve. In the philosophy of medicine, as with philosophy in general, we may assume that words are meaningless and statements false until defined and defended. Intuitive and traditional assumptions about meaning are to

B. Maier, W.A. Shibles, *The Philosophy and Practice of Medicine and Bioethics*, International Library of Ethics, Law, and the New Medicine 47,
DOI 10.1007/978-90-481-8867-3_2, © Springer Science+Business Media B.V. 2011

be rejected as being largely abstractionistic. With an uncritical language we do not know what we are talking about. People, for example, use value terms and do not know what they mean. They use value terms supernaturalistically and as if values were entities somehow beings on their own. (See Chapter 5) Professional scholars as well as everyday people usually assume that they have all the knowledge they need and that they just know what words mean. This is an arrogant use of language. We propose a more modest, which means a more critical use of language. For example, "energy" is used as a scientific term, but on analysis it is found that there is no energy as such.

This misuse is so widely held that this statement should shock even our scientific reader. Many of our beliefs are based on misuses of language. We are so used to the misuses rather than the correct uses that we think it strange to investigate other uses than the familiar misuses. "Infinity" is a pseudo-scientific term and "eternity" even more so. "Being" as used in "human being" is a rather empty term. A human would be more clear and without metaphysical baggage. The philosophy of the natural and social sciences is basically an analysis of the concepts, including definitions, the basic terms and the methods in each science. When this is accomplished the terms and methods are typically found to be unacceptable. If science means true facts, there is no science. There are only more or less well confirmed hypotheses and statistical correlations. It is the task of philosophy, philosophy of medicine, to criticize the pre-conditions in terms of language and methods given in any science. Formal logicians misuse the words "true" and "false" and give empty, contextless, stipulative meanings to ordinary language terms, as will be seen in the Chapter 18. More care, more distinctions are needed. For examples of in-depth definition and clarification of concepts see especially the Chapters 10 and 21.

We typically fail to understand how definition works. Because there are many types of definition we often knowingly or unknowingly argue a case by equivocating between different definitions or by giving a false, biased, or persuasive definition. In conformity with contextualist and ordinary-language views e.g., of pragmatists and Wittgenstein [2] essentialistic definitions are not to be had. To seek a literal definition is to commit the "metaphor-to-myth" fallacy. It is akin to saying, for example, that the medical model is the only truth thereby generating a pervasive fiction. We cannot have absolutely true or literal definitions for anything. We can virtually never find a definition of any term which will be true in the various senses of "true" which will apply to all contexts, times and places whatsoever. That is, "true" also needs to be defined. There is no absolute or fixed definition of person or even of such concrete things as a drug. Furthermore, the definitions given are often circular, for example, that abortion is *wrong* because it is *immoral*. This is a circular statement because it defines a term with its synonym. It is like saying something is bad because it is not good. It is redundant. In addition, the value terms, as will later be shown, are meaningless in themselves. The abortion argument here is simply based on a misuse of language.

Definitions may be rather regarded as perspectival seeings-as. To define non-circularly is to relate different things. To define is to take a model or metaphor. "Humans are machines," or "The world is atomic," are metaphors. We will not, therefore, be able to conclude that the real definition of "medicine" or "cure" is such and such. Because absolutistic, it is not acceptable to say, "x is just y," "x is really y," "x is essentially y," or "x must be y." We may, on the other hand, choose to regard them as such for a certain purpose. An institution is sometimes regarded as a person before the law. This is a stipulative definition. To define, taking a model or metaphor is done in the following example. Medicine is treatment in the context of much uncertainty. Medical decision-making is often like wave theory according to which there are only probabilities.

Other types of definition are possible. When one type of definition is mistaken for another, equivocation results. It is a mistake to take a recommended definition, e.g. "Humanism is evolutionism," as a descriptive definition. One may certainly be a humanist without subscribing to the scientific theory of evolutionism. One may also change one's view of evolutionism and humanism.

Disorders in medicine are usually presented in terms of description, causes, symptoms, diagnosis, prognosis, treatment, and prevention. These create problems of defining in medicine. The most basic terms in medicine are usually undefined or defined in unacceptable ways, such as: bad, bioethics, death, disease, duty, embryo, ethics, fetus, good, meaning of life, medicine, mistake, ought, person, quality of life, value of life, etc. Clarification of these terms will be given in this book. What is a person? Problematic answers have been given which specify that an entity is a person at virtually every stage of development from the idea of conception to only the time if and when one has achieved rationality and a humanistic ethics. We may distinguish between person and personhood. One may be a living being, but not have achieved personhood. So also one may be biologically a female, but not have achieved womanhood.

Each classificatory system and each revision of a classificatory system classifies psychological disorders differently. DSM-III-R, DSM-IV-R, and ICD-10 yield different classifications [3]. In Germany, for example, ICD is used and the International Classification of Procedures in Medicine (ICPM). In DSM III (1987) homosexuality was listed as a disorder, but not in DSM IV. DSM IV is a multiaxial system revised in 2000. The 5th edition will appear in 2010 or 2011 [4]. Ethical terms are almost always undefined or unscientifically defined. Practically in all of the classifications in DSM, "mental" disorders are defined unacceptably. A careful diagnosis would be to merely describe the symptoms reasonably without stereotyping by inadequate DSM labels. The term "mental" is itself a pseudo-psychological term because there is no such thing as a mind. This can serve as a paradigm case of the problem because most think they have a mind, so much so that this statement may strike them as unintelligible. For that matter, ideas, will, imagination and memory as such do not exist either. If this shocks the reader it may then well serve to indicate the severity of the problem. It is called the mentalistic fallacy and will be further discussed in the text.

If there is no diagnosis, there is thought to be no disorder for official and insurance purposes. Yet, there is a significant problem related to the classification of disease entities and the diagnosis of diseases. Some mathematical and statistical approaches used to aid such analyses include cluster analysis, discriminant analysis, Bayesian methods, etc. In ICD-9-CM circumcision is found under "Elective surgery for purpose other than remedying health states" [5]. "V50.2 Routine or ritual circumcision." One definition given was, "Circumcision in the absence of significant medical indication." In same category is ear piercing. (50.3) This is to support an anti-medical and harmful practice of circumcision.

One of the central tasks of the philosophy of medicine is to define and carefully analyze such basic terms of medicine and ethics. This will be done throughout this book. In terms of diagnosis there are, for example, problems with a severely incomplete knowledge base and claims about the degree of certainty one may have. Nosology deals with the attempt to adequately classify diseases. There are also problems with causal definitions because cause is an unscientific concept and it will be analyzed in the Chapter 3. It is often more informative to just describe the individual's state.

2.2 Distinction Between Types of Definition

Abstract or essentialistic definition. An abstract definition is the uncritical use of vague terms, e.g., cause, energy, idea, good, moral, mind, force, etc. This is called also *naming-fallacy, Platonism, essentialistic terms.* General terms can reduce to instances. Abstract terms often or always are not intelligible in this way. Simple everyday terms, like bad and true, are among these abstract terms. Feminists and Women's Studies classes use the word "patriarchy" as a pejorative term, which Elshtain says is "to give a distortion of our society, and it is a dogmatic view" [6]. "For 40 years investigators wrote countless papers about the 'cause' of essential hypertension without wondering whether there was any such entity. 'Hypertension' is an abstraction that omits most of the information on which a sound judgment would be made" [7]. Involved are inflamation, neoplasia, homeostasis, and stress. "The more closely the terms are examined, the more elusive they become" [7].

Activity. Definitions are given as a report of an activity. If the patients can perform certain physical tasks they are said to have quality of life. A term, of course, which is itself open context.

Analogy. Definition by analogy uses the expression: "X is like. . .." What is it like to be without language or in a coma? We cannot know. Wc supposedly can only try to determine what it is like to be us. And this is difficult enough.

Cause. This is definition as giving causes, both true and false causes. "His disease was due to old age." Age is associated with disease, but may not be the cause. Similarly wrong are: "I see wavelengths," and "I hear vibrations." These are not the sorts of things one can sense. Historical statements are typically false ascriptions of causes. Freudianism ascribes false causes, e.g., "I missed the point" is due to breast withdrawal during infancy [8]. Astrology is based on the determined configurations of planets to cause us to be the way we are. In spite of critical philosophical and scientific knowledge people still believe that and some check astrological and full moon charts before doing surgical operations. Now we would think genes do much of that causation. "My mind causes me to think" is giving a false cause because there is no mind. Cause-effect has an infinite number of definitions so we must specify the definition under which a chosen cause is said to have an effect.

Circular definitions or identifications. Synonyms are put together. "Dead" is "no longer alive." "You ought to do your duty." These are fallacies of circularity. The demand for personal medical care, by definition, cannot be given. What is demanded cannot be given. Thus, caring cannot be demanded. This is in one sense a contradiction.

Classificatory definition. Classifications are arbitrary, based on certain problematically selected criteria. Several classificatory systems in medicine are:

ICD-10. International Classification of Diseases. 10th revision. 2006
CPT = Current Procedural Terminology
NANDA *International Classification of Nursing Diagnoses.* nanda.org/html/
McCloskey & Bulechek. *Nursing Intervention Classifications* (NIC) 1992
Omaha Problem Classification Scheme. omahasystem.org/

A standardized classification (Taxonomy) recognized by the American Nurses Association includes an assessment component (Problem Classification Scheme), an intervention component (Intervention Scheme), and an outcomes component (Problem Rating Scale for Outcomes) [9].

Constructivist definitions. These are definitions, which do not claim to realism. All models are created in social settings and so are not perfectly objective. We may construct a strategy or fiction for therapy though not real, but does work. We cannot know all of the causes anyway. Medicine often uses constructive definitions.

Definition in the universe of discourse. This means a description from the point of view of each discipline. There may be a psychological, physical, aesthetic, sociological, mathematical, financial, etc. description of a treatment. Each is a universe of discourse.

Denotative definitions or names. But words do not just name objects by demonstrative meaning. Pointing and sensing are complex activities. The picture of an object is not the definition of an object. To refer to an object or to have an object is a linguistic activity. For Wittgenstein the meaning of a word is just the use of the word in a language context (language game) [2].

Dictionary and encyclopedia definitions. These are summaries of common uses, which we may call the "word-field" of a term. Together with the synonyms of a term we can construct the various different meanings involved and then analyze them. It is one of the significant first steps in doing a philosophical analysis of a term. Such an analysis for example can show that our definitions are full of empty or contradictory meanings and that some may be so vague as to preclude usage.

Eliminative description. For the eliminative materialists, we would replace any mental description with a physical one. On the medical model, all explanations are physical ones.

Endless or incomplete definitions. The possible meanings and definitions of a term are endless.

Exemplifications. These are definitions, which reduce the abstract to the concrete, theory to cases or practice, or give cases or examples. Clinical medicine is medicine exemplified.

Explanation or reason. Definition may be seen as an explanation, like giving a narrative or story. Words resist any fixed or absolute explanation. They just have uses in a language game. The verbal narrative in medicine experiences a present popularity. The patient is to express himself as a narrative, s/he defines him/herself as a narrative, s/he defines him/herself talking about him/herself.

Form of life. Definition itself is regarded as an experiential way of understanding, or an attitude [2].

Humorous definition. Definition is given as a joke, which may or may not contain insight. A circular statement and any informal logical fallacy is the basis also of humor [10].

Literal definition. In conformity with contextualist and ordinary-language views of language [2], literal definitions are to be avoided in order not to be giving a misuse of a literal language. For example, Patrick Suppes wrote, "We do not yet know exactly what mathematics is" [11]. This is not surprising, as we do not have an exact literal definition of anything else either.

Postmodernists hold that words and life cannot be defined. We are in the prison house of language. There is no literal definition. Each history and novel could have been written differently. As soon as you have a definition it becomes something else – other words. In this way also our memory works (as work in progress). Our identity is never defined.

Meaning. Definition is given, as asking for a meaning.

Meta-definition. Definition is itself a meta-activity. It is to be critical, to make comparisons and juxtapositions. Thus the metaphorical forms of rhetoric can be used to explore and analyze a word.

Metaphorical or creative definitions. One may be philosophically dead, intellectually dead, emotionally dead, or socially dead. Thus the metaphorical method is used to define and evaluate. Each technique in the metaphorical method is a way of defining, e.g., by poetry, or informal fallacy, or humor. Defining explores the limits of language. Metaphor, "x is y" is literally false or absurd [12].

"Meaning is use" and "language-game" are also metaphors [2]. Definition and descriptions are themselves metaphors. Health may be seen as involving ethical, consequentialistic holistic, and humanistic reasoning, and all of these as producing the aesthetic value of life. This may be termed "ethical healthcare." That is, holistic health care involves all of one's life.

Negative definition. "X is not a...." This is a limiting definition.

Normative definition. This is what the word is commonly thought to mean.

Operational definitions. We define in terms of procedures rather than abstract existence. Percy Bridgeman [13], Nobel prize-winning physicist, proposed that all terms if they are to be intelligible must be reduced to concrete paradigms, or operational definitions [13]. By rational and intelligible operational definition we remove all un-definable terms. (cf. fallacy of abstractionism) "It is better to analyze in terms of doings and happenings than in terms of objects or static abstractions." Also this is anti-logic: "I can...no longer feel the interest in some of the analyses of symbolic logic" [13] (For a critique of formal logic see Chapter 18).

Pragmatic/Purpose. This is a definition, which may express or state one's purpose. Phronesis is practical medicine.

Recommended definition. A value definition may be intended to promote a certain perspective. The World Health Organization defines health as "a state of complete physical, mental and social well-being and not merely the absence of disease or infirmity" [14]. Holistic medicine should stress the whole environment and prevention. Health promotion is the science and art of helping people change their lifestyle to move toward a state of optimal health. "Optimal health is a dynamic balance of physical, emotional, social, spiritual, and intellectual health" [15]. Here health-care is seen as not just curing diseases, but improving patient's lives. One goal of Philosophical Counseling is to improve the individual and society, not just maintain the present cultural system and state. (See Chapter 17) Bad ethics and inadequate goals are also anti-medicine.

Reductionist definitions. Can thinking be reduced to physiology? Can any universe of discourse be reduced to any other?

Stipulative definitions. The law, for example, stipulates definitions for its own use. It may take the form: "Let the term x be defined as...."

Technical definition. This sort of definition belongs to the type of definitions for a special universe of study.

Use in context definition. Definition is determined by the contexts of its use. The range of its possible uses depends only on the limits of the possibilities of our language. Oh and co-authors in "What Is Health: A Systematic Review of Published Definitions" give the various definitions of "health." They conclude that there is no clear consensus about the meaning of the term "health" [16]. Most definitions implied that theirs was "the" definition. In his later work, *Philosophical Investigations,* Wittgenstein says the way a word is used is what makes it useful in the language [2]. For this reason we have not yielded to the temptation of finding another "better" definition of Health. It has meaning through the perspectives, settings and contexts in which it is used.

Value or persuasive definition. Mercy Hospital (Wisconsin, US) advertises, "We are professional grade." "We are caring."

The Catholic view of the fetus as a person or as an "embryo-person", attempts to be persuasive. "Healthy" is an open value term and not a scientific one. It often only means not presently ill however bad our condition is. Even when one refuses to consider consequences for the lives of people, one claims to be "pro-life."

To define and critically examine given definitions will be a main task throughout this book.

References

1. Schiller, F.C.S. 1930. *Logic for use*. New York: Harcourt Brace.
2. Wittgenstein, L. 1958. Philosophical investigations, 3rd edn. New York: Macmillan.

3. DSM-III-R, DSM-IV-R, ICD-10. 10th revised edition. 3 vols. WHO, Geneva, Switzerland.
4. www.psycweb.com.
5. International Classification of Diseases. 2000. AMA. 9th Revised edition.
6. Elshtain, J. 1981. *Public men, private women*. University Press: Princeton.
7. Murphy, E. 1997. *The logic of medicine*, 2nd edn. Baltimore: Johns Hopkins University Press.
8. Sharpe, E. 1940. Psychophysical problems revealed in language: An examination of metaphor. *International Journal of Psychoanalysis* 21:201–213.
9. Martin, K., and Scheet, N. 1992. *The omaha system: A pocket guide for community health nursing*. PA: Saunders, Philadelphia
10. Shibles, W. 2002. Humor reference guide: A comprehensive classification and analysis. eBook on humor at http://facstaff.uww.edu/shiblesw/ or search: Shibles, Humor.
11. Suppes, P., and Edwards, P. 1967. *Encyclopedia of philosophy*, vol 2, 310. New York: Macmillan.
12. Beardsley, M. 1958. *Aesthetics*. New York: Harcourt.
13. Bridgeman, P. W. 1927. *The logic of physics*. New York: Macmillan.
14. WHO Basic documents. 43rd edn Geneva, World Health Organization. 2001a, p.1. See also Preamble to the Constitution of the World Health Organization as adopted by the International Health Conference, New York, 19–22 June, 1946. URL: http://www.who.int/about/definition/en/ (accessed 2004 July 5).
15. O'Donnell M. P. 2009. Definition of health promotion 2.0: Embracing passion, enhancing motivation, recognizing dynamic balance, and creating opportunity. *American Journal of Health promotion* 24:4
16. Oh, H., Rizo, C., Enkin, M., and Jadad, A. 2005. What is health: A systematic review of published definitions. *Journal of Medical Internet Research* 7:58–60. jmir.org/2005/1/e1/.

Chapter 3
Decision-Making: Fallacies and Other Mistakes

Abstract Medical methods and decisions are often based on traditional, culture-bound practices, rather than on sound, critical thinking considering consequences. As medical language and practice are influenced by the culture, culture must be critically examined. Decision-making is based on language. It mainly involves the making of causal statements, which are phrased in language and can therefore be no more precise than the language used. Frequent causes of irrational medical thinking and decision-making are analyzed especially fallacies from the claim of certainty where only can be probability and medical and societal un-culture of dealing with mistakes. Five levels of decision-making in medicine are presented: 1. Medical, 2. Ethical, 3. Interdisciplinary, 4. Philosophy of Medicine, and 5. Philosophy in general.

Keywords Perception · claim of certainty · captivation by a model · reductionism · consensus · informal logical fallacy · abstractionism · either-or fallacy · rationalization · mistakes

This is a book not just for physicians, nurses, philosophers or bioethicists, but for all those involved in decision-making in healthcare and personal health. This also includes the general public. It is especially for those who suffer and struggle with illness. It is also for management, administration and politics, which directly or indirectly and on all levels, make healthcare decisions.

3.1 Conditions of Decision-Making

Medical methods and decisions are often based on traditional, culture-bound practices, rather than on sound, critical thinking considering consequences. As medical language and practice are influenced by the culture, culture must be critically examined.

Gary Heiman states, "Evidence must be empirical—meaning learned by observation" [1]. This is called "naïve empiricism." The scientific method cannot be ultimately based on perception, because it presupposes language as a basis.

B. Maier, W.A. Shibles, *The Philosophy and Practice of Medicine and Bioethics*,
International Library of Ethics, Law, and the New Medicine 47,
DOI 10.1007/978-90-481-8867-3_3, © Springer Science+Business Media B.V. 2011

Perception presupposes language. Furthermore, there are various linguistic theories of perception. We are not clear how perception works. (For the relation between thinking and language, see Chapter 18) Thus, such important concepts as causality cannot be simplistically based merely on the empirical. In short, the scientific method is epistemologically based on thinking as language-use rather than on mere perception. Perception is also inextricable from language use. The expression, "I see or treat the wound," is not possible without language. Without language we would not know what seeing would be like.

Decision-making, then, must also be based on language. It mainly involves the making of causal statements, which are phrased in language and can therefore be no more precise than the language used. Given that causal theories cannot be based on perception, but rather on language use, the approach to clarifying causal statements would be to analyze the uses and misuses of causal terms, including those dealing with perception. It would be circular to examine perception by means of perception. That is, language must be presupposed in order to do the examining. There is no cause as such, only different models of rendering a situation by means of language.

Physicians, as well as scientists, can take models and methods of treatment and analysis literally, and so be captivated, but there are other models, for example, the psychosomatic approach goes beyond the incomplete classical somatic approach [2]. To avoid becoming captivated and indoctrinated, both principles and methods must be constantly reevaluated. This is why the critical examination of methods and concepts in science and medicine is required by philosophy of science and philosophy of medicine. We must go beyond the given principles and methods of medicine in our society as we must go beyond its enculturated thinking and morality.

Decisions may be rational or irrational. Anyone can make a decision, but surprisingly few can make a rational one without training in critical thinking (speaking) and argumentation. Physicians have training "antagonistic to critical thinking" [3]. Studying medicine at our universities mostly involved and still involves learning by heart and questioning little. Also, making a decision to do something does not mean one will act on the decision. A "command decision" is needed to actually perform an action, e.g., "I will do it now" [4].

3.2 Frequent Causes of Irrational Medical Thinking and Decision-Making

1. The *claim of certainty* is one of the causes of irrational medical thinking although it is not to be had. As a practicing physician, it is my view that physicians, as in many other areas of knowledge, seldom have perfect or certain knowledge, only more or less approximate knowledge. "Cause can never be unequivocally proven in a scientific case" [5]. Wittgenstein thought of knowledge and certainty as language-games, which give us many meanings of these terms, but not an absolute knowledge or certainty [6]. There is only probable, not absolute, cause. Pan-sophism is the pretension to universal knowledge. This is not to be had and it is a fallacy to expect perfection. For example, recognition

of risk factors accounts only for 50% of the causes of ischemic heart disease [5]. We also might find that from three physicians we are given three different diagnoses. Each one tends to take a somewhat different approach, e.g., osteopath, chiropractor, surgeon, nutritionist, general practitioner, etc. The surgeon, J Isenberg, admitted, "A surgical procedure is an experiment even under the most controlled conditions" [7] and "An individual patient is unique and not translatable into a confined and simple system. Any attempt to equate the types of activity that probability theory was developed to evaluate... to the treatment of a given individual with a certain surgical procedure is pointless. They are not equivalent" [7]. The main thing in diagnosing is to leave the diagnosis open for reevaluation and reinterpretation. "Few tests perfectly discriminate between normal and abnormal" [8]. Thus several diagnostic tests are often needed.

Fixed diagnoses and protocols can mean danger for the patient, false categories and diagnoses, wrong treatments.

Charles Eaton, MD (Juniper, FL) gives in his office brochure (2004) the following disclaimer to his hand surgery patients: "No one knows precisely how our bodies work, and no two hands are identical. In the practice of medicine, there is no way to guarantee either accuracy of diagnosis or satisfaction with outcome."

2. The notion that there are *absolute principles and methods* is dangerous and misleading. Concrete cases and individual differences can make the usual principles and methods useless. Methods used may vary from case to case.
3. *Captivation by a model or metaphor.* Several criteria for a term, model, or hypothesis are: What does it mean? What is it like? Does it have explanatory power and practical relevance? As will be seen, even scientific research seldom meets these criteria. For example, the terms energy, person, embryo, disease, cause, etc. are used uncritically and seldom have acceptable definitions.
4. The *Failure to define* terms and words used, including the definitions of disease and health creates confusion.
5. Maintaining the view that there are *absolute or literal definitions* of words is misleading.
6. *Poor management and administration* induces decision-making problems. (See Chapter 8)
7. *Faulty view of the scientific method* cannot provide scientific data.
8. *Lack of knowledge of the informal logical fallacies*, such as, circular definitions, argument from force, etc. creates faulty thinking.
9. *Lack of a sound humanistic philosophy and personal and institutional goals* depersonalizes medicine. Sound decisions can only be made in terms of a well-organized purpose or philosophy. "Cause and effect are not merely legal problems. They are problems...which concern the philosophy of life itself" [9].
10. *Narrow* decision-making fails to consider contextual and holistic perspectives.
11. *Reductionism* of medical treatment to only its physical aspects is referred to as the classical medical model, or body-machine concept. For example, according to this model there is reduction of thought to impulses in nerves or cellular activity. This fallacy is called the "Explanatory Gap" in philosophy. For

example, we do not know how an impulse in a nerve becomes a phenomenal experience.

12. *Lack of knowledge of ethics*, emotions, humanism, and goals for life upon which to base sound decision-making. Thus, for example, decisions can be fair or unfair. It is especially necessary for physicians to know about emotions. Physicians as persons of respect are often said to be like a placebo, a form of treatment, and so part of the recovering for the patient. Thus, their emotions have more than the usual impact on patients. Physicians especially must have balanced emotions in their involvement because of the extreme and demanding emergency situations they daily encounter. (See Chapters 5 and 7).
13. Failure to open up one's decisions for discussion, *lack of open communication* and constructive criticism. (cf. the Chapter 8) Without having all of the reasonably obtainable facts and opinions one cannot make a proper decision.
14. Basing decisions on *dogma,* unscientific beliefs, superstitions or religion. These are blind decisions without concern for evidence, consequences, or humanity. An example is the tenaciously held view that the fertilized egg is a person. Religious dogma also often prevents much needed medical research. Such beliefs are ways of letting people die. Humanistic and ethical management cannot be based on religion.
15. *Lack of critical thinking* (speaking) and the inability to know or present organized and adequate arguments makes rational decision-making difficult or impossible. It may disqualify one from rational decision-making.
16. Decisions made without *consideration of context, consequences and goals* are empty, of no more value than hasty decisions.
17. Decisions can be *prejudicial*. They are often not made objectively in terms of the institutional goal, or task to be performed, but in terms of one's individual belief system and personality. It requires a rational, humanistic person to make a rational, humanistic decision. In the coauthor's questioning of over two hundred classes of about 40 students each, it was found that they thought the average person has about 75% negative emotions. If the average person has about 75% negative emotions and is enculturated and egoistic, one would not, in general, expect rational decisions to be made by them. (See Chapter 7.) Accordingly, many medical as well as managerial decisions are made basically for personal agendas and imposed on the organizational goals, which is often to gain or save money. If this is the only concern it often is to the detriment of the wellbeing of the employees and patients. Defects in one's personality and character are good indications of defective decision-making ability. Arrogance and abusive behavior toward others are predictors of poor management and poor decision-making.
18. Decisions may be based on *mere obedience* to one's superiors, blind obedience, and so the most objectionable acts may be performed in medicine, at the various levels. "Nobody has a right to obey." (Hannah Ahrendt)
19. Failure to realize that one is *oneself responsible* for one's decisions at every level whether pleasing to one's superiors or not. This is, what autonomy, self-determination is about.

20. Failure to have (and to get) the *scientific and practical knowledge* necessary for diagnosis and decision-making. Diagnosis is like detective work. All hidden clues must be uncovered. For example, physicians ask if pregnant mothers smoke, but fail to ask if they are exposed to secondary smoke, which can also be damaging.
21. Decision is a choice and without *rationality* one cannot reasonably choose, and so one cannot genuinely decide. In short, in as much a person is irrational s/he cannot genuinely choose or decide rationally.
22. Decision-making should be based on *adequate information*. In medicine, computer checks of the symptoms and causes of disease can be essential, but must be used cautiously. The recent introduction of evidence-based medicine is an attempt in this direction. (See Chapter 19)
23. *Statistical decision-making* is often used as a "safe" pattern. The physician often assumes that a patient has the common disease for the local area and times. The physician is safer in assuming that the patient has the more common diseases, because they happen a lot, whereas the rare diseases hardly happen at all. This approach is still inadequate. A check for all of the reasonably accessible possibilities is needed, and this can be computer assisted. This is again relevant to evidenced-based medicine. Risk is often given a statistical value, but is based on different conditions and patients in very different situations. Risk is often based on age groups regardless of health status, gender or race, etc. Though useful, statistics are quantitative, vague, and inaccurate abstractions. They always require critical and qualitative evaluation. A sound scientific method requires that one not generalizes beyond the specific data and experiment. There is especially the failure to consider the individual patient when citing general statistics. Even if statistics were not problematic, the court may disregard such data or use it in a biased way. "Laws made in the state capitals and based on generalized data and average ratings of patient acuity and nursing care cannot provide the answers... The judgment of educated, trained, and caring professionals cannot be legislated or averaged" [10]. (See Chapter 19)
24. *Consensus*. The attempt to achieve a consensus may help bring about a decision, but it is based on the fallacy of appeal to the views and emotions of the majority. The decision may be poor. That the majority holds a view does not mean that it is in any sense a rational one. For a sound decision, rational argument and evidence is necessary. One person's sound argument would not prevail over the irrational opinions of all others if consensus were the standard.
25. Nurses are often not allowed enough time to deal adequately with every patient. They then can only *prioritize*. The one who cries loudest, receives attention, those who cannot even cry, remain neglected. This is failed decision-making or none at all.

From the philosophical perspective, Walton stresses rational case methods and dialog between physician and patient [11]. He opposes general principles, and moral universals in favor of a pragmatic approach. Treatment is seen as an ongoing joint negotiation in terms of articulated goals. Here he comes close to a beginning of philosophical counseling, but he is not aware of this approach as it came largely after the book was written.

3.3 Five Levels of Decision-Making in Medicine

Five levels of decision-making may be constructed for medicine:

1. *Medical.* On the medical model, decision-making takes place on the one level of scientific diagnosis and treatment of the physical body. Value terms refer largely to physical well being, for example, "quality of life" means only physical health such as ability to walk, move one's arms, etc. The word "clinical" refers to actual medical practice.

2. *Ethical.* On this level, knowledge of ethics and ethical theory and of the use and misuse of ethical terms is required. Ethics traditionally belongs to the area of philosophy. People usually do not have this knowledge and may accordingly be said to be non-ethical. One needs to know the difference between absolutistic and consequentialistic theories, Kantian deontology and preferential utilitarian theory; and to know the meaning, the uses and misuses of ethical terms such as "good" and if, for example, "intrinsic" value is a use or misuse of that term.

3. *Interdisciplinary.* This includes psychology, therapy and psychotherapy, sociology, economics, history of medicine, political science, law, administrative rules, societal cultural practices, and religion. Interdisciplinary models should not be confused with the ethical, medical, or philosophical levels. Important as they may be: law is not ethics, cultural and normative practice is not ethics, and religion is not ethics.

4. *Philosophical Level.* The *Philosophy of Medicine* is of practical relevance in decision-making. It involves the philosophy of the natural and social sciences, philosophy of law, philosophy of history, philosophical psychology [including the philosophy of the self (person), the philosophy of emotion, the philosophy of therapy], political philosophy, philosophy of religion, aesthetics, etc. as a critique of the methods and concepts used of the various interdisciplinary areas. The philosophy of medicine can resolve the question of the search for a consensus and universal value system upon which supposedly all can agree. It can point out that there is no such universal thing. What is "harmful to society" as well as to an individual must be determined by a sound ethical system, not by cultural or mere consensus norms. On the level of the philosophy of science, one tries to critique, evaluate, and resolve the differences between the various methods, concepts and belief systems. This is the traditional function of philosophy. The analytic task is to critique the concepts and methods of the various disciplines; the synthetic task is to produce new ways of understanding and new methods by means of which to resolve issues. This is where metaphoric models and informed discussing begin.

5. *Philosophical in general.* This is the holistic, integrative and comprehensive modeling of all of the other levels. On this level philosophical models are chosen and creatively developed including philosophies of life. It is also the level on

which the various competing models and belief systems can be critically evaluated and integrated. For example, humanism is a model, which integrates ethics and a pragmatic and practical model with a rational and scientific model. Models of decision-making may be based on such favorable or unfavorable models as the following: altruism, culture, exploitation, family practices, human and natural ecology, humanism, the legal system, obedience, organic wholeness, politics, profit-motive, punishment, religion, rules, science, selfishness, set protocol, the supernatural, etc.

The philosophical level is not a search for one universal principle, or method, but rather an open examination of all principles and methods. Philosophy is not in competition with normative thinking or other models such as religion. It is an attempt to examine, understand and offer theoretical and practical, philosophical perspectives. One of the most important preconditions of critical philosophy is the passion for inquiry, the critical evaluation of each other's views, but also criticism of one's own views. The philosophy of medicine therefore does not allow bioethics and medicine to fall into doctrinaire hands, which unfortunately is often the case. This level may be characterized by the use of all of the methods, theories, concepts and criticisms philosophers traditionally use. Philosophical Counseling (Philosophical Practice) has been established to apply philosophy on the practical level, especially in the 1990s, which is to provide counseling on all of the above areas in which philosophy is involved. (See Chapter 17) It could, and to some extent does, also play an appropriate and essential role in International and National Ethics and Bioethics Committees and Institutional Review Boards.

In sum, the five levels of decision-making in medicine are 1. Medical, 2. Ethical, 3. Interdisciplinary, 4. Philosophy of Medicine, and 5. Philosophy in general.

3.4 Fallacies in Decision-Making

Anything can and often has been said to be a cause regardless of how absurd.

One of the most important methods of rational decision-making is to make sure one avoids informal logical fallacies, that is, misuses of language. Some of the most important fallacies are briefly listed here, and in the next chapter applied to cause.

"All" statements or "none" statements. (Also "always" and "never" statements). We often make "all" statements without checking to see if there are exceptions. For example: "For every cause there is an effect." This is an all-statement, but since it is circular it is empty but nevertheless true. We may say, "You are all wrong." But can one be all wrong, wrong in every way? The same applies to none or no one. Is there ever just one cause, rather are not causes always multiple? Typically, there is not one cause, but many. For example: No one knows the cause. Everything is stimulus and response. Everything will be all right. Everything is chemical. Everything is mathematical. Every physician can be easily replaced. Time heals all. All is relative. Everyone who has this disease survives. No one survives this disease. You are always late. At death we will know everything. Everything you did was wrong. Emotion is female. Reason is male. The deity is all knowing. We can't know anything.

Abstractionism fallacy. Cause is a vague abstraction. One of the most significant errors of human thought and decision-making is the error of using overly vague terms. If we do not, then these terms will not make sense to us. Ordinary language philosophers and pragmatists, especially, wrote against abstract terms. We should be able to reduce all abstract terms to concrete and intelligible terms. If we do not, then these terms will not make sense to us. Waismann [12] and Dewey [13] wrote against abstract terms showing that examples are more convincing than an argument. Abstract terms should be reduced to concrete and intelligible examples. This is one form of the medical case method. We can talk abstractly about prevention of disease, but it makes it clearer to give examples. "Platonism," from Plato's theory of ideas, is the treatment of abstract ideas, e.g., cause, energy, disease, health, mind, the will, the imagination, ideas, etc. as if they have meaning in themselves. They do not. False abstract concepts are often used as a cause in science. If it is surprising that these terms are problematic, it indicates the extent to which they are unquestioned. No terms have meaning in themselves. For example: Our behavior and health are determined by our "genes." We know quite little about our genes, what and how traits are transmitted is generally unknown.

Ad hominem fallacy. The person is criticized rather than the argument. Instead of dealing with what is said or the argument, one attacks the person who says it whether patient or physician. In medicine one often attacks the one who discovered an error rather than deal with the error itself. Isolating out an individual to blame is often an ad hominem fallacy. One may blame an individual instead of a failed hospital policy. This may be partly because it is easier to do so.

Appeal to pity (misdirected sorrow for the failure of others often with blame and contempt). We act because of pity rather than because of the arguments for a rational course of action. We may break a rule of fairness because of pity for one person. We may favor the self-preoccupied person out of pity as well as out of friendship. (For a more full analysis of pity see the Chapter 7) Self-pity can be a form of narcissism.

Argument from ignorance. Because we cannot disprove a belief, does not mean that it is true. People often erroneously say that certain things are true causes because no one has as yet disproved them. In its classical form it runs, "If you cannot prove that x does not exist, it does exist." Of course, we can show that some things do not exist or that they are unintelligible. When we do not know how something works, some come along who claim to know of a mystical cause or cure for it, e.g., in alternative medicine. It is more honest to say, "I don't know" when one does not know. Examples of fallacious thinking are: I don't know how the world was created-it just must have been. Death must be wonderful. I haven't heard a single complaint. The hospital has a low mortality rate and no one has shown that the management is bad, therefore it has good management. But, perhaps it was never evaluated. In my Salzburg hospital situation, physicians continuously and daily cover for bad management so that the patients would have the best outcomes possible. There are in comparison to many managers few physicians who therefore constantly overwork. The harm to and morbidity rate for physicians and patients could be reasonably expected to increase, only taking the management situation into consideration, and a recent independent evaluating team has come to the same conclusion.

Circumstantial fallacy. This includes the failure to go beyond the circumstances, the failure to specify sufficient, adequate circumstances. Evidence and "circumstantial evidence" in law are often proven false. E.g., over 60 people were recently released from death-row because genetic and other tests proved innocence.

Dogma. Dogma is holding a belief without having any or enough evidence. It may even reject evidence itself. Dogma also involves unchanging and absolute views, such as the statement, "This is the only truth," or "This is the only way the illness can be treated." It is a form of the "all fallacy", of overgeneralization, of the abstractionist, literalist, absolutist and other fallacies. Religious causation is fallacious because based on supernaturalism and dogma. Some groups reject science, reason, logic and language; because they claim to have their own true knowledge. This applies to religious belief because it ignores reason and comes by revelation, or in non-rational ways. Reason supposedly is irrelevant to religious belief. One needs no evidence for belief because belief is not knowledge.

Either-or fallacy. The fallacy of thinking something is completely one thing or the opposite. Something need not be true or false, hot or cold, cause or effect. Something may be both a cause and an effect. Cause-effect can be a pseudo-opposite. Epicurus presented the view: *Death means nothing to us, because that which has been broken down into atoms has no sensation and that which has no sensation is no concern of ours* [14]. This is an either-or fallacy. It may not be comforting to a dying patient to tell them, "Do not worry, you are not dead yet." We may not be dead, but we can certainly wish for a long healthy life.

Enmeshment. Sound decisions cannot be made if there is enmeshment, that is, dysfunctional emotional involvement. There must rather be emotional integrity and consistency for good decision-making. One may support an unfair practice or selfish person without realizing the harm done because one is accustomed to do so. One thereby also becomes accustomed to being abused and encourages others to be abusive. Nationalism and family involvement can be also forms of enmeshed ego involvement whereby one supports one's country or family, right or wrong. (See also Chapter 7)

Equivocation is the use of causal terms in different senses in an argument. As there are different meanings of cause (or treatment, etc.) one may equivocate between these in an argument.

Fallacy of anger. People become angry and attempt to win an argument or try to control people by means of some form of negative emotion. Anger is not an argument. It deviates from rational behavior. We may speak of anger (including irritation) as a logical fallacy. Nevertheless, people often regard anger as necessary (absolutistic causation). It is a sign of bad management and bad decision-making. Anger is unjustifiable as a cause of an action. (See also Chapters 7 and 8)

Fallacy of force. Forcing someone into a causal belief rather than presenting an argument. Culture is a strong form of pressure. For example: The patient is always right. The one who complains most gets most attention. Closed, authoritative management is an example of the abusive wielding of power and force. It is a sign of incompetent management. Anyone pressuring another for a certain decision is using an argument from force.

Fallacy of hasty generalization. We "jump to a conclusion" on the basis of a single or only a few instances. For example: Physicians make judgments on the basis of only a few characteristics. Post-traumatic stress need not have one cause, but a lot of constant negativity. It occurs in civilian life as well.

Fallacy of irrelevance. Any kind of irrelevance may produce humor. It avoids the argument. We may prove the wrong point, or continue talking without ever getting to the point. For example: "What is the cure for this disease?" "I don't think I can answer today." Knowledge is sometimes kept from others so as to retain power. It can be a policy to do so.

Fallacy of the majority. Because most people think something is true does not mean it is true. Again, it is the argument, which counts, not merely the fact that experts or people in public debates agree with each other. We cannot vote truth in. The common view that scientific truth is based on consensus is unacceptable. However informative mere consensus cannot be appealed to in decision-making. Also the appeal to majority rule is a fallacy [15]. Examples: Because dialysis machines were limited, one hospital committee decided to not give dialysis treatment to anyone rather than decide who should get treated. The principle of equality was preserved, but not rationality.

False assumption. (Presumptuousness.) We falsely assume something, which was not presented. Causal statements are often assumptions. Common in medicine is the faulty assumption that the patients are telling the truth as they see it. (See Chapter 20.)

False certainty. Claim of certainty where there can be only probability.

False possibility. One asserts that something is possible without evidence that it is in fact possible, e.g. stating, "It is always possible that you will survive the illness," when one lacks such evidence of such possibility. The statement only means, "I do not know for certain if you will survive the illness," not that it is a possibility. It is another form of claim of certainty fallacy. It is important to promote realistic and justified hope, which is partly enhanced by the having of hope itself. Having realistic hope is to some extent a self-fulfilling prophecy. Like a performative utterance, the saying is the doing. It has a placebo effect. Medicine is a science of well-founded probability, constantly open to revision as new evidence appears.

Genetic fallacy. (Fallacy of suspicious origin of an idea.) It is a fallacy to assume that an idea is false (or true) because of the way it was arrived at. It is a fallacy to say, "He cannot be a good physician, because he is the only one who applied for the job." We ask, "Well, where did you get that idea?" It doesn't matter where you got it, or how you arrived at it, as long as it makes sense.

Ignoratio elenchi. (lit. "ignorance of refutation") What is madness? To have erroneous perceptions and to reason correctly from them [16]. This is the proving of an irrelevant conclusion. The arguer thinks one thing is proven, but instead another is. For example: "The journey has been safely performed only we got on the wrong plane."

Intuitionism. (See also legal cause.) This is the claim to have non-rational or supernatural knowledge. It is a form of self-righteousness, not a genuine method of knowing and is unacceptable in medicine and elsewhere. Intuition should not be confused with a judgment based on experience, or an informed guess, which has some basis. Sexual harassment is thought to be a form of intuition and is typically defined as: If one subjectively thinks (intuition) one is harassed one is. It is an unfair and irrational causal notion. It is a collapsing concept, called by Patai and Koertge [17] "accordion concept" because it stretches to include anything one wishes to include, e.g. if a supervisor has coffee with a female employee it is regarded as sexual harassment. Medicine is endangered by such intuitions.

Many question fallacy. (Complex or Double Question.) This is asking two or more questions covertly at the same time. Examples: "Do you always lie?" This asks the two questions: (a) Do you lie? (b) Do you lie always? "Who created the world?" is also a many question fallacy. It assumes that someone created the world, and also that the world was created. But the world was derived from things already here. It did not come from nothing. The question should be asked as two separate questions, which reduces it to absurdity, for example, "If the world was created at all, who or what created it and what can possibly be meant by 'who'?" Of course, each religion has a different and exclusive answer to this question.

Mental causation fallacy. Mentalistic fallacies and abstractionist fallacies use or assume terms such as: belief, cognition, consciousness, contemplation, decision, emotion, feeling, idea, imagination, inner states, intention, mental, mind, planning, sensation, sense, memory, thinking, unconscious, understanding, volition, will, etc. "Psychosomatic" is a mentalistic fallacy if it implies mind and body. "Improve your mind," "He is mental" (a mental patient), "Thinking is an inner process," are mentalistic fallacies. Ascribing cause to such pseudo-entities is a fallacy. Mentalistic terms may therefore be avoided in one's judgment about patients, staff, or in the area of therapy or psychotherapy. Nor would they be justifiable uses in explanations, theories, or practice; for example, they are unacceptable in the statement by Freckelton and Mendelson, "It is necessary to establish that a disease or defect of the mind exists" [18]. There is no such thing as a mind. (For discussion and extensive sources regarding mentalism, see the Chapter 18.)

Potentiality fallacy. The assumption that a fetus, embryo, potential baby are actual babies, or the assumption that it will actually become one, and so to disallow abortion is to commit the fallacy from potentiality. It is an assumption to think the fetus or baby is a person. In any case, let alone the physical chances. The chances of becoming a rational, fully functioning person in an irrational society is minimal as well.

Rationalization. (Compare "wishful thinking.") This is an attempt to justify an unjustifiable act by knowingly or unknowingly finding reasons other than the actual reasons. It is a type of deceit of others and/or oneself. Rationalization is a distortion of truth and falsity. The false is seen as true and the true is seen as false. Euphemism is one form of this. The cause given is often a fabrication or rationalization after the fact. (Cf. circumstantial evidence.) For example, the claim that male circumcision is necessary for cleanliness. If true, this would also then justify the removal of any part of one's body [19]. Often rationalizations are put in the form: The good news is. . ., and the bad news is. . . The bad news is that we will have to amputate one leg;

the good news is that you will only have one foot to wash. Smoking helps me not to gain weight, and keeps mosquitoes away.

Refusal to discuss. I dislike arguments of any kind. They are vulgar, and often convincing. (Oscar Wilde)

Most people are "discussion illiterate" and do not have the ability to present organized arguments about issues. Special courses in speech, debate, and philosophy would be needed to gain an ability to discuss. It is false to assume that just because one can speak one can also discuss. Also, management in maintaining its views and power may stonewall and refuse to discuss.

Take metaphors literally. One of the most common fallacies is to think that one's theory is the only one, or is absolutely true. One is captivated by one's metaphors, be they in science or everyday life. "The only thing we could do was go to war."

You-also fallacy. Attempt to avoid being criticized by criticizing someone else. It is like saying, "I'm guilty, but so are you, so my guilt doesn't count." Example: "I gave the wrong medicine, but this is a problem throughout the system." "The majority rule principle may have its faults, but what system is better." "Management may be bad, but there are also problems with the staff."

Wish fulfillment. This is taking a wish and claiming or believing that it is true or will be fulfilled no matter how unlikely it is. Example: "I'll definitely stop smoking by next week." "I will lose 10 pounds this week." "People will usually act ethically, and rationally."

3.5 Mistakes

All doctors make terrible mistakes [20].

3.5.1 What are Mistakes?

"Mistake" is a value term meaning wrong. By definition, a mistake is bad. Therefore, it is an empty, open context term signifying nothing until given a meaning or context. There are no mistakes, as such. It is false to say that there should never be mistakes, because of the circularity of this statement and because of the equivocation of the term "mistake," and because it is like saying one should never use a value term. If one should never use a value term, and because mistake is a value term, one should never say that there is a mistake. People who say there are mistakes would be required to have education in ethics so as to have an ethical basis for saying something is a mistake. It can be wrong in any dimension: wrong thinking, misunderstanding, wrong action, inexperience, and lack of skill. Even beliefs, e.g. religions, can be seen as errors in judgment. From this it can be seen that we often equivocate with the word mistake, because it is a Chameleon word, which can take on all sorts of meanings. One can say that a mistake (sense 1) is not really a mistake (sense 2). A mistake may be bad because it is a sheer unavoidable accident, or bad because someone is to blame for it and would have been avoidable. If something is a mistake in one sense it may be countered by showing that it is not really a mistake in another important sense. There is a difference between first person and second person mistake, and subjective versus objective mistake. Misunderstanding is not a first person mistake because one can only make a mistake if one knew better. Mistake due to "lack of x (knowledge, skill, etc.)" is not really a mistake in the first person case, but a lack, though another may think of it as a mistake because *they* have additional knowledge. Mistakes are often misunderstandings and so not mistakes except from the point of view of, for example, a failure to cure an illness. Physicians and scientists have only a limited knowledge and, though they are very successful, operate in a world to a certain extent unknown. We therefore do not have complete medical knowledge, but yet must practice medicine. The foundations of science are only in the process of being discovered and understood.

Also, decisions are not self-contained and isolated from contexts of human desires. As a mistake a decision is never justifiable, but the causes may be understandable. Thus, we would never blame, but instead correct for the future. Negative emotions, in this sense, may be thought to be mistakes, which are caused by faulty thinking and are preventable and correctable.

As an accident a mistake was not intended. By strict definition, mistake and error cannot be intentional, or they would not be mistakes. It is contradictory to intentionally do what one thinks is a mistake to do. Scientists often make discoveries by accident, so the result of the "mistake" may be beneficial. In this sense, medicine needs mistakes. We can also turn mistakes and hardships into advantages, e.g. by the establishment of new rules, calling attention to needed changes, etc. Experienced doctors know how little they know and how little is known in medicine. They also know, therefore, how frequent mistakes are. Other doctors do not even realize they are making mistakes. Others rationalize their failures in terms of statistics. One also cannot conclude that if the outcome of patients in a hospital (in an outcome criterion approach) is good, that the performance of management was necessarily good or vice versa. It may, for example, have come at the physical exhaustion, emotional abuse, and exploitation of the physicians or staff. Also, patients are often themselves the cause of medical errors because of non-cooperation or giving faulty information.

Basically, mistakes are due to lack of critical thinking (speaking) and knowledge of the philosophy of medicine, lack of knowledge about one's discipline, about ethics, and about emotions. It is virtually always a mistake to try to correct mistakes by blame and punishment instead of by therapy and education.

3.5.2 What are Indications of Errors?

1. It is estimated that there are 200,000 deaths yearly in the U.S. due to error [21]. A total of 17.7% adverse effects were found in one teaching hospital [21].
2. Treatment errors vary by factors of 1–10 depending upon the location [22].
3. For most operations the mortality rate is 1 in 100 to 1 in 1,000 [21].
4. The National Health Care reported 10% of cases result in errors even death. That is, 850,000 yearly [23].

5. "Upward of forty-four thousand patients die each year at least partly as a result of errors in care" [20].
6. The Institute of Medicine "To err is human" [24] reported that more than 100.000 U.S. citizens die from medical mistakes every year.
7. Lucien Leape of the Harvard School of Public Health, an NPSF board member, estimates that as many as three million medical errors occur in hospitals each year, costing up to $200 billion [25].
8. Inappropriate drugs are prescribed for one in five patients over age 65. In 2003 the RAND Corporation research firm found that the likelihood of receiving care according to accepted medical guidelines is at best 50–50 [26, 27]. It is the third leading cause of death, behind heart disease and cancer. (See Institute of Medicine, 2004) One-third to one-half of antibiotics is mis-prescribed to patients who actually do not need them [28].
9. Each year, approximately 1.3 million patients are injured because of error during their hospitalization [24].
10. Of 5 million patients all will experience at least one preventable adverse event. Approximately one fifth (19%) of medication errors in critical care are potentially life threatening [29].
11. One third of physicians misinterpreted genetic tests for colorectal cancer [30].

3.5.3 Indications that Mistakes are Often Preventable Ones

1. Each year from preventable medical mistakes made in hospitals 200,000 Americans die and many more are injured. Just being cared for can also induce diseases. The Center for Disease Control reported in 2004 that two million people get infections while hospitalized, which may be a form of systematic mistake.
2. Seventy-four per cent of the-adverse operative events in Colorado and Utah in 1992 were preventable [20].
3. Of adverse events 28% are due to negligence [26, 27]. Of 230,000 preventable adverse events, 10.000–14.000 resulted in death [26].
4. "Mortality from lipoplasty [19.1 in 100,000] is higher than mortality from automobile crashes (15.2 per 100,000) or homocides (5.9 per 100,000)" [31].
5. It was demonstrated that measuring blood pressure with the most commonly used type of equipment often gives incorrect readings that may lead to mismanagement of hypertension [32].
6. Prescriptions are usually hastily scribbled. For example, in April 1999, the FDA investigated ten deaths thought to be caused by *Celebrex* (a Cox-2 inhibitor for pain) being confused with *Celexa* (an antidepressant). The issue is compounded because Celebrex in December 2004 was cited as doubling one's heart attack risk.
7. More than 100,000 deaths due to preventable adverse events occur while in the hospital [24].

3.5.4 *What are the Reasons for the Mistakes?*

3.5.4.1 Questionable Medical Treatments

Examples: South African medical students are required to conduct episiotomies even when not required. For example, episiotomies should not be routinely performed but rather only used if there are specific medical indications. Episiotomy persists even where practice guidelines recommend its restricted use. A guideline for use is only 10–30% of the cases. From 1995 to 2003 in Sweden the rate was only 9.7% compared to 100% in Taiwan [33].

Mitka states, "Most of the prostates we remove need not be removed" [34].

As a result, software is being developed for error prevention, cause analysis and reduction.

3.5.4.2 Error is Necessary

As a general principle, if one works, one will make errors. It is inevitable. Rosenthal stated, "Necessary fallibility must be accepted as intrinsic part of the practice of medicine" [35].

3.5.4.3 Uncritical Thinking (Speaking)

A mistake is sometimes metaphorically defined as based on a bad judgment and in this sense may be intentional. Faulty beliefs and uncritical thinking are a main source of error. Mistakes are due to lack of knowledge about critical thinking, ethics, and emotion. Practice does not make perfect if one lacks critical thinking (speaking). To make ethical judgments without studying ethics is like surgically operating without the proper training.

3.5.4.4 Medical Knowledge Is Lacking

Successful treatment is to a large extent based on luck because of our limited knowledge of the human body and disease. "Twenty-seven of thirty clinicians made an error in using the defibrillator" [20].

3.5.4.5 System as a Cause of Error

A system's approach is needed to treat the causes of error as it involves management, and the legal, regulatory and reimbursement sectors, rather than just the individual physician [23].

Australia had 16.6% adverse events in 1992. 15% were due to system error [26]. The notion that mistakes are just caused by "bad doctors" is rejected. Tempelaar concludes that one cannot prove the "bad doctor" theory or that some doctors make more mistakes than others [36]. According to the Joint Commission of Accreditation of Healthcare Organizations, every error should be reported and the cause ascertained especially the systematic causes. Error is not just an individual caretaker's problem although they are usually given all of the blame [37].

In malpractice suits management must be named as one of the involved causes, e.g. "enterprise liability." Liability is shifted from the individual to the company especially where the individual has little control over the situation, thus the term *respondeat superior* [38]. For example, denying a physician access to improving surgical techniques or enrolling too many patients for proper care is an institutional and management problem [39]. However, it is pointed out that the U.S. tort system does not effectively deter malpractice by health institutions or management.

3.5.4.6 Some Mistakes Are Not Mistakes

What one considers not as an intentional or as an un-intentional error, another, e.g., management or the courts may regard it as an intentional or unintentional error.

3.5.4.7 Guidelines Are Not Followed

Dutch College of General Practitioners Guidelines were followed by physicians only 50% of the time [36]. Henry spoke of the Healthcare Commission goal to reduce health care infections, but such simple things as hand washing are only complied with 50% of the time [40]. "Healthcare workers only perform hand hygiene when in-directed only about 40% of the time" [41]. The use of alcohol-based hand rubs (60–90% alcohol) was preferred over washing with antimicrobial soap. They dry hands less and condition the skin. (See also CDC's Guideline for Hand Hygiene in Healthcare Settings).

3.5.4.8 Self-Caused Mistakes

Healthcare workers often do not have a healthful lifestyle. They often do not take care of themselves. Tempelaar found that 47% of physicians in one study had serious stress, 29% anxiety, 27% clinical depression [36]. Dutch physicians had 58% emotional exhaustion (burnout) [36].

3.5.4.9 Patient Errors

One must also speak of "patient errors" which even cause their own death [32].

3.5.4.10 Lack of Sufficient Attention

"Diligence and attention to the minutest detail can save you" [20]. Lack of attention can be disturbed by negative emotions and by those things, which cause them, such as, bad non-supportive or unfair management, overwork, etc. There may be also lack of self-discipline, esprit de corps, and motivation.

3.5.4.11 Misdiagnosis

"We get the diagnosis wrong in two out of five of our patients who die. ... Physicians missed a quarter of fatal infections, a third of heart attacks, and almost two-thirds of pulmonary emboli in their patients who died" [20]. There is 28–57% misdiagnosis of appendicitis in children, almost 100% in infants. Doctors disagree about diagnosis, for example, sending a patient for a hip replacement varies 45%, and in other cases 88%, depending on the city one is in [20].

3.5.4.12 Overwork

The main cause of error was lack of sleep and fatigue [42]. This is an obvious and inevitable cause as healthcare workers are required to work 60–90 h a week. Without sleep one cannot be attentive. (For an analysis of overwork in medicine see Chapter 8.)

3.5.4.13 Limitations of Knowledge in Medicine

"Every day, surgeons are faced with uncertainties. Information is inadequate; the science is ambiguous." Perfect knowledge is not to be had [20]. Protocol intelligence systems have been developed to help prevent error, e.g. for obstetrics. For example, a listing of 6,500 best practice rules and protocols [43]. Real-time best practice and risk management support tool for the OB practitioner was used in over 100,000 births to date. IPROB has demonstrated statistical reductions in the risk and prevention of critical clinical errors [44].

3.5.4.14 Unfair Medical Threats of Malpractice Suits Threaten Physicians

Litigation and punishment do not reduce medical error rates, but rather cause them [20]. Court found that in the case of a patient who irresponsibly failed to take the prescribed medicine before she saw the defendant-doctor, the jury unfairly assessed malpractice. Therefore, the doctor should record if the medicine was taken, but is threatened by an unfair legal system [45].

3.5.4.15 Unfair Blame

Many argue against blame in cases of error and malpractice. Henry gives arguments for the full disclosure of medical mistakes, but without blame. "The new organizational culture must be based on open communication, truth telling, and no blame" [40]. Organizational culture must also be grounded on a system of ethics [40]. Henry bases her view to disclose mistakes on respect for patient autonomy. This may rather lead to lawsuits as the present malpractice situation is based on blame and punishment, not correction. It also presupposes the rationality and fairness of the patient, which is not always given [40]. If a patient felt not informed enough after the treatment then the patient might not be thought by the law to be sufficiently informed. Patients would have to be asked to sign with a witness that they have had sufficient information from the doctor. (The courts may even not accept that and may claim that the witness was coerced.) Myers recommends error reporting to an outside non-punitive state agency [40]. "Medical malpractice suits are a remarkably ineffective remedy" [20].

In Sweden, Denmark, Finland there is No Fault Liability Insurance. Aim is not to punish doctors, but to help the patient. Myers states that a no-fault system of medical liability can be introduced. (cf. Canada's system) The author holds that healthcare is a service, not a right [40].

3.5.4.16 Protocols of Good Management Are Violated

Suppose management is irrational, power oriented, and unfair thereby causing stress among the healthcare workers. If an error is then committed it is the fault of the management, not the physician or nurse. It is also unprofessional for physicians to cover for bad management because this maintains an ongoing risk for the patient. It is irrational to think that error is only at the individual level as if in a vacuum.

3.5.4.17 Unfairness of the Law

Malpractice insurance is needed to cover 1. negligence of a person under your supervision, 2. misuse of equipment, 3. error in diagnosis and treatment, 4. failure to properly inform patients, 5. error in prescribing and administering medication, 6. negligence of care in emergencies outside of working environment, and 7. suit by the employer or collaborating physician. Malpractice claims are self-applicable because, even if baseless, they encourage mistakes [36]. The courts use "reasonable standard of care" as determined by the jury or court rather than a "standard of customary care" as determined by the medical profession [46]. The court trumps medical practice as if it knows medicine better than the medical profession. "The jury makes its own determination of what reasonable medical care requires" [46]. Would you want jury members who are supposedly so knowledgeable to perform your surgery? It may be noted that the law itself fundamentally follows custom and precedent minimizing rational and consequentialistic argument.

As a result of the unfairness, 65% of the physicians surveyed engaged in defensive medicine, ordering more tests than needed, 40% decided not to care for high-risk patients, and litigious patients were to be avoided [47]. Numerous obstetricians and gynecologists in the U.S. have reduced or stopped practice as a result of possible malpractice fears [48]. The average jury award in Georgia has more than doubled since 1995 from $215,000 to $458,000 in 2002. The number of one million or greater awards has nearly tripled, from 4 to 15. 100% of South Florida neurosurgeons have been sued, according to surveys of area physicians [49]. A preliminary survey indicates that as of January 2003, 28 Connecticut obstetricians made the decision to no longer deliver babies. The average payment made by one of Connecticut's major insurers to resolve a claim has risen from $271,000 in 1995 to $536,000 in 2001. Physicians often cannot afford malpractice coverage [50].

The suits are often frivolous. Most suits are withdrawn before trial. 80% with a jury verdict result in no payment to the injured or family [51]. Defensive medicine is not a sound medical practice, but is driven by malpractice threats and unfairness. It involves normally otherwise not recommended tests, for example unnecessary MRI, tomography, biopsy; referrals, refusal to treat certain patients (e.g. litigious ones), eliminating surgery and procedures prone to complications, prescribing unnecessary medication, treatment of little or no medical value [47]. As a result of the threat of malpractice, most physicians have increased the number and frequency of tests to avoid being sued. 39–57% of specialist physicians said they would no longer care for high-risk patients. It is driven by unfair patient claims and an unfair and unreliable legal system. Malpractice insurance rose from about 34,000 in 2000, to 73,000 in 2003 [47]. There are even books giving instructions as to how unfairly win one's case [52].

3.5.4.18 Negative Emotions

The manager who shows irritation and anger with the employee is in violation of professionalism and is causing abuse, which can lead to medical error, e.g. by causing nervousness and tension. Performing surgery after a scolding by the supervisor can lead to mistakes. Anger is a personal and management error and causes further errors on many levels. Physicians and patients are often arrogant and not open to dialogue and help. They are held captive by a metaphor. Many malpractice complaints are also due to bad attitudes of healthcare workers [36].

3.5.5 Case Example: Misleading Diagnosis

A 22 years old woman had unprotected sex on Saturday night and took a morning-after pill on Sunday morning. On Monday, she experienced pain in her right lower abdomen and came to the gynecological ward for treatment thinking her pain was connected with taking the morning-after pill. The CRP (test for inflammation) and leucocyte counts were raised slightly. She was admitted as an in-patient for observation. The next day the pain remained and with no indications of specifically gynecological problems she was sent to a surgeon for suspected appendicitis. She was there given an ultrasound examination, and a clinical check. The surgeon reported no indications of appendicitis. Later in the day she became feverish (39°C). Another blood test was given. CRP and leucocyte counts were raised. In the evening, she was examined again by the gynecologist on attendance. The ultrasound showed no problems in the womb or ovaries. An unspecific sign was the presence of free fluid in the right lower abdomen. This could have come from having taken the morning-after pill and concurrently having had a follicular rupture. She had been constipated for sometime. Clinically, appendicitis was still suspected. She was then again sent to the surgeons who again maintained that there absolutely was no appendicitis present. The same evening at midnight the gynecologist called the surgeon for verification again that it was not appendicitis. She was assured that it was not and that only painkillers were needed. On Wednesday morning the CRP raised higher and the leucocyte count fell. In the afternoon the blood result remained the same. The gynecologist on duty did not react, but simply followed the surgeon's recommendation. On Thursday the fever was still present and there was no improvement of the situation so a laparoscopy was performed which showed appendicitis so critical that the appendix was about to burst. The surgeon now had to come to the gynecological department to operate there. He gave the excuse that the symptoms of appendicitis had not been there, which often is true. But one cannot fully rely on the absence of typical symptoms.

The case shows that there was certainty where there should have been doubt. There was assertion of certainty instead of the open search for a cause. Symptoms were related to causes, even if they did not perfectly fit. There were symptoms and they were guaranteed, not typical ones, misplaced by the surgeons until the gynecologists had to repeatedly challenge. It was held that if there are no indications, there is no disorder. One would more carefully have to maintain that if there are no symptoms that there might still be a disorder. Medicine is an art of probability and uncertainty. There are many clues of symptoms in this case. Appendicitis without specific symptoms was the special challenge. It is not enough to consider only what is in the textbooks, but rather to closely and continuously observe the patient and be open to changing a routine practice.

3.5.6 Personal Experiences: Mistakes

From my personal clinical hospital experience it seems to me extremely problematic how mistakes are dealt with in medicine. The crucial mistake is the way management as well as society deal with mistakes. Making mistakes as well as knowing about the large potential of making mistakes is a tremendous burden and responsibility on healthcare workers. The public and administration expect healthcare workers to be perfect like a pianist who is not allowed to miss one note or one's career is finished. But in the case of health care workers the result can cause great injury or death. It is one of the greatest responsibilities one can have, and whereas in other professions one's performance can be far from perfect, the healthcare worker is unfairly expected to be inhumanly perfect. This is the case even if the patient, management and system are corrupt and negligent. As a result, the physician who makes a mistake can feel incompetent and become guilt ridden and emotionally devastated. Colleagues often fail to properly understand or deal with the problem of mistakes, but rather themselves feel tainted by them and emotionally negativized. Patients unfairly blame the physician before the court using any means to obtain compensation. The court awards for malpractice can be phenomenal. Compensation is sometimes determined by mere subjective accusations against the healthcare workers of the hospitals. Healthcare workers are often punished for making a human mistake. But little is done of what would be needed to correct and prevent such mistakes. Unfair punishment and irrational blame prevent healthcare workers from admitting and discussing mistakes so that errors might be prevented in the future. One of the greatest mistakes and cruelties is to punish when prevention, education and guidance is what is needed. A better understanding of mistakes and a more fair and humanistic treatment in regard to them would help all to honestly confront and more properly deal with mistakes instead of assuming an unrealistic and perfect mistake-free world, which never could exist.

Acting especially in the area of medicine is to make one be prepared to aim at the best, but nevertheless make mistakes. Healthcare workers need support, not blame, need reassurance, help of their families and friends, management, and society to hold in to dealing with ill and endangered people, in spite of the burden of possible and always likely mistakes.

References

1. Heiman, G. 1998/2001. *Understanding research methods and statistics*, 2nd edn. Boston, MA: Houghton-Mifflin.
2. Maier, B., Akmanlar-Hirscher, G., Krainz, R., Wenger, A., and Staudach, A.. 1999. Der chronische Unterbauchschmerz – ein immer noch zu wenig verstandenes Krankheitsbild. *Wiener Medizinische Wochenschrift* 149:377–382
3. Elwood, J. M. 1988. *Causal relationships in medicine: A practical system for critical appraisal*. Oxford University Press, Oxford
4. Shibles, W. 1994. The cognitive-emotive theory on desire. *Journal of Indian Council of Philosophical Research* 11:25–40.
5. Greenberg, P. 2002. In *Causation in law and medicine*, eds. Freckelton, J., and Mendelson, D. Hampshire, GB: Ashgate.
6. Wittgenstein, L. 1969. In *On certainty*, eds. Anscombe, G., Wright, G., and Paul, D. Anscombe G tr. New York: Harper.
7. Isenberg, J. S. 2003. Surgical research and the ethics of being first. *Journal of Value Inquiry* 37:195–203.
8. Lang, R., and Hensrud, D. eds. 2004. *Clinical preventative medicine*. Chicago, IL: AMA.
9. Callinan, I. 2002. In *Causation in law and medicine*, eds. Freckelton, J., and Mendelson, D. Hampshire, GB: Ashgate.
10. Bingham, R. 2002. Leaving nursing. *Health Affairs* 21:211–217.
11. Walton, D. 1985. *Physician-patient decision-making: A study in medical ethics*. Westport, CT: Greenwood.
12. Waismann, F. 1965. *Principles of linguistic philosophy*. New York: St Martin`s Press.
13. Dewey, J. orig.1938/1964. *The theory of inquiry*. New York: Holt, Rinehart and Winston.
14. Strodach, G. ed. 1963. *The philosophy of Epicurus*. Chicago: Northwestern University Press.. See esp. Letter to Menoeceus: 178–195.
15. Shibles, W. 1993. Majority rule fallacy. *Indian Philosophical Quaterly* 20:329–347.
16. Voltaire, F.M. de "Madness" The Works of Voltaire, vol. VI Philosophical Dictionary Part 4 1963/1764.
17. Patai, D., and Koertge, N. 1994. *Professing feminism*, 126–131. New York: Basic Books.
18. Freckelton, I., and Mendelson, D. eds. 2002. *Causation in law and medicine*. Hampshire, GB: Ashgate.
19. Dirie, W. 1998. *Desert flower*. London: Virago Press; Dirie, W. 2002. *Desert dawn*. London: Virago Press.
20. Gawande, A. 2002. *Complications: A surgeon`s notes on an imperfect science*. New York: Picador/Holt. see also www.misdiagnosis.com.
21. Studdert, D. 2002. What have we learned since the harvard medical practice study? In *Medical error. What do we know? What do we do? University of Michigan Forum on health policy*, eds. Rosenthal, M. M., and Sutcliffe, K. M. San Francisco: Jersey-Bass, Wiley. See Harvard Study: Kohn, L., Corrigan, J., and Donaldson, M. eds. 1999/2000. *To err is human: Building a safer health care system*. Washington, DC: National Academic Press.
22. Wachter, R., and Shojania, K. 2004. *Internal bleeding: The truth behind America`s terrifying epidemic of medical mistakes*. New York: Rugged Land.

23. Bogner, M. 2004. *Misadventures in health care: Inside stories*. New York: Lawrence Earlbaum.
24. Kohn, L., Corrigan, J., and Donaldson, M. eds. 1999/2000. *To err is human: Building a safer health care system*. Washington, DC: National Academic Press.
25. Leape, L. in Levy, D. 1997. *Medical mistakes happen too many, AMA poll finds*. USA Today, October 10th, 1997.
26. Hancock, L. 1997. Addressing the "problem doctor". In *Problem doctors: A conspiracy of silence*, 153–170. eds. Lens, P., van der Wal, G. Amsterdam: IOS Press.
27. AARP Bulletin. 2004. 45:1020 www.aarp.org
28. Ropeik, D., and Gray, G. 2002. *Risk*. Boston: Houghton Mifflin.
29. Baldwin, D., Daugherty, S., Tsai, R., and Scotti, M. 2003. A national survey of residents "Self-reported work hours: Thinking beyond speciality." *Academic Medicine* 78: 1154–1164.
30. Andrews, L., and Hibbert, M. 2000. Courts and wrongful birth: Can disability itself be viewed as a legal wrong? In *Americans with disabilities*, eds. Pickering, F., and Silvers, A., 318–330. New York: Routledge.
31. Lapetina, E., and Armstrong, E. 2003. Preventing errors in the outpatient setting: A tale of three states. *Health Affairs* 21:26–39
32. Spath, P. 1999. Reducing errors through work system improvements. In *Error reduction in health care: a systems approach to improving safety*, ed. Spath, P., 199–234. Chicago: American Hospital Association Press.
33. Graham, I. 2005. Episiotomy rates around the world: An update. *Birth* 32:219–223.
34. Mitka M. 2004. Is PSA testing still useful? *JAMA* 292: 2326–2327.
35. Rosenthal, M., and Sutcliffe, K. eds. 2002. *Medical error. What do we know? What do we do? University of Michigan Forum on health policy*. Wiley: Jersey-Bass.
36. Tempelaar, A. 1997. The problem doctor... In *Problem doctors: A conspiracy of silence*, eds. Lens, P., and van der Wal, G., 31–55. Amsterdam: IOS Press.
37. Patton, D. 2001. Legal considerations of sleep deprivations among resident physicians. *Journal of Health Law* 34:377–417.
38. Myers, K. 2001. Medical errors: causes, cures, and capitalism. *Journal of Law and Health* 16:255–288.
39. cf. *Jones v. Chicago HMO* 730 N.E. 2d 1119, Ill 2000.
40. Henry, L. 2005. Disclosure of medical errors: Ethical considerations for the development of a facility policy and organizational cultural change. *Policy, Politics and Nursing Practice* 6: 127–134.
41. Fry, D., and Burger, T. 2005. *Hand hygiene compliance*. OR Insider: 12–15
42. Miecke, S. 2005. Effects of rotating night shifts: Literature review. *Journal of Advanced Nursing* 50: 433–439.
43. www.nejm.org/medicalarticles/medical-intelligence
44. e-and-c.com/corporateoverview.htm.
45. Tenorpe, K.E. 2004. The medical malpractice 'crisis': Recent trends and the impact of State Toll Reforms. *Health Affairs*, January 21, 2004 doi:10,1377/healthaff.WA.20
46. Peters P. 2001. The reasonable physician standard: The new malpractice of care? *Journal of Health Law* 34:105–119.
47. Studdert, D. 2005. Defensive medicine among high-risk specialist physicians in a volatile malpractice environment. *JAMA* 293:2609–2617.
48. IPROB (Intelligent Patient Record for Obstetrics) www.openclinical.org/aisp_iprob.html
49. Medical Liability Reform – Now! Feb. 5, 2008 http://www.ama.assn.org/go/liabilityreform
50. The Hartford Courant, 2002, Nov.17.
51. AARP Bulletin 2006, 47: 11. www.aarp.org, *AARP Bulletin* 2006, 47:11.
52. Babitsky, St., and Mangraviti, J. 2005. *How to become a dangerous expert witness: Advanced techniques and strategies*. SEAK: Falmouth, MA.

Chapter 4
Analysis of Causation in Medicine

> *A central concept in decision-making is causality and few physicians have training in the scientific method regarding causation.* [1]

Abstract Decision-making, especially diagnosis, depends on a theory of causation. In any particular case an unlimited number of models of causation are possible. Cause may be seen as a hypothesis. Cause is an abstract term. Cause is a concept of human understanding. If the patient is to give a report about him/herself, she is likely to construct it according to how she summarizes what has happened in her life, memories, selections, what she thinks to be received. Causes and causation in medicine are to be carefully examined and healthcare workers have to be cautious in ascribing causes too readily in order to avoid circuits and prejudices in diagnosing and treating patients. All causal statements should be regarded as false unless proven otherwise. The critical definitions of causes have an ethical impact on medical thinking and practice, the lack of doing so invalidates it.

Keywords Cause · distal cause · proximal cause · causation · effect · association · correlation · significance · stimulus · response

Epidemiology in practice is said to contain numerous problems in regard to ascertaining causes. Gordis' text, *Epidemiology*, presents a number of them. Data collection is typically incomplete and inaccurate [2]. Classificatory codes and practices change and so invalidate data [2]. Adjustments are seldom made for lifestyle, individual health, stress, other concurrent diseases, and other variables. There are errors in diagnosis, age, race, gender, etc. [2]. Randomizing, e.g., tossing a coin to decide which part of a group to take is not a sound experimental procedure. To distinguish cause as opposed to mere association requires [2]: 1. temporal relationship, 2. strength of the association, 3. dose-response relation, 4. replication of findings, 5. biological plausibility, 6. consideration of alternative explanations (e.g. lifestyle, diet, genetics, etc.), 7. cessation of exposure, 8. consistency with other knowledge, 9. specificity of association to one disease, 10. additional evidence from multiple sources [2]. Deviations from these factors are seen as biases of: selection, unexpected response, exclusion bias, information bias, misclassification, confounding

B. Maier, W.A. Shibles, *The Philosophy and Practice of Medicine and Bioethics*, International Library of Ethics, Law, and the New Medicine 47,
DOI 10.1007/978-90-481-8867-3_4, © Springer Science+Business Media B.V. 2011

(something may be associated with an effect, but not be the cause), inadequate selection of possible causes, failure to consider interactions [2]. We do not have all the information we need to put cause on a firm quantitative and qualitative basis [2]. For example, regarding hormone replacement for post-menopausal women, there are benefits and harms, but the research is unclear and not all of the factors are included in the experiments [2]. The conclusion is that causal decision-making in epidemiology cannot be a statistical or quantative procedure, but rather the critical thinking (speaking) and creative reasoning of the physician in cooperation with the patient who preferably has similar qualities. "Guidelines ...can be of most value where coupled with reasoned judgment in making decisions about causation" [2]. Montgomery [3] presents the notion of "Phronesis" as the use of reason in interpreting particular circumstances [3]. This is like clinical practice. For her, medicine is an interpretative science and includes the social and emotional life. It is not strictly objective, and context and humanity must be stressed.

Mitchell and Benichou in *Encyclopedia of Epidemiological Methods* note that although cohort study, the sampling of a comparison group over time, is the gold standard, bias is found at every stage of research from data collection to failure to review the literature [4]. Cause is only a probability, not something agreed on [4]. Confounding is mixing extraneous factors with the effect of the factor in question. It may not be vitamin E, which has the desired effect, but something in foods containing vitamin E or an interaction with vitamin E. The authors typically follow Sir Austin Bradford Hill's Criteria for causality [cf. 1].

Hill's Criteria for causation are: 1. strength of association, 2. consistency, 3. specificity, 4. prior temporality, 5. biological dose-response credibility, 6. biological plausibility, 7. coherence (does not contradict with what is known), 8. experimental evidence, 9. inference by analogy [5]. The factors discussed are haphazard and lack any philosophical critical depth, nevertheless they are often used as a standard in medical research. Once again, what is needed is critical human reasoning and clarification by philosophy of medicine. A small list of loose characterizations of cause will not do. Evidence-based medicine argues that a "meta-analysis" is needed in epidemiology. This is equivalent to arguing that a philosophy of medicine is needed.

4.1 Decision-Making and Cause

> As general concepts, causality and explanation are far from clear [6].

This quotation illustrates the reasons for introducing the concept of cause here. Decision-making, especially diagnosis, depends on a theory of causation. One of the most fundamental procedures in medicine is the determination of the causes of a disease and/or of the causes, which would lead to successful treatment. Yet, cause is one of the most confused concepts in science and in medicine as well. For these reasons we will continue the analysis of cause begun above. The meaning of explanation, understanding, and reasoning often refer to the informed giving of causes. Professional experience also refers, to a large extent, to the knowledge of causes.

Epidemiology is the study of the elements contributing to the occurrence or nonoccurrence of diseases. It is the study of the cause and control of disease. Causes may be deep or superficial, natural or cultural, or supernatural, religious, philosophical, scientific, political, economic, etc. There is almost never just one cause, but many types of cause. *In any particular case an unlimited number of models of causation is possible.* This means that innumerable decisions are possible. If we are to ask, "What is the cause?" We must first find out what kind of cause or answer is desired, expected, demanded, needed, etc. We may examine the world-field of the term. Synonyms of cause or similar terms are: antecedent, basis, beginning, condition of, desire, determinant, entailment, etiology (study of the causes of diseases), explanation, foundation, ground, inducement, influence, instigator, intention, justification, motive, necessary and sufficient conditions, origin, proof, purpose, reason, root, source, starting point, tendency, underlying principle, etc. Some ethical synonyms of cause are plot, provoke, incite, stir up.

There are many meanings of cause, which can be confused with each other. To ask for the cause is to ask for the answer to "Why?" Many kinds of answers are possible. Cause is an abstract term. One must substitute a specific meaning for it to be intelligible. There is no cause as such. We cannot simply say, "The cause of x disease is y." Things and events cannot cause other things and events. Cause is a concept of human understanding. Without people there are no causes or effects. Causes constitute only one way of looking at the world. It can also be looked at non-causally (See "non-causal events" below).

The distal causes of most all diseases are unknown. This is also true because science has little knowledge about the smallest particles, e.g., quanta, quarks, etc. The proximal causes of perhaps most diseases are also unknown or little known. However, we do often know what will cure or treat many diseases without knowing the causes. The following are formulae for the majority of causes and treatments of diseases. "The cause of x (disease) and also the most effective strategy of treatment is unknown." "X medication/drug is known to be to some degree effective, but the mechanism/reason for it is not known."

What causes people to act as they do? We do not know enough about human personality to be able to genuinely answer this except in a rough way. People often seem to act, "for no reason." Descriptions and classifications of psychological disorders in the DSM IV are, for example, highly problematic [7]. Earlier versions bordered on the absurd [8]. "DSM…is revered too much and doubted too little" [9]. We must know who, what, when, where, how of physical and psychological cause (See "mental causation fallacy" below). Instead of stereotyping patients by pseudo-scientific and derogatory DSM classifications one may rather just describe the specific symptoms and behavior.

All causal statements should be regarded as false unless proven otherwise. For cause as an abstract term, we must substitute a meaning for intelligibility. To "what is the cause?" one must ask first what kind of cause or answer is desired. There is no cause as such. On the pragmatist's view, the meaning of a cause is its consequences. There are no causes as such. It is just a means to bring about a practical result. There is only a cause for a certain purpose. (cf. In operant conditioning "stimulus-response" supposedly only has meaning to the extent that it brings about a certain consequence.) On the ordinary language philosophy view of Wittgenstein causality again may be seen as a language-game [10].

4.2 Synonyms of Cause

The Metaphorical Method explores and analyzes language for clarification, theory construction and analysis for research (See Chapter 1). It involves the analysis of synonyms, antonyms, substitutions, reversals, associations, metonymy, analogy, etc. In the analysis of synonyms it is seen that cause takes a number of different seeming forms and meanings. Causal statements are only as precise as the language in which they are put. We must explore the causal aspects of each sentence. One synonym of cause is "agent of responsibility," though it is not clear how much an individual can be genuinely responsible for his or her actions. The causes of emotions are ultimately one's own assessments, not things, others or situations (See Chapter 7). Other synonyms are: arrangement (order, regulation, structure, direction), ascription to: claim of a cause, attribution to, association: as a type of association, constant contingent conjunction (David Hume), summary of past associations, basis: the conditions or reason(s) for something happening, be-cause: since, on account of, in view of, blame as causal word (you could hypothetically have done otherwise than you did do), cause-effect. "For every cause there is an effect; for every effect there is a cause." This is circular and true by definition. Cause without effect is a contradiction. Effect without cause is a contradiction. Other synonyms are: change, transformation, potential cause versus actual action, circumstantial cause, clarification, commit, conditions for, connection, consequence (when x happens, y happens.), construction (cause as construction: there is no cause in reality), fabrication. In therapy, we cannot construct all the details of all thought and action, but we can construct a likely story or narrative as to what happened, what the problem is and what the solution might be. If the account is accepted by those concerned it may or may not lead to effective solutions. (cf. circumstantial evidence, decisions of administrative meetings, negotiations, debates, arguments, discussions, etc.) Cause in terms of control viewed in terms of controlling the situation, detection of causal factors or effect, correlation. To cause as to create, to do, to make, or to entail. Cause as dependence, disappointment as faulty expectation, drive (a pseudo-psychological term). Causing as being due to. Cause as an ending (cf. $c > f > c > f$), cause as an ethical impulse. An event as cause (event is future consequences), eventuality (conformity), explanation, force (cause is supposedly not a force), form, ground, if-then hypotheses, impetus, induction, influence (if one claims an influence it must be spelled out), power, prediction, relevancy, reliance, result, root, and interdependence. To know often means to be able to give a cause (causality and epistemology), to give motives, to produce a narration: cause as a constructed artificial story, cause as a necessary condition (required before something happens, but it may require other sufficient conditions; only that which is useful to cure or change), origin (not absolute origin), past cause as just a present configuration of events premise, provocation, human reasons versus material causes [11], source, step by step procedure, stimulus, sufficient conditions as full and adequate conditions needed for something to occur. To cause as to perform, perpetrate, persuade, follows that, reform, inform, impress, seek, shape, teaching as cause, to explain, x correlating with y (without causal force a x could be interchanged with y).

4.3 Antonyms for Cause

These are in a sense also synonyms: consequences, effect, purpose, result (as effect), and success.

4.4 Metaphorical Models for Cause

Cause is seen in terms of any other term, e.g. cause as association. Events can be associated without being causal. Cause may be seen as disappointment, as explanation, as instruction, as a recipe, as a riddle. Cause may be a form of our own creation and understanding. But we also can look at the world and what happens non-causally. The effect can occur without the causal treatment factor, e.g., the body often heals itself in spite of treatment. We ascribe cause where there is none. We can often pragmatically provide treatment without knowing the causes.

Cause is given as an answer to "Why?" Many kinds of answers are possible, e.g., "For no reason."

Dialogue can cause, not solve problems if people are uncritical: in cause x, e.g., cause *problems.*

Cause-effect can be seen as stimulus-response, and mere correlation. Critique: But the correlation must be relevant so causality is still not avoided. What is to be the difference between the stimulus and the response if there is no time or causal factor? One can put effort-success in place of cause-effect.

4.5 Substitutions for Cause

In the following we may give various ranges and specific characteristics of substitutions for x:

x (caused) y to happen, e.g., negotiated, influenced, etc.
x happened (because) ———.
The ——— (cause, etc) of x is ———.
Who or what (caused) x?

4.6 Temporal Factors in Causality

They can be viewed as present or past association, as expectation, as timeless stimulus response. Time is involved in numerous causal words and phrases, e.g., "I will go" implies cause.

Cause is involved in the hypothetical conditional: If you do x, then y will (in future) happen.

Post hoc ergo propter hoc. Temporal implication is found in: to become, results in, etc.

4.7 Types of Causality

Causal theories are not based on perception, but rather on language use. Perception presupposes language and there are various theories of perception. We are not clear how perception works. Thus causality cannot be simplistically based merely on empirical experience. Hume's "sense ideas" should be replaced by language and "sense language." The approach to clarifying causal notions would be to analyze the uses and misuses of causal terms, including those dealing with perception. It would be circular to examine perception by means of perception. There are physical causes (who, what, how, e.g., epidemiology), psychological causes, statistical causes, mathematical causes, and cause as it is used in each discipline. Case distinction may be made between I, we, they, you, or it caused. Causality may be primarily subjective or objective, supernatural (fictional) or natural. The type of causation should be specified to avoid equivocation and evidence for it provided. Cause can be an inert object, active object, abstract concept, or a human agent. Because religion is supernatural and absolutistic, rather than consequentialistic, it has a lack of concern for naturalistic causes and consequences.

The types of definition also generate different types of cause (See Chapter 2). A distinction is made here between the following types of definitions of causality using some types of informal fallacies and the metaphorical method:

Absolute versus hypothetical cause. Over-abstract, and absolute definitions are rejected because they are fallacies. Cause may be seen as a hypothesis. On Hume's view cause is a constant, contingent conjunction based on past experience [12]. It may not hold true in the future. To consider an absolute cause is unscientific. There is no absolute or necessary causation though logicians and scientists misleadingly speak of "necessary and sufficient conditions" as criteria of causality. "X would not have happened 'but for' the action or inaction of the healthcare worker." "But for" cannot establish cause generally. "If only I had left home three seconds later, there would not have been an accident." On this argument the cause of the accident is thought to be just my being early. "Most situations in health and disease do not fulfill the criteria either for necessary or for sufficient causation" [13].

Beg the question. This is a statement, which assumes what it is supposed to prove. "For every cause there is an effect" is a circular statement.

Category Mistake. The concept of, and evidence for, cause is often different in different disciplines. To use the same term, "cause," for different disciplines creates a category mistake. Also, to use causal terms in different senses in the same discipline creates equivocation. J. Stapleton speaks of "the chameleon-like ability of that concept [cause] to mean vastly different things to different people" [14]. Causal beliefs of the average enculturated person are typically false.

Cause in medicine is not the same as cause in therapy or in theology. For the symbolic logician, cause is implication: "If x, then y", "x implies y" whose meaning of implies is determined by truth tables, or simply represented by a horseshoe sign. It is only a "formal logical cause" or "symbolic logic cause." It is a false and simplistic reductionism of the ordinary notions of causality. Its presupposition of truth tables for its meaning is unscientific. This makes such logic stipulative and irrelevant for rational discourse. It becomes a mere game played by logicians. Symbolic and formal logic can have no acceptable use in medicine as the critiques of the pragmatists and ordinary language philosophers have shown (See full critique in the Chapter 18).

Cause as association. Events can be associated without being causal. (Cf. metonymy)

Cause in each universe of discourse. The kinds of causes and evidence for them differ in each discipline. Chemistry, medicine, religion, sociology, statistics, etc. each have different notions of cause. We may have mathematical causation, medical causation, economic causation, supernatural and mystical religious causation, qualitative causation, quantitative causation, metaphorical causation, etc.

Circumstantial cause or evidence. In criminal law and administrative law, personal guilt is often determined only by circumstantial evidence. This is a significant embarrassment for legal practice regarding its methods of determining causal evidence. Similarly, healthcare workers may be found guilty of malpractice on the basis of circumstantial or legal, rather than actual evidence. In medicine, the main questions about a fatal event could be who initiated, or promoted, or could have prevented, or was the last person who presumably could have turned it around? It makes a great difference which question is asked in ascribing blame.

Constructivist definitions of cause. These are definitions, which do not claim to realism, but rather to present structures, organizations and configurations. Circumstantial evidence, administrative committee decisions, negotiations, debates, arguments, discussions, etc. are often mainly constructivist. We speak about, "What will fly," or "What the traffic will bear." That is, a hypothetical narrative is created by consensus and this is then taken as the basis of the decision. This obviously has grave shortcomings. If the patient is to give a report about him/herself, she is likely to construct it according to how she summarizes what has happened in her life, memories, selections, what she thinks to be received.

Circular definitions of cause. Example: "Causality is producing something or that which brings about an effect." "For every cause there is an effect; for every effect there is a cause." These are circular and true by definition. Cause without effect is a contradiction. Effect without cause is a contradiction because the two terms are defined in terms of one another.

Conceptual confusion in causation. False concepts are often used as a cause in science, e.g., energy is a pseudo-concept.

Contradiction. If one tries to avoid causal notion by reducing it to correlation of events (or of stimulus-response), it is nevertheless brought back by speaking of statistical *significance*. But to know if something is significant one has to know about cause, so it does not avoid causality. There is also circularity here because stimulus and response are synonyms of cause-effect so it is blind to think one is avoiding causality by substituting stimulus-response.

Correlation. Bertrand Russell's philosophy bans *cause* replacing it with correlation. In psychology, for example, cause-effect is seen as stimulus-response, mere correlation. In attempting to avoid the fallacy of cause as a mysterious force, correlations are used. To avoid saying x caused y, it is said that x has a significant correlation with y. The problem with this is: The reduction of cause to correlation assumes what it is supposed to prove because it is said that the correlation must have "relevance." Not everything may be correlated with everything else. But to determine relevance is to assume causal significance. Thus, correlation does not circumvent the notion of cause, but rather presupposes it. Also, stimulus (S) and response (R) are already synonyms of cause-effect and so do not avoid the notion of cause and its problems. We can instead speak only of "event x *and* event y." S and R would be then causally neutral. Instead of stimulus-response one could speak of event 1 and event 2 regardless of which came first. We could then find a correlation between them. On such a view R could even precede S. One could say the response "caused" (as correlation) the stimulus. The response could also serve as a stimulus in such a way as to have S-R-S-R-S etc.

Also, there is virtually never a single stimulus, S, but S1, S2, S3, etc. and R1, R2, R3, etc. S-R leaves out thinking and language use (S-language use-R), which are the most important factors for human behavior. To exclude the thinking between stimulus and response is false reductionism. The notions of response, stimulus, and event are vague. Stimulus and response are determined by the language they are described in. Which parts of a situation are we to label as the stimuli and which can we exclude? Also, the statistical correlation is set very low. The statistical significance of .05 is problematic. But even high correlation may provide no evidence for being a cause. If nearly everyone who has cancer drinks milk, it still does not show that milk causes cancer. Cause as statistical correlation is often vague, circular, over-abstract, and therefore highly problematic. Statistical relevance does not mean significant, qualitative relevance. Every experimental and statistical study needs to be evaluated from the point of view of a sound philosophy of science. For nearly every experimental study there is a counter study, which undermines it.

Defensive self-justification causation (cf. Chapter 8). The physician is not required to give treatment, which is intrusive, burdensome as well as futile, but sometimes law or policy requires it. It may require the healthcare worker to do unneeded tests in order to prevent a possible legal problem. However, defensive medicine is said to be of marginal or no medical value to deter a patient from filing a malpractice claim. In a survey of 824 physicians working in emergency medicine, general surgery, orthopedic surgery, neurosurgery, obstetrics-gynecology, and radiology all but 7% reported having engaged in some form of needless defensive medicine. One third of respondents had frequently prescribed more medications than were medically indicated. Nearly 40% of respondents intend to avoid caring for high-risk patients or have already begun to do so. Occasionally certain high-risk procedures were avoided. Patients with complex medical problems and those seen as litigious also were avoided as patients [15].

Degrees of cause. There are degrees of causality: Adequate explanation, causal guess, circumstances (cf. law), conditions, expectation, foresight, hope, intuition, possibility, probability, and speculation. Most distal causes of diseases are unknown, especially for any particular person. Who and what caused the crime? Was it the individual, trigger finger, society, poor education, lack of critical thinking (speaking), etc. in the schools? It is a faulty question to ask what the literal cause is or assume there is an absolute cause, e.g. "What really caused you to do x?" "What is the genuine cause of the disease?"

A common medical error is to claim certainty when there is only some degree of probability based on incomplete knowledge. "Absolute safety of any substance can never be proven" [16]. "Just about any food has the potential to cause illness" [16]. The statements made in nutrition journals and the recommended daily requirements are among the most unreliable in the area of medicine.

Because even the smallest event or detail may result in disastrous consequences, it is an argument for truth-telling on the part of both patients and health care workers. It is also an argument for disclosure and full, open communication in decision-making. That is, we cannot always know in advance what consequences even the smallest detail will have. Even personality traits of the staff may result in harm to the patient, e.g. arrogance or envy, may cause one to fail to discuss a problem diagnosis with a more experienced senior staff member. The doctor often has a special kind of emotional influence in medicine. Each member of the treating team in terms of their own personal and professional structure potentially might influence the whole team as well as the attitude of patients and therefore each member might be a potential source of conflict.

Descriptive cause. This is not to imply that there are absolutely objective descriptions, but we can find inter-subjective characterizations of an event. This event would not have happened unless the alleged causal one occurred. Evidence-based medicine (EBM) claims to provide objective description, but does not do so (See Chapter 19). EBM trials do not determine causation. We especially need observational clinical studies [13]. Confounding in statistics is a false causal result due to a third factor.

Desire or intention. A desire or intention is not a proper cause. Cause, desire, and intention differ in each discipline.

Dictionary and encyclopedia definitions of causation. There are various definitions of causation found in reference sources.

Emotional causation. Each theory of emotion yields a different model of causation. On the philosopher's Cognitive Theory of Emotion [in therapy called the Rational-Emotive Theory (RET)] our emotions are caused by our assessments. Irrational and unrealistic thinking and decision-making leads to negative emotions; and rational and realistic decision-making leads to positive emotions. As humans are involved in virtually all decisions, one cannot make humanistic decisions without a sound knowledge of emotions.

Evidence of cause. Virtually, all causal statements may be regarded as false unless proven otherwise (Fallacy or false causality). It is usually the burden of the asserter to defend the causality, e.g. "What view of causality are you wishing to use here?"

False Authority. It is not an argument to say that something is the cause or effect because someone says so. Because a person is an expert or famous authority does not in itself make anything they say true. Peer review is often set up by those merely sympathetic to a certain viewpoint and is often biased. Rather, the specific arguments and evidence are needed. It is not an argument to say that a causal assertion is true because being presented in a medical resource, dictionary, encyclopedia, or by a scholar. The healthcare worker's decisions are often unjustifiably limited by religion, law, protocol, superiors, government, and by associations such as the American Medical Association or nursing associations. All of these sources must, however, be questioned. We ourselves must know the arguments and reasons for our actions. In medicine, for example, as patients it is important to know as much as we can about the disease we have. One may, however, provisionally accept advice from an expert or specialist in areas to the extent that one is not able to determine the facts or arguments for oneself, e.g. law, medicine, economics, etc. Even then, other experts may be consulted and compared. The philosophy is the critique of the concepts and methods of the various disciplines and so tries to avoid the appeal to authority fallacy.

False blame (cf. defense mechanisms, false cause, legal causality). One may be said to be guilty of something he/she did not do or could not have prevented. We cannot, for example prevent all mistakes. Our legal system is found to be permeated with illogical reasoning, faulty rules of evidence, and inadequate or outdated understanding of psychology [17]. Does it ever make sense to blame anyone? In law, Clarence Darrow often obtained mitigated sentences for his clients on the basis of extenuating circumstances. If such circumstances were fully understood, it may be that no one would ever be to blame [18]. If, for example, we only do self-defeating or harmful things out of ignorance, why should one be blamed? Certainly harmful acts are usually or always due to lack of education, lack of knowledge of cause and effect, one's societal influences, and one's own level of understanding. We cannot blame people if they could not in actual fact have done otherwise (The section on clarifying understanding in the Chapter 8 would be also useful to understand this point). There are always further causes for one's actions: We blame others when, in fact, we may have largely caused the person to act that way because of our negligence or failure to teach about critical thinking (speaking), ethics, rationality and emotions, etc. Ultimately, society and the court itself may be the cause of a person's crime, because it brought about the conditions for it, e.g. by poorly educating and teaching its citizens or by punishment, rather than rehabilitation. It is a militant, enculturating society, not just an individual that pulls the trigger. Society murders itself.

False cause fallacy. Post hoc ergo propter hoc. The fallacy of assuming that if something happens before an event it must be the cause of it. Other fallacies relate to this one in many ways, e.g. genetic fallacy. Because A happens before B happens, does not mean A causes B. For example, energy is not a cause, but a description of what happens. Energy as such does not exist. It is a property of events, that is, a reference to an action or what happens, not a thing or substance. Conflation of causes in experimental research trials is the failure to isolate proper causes. Often relevant variables are excluded. If one takes a drug or gets treatment, and improves, it may not be because of the treatment or drug, and even be in spite of them.

False reason. Cause as the reason for an event. Reasons are explanations or justifications. The effects are consequences. Dogmatic, supernatural and thoughtless decisions are typically made without concern for consequences or reasons. This is true also for medicine.

Reasons for events are often relative rather than absolute. Of about half of the diseases listed in the *Merck Manual* the cause is listed as being unknown. When the cause is known, it is often a proximate or immediate cause. The full causes of "thought," perception, and virtually ultimately everything in our environment are fundamentally unknown. About any cause, we can ask for a further cause. It is not completely known why we perform any act. "Our basic medical models rarely account for more than 30% of whatever outcome we are investigating." Physicians are often faced with the unexpected [19].

Faulty question fallacy. A faulty or meaningless question is asked such as, "What causes disease?" "What do patients really want?" "What is the goal of life?" But these are the questions people often ask. They are over-abstract and cannot be answered.

Indoctrination or "cultural fallacy." Our view of causality is largely based on indoctrination of our cultural beliefs and language. Culture is often an enemy of rational inquiry. (Folk or lay causality, in medicine "lay etiology" for symptoms, diseases, disorders.) Thus, the causal beliefs of people are typically false. Even legal causality appeals to intuition and common sense, which can be arrogant (See "legal causality" above).

Language-game of causation. Causality may be seen as a language-game [10]. There are no causes as such. Rather we have to find out what cause means in a particular case and we must look at the particular language situation and the context in which it is being used. In "x may cause cancer," we need to know the quantity, detailed description of the substance and the full context.

Legal causality. In the legal area, causation is seen in terms of blame, punishment and values. It often merely stipulates or assigns a cause even when the facts indicate clearly otherwise. In the courts, circumstantial evidence and other fabrications are employed in the determination of a cause. Cause is often an *ex post factum* story lawyers tell to win their cases. That over 100 death row inmates were recently released due to genetic testing is one of the greatest embarrassments showing of the failure and inadequacy of legal procedure and evidence. Such cases required the most solid legal evidence, yet failed. That 25% do not return to prison does not show success. The failure rate is not 75%, but somewhere between 75 and 100%. The existence of up to 100% recidivism and failure rate shows also that the criminal system has failed to correct and failed to protect the public, and has done so at great expense to the people. That the public and the lawyers are not at all embarrassed about this shows their insensitivity and lack of critical thinking. As with war and bad management, the criminal system thrives on failure. It is self-contradictory for physicians to merely blindly serve society, unmoral cultures, unethical moralities, and their unethical institutions as they have been doing by neglecting more aggressive political and societal engagement. The embarrassment falls also on the medical profession for not doing so. Several exceptions are: Physicians for Social Responsibility, Physicians for Human Rights, International Physicians for the Prevention of Nuclear War, etc. A philosophically enlightened medicine would not be a mere medical slave of society.

Court data are mostly untestable [20]. Decisions are based on "intuition" and "common sense." There is a "resort in desperation to idealized notions of intuition and common knowledge. Where recourse is had by the courts to such vague formulations of common sense... [it] leaves the role of proof...problematic and uncertain" [20]. The jury's notion of cause is arbitrary. The jury can determine causality to be anything it wishes. Jury decisions are based on intuition and uncritical common sense and then the reasons given for the decision are rationalized on these bases [20], "As long as the law seeks refuge in such unscientific notions such as intuition and common sense and intuitive assessment of causation...proof of causation...will remain the unsatisfactory handmaiden of such phenomena as the aura of persuasiveness of an individual witness on a given court day. This is a recipe for inconsistency and error in decision-making" [20].

Hollingsworth and Lasker claim that Daubert versus Merrell Dow Chemicals shows that scientific and epidemiological medical evidence is now to be the new basis of courtroom evidence, not legal priority [21]. A critical analysis of Sir Bradford Hill's criteria of causation is also used: strength of association, consistency of findings, specificity of dose-response relationship, temporal relationship, biological credibility, exclusion of alternative explanations [21]. Clinical reasoning about causation is not acceptable unless based on scientific evidence [21]. Ironically, the law requires scientific evidence for testimony, but its own methods of the religious oath, a problematic appeal to precedent, and to an arbitrary jury would not pass as good scientific methodology. "Courts that ignore the scientific method when reviewing medical causation opinions do a disservice to the legal system." [21]

Tamanaha argues that according to postmodernism, there is no true meaning in law. All understanding is interpretation. Postmodernists oppose law as claiming to be authoritative, neutral, good or right [22].

Formalism is fallacious and mechanistic deduction from alleged absolutistic laws. On the other hand, Oliver Wendell Holmes was a pragmatist holding that law should serve human purposes. Roscoe Pound and legal realists saw law pragmatically as instrumental [23].

The patient must be given information even about rare side effects, benefits and alternatives [24].

This is true even if, as in one case in Australia, the risk is 1 in 14,000 [24]. If the physician performs an operation without consent of the patient, even if it benefits the patient or saves the patient's life, there is no need to even prove harm done to establish the physician's guilt in English law [24].

In a strong sense, the courts, the politicians, business people, and the religious try to usurp medical decision-making. "It is for the courts to decide what is the required standard of care, and not for medical practice...to do that." [20] The jury can decide causation without or in spite of the assistance of expert testimony. Introduction of medical evidence and literature may be rejected by court as hearsay. Allowed are only statements delivered in person. So-called medical experts are given absolute immunity in testifying and often are not especially qualified to ascertain the standards of practice. Hoffman suggests that they should not be given such immunity, but be held responsible for their testimony and challenged on it [25].

Standards of practice should not be determined by the judge or the jury, but rather by the medical experts [25].

Legal decisions made by legislature and methods of law: precedent, presumptions, adversarial method, etc. often prevent causes from even becoming known. So-called "expert witnesses" are often biased. "There is a need for clearer tests of causation as a matter of law and for greater synchronicity between scientific/medical criteria for proof of causation and those used in law." [20] I. Callinan states, "I sometimes think that theory [chaos theory in science] may have relevance to the legal theory of causation." [20] I would suggest that the present concepts in law be severely critiqued and reexamined by philosophers of law and philosophers of medicine (See also discussions of law and medicine in the Chapter 12).

The American College of Obstetricians and Gynecologists (ACOG) reported, that the fear of being sued is driving obstetricians-gynecologists' to stop delivering babies [26]. Medical liability reform is ACOG's top priority. One in seven ACOG Fellows report that they had stopped practicing obstetrics because of the high risk of liability claims. Services were curtailed in many areas because of the risk of liability claims or of being sued, e.g. decrease in amount of high-risk obstetric care – 25.2%, decreased gynecologic surgical procedures performed – 14.8%, etc. In 2003, one in two Fellows had been involved in a claim in the last 4 years. Over 76% of ACOG Fellows reported they had been sued at least once. 57% had two or more claims filed against them, and 41.5% had three or more claims. Ob-gyns have an average of 2.6 claims filed against them during their career. Obstetric claims accounted for 61% of claims against ob-gyns; 38% were gynecologic claims. From 1999 to 2002, the top primary obstetric allegation was the neurologically impaired infant (34%). Almost half (49.5%) of claims against ob-gyns are dropped by plaintiffs' attorneys, dismissed, or settled without payment. Of cases that do proceed to court, ob-gyns win eight out of ten (81.3%). From 1999 to 2002, on average, the length of time from occurrence to closing of the claim was 4 years.

Hoffman noted that in medical malpractice cases lawyers tend to represent the highest income cases and ignore the cases less than $100,000 even if the latter are more meritorious. Furthermore he argues that, "Neither judges nor jurors are particularly well equipped to gauge...whether an individual physician's conduct conformed to the accepted standards of practice" [25]. Law (judges, lawyers, jurors) should not be determining the standards of care of the medical profession. The law takes over decision-making for the medical profession just as religion tries to do. Now there are "professional physician witnesses" serving as "experts" and being paid five figure fees and often they never even practiced in the relevant specialty. He concludes, "We need to take the job of evaluating compliance with standards of care out of the hands of judges and jurors" [25].

Literalist fallacy. One of the most common fallacies is to think there is only one cause or solution. "The only thing we could do was going to war." The sociologist typically thinks all causes are social ones. This is the same model captivation held in each discipline, e.g., the individual medical model according to which all diseases are caused and cured by physical operations, drugs and medicine. Therapists e.g. are usually captivated by questionable models of therapy. Physicians and patients are often held captive by a metaphor, not open to arguments and discussion [27, 28].

Logical cause. Cause is reduced to a basic connective or formulation in formal symbolic logic, e.g. logical implication between propositions, if-then propositions, or conclusions such as "therefore...," or "It follows that..." Logical cause is a stipulative and constructive cause as those found in mathematics without having to have any basis in reality or outside of logic. Going from premises to conclusions needs reason outside of logic. Logic cannot establish the truth-value of the truth of a sentence in the first place. Deductive logic is inadequate in medicine. "Logic cannot establish the context of any factual truth" [29] (For a strong criticism of formal logic see the Chapter 18).

Mathematical causation. Like for statistical and logical causation, cause is reduced to a mathematical construction. Cause in science is often only a mathematical cause, irreducible to empirical evidence. The problem is that science and cause rest on language and not numbers. Chaos theory, particle theories and quark theories, for example, are largely mathematical theories.

Metaphorical causation. Causality may be seen in terms of metaphors such as: a significant correlation, an instruction, or a recipe based on experience [30], a problem to be solved, an irreversible succession, an influence, etc [31]. Creative new models of cause may be given. Cause may thus be analyzed in terms of any other term.

Negligent causality (See also "false blame"). Awards of millions of dollars in malpractice cases have been out of all proportion to the harm done. Cases are adjudicated by unacceptable legal methods, which often discount medical evidence (See legal cause above). Medical negligence may not involve a positive action, but only a failure to treat or warn a patient. The question arises as to how much one must be responsible to warn against. Is a doctor negligent if he/she does not warn the patient about everything that might shorten the patient's life? There are an infinite number of risks, which one could be warned against and the physician cannot be expected to give them all. Nor could any healthcare worker know of but a few of the typical risks of all possible occurrences. Should the physician find out all about the details of each patient's life and point out all of the risks due to the patient's lifestyle and habits as well? Patients have a responsibility to learn about their own lifestyle risks and preventative care by means of their own research. They have autonomy. They are often the ones who choose which sort of physician to visit in the first place. Different physicians would warn of different risks depending on their specialty.

The statistical amount of risk must also be understood to be in need of much interpretation, which the physician cannot be expected to elaborate on, or which is not even known yet. 1 in 14,000 risks may turn out to be 1 in 10 in some cases and 1 in one million by other qualitative analyses (See section on statistical cause). Statistics are abstractly quantitative and in need of much qualification. They are falsely assumed to be perfect. Suppose an operation has 50% success rate. The patient may sue for malpractice whenever the result is negative. The surgeon may be accused whenever an operation fails. The higher the risk, the higher is the likelihood of a lawsuit. This discourages a great deal of desired, but risky surgery. It also discourages reporting of mistakes. In the U.S., it was only as recently as of July 1, 2001 that hospitals were required to disclose to patients all unexpected results, that is, errors. It was found that they tend to not report preventable harms, one reason for this being the fear of malpractice suits [32]. Malpractice insurance is so high in the U.S. that many obstetricians have given up the practice of obstetrics. The American cost is over $200,000 per year in obstetrics and the number of deliveries must be limited. Physicians must by now be familiar with medical court decisions. Irrational court decisions have undermined sound and available medical practice.

Noncausal events. In contemporary physics some events are said to be uncaused. Some scientists say the law of cause and effect must be given up [33]. Quantum theory is basically a mathematical theory of causation. "Quantum theory can't be explained" [34]. Light is not a particle or wave, but can be both or either at different times. "The microscopic phenomena described by quantum mechanics are intrinsically random. If, as Einstein believed, their randomness is a manifestation of a deeper level of determinism, that level has not yet been found" [35]. Some scientists think scientific concepts of cause no longer apply regarding cosmology. We cannot explain what we need to explain, e.g. the nature of "black matter" and "black holes" both of which are said to dissolve the notions of space-time continuum and the laws of cause and effect. One can see the world non-causally, e.g. in "black holes" the concepts of time and space supposedly no longer apply.

Normative common views of cause. There are numerous false ascriptions of cause based on normative belief. For example, the cause of emotions on the cognitive theory of emotion is ultimately you, not as is usually thought, things or others. To say someone or the situation caused you to be angry is false. You are the ultimate source of your anger because of your negative assessments. Blame suggests that one could have done otherwise than one did. But this is a faulty assumption, because if one could have, one would have. It is only from an idealistic and unrealistic viewpoint that one could have done otherwise and it ignores the reality of one's actual abilities (See also Chapter 7). Callinan speaks of the idealistic "tendency for lawyers to believe that all is controllable" and so one is to blame [20]. Buetow and Elwyn wrote, "These patients could have acted otherwise" [36]. This is incorrect. But, of course, harmful behavior needs to be corrected. Psychological disorder and need of therapy or education are almost always ignored in the U.S. courts because blame and punishment is demanded in their place. It is in only the most extreme cases that psychological illness might be admitted as a cause.

Operational definitions (See discussion in Chapter 2). Cause is defined as what in fact happens, e.g. there is no energy in magnets as such, but it may only refer to the fact that iron filings go to the magnet. We never find energy in itself. It is only an attribute, descriptive of an action. The term "energy" is misused in both in science and in medicine.

Oversimplification is often due to the assumption that there is only one cause (or answer, model for cause, etc.) or one effect. A simplistic cause is given. This is perhaps one of the most frequently committed errors. One or two causes are given when numerous others are relevant or needed. We may then speak in the plural of the causes, rather than of a single cause. It is also a faulty question fallacy if we ask what *the* cause is, or assume there is an absolute cause, e.g. "What really caused him to do x?" "What is the genuine cause of the disease?" These are faulty questions. It is oversimplification to say, "The doctor cured me." "Nature" cures many diseases, e.g. colds.

Pragmatic definition. The meaning of a cause is its consequences. (e.g. John Dewey) There are no causes as such. There are just means to bring about a practical result. There is only a "cause" for a certain purpose (cf. In operant conditioning, "stimulus-response" supposedly only has meaning to the extent that it brings about a certain consequent change of behavior). This avoids faulty incomplete or fixed notions of causes and therefore avoids unfair blaming. The emphasis is to humanistically treat, correct, and educate regardless of the harm done by someone.

Quantitative cause. Quantitative statistical-mathematical determination of causes is mainly used in the sciences of the Western World. Research in social and natural sciences, including medicine, usually uses such quantitative experimental design. Besides being a form of the reduction of thinking and language to mathematics, the approach is based on a faulty view of the scientific method, which is based on language use, not mere perception (See extensive discussion of this in Chapter 18). The philosophy of science has shown that this method as typically used is in fact unscientific. The concepts of the various disciplines are typically not well defined or undefined, unexamined and confused, yet they are used unquestioningly in experiments. The notion of "cause as correlation" is highly problematic. Experiments of lab animals are illegitimately generalized to apply to humans (personification), etc. For example, in toxicology, there is imprecise extrapolation from results with animals to humans. Animal studies as the basis of causation are not uncritically accepted in the courts [21]. Saccharine, for example, was found harmful to rats, but not to people. Also animals develop different kinds of cancer [21]. However, a substance still might be harmful to people.

For each experiment proving one thing in medical research there are often others proving something else or the reverse. Experimental results and techniques must always be subjected to the critique of critical philosophy of medicine, but they virtually never are.

Alternative medicine must be critiqued as closely as is medicine. Drug testing takes many years, 10–20 in the U.S., and the drugs are still inadequately tested. Once the drug is released for general use a great deal of experience is obtained about the drug, but it is usually not adequately integrated into the drug approval process. However, the Medwatch website does collect such information on a voluntary basis. In 2004, around 1,000 voluntary reports were sent each day to Medwatch regarding the dangers and side effects of drugs. 300,000 reports are received each year. The agency, however, has too few people and no safety system to deal with these reports. Furthermore, we cannot tell what lower doses of a toxic substance will do. We often do not know the mix of toxic substances with other compounds and foods and also with the environment. In addition, the medication must be tailored to the individual.

Recommended cause. In a strong sense, many causes in medicine are hypothetical and so more like recommended causes than actual ones.

Religious causation. Religious causation is frequently erroneously confused with medical causation. Disease is thought being caused by sin and God's punishment, etc. Often it is thought to be a miracle if someone unexpectedly survives. Miracle only means that we do not know how something happened and think the result is desirable. Miracle is a religious value term. We do not say it is a miracle if a friend unexpectedly dies. Miracle is not a type of cause.

Self versus other causation. We must distinguish between I, versus we, they, you, or it caused something. The evidence is different in each case. I have first-hand knowledge about my lying, but others do not. It may be difficult to determine if others are lying. A different kind of evidence is needed (See Chapter 20). We may, however, be mistaken or unclear about cause whether it refers to oneself or others. Suffering and pain may be different for different people.

Self-cause. Cause can be seen as being self-caused. Who and what caused the crime? There are consequences of not inquiring. There are proximal and distal causes. What caused the medical error? Was it the nurse, physician, policy, bad management, the fact that the hospital was unfunded, the lack of education in critical thinking and ethics, etc.? We often proximally or distally cause our own illness or inadequate treatment (See Chapter 11). For example, one causes one's own health problems due to one's unhealthful lifestyle, opposition to medical research for religious or other reasons, refusal to contribute organs after death, opposition to stem cell research, etc.

When is one a victim? When is one innocent? Some even claim being victims because made dependent by being supported. If we vote for an aggressive military, or against health care funding we cannot claim innocence.

Statistical fallacies. Often cause is only numerical probability. "The practical issues are qualitative" [37]. There is no linear relation between cause and effect. An absolutely regular heartbeat is a prelude to health problems [37]. Statistics is only as good as the prior experiments and the clarity of the terms used [38]. "None of the contributors are prepared to endorse the statistical methods of causal analysis most widely taught and employed in quantitative social research" [38]. By relying on statistics there is often failure to consider the individual patient (e.g. when citing general statistics). There is also often a failure to give the correct application of the statistics, e.g. to point out that they may apply only to those who are overweight. We may not just tell the patient, "The risk of the operation not being successful is x% of the time," or "Only one in 20,000 dies from taking this medicine/procedure/drug/operation/injection/anesthesia/etc." Rather we must know how this figure was arrived at and relate it to the particular health of the individual.

Similarly we can't simply maintain that x is the standard dosage. We need to know further conditions like age, gender, kidney and liver function, etc. In one case, the patient was given the standard dosage of a drug of which soon was seen to be twice the dosage "needed," and, after consulting another specialist, it was determined that the drug was not necessary at all. It is important to avoid making certainty statements. The use of uncritical statistics in medical research, as in psychology and other subjects is virtually always faulty in significant ways.

Stipulated cause. Cause may be merely stipulated, e.g. in law.

Substitutions for cause. We may substitute different meanings for cause and cause-effect. For example, we may substitute "effort-success" in place of cause-effect. This is another form of recommended cause. Sometimes one cannot be successful because, for example, the cancer is too widespread.

Supernaturalism fallacy. This involves abstractionistic, and supernatural causes. Causal explanation is based on astrology, religion, unexamined alternative medicine, mentalistic psychotherapies, or unfounded and unscientific belief systems. The causal beliefs of the average person in each society are typically supernatural. Religious and mystical beliefs can predominate over medical and psychological findings and practice in the various cultures. Again, there is the attempt of some churches and individual healthcare workers to determine medical decisions as dictated by religious belief. Dogmatic belief system holders often try to usurp medical and humanistic decision-making.

Teleological fallacy. The notion of a "final cause" (teleology, fate, determinism) is a false notion. Things have potential and abilities not ultimate meanings or reasons for being, etc. This fallacy often used by religious people tries to establish the existence of a god.

Unknown cause. (cf. noncausal events) Sometimes the cause is unknown. One cannot claim that a nutrient or food substance has no negative side effects. In so many cases high blood pressure has no known cause [39]. Two men could have been the father but neither is available for testing. Which is to be regarded as the father? In Italian law, if a wife gives birth it is assumed and stipulated that the husband is the father even if he has not touched the wife in 10 years and never contributed to a sperm bank.

Value or persuasive cause. Blame ascribes wrongness to the cause. Here is, then, vindictive causality and persuasive causality. "This has no effect," means only not the *desired* effect. It is not descriptive, but a value judgment. Cause can be used as an uncritical moral term or as a critical ethical term. "The outcome was bad," does not necessarily mean that the operation was not a success. It may mean that a desired outcome was not achieved, though the other outcomes might have been achieved.

4.8 Summary

Causes and causation in medicine are to be carefully examined and healthcare workers have to be cautious in ascribing causes too readily in order to avoid circuits and prejudices in diagnosing and treating patients.

The critical definitions of causes have an ethical impact on medical thinking and practice, the lack of doing so invalidates it.

References

1. Hollingsworth, J., and Lasker, E. 2004. The case against differential diagnosis: Daubert, medical causation testimony, and the scientific method. *Journal of Health Law* 37:85–111.
2. Gordis, L. 2005. *Epidemiology*, 3rd edn. Philadelphia, PA: Elsevier.
3. Montgomery, K. 2006. *How doctors think: Clinical judgment and the practice of medicine*. Oxford: Oxford University Press.
4. Mitchell, G., and Benichou, J. 2000. In *Encyclopedia of Epidemiological Methods*. Chichester, NY: John Wiley.
5. Hill, A.B.. 1965. The environment and disease. Association of causation? *Proceedings of the Royal Society of Medicine* 58:295–300.
6. Salmon, W. 1998. *Causality and explanation*. New York: Oxford University Press.
7. *Diagnostic and statistical manual of mental disorders*. 4th edn. Text revision (DSM-IV). 2000. Washington, DC: American Psychiatric Association.
8. Shibles, W. 1974. *Emotion: The method of philosophical therapy*. Whitewater, WI: Language Press. See Chapter on DSM II.
9. Andreasen, N. 2001. *Brave new brain. Conquering mental illness in the era of the genome*. Oxford: Oxford University Press.
10. Wittgenstein, L. 1958. *Philosophical investigations*, 3rd edn. New York: Macmillan.
11. Melden, A. 1969. The conceptual dimension of emotions. In *Human action*, ed. Mischel, T. New York: Academic.
12. Hume, D. 1993. In *Enquiry concerning human understanding*, 2nd edn. Section 4–7, ed. Steinberg, E.; Hume, D. 1978/1888. In *Treatise of human nature*, ed. Selby-Bigge, L. Oxford: Clarendon Press
13. Elwood, J.M. 1988. *Causal relationships in medicine: A practical system for critical appraisal*. Oxford: Oxford University Press.
14. Stapleton, J. 2002. In *Causation in law and medicine*, eds. Freckelton, I., and Mendelson, D. Hampshire, GB: Ashgate.
15. Studdert, D. 2005. Defensive medicine among high-risk specialist physicians in a volatile malpractice environment. *JAMA* 293:2609–2617
16. Goldstein, M., and Goldstein, M. 2002. *Controversies in food and nutrition*. Westport, CT: Greenwood Press.
17. Mitford, J. 1998. *The American way of death revisited*. New York: Knopf.
18. Shibles, W. 1987. Reformatory blame. *Journal of Rational-Emotive Therapy* 5:266–281.
19. Katerndahl, D. 2005. Is your practice really that predictable? Nonlinearity principles in family medicine? *Journal of Family Practice* 54:970–977.
20. Freckelton, I., and Mendelson, D. eds. 2002. *Causation in law and medicine*. Hampshire, GB: Ashgate.
21. Hollingsworth, J., and Lasker, E. 2004. The case against differential diagnosis: Daubert, medical causation testimony, and the scientific method. *Journal of Health Law* 37:85–111. 509 U.S. 579. 1993.
22. Tamanaha, B. 1997. *Realistic socio-legal theory: Pragmatism and a social theory of law*. Oxford: Oxford University Press.
23. Summers, R.S. 1982. *Instrumentalism and American legal theory*. Ithaca, NY: Cornell University Press.
24. Hope, T., Savulescu, J., and Hendrick, J. 2003. *Medical ethics and law: The core curriculum (for medical students)*. London: Churchill Livingstone/Elsevier.
25. Hoffman, D. 2005. The medical malpractice insurance crisis again. *Hastings Center Report* 35:15–19.
26. communications@acog.org July 16, 2004.
27. Schneider, St. 2005. *The patient from hell: how I worked with my doctors to get the best of modern medicine and how You can too*. New York: Da Capo Press.
28. Hanson, N. 1958/1961. *Patterns of discovery*. Cambridge: Cambridge University Press.

29. Phillips, C. 1995. *Logic in medicine*, 2nd edn. London: BMJ.
30. cf. Honoré, A. 2002. In *Causation in law and medicine*, eds. Freckelton, I., and Mendelson, D. Hampshire, GB: Ashgate.
31. Gasking in Taylor, R. 1967. In *Causation. Encyclopedia of philosophy*, ed. Edwards, P., vol 2, 56–66. New York: Macmillan.
32. Lamb, R.M. 2003. Hospital disclosure protection: Results of a national survey. *Health Affairs* 22:73–83.
33. Bridgeman, P.W. 1927. *The logic of physics*. New York: Macmillan.
34. Moring, G. 2000. *CIG to understanding Einstein*, 100. Indianapolis, IN: Alpha Books.
35. Tavel, M. 2002. *Contemporary physics and the limits of knowledge*, 214. New Brunswick, New York: Rutgers University Press.
36. Buetow, S., and Elwyn, G. 2006. Are patients morally responsible for their errors? *Journal of Medical Ethics* 32:260–262.
37. Jenicek, M., and Hitchcock, D. 2005. *Evidence-based practice: Logic and critical thinking in medicine*. Chicago, IL: American Medical Association Press.
38. McKim, V., and Turner, St. eds. 1997. *Causality in crisis? Statistical methods and the search for causal knowledge in the social sciences*. Notre Dame: University of Notre Dame Press.
39. Ropeik, D., and Gray, G. 2002. *Risk*. Boston: Houghton Mifflin.

Chapter 5
Ethics and Non-ethics

Ethics must fight with three opponents: thoughtlessness, egoistical selfishness, and society. [1]

Abstract A distinction is made between morals and ethics. Morals are not critical, but enculturated, more like learned habits or rules than thought out behavior. Ethics is a critical examination of morals and cultural practices. The medical system of a culture is also based on such morals. Common morality is basically self-contradictory. There is ignored moral inconsistency and searched ethical consistency. According to the ethics versus moral view, something can be moral, but unethical; immoral, but ethical; both immoral and unethical; both moral (by chance) and ethical (by reason). As to suggestions for a specific ethical theory, which is consequentialistic and combines the scientific basis of medicine with the philosophy of medicine one may recommend a naturalistic, humanistic theory of ethics. It also benefits from the philosophy of science and pragmatic Philosophical Practice.

Keywords Culture · enculturation · ethics · morals · non-ethical · Naturalistic Theory of Ethics · Humanism · ethical terms · misuse of ethical terms · universalization

5.1 Introduction

Socrates said, "The unexamined life is not worth living – that you are still less likely to believe" [2].

Socrates is certainly correct that the average person, including the professional, is not "likely to believe" in an examined life. The statement is inaccessible to the non-philosopher. Such examination and critical thinking (speaking) are not tolerated as Socrates fatally found out especially by people who want to comfortably remain enculturated without asking any questions. Religion and culture oppose critical thinking. They must oppose critical thinking (speaking) if they are to survive as they do. Criticism would ruin them. By "culture" is meant traditional, and uncritical and commonly accepted societal practices of a nation or race no matter how they are. It does not here include those institutions critical of culture such as open education,

B. Maier, W.A. Shibles, *The Philosophy and Practice of Medicine and Bioethics*,
International Library of Ethics, Law, and the New Medicine 47,
DOI 10.1007/978-90-481-8867-3_5, © Springer Science+Business Media B.V. 2011

and philosophy. "The teaching of medical ethics was traditionally…often skimpy and formulaic… [or] barely mentioned" [3].

Today we can say that critical philosophy of medicine is not mentioned at all. Even *Medical Ethics Today* (2004) supports a medical system following culture. On the other hand, they also state that medicine should not just support community values [4].

If one wants to be healthy and promote health, one will often have to go against culture and experience the rigid criticism of its members.

Albert Schweitzer wrote, "The progress of ethics consists of our decision to think pessimistically of the morals of society" [1].

Cultures create the familiar and the strange. They like nearly everything else are changeable, dependent upon geography, place and time, and are socioeconomic products. The more dualistic, and metaphysical, the more alexithymic people become, that is, being unaware of their emotions, being irrational, and having a paucity of imaginative thought. You only think if you think yourself [5].

This is similar to John Dewey's pragmatic view that you are only ethical if you yourself make rational decisions [6]. It is little known that people do not understand ethics. One cannot break ethical rules if one has none. People, including professionals, think ethics is innate and spiritual qualities. It is not.

One main problem is unquestioned enculturation. People blindly accept tradition and culture, regard it as a standard of morality, and reject anything, which deviates from their particular belief systems. People think normative common and traditional practices are the standard for ethics because they are used to them in their culture. Culture can be a pejorative term, like prejudice, and dogma, and enculturation. "Speaking differently, rather than…arguing well, is the chief instrument of cultural change" [7]. We must, therefore, deprogram its irrationality, superstition and common (un-inquired) (non) sense. Dewey and Tufts state, "The intellectual distinction between customary and reflective morality is clearly marked" [8]. Reflective morality favors criticism of culture. In terms of change we need to go from customary to reflective morality [8]. Nietzsche states, "What is needed above all is an absolute skepticism toward all inherited concepts" [9]. Desires are habits and cultural tendencies and must be remade in terms of consequences [10]. GH Mead agrees that culture can and must be changed. It is in communicative evolutionary process [11]. Nietzsche wrote in *The Gay Science*, "Not to question, not to tremble with the craving and joy of questioning… that is what I feel to be contemptible, and this feeling is the first thing I seek in everyone: some foolishness persuades me ever and again that every human being has this feeling, as a human being" [12]. Schweitzer held that we must be highly critical of society and its past and present practices and through fundamentally individual critical thinking (speaking) renew it by making it ethical again [13]. This is also one task of the healthcare worker which cannot continue to be ignored. "Often ethics prescribes higher standards of behavior than does the law, and occasionally ethics requires that physicians disobey laws that demand unethical behavior." [14] Physicians may have to promote their positions "forcefully" against those of governments, health system administrators, and/or commercial enterprises [15].

Children, like philosophers, are full of curiosity and questions, but they soon learn not to ask them. Questions irritate. We may wonder why by the time they are in high school there can be little curiosity left. They are captivated by the common beliefs by then. Philosophy is rarely taught or taken as a subject. (See Chapter 18)

The result is an indoctrinated populace, which is largely emotionally dysfunctional (with anger, hatred, revenge, selfishness, etc.). On the common view, it is only the unexamined life, which is worth living. Cultures and religions are treated as being sacrosanct and enjoy absolute protection and tolerance regardless of the irrationality or harmfulness of the naturalistic consequences, which range from opposition to medical treatment and to medical research to favoring female and male circumcision and war. Philosophers try to show how and why people, including professionals, are so irrational and in need of education in the attempt to correct it. Also, bioethics can be no better than the rationality, emotional stability and moral soundness of the culture in which it is practiced. It will be shown why people are not and cannot be ethical. Bioethics shows in miniature what is wrong with society in general. *We will not solve problems in bioethics until we solve them in the other areas of society as well.* If, for example, money continues to be valued more than people, war more than peace, etc. medical care and advances in medicine will continue to suffer. In this sense we do not have an ethical right to be superstitious, irrational or dogmatic.

Philosophy is a critique of culture. Virtually all cultures are based on absolutistic, traditional, supernatural and irrational thinking. A sound philosophy is an attempt to bring humanistic and practical critical rational reasoning into play in order to better society both individually and holistically and its institutions including medical practice. *Bewußte Lebensführung* (rational conduct of life) refers to a lifestyle, which is conscious, aware, critical and concerned with the social, environmental, ethical, and with learning as much as we can about our world. This is also what philosophy is about. Medicine does not exist in a vacuum. It is part of society. If the society is immoral, medicine is immoral, unless it takes on the task of influencing and guiding society in a more ethical direction. We can speak of normative uncritical morals as moral contamination. In this sense, medicine cannot just follow and be the servant of society, but must guide and lead it. It must set policies, which genuinely promote health and prevent death and change the culture to conform to this professional medical goal. Medical ethics should no longer be an aberration and contradiction to society. One American Medical Association Principle of Medical Ethics is: "III. A physician shall. . .recognize a responsibility to seek changes in those requirements, which are contrary to the best interests of the patient" [16].

A kind of dilemma is created. Medicine is practiced in a certain cultural environment. Should medicine continue to blindly comply with and encourage each culturally based or religious medical practice regardless of the naturalistic consequences? Physicians want to heal everyone adequately to what is clinically needed to continue one's life with as much health, capabilities, insights and enjoyment for life. Physicians do not want to have to make decisions imposed on them by an unethical society which does not want to provide needed money and staff, but rather puts its wealth into areas which oppose medicine such as war. Only when ill themselves, people start to realize what healthcare provision is about, before that they have somehow the idea to never be getting ill themselves and not to have to contribute to others who are already ill. Medical decision-making is not to be taken over by uncritical societal issues such as putting economic, legal and religious viewpoints over the medical issues.

The principles of *autonomy* and *equality* require that all be served equally according to the subjective desires of the individual. These principles, however, contradict the principles that one should base one's decisions on a sound ethical theory, reason, and a concern for the naturalistic consequences for others as well as for the individuals involved. Should those who are destructive, selfish, uncaring, unwilling to follow physician's orders, etc. just be treated without ethical considerations? If not, which ethical considerations apply?

A solution to the above dilemma. Both the principles of autonomy and equality, and an altruistic ethical theory may be used to resolve the dilemma. (See also Chapter 12) We can have rational equality rather than blind equality, caring rational understanding rather than absolute autonomy. Destructive people can be treated, but an attempt must be made to correct their negative behavior through therapy or education in addition to the physical medical care given. In this way, all may be medically treated regardless of the destructiveness or negativity of the personality, irrationality of the belief system or harm they do to others, if we can understand them and not blame them for their condition. The psychotic cannot be blamed for such condition, but recommended for additional treatment. In the typical American prison system the prisoner is blamed and punished, but virtually no needed education, therapy or genuine correction is given. On release, the prisoner returns uncorrected to the street to commit crimes again. Instead of this, education, correction and therapy are needed. Similarly, a limited, physical centered medical model should go beyond the blaming and physical approach and see that education and therapy is given at the same time. In short: Autonomy becomes rational understanding and concern. Blind equality becomes fair and rational consequentialistic distribution. Physical medical treatment becomes holistic psychological and physical medical treatment. Blame becomes blameless understanding, but offering correction of personality negativity, destructiveness, faulty beliefs and enculturation. Partial local health coverage becomes worldwide universal health care. Medical health becomes medical, psychological, and societal health. (cf. WHO-definition of health) Medical health care cannot be effectively separated from psychological and societal health care. (cf. the psychosomatic approach)

The physician has a direct and indirect task to see that the destructive patient obtains the needed psychological help. This may be provided through therapeutic referral or philosophical counseling and preventative measures. Stress here may be placed on preventative medicine and preventative therapy.

"Non-ethical" is used here in the sense of not knowing about ethics. It means that one is neither ethical nor unethical. They simply do not know about the use and misuse of ethical terms or about ethical theories. This may also be termed "ethical illiteracy," non-ethical thinking and behavior. One may also be unclear about morals and so be "morally illiterate". One may not be aware of morals or not follow normative cultural practices. If one has not studied ethics, there is nothing upon which to base a judgment. 22% of Americans are functionally illiterate [17].

People at all levels of education and society are usually "discussion illiterate," that is, they are not able to give arguments for or against most subjects. When asked for arguments regarding war, god, abortion, etc. they do not know any and cannot give any, but they "know" what is wrong or right according to mostly common or societal views. They show little knowledge of critical thinking such as giving arguments. If they are capable of giving one or two they are incapable of critically evaluating them.

The philosopher's task is honest, open inquiry and critical thinking. Non-philosophers oppose them and are therefore against rational thinking. There is widespread aversion to any kind of deeper inquiry. People turn their heads from it. It is regarded as being anti-social and improper if not insulting. It is regarded as an attack on one's cherished beliefs. Do not ask too much, do not inquire – it might make you feel uncomfortable. There is resistance to thought. Beliefs are held in spite of evidence.

This irrationality prevails in and outside of the academic world. Scientists are often captivated by their models in spite of the lack of evidence for them. Scientists seldom have a background in the philosophy of science or ethics of science. Medical practitioners typically know little about the philosophy of medicine or ethics of medicine. Dewey and Tufts wrote, "Ignorance is the root of all evil" [18]. Ethics and philosophy courses are beginning to be taught in relation to the medical field. For example, Baylor College of Medicine and Rice University Center for Medical Ethics and Health Policy offer a Graduate Program in Philosophy with a Specialization in Bioethics. The final thesis is written in philosophical bioethics [19]. Students may engage in clinical ethics work, attend meetings of ethics committees and institutional review boards, participate in clinical teaching experiences, and observe ethics consultations. They are also given the opportunity to participate in Baylor training programs for those wanting to do empirical research.

Advanced degrees are no guarantee of being a critical thinker. Academics, healthcare workers and scientists of all sorts usually still have religious beliefs, which would outrage the rational person. Their enculturation may prevail over their reason. Even philosophers are often no better than anyone else in this regard. For example, they often believe in metaphysics and the supernatural, are captivated by symbolic (or formal logic) or Aristotelian logic neither of which bakes any bread. They may not have any use for the pragmatists (e.g., John Dewey) and ordinary-language philosophers (see bibliography for Wittgenstein) who have extensively and convincingly argued against the myths of formal logic. (See Chapter 18)

5.2 A Naturalistic Theory of Ethics

John Dewey and the pragmatists pointed out that an ethics based on supernaturalism, including religion, is unethical [6]. To correct the unacceptable substitutions people usually give for open-context ethical terms we could substitute a naturalistic theory of ethics. On such a system, by ethical terms, e.g., good, right, etc. we may mean or refer to the following: To bring about our (a) informed, (b) wants and likes (c) deliberately (d) on the basis of inquiry (e) with as adequate and full consideration (f) as reasonably possible (g) of the naturalistic and global naturalistic consequences of bringing about the informed wants and likes of everyone including concern for animals and nature. In a word, it is humanism or human and natural ecology on a world-wide, all-encompassing level which includes the knowledge found in the various natural and social sciences including philosophy and aesthetics, mathematics, and astronomy. It is in this sense that we may speak of an adequate, holistic ethics. *Ethics in medicine could be based on such a rational, holistic consequentialism, on humanism* [20].

Humanism means simply that it is in my informed and sensitivity-based interests to help others. "A workable biomedical ethics is humanistic" [21]. "Humanism and the humanities are essential to the fullest maturation of the physician" [22]. Elliott, a physician and philosopher of medicine presents a holistic, naturalistic, pragmatic view combined with ordinary language philosophy [23]. Joseph Fletcher proposed an utilitarian "situational ethics" which is like clinical ethics stressing the concrete circumstance rather than the imposition of set apriori rules, or universal theory. Though a theologian he favors humanism over theism. He stresses the rational over the revealed and authoritarian [21].

On this theory, nothing is good or bad in itself. The theory is grounded on naturalistic wants and likes [24]. Ethical questions take the form: (1) What do I want? (e.g. goal development, career guidance) (2) How do I resolve conflicting wants? (3) How can I bring about my wants? These wants must, however, be carefully considered and based on inquiry. Aristotle thought that people are basically beings of desire. Virtue ethics arises out of desires, not out of duty. Duty and rights are basically denials of desires. If there are no desires and goals there is no ethics [25].

Goals influence all of our lives even when we are unaware of them as doing so. Without a clear and adequate ethical, rational and critical knowledge of our goals we cannot begin to establish or critique a bioethics.

The U.S., similar to many other nations also in Europe, has more or less questionable moral goals. It has no ethics. Decisions are based on enculturation, power groups, economic self-indulgence, an anti-philosophical and anti-critical public, consensus, problematic legal methods, and supernaturalism. Bioethics and medicine are subject to these factors.

This morals based on bringing about one's wants may seem selfish. But selfishness involves bringing about one's own wants at the expense of those of others. It is to do good for oneself, but bad to another. However, by definition, bad is nothing one would want to do. The naturalistic theory is also not for selfishness because this humanistic ethics is based on inquiry and naturalistic consequences leading to the understanding that one of one's wants is to live harmoniously and ecologically in the world and society at large. Naturalistic ethics, more than any other ethics, has led to humanism and altruism. It is the opposite of selfishness. Basically, the view of John Dewey, the humanists and pragmatists is that ethics is consequentialistic involving enhancing human wants and abilities to the maximum in harmony with others and with nature now and in the future. The value of life constantly changes with our knowledge, and ability. In regard to ethics being based on bringing about our naturalistic wants and likes, Dewey even argues that these can be changed. We can change our desires [26]. The value of our life is accordingly not an absolute, but is in process. On this view, it is a teleological fallacy to look for or state *the* purpose or the goal of humans. The value of life reduces to naturalistic, consequentialistic criteria as opposed to non-naturalistic ones. Similarly we cannot ask for *the* purpose of medicine, but only how medicine may best be used to naturalistically, holistically and consequentialistically help humans and humanity. If this method were applied, it would cause a revolution in present medical practice. There are no moral reasons in medicine, only reasons. In a strong sense, this naturalistic ethics is to be scientific and efficient. Aristotle said ethics cannot be a science. Why not? Medicine also benefits from the philosophy of science and pragmatic philosophical practice [27]. According to the "naturalistic fallacy" held by G.E. Moore and people in general, one cannot reduce ethical terms to naturalistic terms because ethical terms supposedly are not scientific terms. People identify ethical terms with normative and abstract terms. But if so, then ethical terms have no relevance to our naturalistic and scientific lives. If they cannot be reduced to naturalistic terms then they must remain meaningless. Thus the "naturalistic fallacy" is not a fallacy at all, but the metaphorical technique of reversal: the fallacy is to think that ethical terms have meaning in themselves. People falsely use ethical terms as if they were "non-natural [supernatural or religious] properties" as G.E Moore held [28]. People falsely think that ethical terms are really religious terms and so ethics should not be taught in schools except as a special kind of religion. Rather ethical terms have naturalistic, scientific and descriptive uses and misuses. For example, Molewijk does not define ethics so he does not know how to integrate ethics and medicine or resolve Moore's "naturalistic fallacy" [29]. It is simple just adopt a naturalistic theory of ethics. The use of ethical terms is not mystical.

It is these commonly held normative views, which need justification.

Present political priorities	Humanism priorities
culture/chance based	reason/ethics based
inconsistent values	consistent values
uncritical normative values	critical ethics
private interest groups	concern for all
egoism	altruism
priority for war	opposition to war
supernaturalism	opposed to supernaturalism
power play	support of people
anti-inquiry	pro-inquiry
competitive/extreme sports	recreational sports
pro-punishment	anti-punishment
anti-critical education	pro-education
anti-philosophy	pro-philosophy
negligent lifestyle	responsible lifestyle
unhealthful nutrition	healthful nutrition
rich over poor	concern for all
nationalistic	internationalistic
against national healthcare	pro-international healthcare
money over humans	humans over money
minimal medical research	maximal medical research
minimal healthcare	maximal healthcare
anti-environment	pro-environment
minimum social welfare	maximal social welfare

A problem is not a problem as such. A moral problem is not an ethical problem. Should we bomb the village or the city is a moral problem based on normative acceptance of war, not an ethical one which questions war and comes to the conclusion to oppose it.

We may now contrast the present policies with a humanistic policy. On a humanistic philosophy the following would have priority:

Prevention, education and correction in place of punishment in prisons.
A preventative Peace Corps would be as large as the military.
The military would be internationally organized and kept at the minimal level necessary, have non-lethal weapons, and be used only for stopping war wherever it may occur in the world.
The U.S. would be a model for rationality.
Europe would be a model of critically challenging tradition.
Support for worldwide healthcare.
Provision of food and medical care for all endangered people of the world as a major goal of politics.
Provision of education for all of those requiring it.
Exposition and elimination of all harmful belief systems, of ideas, which harm humankind.
Worldwide regulation of organ donation and transplantation policy: only those who have chosen to donate organs are to receive organs, but all must choose to do so or to opt out.
Education to stress critical thinking (speaking), emotion, ethics, humanism.
Medical research provided and care tax exemption and thus heavily funded.
Everyone should be given a right to a college or technical school education.
Protection of resources and the environment and animal life.
International and World-citizenship.
Agency established for international welfare and social work.
The least well off are also cared for, not just most as on a utilitarian view.
Fully funded, secure retirement plans for all.
Restructuring and humanization of the legal system.
Restructuring and humanization of the tax code.
Elimination of all special interests.
Restructuring and humanization of the political system.
Ethical Commission to constantly evaluate and ensure humanistic national values and practices.
Stated and well-founded value system.
Change of the constitution to reflect more rational and humanistic values.

5.3 What Is Ethics in Actual Usage?

Those who have not studied ethics or philosophy are not in a favorable position to make ethical decisions. The ordinary language approach in philosophy deals with what the healthcare worker and others actually say. Davies and Hudson [30] presented actual statements of physicians who were in important decision-making positions as follows: "'I'm the judge and jury...and that may be the patient's bad luck'" [31]. "'At the risk of sounding arrogant, I really don't care what they [ethics committees] think'" [31]. "The physicians often defined ethical *dilemmas* as situations with *'no real answer'*" [32]. This last statement is both circular and false. "Dilemma" means "a situation with no real answer." "'I think it's a matter of finding something that a family can live with and you as a physician feel is justified'" [32]. This statement stresses a mere consensus rather than informed argument. Rather, we need to know what the basis is for the justification and if one has knowledge of ethics upon which to make such judgments. One physician stated, "Medical ethics is not a useful field of study" [32]. "'I don't know what an ethicist is, and I think that's a made-up jargon term. It was noted that opinions about medical ethics were extremely varied'" [33]. "Doctors are no better qualified to make ethical decisions than most people" [34]. Nevertheless, the American Medical Association and other medical bodies set ethical standards.

People are not able to specify the difference between a descriptive statement and an ethical one. This is also true of members of nearly every profession including many philosophers. They have the view that values are somehow within us, and that one just knows what is ethical without having to have any training or education about it. The same is true of religion. One virtually never reads the literature in the philosophy of religion about the arguments for and against the beliefs in religion. Yet, it is erroneously thought that religion is a good basis of ethics. Religion is in fact not a proper basis for ethics at all [35]. John Dewey wrote, "Religionists disparage...intelligence as a force. They properly feel such faith to be a dangerous rival" [36]. "Religion and ethics are different categories of human enquiry. Religion is as different from ethics as it is from mathematics. Religion is about faith; ethics is about reason" [37].

It is also thought that ethics is characterized by being judgments or opinions. But many judgments do not at all appear to be ethical ones, e.g., "I think he has cancer." Some issues are said to be moral issues or ethical questions, so as to suggest that ethics might be characterized by the subject matter in question. Accordingly, if an issue is about abortion, killing, lying, cheating, sex, pornography, cancer, etc. it can be characterized as being an ethical one. But, "He lied about his illness," is a perfectly descriptive statement, so an ethical statement cannot be characterized by just being about a certain topic. What makes a statement ethical is if we add that something is good/bad, best/worst, better than/worse than, right/wrong, or that one should or shouldn't do something (duty). "He lied," is descriptive, but "He lied and that is bad," is an ethical statement. An ethical statement is like a meta-statement, a statement about a statement. Any descriptive assertion can be regarded as good or bad, so there are no ethical questions or topics as such. Contrary to what people say, nothing is as such an "ethical issue." Thus, one reason why one cannot be ethical is that one does not know what an ethical statement is.

Thus far we have only identified what an ethical statement is. Another reason why one cannot be ethical is that one does not know what ethical words mean. What is meant by "good"? An examination of its use in everyday language will show that it is in itself meaningless. Although people think ethical statements are the most important statements there are, *they neither know what they are nor what they mean*. People were taught everyday moral rules, but virtually never taught ethics in the schools. One has absolute rules such as "abortion is always wrong," to avoid thinking about the naturalistic consequences. "Some think that all we can ever do in ethics is state our own position." Or that ethics is just a matter of taste [38]. William Bartley wrote, "Words like 'nice' and 'good man' are virtually meaningless in most contexts" [39].

If a physician says she or he will give you the best care, you have no idea what kind of care you will receive, because "best" means nothing. It could mean that they will try to find you an organ for transplantation, but in the U.S. roughly half of those on the waiting list do not receive such organs. You only know what "best" means if a meaning or reason is given which specifies what is in fact signified by it. Rachel's statement reflects this view when he writes, "Whether something is good or bad depends entirely on the reasons that can be given for or against it" [40]. Such empty ethical terms as "helps," "quality care," and "best treatment" are frequently found in advertisement claims for medications, healthcare products, drugs, etc. In short, ethical terms have no meaning until one is given to, or substituted for them.

What people do in practice substitute for ethical terms are usually fallacies. (e.g., circularities, wrong in-itself fallacies, or absolutisms.) Good-in-itself or duty-in-itself are unacceptable substitutions. They say nothing and even preclude reasons or meanings from being given. If "good-in-itself" did have a meaning or reason it would no longer be good-in-itself, but good for that reason. The substitution for "good" based on cultural or normative practices, supernaturalistic, dogmatic, relativistic theories, would also be unacceptable, because they are irrational [41]. Good in-itself falsely implies there is an intrinsic good. "'Intrinsic nature' is...an expression which has caused more trouble than it has been worth" [42].

The following are examples of the misuses of ethical terms. Most everyday uses of ethical (moral, value) terms are misuses especially in the sense of being circular or absolutistic (e.g. wrong-in-itself). Examples:

Morally justified. (circular, redundant) This means not justified by any reason, but enculturated and so not even to question.

Utilitarianism is to produce the greatest good for the greatest number but it does not say what good is. To find that out we would need an ethical theory, which utilitarianism is not. Utilitarianism is based on agent-neutral intrinsic [that is, good-in-itself] good [43]. This treats utilitarianism as an absolutistic view, which it need not be. One could say that a dogmatic utilitarianism produces the greatest dogma (fixed belief) for the greatest number. We usually call this an absolutistic morals or non-ethics. It does show that utilitarianism is not an ethical theory, but rather pre-supposes one such as a naturalistic theory of ethics. As such, utilitarianism is just an empty mathematical-like formula, which does not allow us to decide between the greatest good or the largest number to try to produce the greatest good for the greatest number. Utilitarianism is also a limit of the good one can do otherwise it would state, "Produce the greatest good for all, not just for the greatest number." Even universal utilitarianism, then, is not universal. In any case, most people are at best limited utilitarians, limited to very few, immediate people, or often only to oneself.

Circular:

"The *wrongness* of an act is just its moral *objectionableness*" [44]. (circular)

We *should* have *respect* for all humans. (circular)

There are *morally right ethical values.* (This is four times redundant or circular.)

"Regardless of how their disease developed, it is morally wrong not to attend to their needs when they arise" [45]. "Morally wrong" is redundant and an in-itself fallacy. We do not know why he thinks it is wrong. It is just a misuse of ethical language.

It is *wrong* because it is *against my conscience.* (circular) Furthermore, conscience is not an acceptable criterion because it is enculturated and based on whatever one by chance happened to learn. Nevertheless, it is held for the usual orientation not challenging this concept.

It is wrong because against the consensus. (Appeal to majority and appeal to authority fallacies.) Consensus does not determine ethics.

Ethical fairness. (circular, redundant)
"To *prefer* to die is just *wrong*" [46].
"It is *immoral* and *wrong*" [47].
Justice should be based on *fairness.* (circular)
Morally right action. (circular, redundant)

To speak of ethics as bringing about naturalistic consequences-as-such is also an in-itself fallacy. We need to know which naturalistic consequences are meant and if they are good (in some specific sense) consequences.

Each person has a *right* to health care. (Right is open context and can be neither true nor a reason.) It is unacceptable to base ethics on rights because rights is an ethical term and this just generates circularity. For example, "X is ethical because it is a right," is circular. Secondly, to claim a right in itself is an ethical fallacy. A right is like a law, both of which may be unreasonable and unfair. One may have a right to inherit, but be totally undeserving. Consequentialism opposes the idea that some things are just wrong. War is wrong, but only for reasons. Because people have the thought (or intuition) that war is just right, their thought (intuition) cannot be trusted about anything.

The blastocyte or the fetus has a right to life. (There is a right to life only given by the pregnant woman.)

Intrinsic rights. (This is unintelligible and an absolutistic fallacy. Compare "intrinsically guilty" or "intrinsically true.") Abortions, cesarean sections, for example, are neither intrinsically good nor bad. Nor is medicine intrinsically good or bad. It is good only in a certain sense or for certain reasons, for what it is capable of doing. To say money is intrinsically good or good-in-itself is to value the currency itself, not what it can buy. If anything is intrinsic in ethics it is intrinsic ignorance. Singer holds that nothing has intrinsic value if it has no awareness, desire or "will" [48]. Thus, if one has no knowledge of ethics, one cannot be ethical or unethical, one rather is non-ethical.

Ronald Dworkin in ethics stresses rights over consequences [49]. But rights are open context terms. There are no rights as such. Rights easily become rigid absolutisms and ignore consequences. Countries claim a right to go to war or a right not to help people in need. A person has a right to autonomy regardless of how irrational one is. To insist on one's rights is like a form of principlism, which is a way to prevent healthcare workers from making decisions on the basis of reason. Stress on rights would substitute rights for reason and consequences, exclude consequences and rational decisions. It is like an egoistic demands-based ethics. Rigid rights, principles and autonomy tend to disallow the reasoning and expertise of physicians and healthcare workers.

We need the courage to open ourselves up to questioning. Students often think that questioning and education are tricks to make one think. Critical thinking (speaking) goes beyond the level of comfortableness and is thought to be antisocial and smash one's comfort zones. There is a fear of questioning. Open-mindedness may make one vulnerable. When Shakespeare says that we die many deaths before our own, he is partly saying that death is the uncritical, emotional dullness of not using one's reason.

It is not just the average person who is non-ethical. Professionals in any field, including many philosophers and ethicists themselves, are often quite confused about ethics and morals. Ethical theories themselves need evaluation. The literature on bioethics often refers to Kantian principles for guidance, but who is really clear about Kant's view of ethics? Onora O'Neill, a Kant scholar, wrote, "Kant's *Groundwork [for a Metaphysics of Morals]* is the most read and surely the most exasperating of his works on practical philosophy" [50]. There are many interpretations of Kant's philosophy. Fletcher interprets Kant as holding that *we should just obey rules because we should.* Kant does say *we should do our duty because it is a duty.* [*Pflicht an sich.* Duty in itself] [51]. This is circular as well as an absolutist or in-itself fallacy. The literature on bioethics stereotypes philosophical views and uses them as slogans without needed criticism and clarification [52]. Philosophers are needed to clarify such views to show their meaning as well as criticisms of them.

Kant says we should act out of a "good will," but this is an obscure notion and assumes also the mentalistic unscientific notion of a "will". To say we *should* act out of a *good* will is also redundant and circular. The Yale philosophy professor, Allen Wood states, "Good will" is "unknowable metaphysics" [53]. It is also circular to say, "The best will in the world does the right thing" [54]. On another interpretation we may briefly characterize Kant's ethics as follows: Ethics is created by humans. It is a form of our understanding imposed on the world. There is moral law within us. According to Kant's theory of knowledge concepts without sensations are empty, sensations without concepts are blind. Similarly one might suggest, law/ethics without action is empty, action without law/ethics is blind. The ethical form is law-like for ethics to have a standard, which is assumed to apply to all alike. His formal principle is: "Act only on that maxim through which you can at the same time will that it should become universal law" [55]. It is hard to know what Kant means by universalization here. In bioethics it is assumed that one can just somehow universalize, but it is not clear how to do it especially as, on one view of Kant, he cannot primarily consider consequences or wants. On what basis could one then universalize? O'Neill does say that the basis of the principle is non-contradiction [56]. It will not help us then to know what to do, just to know we should not contradict ourselves. Universalizing will also be further discussed below.

On one view, the Kantian categorical imperative tries to approach a "universal" decision making program, aiming at inter-individual exchangeability in the sense of consequences for the decision-maker as well as the ones concerned by the decision. The definition of the quality of consequences remains unsolved. Rational decision makers are required to bring about rational consequences. On the other hand, he tells us not to primarily consider consequences, wants or likes.

The principle of non-contradiction is used to oppose lying (because it undermines assumed truth telling and confidence in each other), to oppose suicide (because contradictory to remaining alive to be the ethical law-giver), to oppose the development of one's natural talents and capacities (because as a rational person it is self-contradictory not to or to instead seek only pleasurable indulgence). For Kant the goal is a community of ends of everyone, a shared commitment to universal principles. One could understand this as a consistency principle: If we are to live in an ethical world all must have ethical principles. There is certain logic in the non-contradiction principle. We ought to know that if we go to war it supports the institution of war and is contradictory to staying alive. "Reason" is given special interpretations such as a "faculty of principles." We supposedly must ground morality on our rational "reason," even if we do not know what that is. Unlike others, Wood sees Kant not as a deontologist, but as a consequentialist [57]. We must follow law-like duties because the natural desires and behavior of people are untrustworthy. We must therefore try to make our culture and society more reasonable. If we think about law, it is something one should follow and so one might conclude that if there is to be law there must, by definition, be the duty to obey it. One reason for following law is because people are not very rational and their wants and inclinations depraved. Accordingly, Wood refers to "Kant's picture of human beings as arrogant, antagonistic, deluded, and unhappy" [58]. This account is not meant to give the correct interpretation of Kant, but to indicate the problems with even trying to interpret Kant.

Kant spoke of the forms of understanding, and other "faculties" creating an outdated mentalistic "faculty psychology." His follower Ernst Cassirer, in *The Philosophy of Symbolic Forms*, replaced "faculties" with symbols and language such that experience is transformed into language, which then constitutes "reality" [59]. Wittgenstein [60] brought the trend full circle by holding that language has epistemological primacy and that the limits of our language are the limits of our world. On this view, ethics is a linguistic, rather than mentalistic construct. Ethics becomes a use of language. To investigate ethics is to investigate the uses and misuses of language. This is the position of ordinary language philosophy and the one stressed in this book. (See also critique of Kant in the Chapter 17)

What could it mean to seek universals in ethics? Universal principles are like algebra of morals. Do not look for a common morality. If rules are absolute, absolutism is a way to avoid deciding. Carse wrote that universals and vague appeals to an absolute justice are ways to block rational discussion [61]. Pellegrino argued that it is futile to look for one unified theory of all medicine [62]. It is as if to say we are not intelligent enough to make decisions, therefore we must appeal to an absolute commandment, law or fixed mathematical formula such as utilitarianism. Universal can just mean a plurality. A particular can also be considered to be a universal in many ways, e.g., in that many can understand a particular, or perceive a particular and that language involves such universal principles.

Laws are statements using or implying empty value terms such as commands, should, should not, duty, right, wrong, good, bad, better than. Law as codes or canons of morals or wants is neither ethics nor as such ethical. Law as rules, statutes, regulations, ordinances is empty of content and so not ethics. If the legal statement stresses the value terms, ought, should, duty, responsible, negligent, and their negations it places stress on action. Law as commanding action is similar to concepts such as: proselytize, indoctrinate, promote, persuade, enculturate. On this view, the rules *should* be followed. In this sense, it means no more than that something or other should or ought to be done or the command to do or not do something. Again, this is not an ethical theory nor is it based on one. Similar value terms are used in every area of life, but it does not mean we should follow them.

The law, however, can be based on an ethical system, but it is typically based on an unacceptable cultural belief system instead. Law based, for example, on a naturalistic and humanistic ethics would be such an example. As the law often punishes instead of providing education, therapy or correction of the offending cause it is especially unethical. It is not an ethical system when action is based only on threats of punishments. In sum, the law is neither theoretical ethics nor practice of an acceptable basis or standard of ethics.

There is no duty in itself, no free-floating duty. On a naturalistic theory of ethics absolute or abstract duty- and ought-statements would be reduced to consequentialistic or hypothetical if-then statements. "It is your duty to do x," becomes, "If you do (not) do x, y will (might) happen." We may refuse to obey an unfair law because the consequences are thought to be unacceptable. Even some religious physicians do not follow their religion regarding certain medical practices because the naturalistic consequences would be too disastrous. On a naturalistic theory, we would always first ask for the reasons for and consequences of doing an alleged duty or obeying a rule or law. What will happen if I do (not) obey this law? This theory would oppose blind obedience of the sort taught in society, schools, in the church and in the military. To obey blindly, and to have blind belief, faithfulness or unquestioning loyalty is unethical.

In science there are no absolute laws, truths, or absolute facts, there are rather hypotheses. For some reason, ethicists have sought to try to create or discover universals. It is not clear what universal means in this context. Universal is, in the first case, not a substantive but a modifier. "All life dies" is a universal statement, but it is a descriptive and quantitative, not a value statement. It is a descriptive statement, which is universal. We may desire universal healthcare. This often refers only to a certain country, e.g., hoped-for universal healthcare for the United States. Universal really would be universal world healthcare. Here universal is a modifier of healthcare. We may say "All people in a society act in such and such a way." Universal action is limited to the society and only gives custom, practice and morals, not ethics. To seek to universalize action is not an ethical principle. Universalizing is merely a quantitative notion. To merely apply a rule to everyone is not an ethical principle. It is only to include everyone.

In ethics, universalization means an ethical term is combined with a form of the term "universalize." In order to determine what to universalize specifically means, a theory of ethics is needed. In ethics we may speak of universal rights, universal good, universal duty, etc., that is, to apply universal as a modifier of ethical terms. But what is attempted to be universalized, can be beneficial or harmful.

Kant gives us the principle, "act only on that maxim whereby you can at the same time will that it should be a universal law." Universalization is, for example, to have rules that affect everyone without exception. As that demand is too stringent we can say "try to" affect all. It is like seeking a law in science. One problem is that it commits the "all-fallacy." No law is absolute or applies to everyone, nor is there universal agreement. How is one to determine universality – by what criteria? Again an ethical system would be needed to determine that. The non-ethical utilitarian principle of producing the greatest good for the greatest number is a form of the principle of universalization. Universality as equality suggests equality of treatment, but equality is not an ethical principle. There are also many different interpretations of what Kant meant by universalization, but whichever interpretation is used, the above points should be considered. If law and universalization are used as principles in bioethics their full and detailed meanings would first have to be clarified much more than they now are.

In view of the above we can make a distinction between an ethical (critical) ought/duty and a moral (uncritical) ought/duty. There can be rational ethical and irrational moral duties. A rational medical obligation is much different from a supernatural religious obligation.

Ethics is not consensus. Such misuses of ethics dehumanize humans. They are contradictory to being human. "Non-decision is subhuman" [63]. Consensus is regarded in philosophy as a fallacy. Consensus, like democracy, presupposes the existence of educated participants, and those interested in the interests of all, not just one narrow group of people or political unit. A broad democracy or consensus would include the representation of all people, e.g., as represented by international bioethics organizations. It is not merely a survey of uninformed individuals or local preferences [64].

We cannot uncritically base ethics on conscience. Conscience is whatever we happen to believe or have been enculturated into. It would have value if it were based on a sound ethics, however. Thus, the following position of the British Medical Association (BMA) is not easily acceptable. The BMA says physicians should act within their own conscience [65]. Conscience may err. Conscience is often what we are "moralized" into.

This reduces ethics to morals.

Rather, on a naturalistic, consequentialistic, humanistic theory of ethics, we can and must use our intelligence to determine what ethical terms mean and how to be ethical. *Phronesis* is intelligent deliberation, practical reasoning and ability to figure out the most effective and useful action in each specific situation. Ethics is not the acceptance of a fixed theory, which will apply to all cases, times, and places, but the subjection of each ethical view to scrutiny. Our moral decisions are based on diverse relativistic belief systems and practices such as party politics, cost-benefit, punishment, war, rituals, etc. and thereby lack adequate overall consequentialistic ethical direction. We are literally and metaphorically nationalists, not internationalists. Most people have not reached the level of qualifying as world citizens or critical thinkers able to use their reason for the benefit of humankind. Accordingly, Kilner states, "Western ethics is impoverished and ultimately unconvincing to the extent that it lacks a story to explain and ground its concepts" [66]. The same would apply to other peoples of the globe.

Brown and Singer state, "The idea of living an ethical life offers a revolutionary alternative to our present way of living" [67].

5.4 Ethics and Morals: An Unethical Society

> In life we are surrounded by death, so too in the health of our intellect we are surrounded by madness [68].
>
> If an unfriendly foreign power had attempted to impose on Americans the mediocre educational performance that exists today we might well have viewed it as an act of war [69].

Michael Kline criticized that we do not question customs, which become deeply held and extremely difficult to change over time [70]. Smoking, drinking, drug taking have become norms and "any attempt to eliminate or modify or reverse [such] behaviors. . .often provokes resistance" [70]. Kline nevertheless recommends trying to change such behavior and customs for health reasons. If people are non-ethical, we must appeal to their irrationality and uncritical morality, but preferably help them to become ethical through discussions, courses, in-service programs, and providing models. Kossek and Block state, "Morality comes about as a result of the codification of traditional behaviors, conventional wisdom, particular familial or social orientations, and current public opinions. Morals are not subject to intense scrutiny, do not require a sound philosophical foundation (or sometimes any particular foundation) . . .. Ethics on the other hand, demand a supportable philosophical foundation" [71].

A distinction may be made between ethics and morals [72].

Ethics is critical thinking (speaking). Morals are not critical, but just uncritically taking over the unreflected values of society. Morals may be used to refer to the uncritical usual customs, rules, beliefs and practices of a society. They are not based on rational justifications or naturalistic consequences. They are enculturated and indoctrinated practices, more like learned habits or rules than thought out behavior. The medical system of a culture is also based on such morals. They are only descriptive of practices, customs, traditions and rituals. The popular, normative and common enculturated morality is often played off as ethics. People have familiar beliefs, which, regardless of how absurd, set their standard for what is right or wrong. They hold on to their beliefs tenaciously to their death simply because they are familiar. This is the fallacy of argument from familiarity and argument from tradition.

Ethics, by contrast to morals, may be used to refer to an analysis of what ethical terms mean, and stresses the understanding of ethical terms, their uses and misuses and consequences. Ethics is a way of decoding and dismantling morals in order to evaluate outcome when morals are applied. It also involves the creation of rationally justified and sound theories of ethics with special concern for the naturalistic consequences of such theories in actual practice in specific contexts. Ethics is therefore not at all the same as morals, but a critical examination of morals and cultural practices. Ethics is in this sense a critique of culture. According to Albert Schweitzer, ethics must fight with three opponents: thoughtlessness, egoistical selfishness, and society [73].

To take one example of how the culture is harmful: 80 million to 114 million women were brutally circumcised including 80% of the girls in Alexandria. Sudan circumcision or infibulation of young women ages 15–19 in 1990 totaled about 90%. There are four levels of circumcision: 1. excision of the prepuce, 2. plus all/part of the clitoris, 3. plus all/part of the labia minora, 4. plus infibulation which includes the stitching up of the vaginal opening. The operation at levels 2–4 guarantees that the woman will never experience sexual satisfaction. Roughly 15–20% of the women have infibulation. It is a cruel, often fatal and totally unnecessary operation based on cultural morals, such as tradition and religion. It was performed usually by women without anesthesia or sterile instruments. Hemorrhaging, shock, infection, pain, lasting complications, and high maternal mortality result. A circumcised woman is so scarred that her genitals could easily burst during childbirth and cesarean section is usually recommended and normal birth prohibitive. This means that in areas without a hospital birth-giving women can easily die.

Circumcision is not associated with one special religion [74].

Popularly and politically accepted are just the uncritical and enculturated morals of the society regardless of naturalistic consequences, or harm done. It is blind conformity, not ethical practice. Ethics is critical of one's beliefs and of one's familiar and comfortable practices. Ethics is therefore found by people to be incomprehensible, inconvenient, irritable, or outrageous. Ethics is often rejected out of hand. Ethics is regarded as immoral, and in a sense it is deliberately so. Often to be ethical one has to be immoral. In morals, as with the acceptance of some cultural practices and war, harmful consequences are totally accepted, rationalized, or covered.

Something may be ethically right, but morally, normatively, traditionally, and commonly, as well as, legally "wrong." Legal wrong is used in a non-ethical sense. It could instead be called common or legal practices and beliefs. It is often even thought immoral to question morality, to critique it, and so ethics is avoided or opposed. This was the problem with the questioning by Socrates. Ethics is usually not taught in schools as ethics, only as morals. To take the reverse view one could in disagreement say that it is unethical generally not to critique morals and the traditional practices and beliefs of a society. Ethics requires constant reevaluation of our values and has the practical task to oppose those views, which block it. It is for this reason that healthcare workers must constantly critique society and existing medical practices in order to be ethical.

Bioethics is not just a critique of medicine and scientific findings, but is also necessarily a critique of the basic beliefs and practices of our society as these affect medicine. Bioethics' problems stem from problems with society. Society's "philosophy" is not the same as the "Philosophy of Society." The former is the uncritical perspective of the society the latter is a critical one in the life we could lead. Society, as such, does not have a coherent philosophy rather a random contradictory collection of beliefs, practices, dogmas, and customs. Bioethics is not mere bio-morality, it is a critique of morality. The pragmatist-humanist John Dewey argues, that "philosophy is inherently criticism" [75]. It involves criticism of culture and humans in all ways as well as critical thinking about ourselves.

We "live" in a world at war with ourselves, with billions of starving, in need of medicine, without jobs, punishing others or being punished, with illiterate people having, in addition, numerous negative emotions. The United States is a nation with the greatest technology, but not with adequate morality. In the area of morality there has been, in general, questionable progress. To the extent that the United States and European nations are in the best positions to improve the world, they have rather chosen to annihilate it and let billions die. Technology has advanced remarkably, but ethical behavior and critical thinking have not. Those who are in most ways significantly able to help those in need let them die instead. The morals of the major and minor countries consist of selfishness, of killing as a foreign and/or domestic policy, of the use of cruel punishment of prisoners and violators of the law, on the basis of cultural beliefs. This is the lowest form of morals and is distinctly a paradigm of being unethical. It is this unethical societal background, which imposes itself on the medical profession.

Society believes strongly in moral retribution and that criminals should be punished for their crimes. They only in the most extreme cases allow psychological problems or similar reasons as a defense. "Not guilty by reason of insanity" is seldom allowed. Prisoners, however, are also products of an unethical society. The McNaughton rule became the standard for insanity in the United States and the United Kingdom, and is still the standard for insanity in many states. The "McNaughton rule" (1873) for insanity is not knowing right from wrong and was a standard to be applied by the jury, after hearing medical testimony from prosecution and defense experts. The rule created a presumption of sanity, unless the defense proved "at the time of committing the act, the accused was laboring under such a defect of reason, from disease of the mind, as not to know the nature and quality of the act s/he was doing or, if s/he did know it, that s/he did not know what s/he was doing was wrong." If sanity requires knowing about the difference between right and wrong, and if people do not know about ethics, then they would to that extent also not be considered sane. Also in disease of the "mind," mind is a pseudo-psychological notion. The definition just has the meaningless terms right and wrong as criteria.

In 2005 the U.S. was thinking of drafting doctors. Doctors cannot complain as many of them seem to believe in war. Sidel and Levy state, "War is clearly antithetical to public health" [76].

Why does this even need to be said? Is it not obvious? They state that there were 45 million deaths caused by the military in the twentieth century [76]. They state that participation of the healthcare worker in the military may be contradictory to their profession and unethical [76]. When the budget is exhausted for war, it does not allow funds for healthcare and the U.S. is not able to give adequate aid to all people who would need it [77]. War is conventional culture. Culture is the uncritical eye.

5.5 Value Contradictions

Common morality is basically self-contradictory. One could say that holding contradictory beliefs is a sign of uncritical morality rather than critical ethics. There is ignored moral inconsistency and searched ethical consistency. According to the ethics versus moral view, something can be moral, but unethical; immoral, but ethical; both immoral and unethical; both moral (by chance) and ethical (by reason). "People can and do frequently hold mutually exclusive moral beliefs, meaning they are in contradiction with themselves, yet most individuals holding such beliefs normally don't see the conflict" [71]. As morals do not involve critical ethics it is understandable why there is contradiction and inconsistency regarding morals. The people of any nation are typically anti-inquiry and anti-philosophy, yet become puzzled and angry when they have to accept the consequences of that. The Chinese and Khmer Rouge and others purposely killed intellectuals and university people and in 2005 the U.S. government is known as an anti-intellectual government more interested in faith and business than philosophical, scientific and medical research. The anti-intellectual priorities are shown when a baseball contract for one single player was written for 100 million dollars in 2006. As an example of anti-inquiry R Moser, editor of *JAMA* and Director of the Division of Scientific Publications AMA, in "An Anti-Intellectual Movement in Medicine" [78] stated, "I have detected a subtle anti-intellectual, anti-specialty, anti-research movement developing" [79]. "I see a vast swampland of intellectual impoverishment in medicine ahead" [80]. This was in 1975. The anti-inquiry, anti-philosophy nature of people generally has never been subtle. His statement characterizes the situation also in 2009.

People everywhere are opposed to at least some killing under some circumstances, for example, if someone intentionally runs over a child. One might then try to show that it would be a contradiction for one to oppose killing in these circumstances, yet not oppose it in war. This is not to say that it will convince them, for people, including professionals, are quite prepared to ignore or rationalize contradictions. The result is again that, in the end, they will probably not be embarrassed regardless of the number of people killed. Whether one killed 1,000, 10,000, 100,000, 1,000,000 in a war are just figures for which people have no concern. They support policies, which can kill nearly everyone alive immediately or over time. *If one agrees to kill one, the actual number killed is of little concern.*

5.6 Examples of Contradictions

1. We regard illness and early death as natural phenomena. Yet many illnesses people bring on themselves [overeat, anti-nutritional food, little exercise, failure to follow physician's instructions, putting church and military over medical research, high risk behavior (AIDS, STD, extreme and Olympic sports), unsafe working conditions, lack of preventative medical care; unsafe autos, drivers and roads etc.]
2. As a patient you wish autonomy regarding medical decisions, but do not or cannot have the background required to make such decisions either in terms of medical knowledge or in terms of ethics, emotions, belief systems, or critical thinking. Autonomy is virtually always a preference rather than an ethical decision.
3. Physicians are dedicated to restore health and save lives, yet medical errors are the fourth or fifth cause of death. In these cases a few are sacrificed to save many.
4. You fell asleep while driving home after a long and exhausting night-shift and caused the death of someone very close to you. You need intense therapy to overcome the guilt. Yet you have no guilt regarding anyone else in the world who is dying because of your failure to help him or her.
5. Your cat dies. It is a tragedy. You mourn for months. Your country kills two million in an unwise and unnecessary war. You pay no notice. "Few could stand by and watch a child drown; many can ignore a famine in Africa" [81].
6. In court cases, people sue for the slightest injustice. The millions killed in war have no rights or appeals.
7. You read about arbitrary and unnecessary carpet-bombing of another country. You go to the beach. Compare: "They are killing your family now." You go to the beach.
8. Roughly a billion people do not have enough to eat, and roughly 10 million die of starvation each year (UN study 2002). One of the major problems in the Western world is overeating which also causes earlier death and unnecessary expense and overloading of the health care system which prevents the needy from being cared for.
9. Six million children die yearly due to preventable diseases.
10. You refuse to donate organs, but demand to receive them if you need them.
11. The U.S. has no national healthcare plan and a bankrupt Social Security System and an inadequate Medicare/Medicaid System. U.S. citizens vote to spend billions on a bloody, unnecessary war, increase spending on the military, and give tax breaks to the rich.

12. If your child were shot for no reason, you would demand the ultimate death penalty. If you vote for an unreasonable war should you not also be given the death penalty?
13. You would never kill your mother or near neighbor, but people feel great pride about killing people far away, other mothers and distant neighbors.
14. The U.S. Attorney General Ashcroft was enthusiastic about war, but outraged about statues, which revealed women's breasts. People often feel more strongly against activities such as pornography or nude beaches than about mass slaughter. The Gulf war produced pictures of severely wounded children with their clothes blown off. Is it the nakedness, not the blood, which would offend?
15. Political leaders claim to be moral, civilized, and intelligent, yet kill instead of preventing, communicating, or problem solving.
16. If the killing of others is accepted, as it is a normal foreign policy, then caring, love, kindness, trust, and honesty have no meaning. To pretend otherwise is hypocrisy. In this sense, virtually no one values anyone. People will kill to protect self, family, country, religion, beliefs, but give little or no concern or protection to other human beings. People are insensitive to killing and humanity other than to those closest to them. Noddings supports caring especially one's immediate friends or family [82]. This is a too narrow concept. Reports of deaths on the evening news people regard as a source of entertainment as pointed out by Kuhse, "destruction of human life as a matter of amusement" [83].
17. People sometimes show extreme care in being polite, smiling, or mechanically following the trivial rules of everyday life, yet in matters of humanity and killing, there is little or no sensitivity or concern. There are ironically "rules of war" and the Geneva Convention, which are like the etiquette of killing.
18. If your life were so valuable to you, why would you not recognize that the lives of others could be equally valuable to them?
19. Some would accept killing, shooting, blowing up, or setting people on fire, but object to strangulation or use of gas or biological weapons. It seems to be like preferred methods of cooking food.
20. Your church opposes medical research, but you demand the best medical care.
21. Because people are non-ethical, laws proliferate, egoistic selfishness prevails, and watch groups such as unions, ethics committees, hospital and physician evaluation organizations (e.g., hospitals are graded at healthgrades.org), medical ethics boards, human rights groups, etc. must constantly monitor and try to correct violations. They can never do so because they can never take the place of people and institutions being sensitive and ethical in countless specific ways in the first place. Also, the violators often have the greater amount of power. In addition, the watch groups themselves are typically non-ethical, e.g., unions can be corrupt.
22. Family values. Cultures and medicine usually support what is called "family values". Family help can be and often is among the greatest help one can have. On the other hand problems can arise. The family is not a value or valuable as such, nor is it a keeper of values or ethical. Each family member has his or her

own morals and so may the family. It is a cultural institution, which varies from culture to culture. It involves the support and perpetuation of the culture and its usual practices (morals). The family, like culture, is basically non-ethical, and many of its practices are also harmful, explicitly anti-humanistic. For any particular family may be degrees of good or bad, functional or dysfunctional. Thus, it is better to speak of each individual family, rather than family in general. Also, families from one culture bring their "family values" into other cultures. Nations often consist of many contradictory values. Furthermore, in the West, roughly half of the marriages end in bitter divorces and the other half often remain problematic relationships. In order for the family to have ethical as opposed to only cultural values or moral values, the partners need to know about ethics, yet only a tiny fraction of any population has the required philosophical and ethical knowledge. They are rather led by dogma and are married in the Church or by the prevailing religion, and also subject to state and local cultural moral laws. (See also the Chapter 6). As a consequence of the above: (a) Families are often "enmeshed" with each other, that is, trapped in negative and abusive relationships calling for "family therapy" but is seldom obtained. (b) Families develop into egoistic power groups, whereby their members defend each other against all others. Unfair family preference is given in the various areas of society. For example, Germany allows organ donation only by family members (relatives) or close relationships. But this often allows donations for the undeserving, but not for deserving others. (c) Estates are often left to selfish and dysfunctional family members just because they are family members and this practice is fixed into law by the state. Often women were and are not allowed to inherit at all. (d) In medicine, family members of the healthcare worker are often or usually given preferential treatment. It is also a factor giving one preferential consideration in organ donation and triage. (e) Families, often unlike others, are allowed to have children without regard to their qualifications, education or ability to care for such children. (f) In medicine and other areas, relatives are often asked to decide for the patient or other relative when they are not able to decide for themselves. But the relative may be unethical or disliked by the patient. Although some family support can function well, so also can non-family support.

There is also another egocentric fallacy at the heart of the concept of the family. It is the view that the child is one's own flesh and blood, one's own genetic make-up. However, against this view are the following: (a) People typically know almost nothing about their own genetic make-up. (b) Each individual is different. (c) There are millions of possible egg-sperm combinations for a couple. (d) If the child is seriously retarded or disabled one may not so readily claim identity with it as an offspring. (e) There is a misplaced desire to have a child as if one were duplicating oneself. One does not duplicate oneself. Human cloning is not allowed and even that would not create a "perfect" duplicate. Any member of one's family may have a similar gene pool and so produce a similar child. Should a husband be impotent his brother may instead contribute the sperm should a similar gene pool be desired. Many do not hold the genetic myth, so feel instead free to adopt a child, or to obtain carefully selected sperm from a sperm bank.

5.7 On Being Non-ethical and Anti-Inquiry

There is literally no personal value of life unless one knows what ethics is. It is not just that people do not know about ethics, they do not want to know. Enlightened management and personnel "conflict resolution" presuppose that one knows about ethics and emotions, which is usually not the case. Many people in virtually all cultures are anti-inquiry. Statements to this effect are abundant:

"It is hardly possible for health care professionals – most of whom are either ignorant of or in disagreement with moral theories – to apply theory to their own very real moral problems as they arise in the specific cases for which they are legally, medically, and morally accountable" [84].

Rendtorff and Kemp under the principle of autonomy regarding bioethics refer to the necessity for an individual to have the capacity to create ideas and goals for life, and the capacity for moral insight, and self-legislation [85].

This assumes that one is only a moral/ethical person if one knows about ethics. John Harris even defines "person" as one "capable of valuing its own existence" [86].

"Value decision-making is frequently based on morally irrelevant grounds, is inconsistent and idiosyncratic, and results in much unnecessary suffering and the wasting of limited resources" [87].

"How little anyone cares about living philosophers" [88].

"Bioethics professors were trained in the arcane field of philosophy" [89]. Opting for a transcendental, supernatural, religious basis for bioethics, Smith's main objection is that bioethics now begins to base its morality on critical thinking and "rational analysis" [90].

Anne Maclean asked, "Why should we attach more weight to the pronouncements of philosophers on moral issues than to those of other people?" [91]. The answer would be the same as to why we should attach more weight to the pronouncements of physicians on medical issues than to those of other people.

"As long as there has been such a subject as philosophy there have been people who have hated or despised it" [92]. "Clinicians consistently argue that they cannot see how philosophy is clinically useful" [93].

"Ordinary people pay little attention to theories when they make their moral judgments" [94].

"The disadvantages of ethics are fairly obvious: they do not provide the answers. . .Some ethicists are accused of complicating matters further. On the whole, philosophy is better at asking questions than providing definite answers to them" [95].

Harding wrote, "I can make no sense of the claim that someone has a duty to die if the person has never been able to understand moral obligations at all" [96].

In short, it is as if ethics does not apply to medicine. The physician merely treats so as not to violate the existing laws without having knowledge of or concern for ethics or the philosophy of medicine. People are basically enculturated or indoctrinated into their culture or religion regardless of profession. Beliefs and practices are determined geographically. Where one is born determines what one believes. If the world is primarily non-ethical and children are enculturated into it, one may ask if it is justifiable to give birth. People are non-ethical because they non-ethically (without knowledge of ethics or morals) give or take rights away.

Medicine is therefore practiced differently in the different cultures, religious communities and organizations. Even if the morals were taken over from a culture, it would not thereby be ethical. On the contrary, as was earlier mentioned, for a naturalistic theory of ethics one must be deliberate and the ethics must be based on rational inquiry. Fletcher says, "Whatever we are compelled to do [e.g., by culture, dogma, military, or religion] is amoral" [97]. Compelling and expected customs are no theories of ethics.

In a strong sense, ironically, for most people, ethics, morals and values are valueless. Joseph Margolis even argues that life is meaningful (significant) or has value only if one acts morally [98]. This statement is supportable unfortunately being also circular. Metz notes that life is not meaningful independent of the choices one makes [99]. Just following rules or culture is not to make choices, but enculturation. Choice or deliberation involves our positive action, but choice also requires critical inquiry. Against the existentialist position of the philosopher Sartre, mere choice cannot be the foundation of ethics. Loewy maintains that one is a person *to the extent that one can understand values*, possess creative intelligence, have reflective inquiry [100].

Even in a society where killing is not directly justified, it is carried out and justified indirectly in daily practice. Moral choices are made even if one does not know about ethics, although such "choices" are blind ones. When one decides that it is acceptable to harm and kill others, to what extent should this nullify one's right or preference to medical treatment and life, for example, when resources are scarce? It is clear that they put a low value on human life.

"The capacity for and exercise of self-determination [autonomy] can be... the – fundamental ideal of the person [patient] within medical ethics" [101]. Firstly, what is the self to be here? Secondly, why is autonomy, e.g., of an uninformed or selfish person to be the ideal? And thirdly, it would seem especially an ideal in medical ethics and is often not realizable. According to the principle of autonomy the patient has the ability and a right to determine his/her own treatment. Some say the ability to be autonomous is the characteristic of being a person, i.e., have the rational and psychological abilities and behavior and knowledge of one's own wishes and values and social abilities [102]. There are serious problems with the principle of autonomy. Patient centered therapy is not advisable where the patient lacks knowledge and is unable of critical thinking and finding out about the best, adequate decision. We must consider one's ability to judge just as we consider the patient's medical disability. Anti-inquiry is anti-medicine. The so-called "autonomy of the patient" means respect for the patient as a rational agent acting freely and not under constraint. But what, if a patient is not rational, and under cultural constraints and cannot evaluate his/her quality of life? Abandoning patients to their autonomy is all too easily done. Respecting autonomy in the competent person presupposes beneficence on the part of physicians (For further analysis see Chapter 9) [103].

It is also the assumption made in Rogerian therapy and Socratic oriented views of Philosophical Counseling, that the patient can solve his or her own problems and that the therapist need not actively guide and challenge the patient [104]. In opposition to this view, the patient should not be regarded as the expert in the areas of ethics, decision-making, emotion, or medicine, however patients should be encouraged to participate as much as possible in learning about all of these. The more the patients know about their condition and participate in the treatment, the better. The physician also depends on the patient to report the effectiveness of the treatment given, e.g., reactions to medications and drugs. The informed, not the uninformed, patient is necessary for successful medical treatment. One may be said to have autonomy only to the extent that one is qualified to have it. "The comprehension by patients of medical information is not outstanding" [105]. The information given is not as well. "Our own moral beliefs are often genuinely unclear, uneconomical, incomprehensive and incoherent" [106].

5.8 Brief Conclusion

Philosophy may be defined as a critique of the concepts and methods in the various disciplines, e.g., of the concepts and methods of medicine. It is also therefore, *a fortiori*, a critique of one's culture. Medicine is usually practiced without concern for ethics. Decisions are usually made intuitively, politically, culturally, administratively and according to law. There is little concern with holistic treatment or treating the patient as a whole personality in society.

As to suggestions for a specific ethical theory, which is consequentialistic and combines the scientific basis of medicine with the philosophy of medicine one may recommend a naturalistic, humanistic theory of ethics. Ethics in medicine would refer to a rational, holistic consequentialism, be scientific and efficient. It also benefits from the philosophy of science and pragmatic Philosophical Practice [104].

In many areas people are working to promote death rather than to prevent it. Culture stands against medicine and philosophy. The armies, not medical care, lead the world and medicine helps them to do so, though the two are fundamentally opposed. Medicine, like philosophy, goes against culture and the beliefs and practices of the vast majority of people. It is understandable, then, that bioethics and the philosophy of medicine will clash with culture. *It is understandable also that culture, religion, politics, economics, and law have succeeded in controlling and domineering medicine.* Medicine is a mere disenfranchised servant of the society. Medicine must be made rational, whole, and humanistic again. It must lead, not follow, the culture. To do so it must hold its principles against the uncritical culture and society, and ironically against the average person, whom it helps and heals. In the words of Albert Schweitzer, *Ethics must fight with three opponents: thoughtlessness, egoistical selfishness, and society* [107]. *Renewal of the culture is only possible if ethics becomes once again the domain of thinking people* [108]. Medicine has the job not merely to serve society and culture to advance their welfare. It has the task of taking a leadership role in helping to develop their ethics and character. The AMA Principles of Medical Ethics [16] state, a physician shall recognize a responsibility to participate in activities contributing to the improvement of the community and the betterment of public health [109]. Physicians and healthcare workers must lead society in medical decision-making, not merely blindly serve existing culture and political and militaristic programs.

5.9 An Ethics Text for British Medical Schools

The General Medical Council requires ethics and law as part of the core curriculum for medical students. The text, *Medical Ethics and the Law*, is a consensus outline of the core guideline [110]. Because of the importance of this book as official required reading for medical students it would be well to offer a critical review of the book. It is an attempt to provide a guideline and principles, which medical students can follow without having to engage in their own reasoning. A similar, but less official book in the U.S. is Jonsen's Clinical Ethics [105]. An alternative to such protocol oriented books are bioethics anthologies such as Kuhse and Singer [111] whereby specific issues can be presented in depth for discussion. The following is a review of the book by Hope and his co-authors. Consequentialism is unfairly represented in this book as considering only partial consequences [112], utilitarianism is superficially critiqued, naturalistic and humanistic ethics of John Dewey – perhaps the most important and workable ethical theory – are not even presented, Kant's deontology is uncritically stereotyped and as presented of little application in medicine [113], the four unacceptable principles (principlism) of Beauchamp and Childress [114] are uncritically presented (see extensive critique of these principles in the present book), the account of virtue ethics is also inadequate. Hope and his co-authors do not clarify the use and misuse of ethical terms and we never learn what ethical terms mean. Communitarianism is presented, but it is not an ethical theory and based by them on the fallacy of consensus [115]. The account of feminist ethics misses the central basis of the feminist approach, which is the anti-patriarchal model and fails to analyze the concepts of caring, and narrative. (See critique in the Chapter 9) They state that one should clearly distinguish between medical and ethical facts [116]. First of all, there are no absolutely true or objective "facts." Secondly, on a naturalistic theory of ethics, both ethics and science use the same method of inquiring to find ways to achieve specified goals and consequences. Ethical methods and scientific methods are similar. It is a myth to think that ethics is separate from or cannot be reduced to or taught as a science. Such a myth is based on the view that ethics is religion or that we cannot derive ethical statements from factual statements called the "naturalistic fallacy," or "is-ought fallacy." But this so-called fallacy is not a fallacy because if ethical terms are to mean anything at all they must be reduced to naturalistic terms. Otherwise the abstractionist, empty ethical terms will have no meaning whatsoever [117]. Emotions are said to play a part in reasoning and ethical argument [118]. However, no clarification or theory of emotions is given. Formal and symbolic logic are given by Hope as methods of obtaining valid arguments and reasoning. This is a myth. Formal logic is abstractionist metaphysics and of no use in practical or any reasoning. (See Chapter 18) Informal logical rhetorical fallacies such as circularity and appeal to authority fallacy are necessary to be aware of for sound reasoning, but the ones given in the text are formalisms and do not represent the main ones and the ones given are trivial, e.g., the "No-True Scotsman Move" [119]. A discussion of desire-fulfilment theories is briefly presented, but again no definition or clarification of desire is given or explanation of how desire may or may not lead to action [120].

"Intrinsic goodness" is discussed without mentioning that it makes no sense to speak of such an empty category. Nothing is good in itself, or intrinsically good. "Best interests" of the patient or infant is discussed, but "best" is an open context, empty term. No attempt of clarification is given, but instead an appeal to culture, to religion, to a mentalistic "freedom of the will" [121].

In the discussion of ethics as rights it could have been pointed out that right is an open-context term and meaningless in itself [122]. There are no rights in themselves. In this case the authors did at least point out that rights need not be absolute and do depend upon consequences. Autonomy is appropriately but briefly criticized and it is pointed out that to be autonomous must rest on rationality and there may be higher order desires and even a life plan [123]. These are important points, but it would have been good to expand the higher order desires to conclude that autonomous decisions should be made holistically and in terms of one's life plans. This would mean that to be autonomous would require philosophical and critical thinking. But it is part of a naturalistic theory of ethics to consider consequences on the highest level possible, something the enculturated and philosophically untrained person cannot do (See Chapter 12). Law is presented as required, but a critique and evaluation of law is not to be found [124]. The philosophy of law is an examination of the methods and concepts in law. *Stare decisis* Latin "to stand by which is decided" is the fundamental jurisprudential principal that the precedent decisions are to be followed by the courts even though the case, if considered anew, might be decided differently by the current justices. This commits the fallacy of appeal to tradition and precedent. Physicians are often treated extremely unfairly in the courts and it is negligence not to include such issues (See critique of the law in the Chapter 3). To be a lawyer or even Supreme Court member does not mean one is ethical or knows anything at all about ethics. Law and jurisprudence are not ethics.

5.10 Case Example: Medicine and Dysfunctional Culture (Made Available by Dr Wolf Michael Luetje, Head of the Women's Hospital Viersee in Germany)

S is one of the two daughters of D. (D who has several shops in the Katmandu Valley. Katmandu, capital of Nepal). The family is of royal origin and wealthy. S was age 20, married for 2 years and was attending a business school. She was pressured by both families to have a child. They thought it abnormal not to do so. As a woman without children is without honour and she may lose her husband. The woman was desperate. Her father then sent her to a fertility clinic in India where she underwent IVF (in vitro fertilization) because her Fallopian tubes were blocked due to an infection. In her sixth week she had a miscarriage. Her husband threatened to leave her. She had known for a long time that he was unfaithful to her. Her father then arranged a second IVF in Germany. This also failed. After her return to Katmandu she became an "untouchable," meaning that she had to serve all family members and that her husband would not touch her also. She also had to stop her studies. As she was infertile and "untouchable" her husband had no use for her and had other women instead. Her father did not give up and 2 years later sent her and her husband to America for fertility treatment at a cost of $10,000, which is the sum of healthcare costs for 10,000 women in Nepal per year. He had lost his only son in an automobile accident and perhaps in the hopes of having a grandson to "replace" his lost son he had sent his daughter to America for fertilization. She was successful this time and delivered a son in America. A caesarean section was performed at an astrologically favourable date selected by the father. Upon her return she was welcomed as if she were a queen and her husband took especially good care of her and the child. She completely regained her original social position and was also allowed to continue her education. Reproductive technology saved this woman from being destroyed by her culture.

Along with humanism and the naturalistic theory of ethics comes an ethical concern with global medicine and concern with all those who are in need of food and medicine.

5.11 Case Example: Military Medical Service as Contradictory to Medical Practice

The editors of the volume containing the article by Sidel and Levy [125] felt compelled to give the following caveat regarding challenges of the very morality of physicians serving in the armed forces. "The following chapter is controversial. The field of ethics is a discipline of logical and philosophical analysis that requires debate. For true debate to occur, opposing viewpoints must be advanced forcefully and analyzed rigorously. The editors recognized that examining opposing viewpoints could challenge even our most basic presuppositions and that these challenges would cause discomfort. Were we not to include the challenges, we would fail to generate the required thoughtful analysis and debate."

"Conventional war has largely disappeared. . .replaced in recent years by fierce ethnic or religious rivalries" [126]. The requirements before going to war are not known or considered. All wars are in the philosophical and rational sense unjustifiable and crimes against humanity [127]. Now hundreds of thousands of children are used as soldiers and killing is in the millions. Gelfand says it is not a contradiction for a physician to develop bacterial weapons [128]. But the military has the job to kill, healthcare workers have the job to heal and keep alive. No two professions could be more contradictory. But is the physician's helping to heal soldiers not contributing to killing again? War is totally contradictory to the physician's profession. Also, military healthcare workers may well be tried for war crimes just as would be appropriate for anyone serving in the military. Thus, we may ask, "Should physicians serve in the military service and so contribute to war and killing?" This is a contradiction to the physicians' oath to heal and save lives and is not justifiable. It contributes to more killing, injuries and death. Just as the physician has the duty not to support any criminal activity s/he has the duty not to support the military.

This issue was addressed by Michael Gross who wrote that, every soldier is ruled by the military and state and has no rights to life or medical treatment except for the purpose of fighting for the military [129]. "War fundamentally abridges an individual's right to life" [130]. "War. . .undermines each actor's right to medical care" [130]. "Combatants lose their right to life as they gain the right to kill" [130]. They have no individual right to refuse medical treatment or to die. They can, however, daily be commanded to sacrifice their lives. Combatant risks of 50–100% casualties are usual even for minor military gains. Soldiers can also be required to take experimental and harmful investigational drugs [131]. Soldiers do not have identity except as part of a fighting force. They have no use, or patient rights except as military fighters. "Right of one's own wounded soldiers to receive medical care is contingent upon their 'salvage value'" [129]. Salvage value which is the likelihood of returning to battle, replaces the "quality of life" criterion. "Those beyond salvage. . .may not appeal to any right to life to secure medical treatment when resources are scarce" [129].

"The good of the self is not a concern of anyone in the military" [129].

"Utility allows military necessity to trump other military constraints on military action" [130]. War overrides civil liberties, autonomy and the individual, whether civilian or combatant.

"Enemy soldiers have no intrinsic right to medical care" [129]. Healthcare workers may treat them only if they are no longer a threat. Civilian's and non-combatant's rights to life or treatment are secondary to military concern and may therefore be denied altogether. "During war the state rarely sacrifices a few lives to save many. Instead, it sacrifices the lives of many to save some intangible national asset." (e.g., freedom, democracy, way of life, military interests, revenge, religion, culture, race and ethnicity reasons, power, economy, pride, legality, treaties, etc.) [129]. It is supposedly for the interests of the state, but the interests of the state may even be opposed by the majority of citizens [132]. "Non-care giving arise[s] as physicians are asked to contribute to the practice of war and the development of weapon systems (e.g., biological warfare) rather than healing the sick or injured" [126]. Physicians should not even help to develop non-lethal weapons. They are still weapons.

Medical units and personnel are not respected by combatants and are directly attacked or killed. This happens also as a result of the inevitable collateral damage. Medical units are used for military disguise and so are military targets [133]. Military blockades also block medical help.

Basically, at present members of the medical profession are treated as mindless collaborators of the military establishment. Medical ethics and practice is taken over by the military. There is no longer medical ethics and professional medical practice. It has become military medicine, which is a contradiction in terms, similar to saving life versus killing. "It is an inherent moral impossibility to be a physician-soldier" [134]. By killing, the military thinks it is saving lives, that the military is a kind of medical corps. The prevailing principles of military service are obedience and support of the fighting force. Sidel and Barry state, "In our view, the ethical principles of medicine make medical practice under military control fundamentally dysfunctional and unethical." "We believe the role of the 'physician-soldier' to be an inherent moral impossibility" [135]. They argue that: 1. The best interests of the patient are subordinated to the goals of the military. But according to the Geneva Conventions the wounded must be cared for in war equally among civilians and the enemy. One cannot give priority to one's own troops or deny care to others although that is the actual practice. In commenting on the authors' objection here in a sense the obligation to provide medical treatment seems moot because the soldiers are sent into killing fields anyway. It is in some ways like sterilizing needles before executing prisoners. To be able to kill, but then treat with care makes war into a game. 2. Medical research is conducted on soldiers without their informed consent. FDA allows wavers for the military to use drugs on soldiers without requiring informed consent. 3. Triage treatment favors soldiers who can return to duty rather to the exclusion of all others. The authors believe the demands of the military have an "inappropriate" priority [136]. 4. Poor medical records are kept, or not kept at all, e.g., of the 150,000 U.S. troops, which received controversial anthrax vaccine in the Persian Gulf War. Bad batches of vaccine and other adverse effects could therefore not be identified. The anthrax trials were 40 years old and more recent ones used in animals. Those refusing the vaccine were threatened with punishment. 5. The military physician has always absolute decision power over which patient is treated. 6. The military can require immunizations or medications or exposure to unknown hazards (e.g., Agent Orange), which have a short-term value for immediate battle plans, though the soldiers may be seriously harmed in the long run. 7. The healthcare worker's concern for life should be in terms of helping everyone in the world. It does not stop at the nationalistic borders. Sidel and Barry recommend a global perspective of medicine [137]. 8. The killing of citizens by the military clearly opposes medical practice. 90% of deaths in recent wars were among civilians, including women, and children [134]. 9. Healthcare workers are theoretically given immunity from attack and should not be used as combatants, yet they are being used that way. 10. Physicians should always do no harm. They should therefore not engage in military biological, chemical, and nuclear weapons warfare research. There should be no involvement with torture. Medicine is used by the military and prisons as a weapon [138]. Therefore, physicians become combatants, torturers and poisoners. 11. The military does not allow physicians to exercise ethical judgments. Healthcare workers must first obey orders then act as healthcare workers. In 1967 Howard Levy a drafted U.S. Army Medical Corps dermatologist refused an order to train special combat forces in dermatology because it undermined physician-combatant role. He was given a dishonorable discharge and 3 years in military prison [138]. Military physicians are not allowed to protest unethical military actions, or an unjust war, e.g., the U.S. bombing of a North African pharmaceutical plant, which supplied half of the medicine needed for the region. Conscientious objection (CO) in the U.S. is based only on sincere religious training and belief [139]. The military is anti pacifists. Healthcare workers should be allowed as COs. Three hundred American medical students refused to serve in the Armed Forces in protest of the Vietnam War [137]. Yolanda Huet-Vaughn, captain and physician in the U.S. Army Medical Reserve refused to serve in Persian Gulf War in 1990, because she thought it was an immoral war and violated the goals of medicine, and her concern for humanity. She was convicted and imprisoned at Fort Leavenworth in Kansas. The authors recommend physicians not be required to serve in a doctor draft and furthermore help end war by their refusal to participate in it in any way [140]. They further recommend that healthcare workers should take the leadership in preventing and opposing war [137].

In sum, the military and belief-systems not objecting to war and even promoting it are contradictory to medicine and keeping people alive and the professional physician should not participate in the military in any way and take the leadership in doing all possible to oppose all military action.

War is a cultural metaphor the populace has accepted and is held captive by.

5.12 Insensitivity to Killing: The Failure to be Embarrassed

Where the intellect is absent it cannot be embarrassed.

Unembarrassed goes together with terms such as defensive, apathetic, unquestioning, dull, uncritical, enculturated. However, we can at least show how one is embarrass*ing* by examining the arguments and pointing out contradictions and fallacies in one's thinking. People, including professionals, just follow commands and so perform the most horrible acts in war, business and in everyday life. Because there are commands they are not affected or embarrassed by the acts. Canetti wrote, "'That's not me. I could not have done that'. . .How undisturbed they remain. . .They feel no guilt and regret nothing. The [horrible] act did not get into them" (Translation of the author) [141]. They cannot evaluate their own actions without having insight into ethics. There is blindness to arguments, which is without guilt. Only the gracious have sensitivity and can be embarrassed.

How can one possibly get an enculturated, insensitive, uncritical, and uncaring person to understand the fatal consequences of acts in war? Absolutists have failed to understand the very concept of consequences. They rationalize that they are doing good. So perverse is the understanding that they will say they are not killing, but preserving peace, neutralizing the enemy, saving lives, eliminating an evil force, or making the world safer. Citizens will be enraged to be told that they are killers. Nevertheless, they are by their votes. We possess no intelligible argument to conclude otherwise. To support war in any way is to participate in war. Insensitivity to human life is combined with irrationality and unethical behavior to the detriment of humanity. It is what is meant by ethical depravity.

People are concerned with and sensitive to the immediate. If my pet dies I perceive that as a great tragedy. Ten thousand killed by a bombing raid is considered a great success. Even if it were not, it is just a number. There can hardly be a greater insensitivity than this. One reason for this is perceptual distance. We know what we can immediately relate to. We feed our pet, run with it, pet it. That millions are starving in the world is only a concept, if that. We sometimes become affected if we meet the child or see a picture or video of children starving. But that too we become used to. We have seen too much of it in the comfortable setting of our living rooms to think and feel it unacceptable. We become insensitive to killing. It is thought to be normal. It is even thought to be moral. But those who actually experience war, the soldiers who did not really know where they were going into may suffer from posttraumatic stress disorders. Their bodies know more than they do. But bodies can break down under the stress as well (See Chapter 13).

Physicians may never be able to adjust to the immediate perception of continually seeing and treating people severely ill and dying. Each death can be experienced as a profound tragedy and occasion for grief. One has to sensitively experience it to understand the trauma daily observed and the responsibility for other people's lives. The immediate holds us captive and prevents us from understanding even the tragedies in the lives of our own physicians. We remain relatively insensitive to anything beyond our direct experience. We are perceptually egoistic. Rather we may try to see the world as our family.

We may approach the issue from the ordinary language philosophy perspective: What use does "kill" have? "Kill" objectively means to cause death. "Kill" as an ethical term means to destroy, devastate, damage, demolish, ruin, to cause a body to fail and a person to die. On this meaning, killing by definition is wrong. Killing another is not like killing oneself. Another's death is not like mine. But do we know what death is? [142] (See also Chapter 21). We are unclear about death – the complete destruction of our thought? What does one think (say) here? When we kill can we genuinely know what we are doing? We use the word "killing" as we use other verbs. But is it like them? Our words are living words used in practical everyday life. Such words as killing people, dying and death go beyond the use of such conscious words. What are the naturalistic consequences of killing? Does killing reach beyond our living language uses? Does killing somehow take us beyond the limits of language – into death itself? What fallacy is this, death-in-itself? We speak of death as we speak of other things and so prevent ourselves from grasping its meaning. We remain insensitive to killing because we are unable to understand its meaning.

To find out exactly what something is in a particular case, we must discover the actual assessments and actions relevant to it. Typically, we cannot imagine our own death and what it might mean to die. What we say about death is *said* – a metaphorical, linguistic construct. We constitute ourselves by language. We are linguistic constructs. Our grief is typically not because of the death, but because of one's own losses or because of other self-talk (thinking).

One of the main reasons for the incredible insensitivity to killing is that it is regarded as a common word like swimming, winning, running. But it is not. This exposes another reason why people cannot be embarrassed about his or her killing of others. The government may be shocking, but it truly represents the citizens who have voted for it. Roughly 90%, for example, supported the first Gulf War in 1991, most oppose universal health care, yet are religious, etc.

Another ordinary language view is the observation that "life" is only a symbol and so not the sort of thing that can be meaningful [143]. "Life" only has meaning in various language games or contexts. Thus, we would not look for *the* meaning, purpose, or value of life as such. "It's a good life" can mean something like "I am happy," with no actual assertion about life or what it may be. "Meaningful" in "Life is meaningful" can indicate, "Life is good." "Meaningful" functions as an open-context value term here. The synonym, "significant," especially connotes value. That is, each term has no literal meaning, but gains meaning by its use in a language-game. Additionally, there is no life, meaning, value as such but only linguistic usages. They are language constructs not mentalistic ideas or "objects" independent of language. Language, not perception or objects, has epistemological primacy. Medicine uses the scientific method, but the scientific method presupposes and is a linguistic construct. Every theory in science is a linguistic construct.

5.13 Case Example: On Sensitivity

The following is a narrative of what it meant and means to me, the author of this book to be a physician. What is it to experience genuine sensitivity for others? It is part of my profession and to a large extent defines it. Having devoted my life to helping people, much of it voluntary, their death can become a threat almost as overwhelming and intense to me as if it were my own, if I would know what my death will be like. Having witnessed someone whose death could perhaps have been prevented or been even remotely involved in such a case is a challenge to all of my professional thinking and personal emotions. What such sensitivity does to me is that it shows the great value of life and tragedy of death. Every death of a person is the death of a world – including my own. The confrontation with death stops life. My life becomes paralyzed. How could I ever love, enjoy, have fun, knowing how those must feel who have lost their beloved. Death is what the physician must prevent. If I could not prevent a person from dying I would feel it as a failure as a physician. I ask why I am still alive after the death of a patient whom I have taken care of and felt responsible for when this other person is not? This experience tares me into a sort of life in death, a living death. Life's fragility and borderline struggle is more evident than ever. It is almost impossible to find my way back into my everyday life or let myself be concerned with the little or large sorrows of those around me. They cannot understand I am in pieces. As a physician one always has to be aware, one little mistake and someone's life is gone. You strive to get back to your normal life before, but there is no way back. Why did I become a physician? To find out about that? No. You kind of knew, you might have to, but you hoped, you might not. My motivation for becoming a physician is similar to that of A Schweitzer [144]. I have to decide what from my life, I have to sacrifice, how much I can keep for myself in all the areas of my life as a human being. This is my very personal decision. It depends upon my understanding and care. Common morals and societal guidelines would not work for that. They are too superficial, not sensitive enough for specific contexts (See also the Chapter 9).

Schweitzer takes an existentialist position. The genuine ethical challenge can only be reduced to a personal subjectivity and sensitivity, not to moral or societal imperatives. "My personal ethics is basically subjective, because of acknowledging for every one of us the liability to decide how far s/he wants to go with sacrifice" [145]. The humanistic point is that we can do all we can for others, as a person and physician, but each one of us has to guard that we do not destroy ourselves and our lives in the process. Altruism should not exclude the life of the altruist (See also Chapter 10).

5.14 Case Example: Tsunami Disaster and Cultural Irresponsibility

"Tsunami" is a Japanese word from the 1960s used to refer to large and devastating tidal waves. On December 26, 2004, the catastrophic tidal wave took the lives of eventually around 200,000 people in 12 countries bordering on the Indian Ocean. Additional millions are left hopelessly without shelter, jobs, income, health care, food or water causing thousands of additional deaths. The initial reaction of the U.S. President G W Bush was none. He waited 3–4 days before responding. When he did make a statement to contribute to one of the largest natural disasters of this sort in known history, the U.S. government initially offered to contribute only $15 million, considerably less than even the tiny country of Austria gave (100 million). The Pfizer company alone pledged $35 million. Around 3,000 lives were lost in the New York 9/11 attack, not 200,000, yet the U.S. reacted to the attack as if it were the greatest disaster in the world, totally revised its entire government into an armed camp called "Homeland Security," and mourned as if it were the greatest victim known to humankind. However, every life lost is a tragedy, here but also everywhere. Not only here. Hundreds of billions were given for the 9/11 attack, and the average payment from the 9/11 Fund to affected families was over seven billion dollars. The tsunami funds are measured in millions not billions and the families for the most part will receive virtually no payment at all and return to starvation conditions. Noam Chomsky in his book, 9/11, wrote that our anti-humanistic and aggressive militant policy caused the attack on the U.S. He said, to find out who caused the attack, look in the mirror [146]. By Bush failing to act immediately to the tsunami news he should have washed out all of the compassion he seemed to have regarding 9/11. Suppose the Bush administration reacted as slowly to 9/11 as it did to the southeast Asia disaster?

It was only after Bush returned to Washington from his vacation the donation the U.S. pledged eventually raised to 35 million and then later to 350 million toward the tsunami disaster. This was presumably because of national and international criticism that the U.S. was giving too little.

Compare: A 6 year old in China gave 22 Euro, all of his earnings, and Michael Schumacher, a private person gave 10 million.

The U.S. gives a miserly .1% of its GNP instead of the .7% recommended by the U.N. Even if it were to give generously to help with the tsunami crisis, it still will have an un-humanistic reputation because it does not do its share to deal with the more major problem of all the dying people of the world.

Millions die each year from preventable diseases. One billion people in the world are starving, seven million in Zimbabwe alone. Sub-Saharan Africa has the highest number of starving, about 50%. Two million children under the age of five needlessly die of pneumonia each year. One million people contract malaria each year. 240,000 die of AIDS each month. 13 million children die of diseases caused by malnutrition each year. From January to April more people will die in the eastern Congo than were killed in the tsunami. 12 million each year die of lack of water or polluted water. 20% of the world's population lives on less than $1 per day. Statistically there is enough food to feed the world but the rich nations let the starving die. *The real tragedies are not just the needless, fatal wars or the preventable natural disasters, but the societal and cultural opposition to humanism, ethics and critical thinking (speaking).* There is a failure to be able to think of ourselves as global, world citizens. Those engaging in lethal war should be regarded as war criminals. War is terrorism. There is no justified war [127].

As stated in *Der Spiegel*, there is a lack of ethical sensitivities (*ethische Empfindungen*) [147].

On the other hand, private donations of people were extremely generous. This seems to indicate the contradiction we inevitably find on the level of normative morality. People sometimes respond to a well-publicized emergency, which involves their own people (tourists) but have no real awareness or concern about the fact that people are needlessly dying every day all over the world.

A Peace and Humanitarian Corps is needed, which is as large as the Military consisting of volunteers from each discipline (agriculture, communication, ecology, economics, education, medicine, philosophy, philosophy of religion, politics, sociology, therapy, science, etc.) to prevent war and meet the physical and psychological needs of all people of the world. Organizations such as Doctors without Borders, the Red Cross, OXFAM, etc. are excellent examples of what a peace corps and genuine heros should look like.

The tsunami was not a new phenomenon. In 1883 there was one from the explosion of Krakatau Volcano in Indonesia, which drowned 36,400. In 1992 there was a proposal in Jakarta, Indonesia to develop a high tide warning system for the Indian Ocean, which would cost two million dollars, but it was not adopted. This is a massive form of letting die.

The members of the Peace Corps can be basically volunteers and each member of each society would be expected to participate for at least a year. Those receiving welfare can continue Peace Corps service until they are qualified for other work. This would help to guarantee total employment for each country and reduce welfare. The military should be as minimal as necessary and for defense alone, not for aggression as it is at present. Its weapons should be completely non-lethal and environment friendly. All nations could develop their own Peace and Humanitarian Corps, which coordinates with all others. The large nations, including the USA, could especially reverse their policies to become aggressive models for humanitarian aid and peace. Altruism, rational problem solving and helping all others is also in one's own self-interest. There is a global ethics. We can learn from it how to save millions of other lives. The island we all live on is the earth. The healthcare profession could thus be part of and defined by taking a leadership in this humanistic goal-directed, universal medical philosophy. It is a universal Hippocratic Oath.

5.15 Case Example: Culture and Family as Anti-Medicine: Female Circumcision

Savulescu gives the example of the request of an African woman for infibulation – the excision of her clitoris and sewing together of her labia. He says there is no reason to comply with religious requests that are based on irrational beliefs [148].

The following is a review of the two books by Waris Dirie who details her cultural and familial life as a Somalian Nomad who was totally circumcised (infibulation) at age five [149, 150]. The circumcision was thought to be required by the Muslim religion as well as by the culture. All of the women she and her mother knew were circumcised. If a woman in Somalia (and typically in other Muslim countries) is not circumcised she is regarded as unmarriageable and "unclean" and "impure" [151]. Waris Dirie therefore also wanted the procedure and had the prevailing belief that she would thereby become a woman.

In many countries, women traditionally are taught to think that they are more of a woman if they become a mother.

A traveling Gypsy woman at high cost performed the circumcision. Sex and circumcision are basically taboo topics, so are virtually never discussed. Thus, no information about such things can be generally understood or evaluated [152]. The circumcision involves cutting out the clitoris, and lips of the vagina (labia maiora and/or minora) and then sewing it up so that only a scar will eventually remain where the entrance to the vagina once was. A small opening the size of a matchstick is left so that fluids can come out. As a result urination for Dirie was painful as only one drop at a time could come out. Every step of the procedure was unbearably painful and life threatening. The legs were bound together for over a month and the child left alone in a separate place until she either died or was healed. The cutting was done without anesthesia and in unsterile conditions. The cutting itself was done with a regular or jagged razor blade, broken glass, sharp rock, scissors, or one's teeth to bite out the genitals [153]. The sewing up of the wound was done by making holes with thorns and then threaded. Dirie's operation was done on a stone, which was afterward drenched with blood with her genitals lying on top of the rock, and would be most likely eaten by animals. Girls often die from the procedure, as did her sister, with no one speaking about it or saying why they disappeared.

The result of infibulation is that the woman is deprived of virtually any usual sexual feelings [154]. Ironically, the procedure, which was culturally thought to make her into a woman, was the very procedure which deprived her physically and erotically from becoming one. When years later extremely painful menstruation began, she had to be cut open although it was a violation of her culture, to allow the fluids to pass through [155]. The pain of menstruation was so great that Dirie, usually extraordinarily courageous and uncomplaining, reported that she wanted to die [156].

The results of female circumcision are: bleeding to death due to the operation, infections, fever, extensive scar formation, tetanus, hepatitis B, chronic urinary and bladder infections, pelvic infection, cysts, abscesses, neuronomas (tumor of the nerves), dysmenorrhea, frigidity, depression, death. One could add to her account that massive keloid scarring can interfere with walking during one's whole life. One cannot usually give birth vaginally in these cases without disruption of the scarred tissue, or even bleeding to death [151]. The smallness of the opening and the scar tissue prevents normal birth. Birth is often self-birth and there is no possibility of cesarean section. On the wedding night as before giving birth she must be cut open [157]. In the hospital such women must typically be given cesarean section deliveries. But in their countries this is often not available.

The United Nations estimates that 130 million girls were circumcised, two million a year, most in Moslem areas, including Somalia (80% of the women), Egypt (84% between age 3–13 [158]), Sudan; and one could add to their list: Ethiopia, rural Saudi-Arabia, Kenya, Chad, Malaysia, Indonesia, and United Arab Republic. (80% of the female population in 28 countries in Africa.) 27,000 New York State women were or will be circumcised [159]. Admittedly, statistics of these sorts can only be estimates.

Many years later Dirie tried to explain her circumcision to a friend in New York, but as she was unsuccessful she simply showed her the scar where once her genitals had been. She wrote, "Tears poured down her cheeks as she turned away." The friend said, "Its horrible, Waris. I can't believe that anyone would do this to you" [160].

What were the forces, which caused the circumcision? Most immediately was the family and, in general, the culture. Dirie's mother ruled her own life by the Koran and thought her religion required circumcision. Her father required total obedience, ruled the household strictly, beat both wife and children, and he thought circumcision was a necessary condition for her to get married. When she was 13, Dirie's father demanded that she marry an elderly, crippled man in exchange for which he would be given a few camels. Dirie escaped from having to submit to this arranged marriage by running away without food or water across a burning desert and not knowing which was the right direction to Mogadishu where relatives lived. She carried nothing with her. Without shoes she walked over burning sand, sharp rocks, thorns, scorpions, snakes, and in the dark as well as heat of the day with cut and bleeding feet until she was exhausted. She loved her family, but it was also the source of great harm to her both physically and psychologically. She stated, "My parents were both victims of their own upbringing, cultural practices that have continued for 1,000 of years" [161].

In spite of her courageous, rational and humanistic opposition to female genital mutilation (FGM), she nevertheless had her own son circumcised as a baby, though there were no medical reasons to do so. She said it was for medical cleanliness, which ironically was one of the reasons people gave for female circumcision, but the latter involved religious "cleanliness" [162].

Religion, often almost identical with the culture, was also a cause of circumcision and dysfunction. "In'shallah" is said to mean that everything that happens is a decision of Allah. If so, then female genital mutilation must also be a decision of Allah. Dirie is a believer, yet does not think Allah decided for female circumcision otherwise he would not have given women genitals in the first place. She believes that Allah's plan for her, her destiny, was to campaign against female circumcision and help the Somalian people change their dysfunctional cultural ways. Still she believes that Allah will take care of her mother and father who were in the most desperate conditions when she visited them after being gone for 20 years [163, 164]. In addition to superstition, and fate, she believes in prayer for rain and for other things [165]. Nevertheless, she says she opposes superstition [166]. It is hard to know that one still has superstitions when they are still bound up with enculturation. She believes in devils that inhabit the sick and stand at crossroads to confuse travelers [167]. Religion thus permeates the culture preventing rational education, thinking and medical care. It may be noted that in 1997, due to Islamic pressure, the Egyptian court overturned the ban on female genital circumcision and earlier Muslim Fatwas (e.g., Jan. 29, 1981) through Sheik of Al-Azahar who defended female circumcision as being necessary. Medicine is largely folk medicine and based on religion. Dirie mentions the word "ummi" meaning untouched by knowledge from any source other than Allah [168]. Here is pride in ignorance.

It may be noted that the Islamic denial of sexuality and information about it in the Moslem world is similar to the Western Catholic, and some other Christian religions, which deny and undermine human sexuality and may be viewed as a form of psychological circumcision. Dirie wrote, "Families will claim it is their 'religious right to mutilate their daughters'" [169].

With continued love of family, her culture and religion she found she simply had to object to some of their practices. She states, "Stand up for yourself and not let people push you around for no reason" [170]. By both her enmeshment with and yet opposition to her cultural practices, she was caught in a double bind leading her to be confused due to the contradiction in values [171]. She felt guilty because of her opposition as she says, "I could imagine them [Somali people] saying, 'How dare you to criticize our ancient tradition!'" [170]. She thought she would be attacked or killed if she returned to Somalia [172]. She wrote, "I denounced my family and a tradition.... It made me an enemy in the country!" [173]. Her own religion opposes her speaking about and opposing female circumcision, and her occupational career in modeling is against Moslem beliefs. She faced the experience philosophers and other critical thinkers have when they inquire into the customs of culture and religion or try to establish a more humanistic society. The process of her breaking out of society to become a more humanistic and critical thinker was courageous, painful and slow. She wrote, "My visit showed me how difficult it will be for people to change" [174]. Physicians, like philosophers, must similarly have the courage to critique society, religion, and family beliefs in order to provide sound medical treatment in spite of the criticism or rejection they will inevitably experience. The U.N. organization in Bosasso tried for over 6 years to lessen female genital mutilation, but without success. They tried to teach mothers to use circumcision as occasionally practiced in Saudi Arabia which is just a ritual without cutting [175]. Perhaps because of the failure to be able to change adult thinking as proposed in her first book (1998) in her second book Dirie set her task to influence the Somalia children by improving their health and education.

Dirie's books about her experiences show the gradual development of critical and humanistic thinking and that to do so one is required to go beyond and critique one's family, culture and religion possibly even at the expense of one's life. This is true of philosophers as well as of scientists, as was the basis of the scientific revolution which freed physicians to gain medical knowledge. Medicine cannot just be based on family values, culture and religion and have it be scientific and successful. It must go beyond them.

By critiquing society, people will suffer rejection by the indoctrinated and uncritical members of society. The questioner must be like a "desert flower" which can endure adverse conditions for long periods of time. Dirie's first name is *Waris*, meaning "desert flower" in Somali. Dirie had to try to survive the dysfunctions of her family, culture and religion in order to become a more humanistic and critical person.

References

1. Schweitzer, A. 1923/1996. Kultur und Ethik. Beck'sche Reihe, München: 338: *Mit drei Gegnern hat sich die Ethik auseinanderzusetzen: mit der Gedankenlosigkeit, mit der egoistischen Selbstbehauptung und mit der Gesellschaft.*
2. Plato, *Apology* 38a: Hamilton, E., Cirnes, H. eds. 1961. *The collected dialogues of Plato.* New York: Bollingen.
3. BMI. 2004. *Medical ethics today: The BMA's handbook of ethics and law*, 2nd edn, 651–652. London: BMI.
4. BMI. 2004. *Medical ethics today: The BMA's handbook of ethics and law*, 2nd edn, 16. London: BMI.
5. Heidegger, M. orig.1951/52/1992. Was heißt Denken? Reclam, Stuttgart: 3: *In das, was denken heißt, gelangen wir, wenn wir selber denken. Damit ein solcher Versuch glückt, müssen wir bereit sein, das Denken zu lernen.* cf. Maier, B. 2003. Vertrautes und Fremdes in ethischer Perspektive. Zwei wert- und emotionsgeladene moralische Begriffe. In *Psychosomatische Gynäkologie und Geburtshilfe. Beiträge der Jahrestagung 2002 der DGPFG*, 291–296. Gießen: Psychosozial-Verlag.
6. Dewey, J. 1939. *Theory of valuation*. Chicago: University of Chicago Press.
7. Rorty, R. 1989. *Contingency, irony and solidarity*. New York: Cambridge University Press.
8. Dewey, J., and Tufts, J. 1908/1932 rev. Holt, H. *Ethics*. New York: Holt.
9. Nietzsche, F. 1967. *The will to power*, 221. (Kaufmann, W., and Hollingdale, R. trs. New York: Vintage.
10. Dewey, J. 1939. *Theory of valuation*, 29. Chicago: University of Chicago Press.
11. Mead, G.H. 1934/1977. *On social psychology*. Chicago, IL: University of Chicago Press.
12. Nietzsche, F. 1895–1901. Gesamtausgabe in Grossoktav. In *A Nietzsche reader*, vol 1–15, ed. Naumann, C., Leipzig, G., and Nietzsche, F.. Tr. Hollingdale, R.J. New York: Penguin.

13. Schweitzer, A. 1923/1996. Kultur und Ethik. Beck'sche Reihe, München. Schweitzer's thought is radically individualistic and critical of society. *„Der Zusammenbruch der Kultur ist dadurch gekommen, dass man der Gesellschaft die Ethik überließ. Erneuerung der Kultur ist nur dadurch möglich, dass die Ethik wieder die Sache der denkenden Menschen wird, und dass die Einzelnen sich in der Gesellschaft als ethische Persönlichkeiten zu behaupten suchen*"(351f) cf. Maier, B. (2003). Vertrautes und Fremdes in ethischer Perspektive. Zwei wert- und emotionsgeladene moralische Begriffe. In: Psychosomatische Gynäkologie und Geburtshilfe. Beiträge der Jahrestagung 2002 der DGPFG, 291–296. Gießen: Psychosozial-Verlag.
14. Medical Ethics Manual. 2005. *World medical association*, 12. France: Ferney-Voltaire Cedex.
15. Medical Ethics Manual. 2005. *World medical association*, 26. France: Ferney-Voltaire Cedex.
16. American Medical Association. 2001. AMA Principles of Medical Ethics. Adopted by the AMAś House of Delegates, June 17th, 2001.
17. National Institute for Illiteracy. 2000. nifl.gov/reders/lintro.htm#D.
18. Dewey, J., and Tufts, J. 1908/1932. *Ethics. rev*. Holt, H., 288. New York: Holt.
19. bcm.edu/ethics/phd.htm.
20. Shibles, W. 1995. Analysis of the definitions of Humanism. *Scottish Journal of Religious Studies* 16:51–61.
21. Fletcher, J. 1979. *Humanhood: Essays in biomedical ethics*, 18. New York: Prometheus.
22. Pellegrino, E. 1979. *Humanism and the physician*, XI. Knoxville TN: University of Tenessee Press.
23. Elliott, C. 1999. *Bioethics, culture and identity: A philosophical disease*. New York: Routledge; Elliott, C. ed. 2001. *Slow cures and bad philosophers. Essays on Wittgenstein, medicine and bioethics*. Durham: Duke University Press; Elliott, C. 2003. *Better than well: American medicine meets the American dream*. New York: W Norton.
24. Shibles, W. 1995. *Emotion in aesthetics*. Dordrecht, The Netherlands: Kluwer Academic Publishers; Shibles, W. 1995. *Unsere Gefühlswelt*. Mainz: Lermann Verlag; cf. Shibles, W. 1995. Analysis of the definitions of Humanism. *Scottish Journal of Religious Studies* 16:51–61.
25. Shibles, W. 1994. The cognitive-emotive theory of desire. *Journal of Indian Council of Philosophical Research* 11:25–40.
26. Dewey, J. 1939. *Theory of valuation*, 29. Chicago: University of Chicago Press.
27. Shibles, W. 1980. Ethics as a science: Going from is to ought. *Iowa Science Teachers Journal* 3:26–32.
28. See Edwards, P. 1967. *"Naturalistic fallacy" and "G.E. Moore". Encyclopedia of philosophy*. New York: Macmillan.
29. Molewijk, B. 2004. Integrated empirical ethics: In search for clarifying identities. *Medicine, Health Care and Philosophy* 7:85–87.
30. Davies, L., and Hudson, L. 1999. Why don't physicians use ethics consultation? *Journal of Clinical Ethics* 10:116–125.
31. Davies, L., and Hudson, L. 1999. Why don't physicians use ethics consultation? *Journal of Clinical Ethics* 10:119.
32. Davies, L., and Hudson, L. 1999. Why don't physicians use ethics consultation? *Journal of Clinical Ethics* 10:121 *my italics*.
33. Davies, L., and Hudson, L. 1999. Why don't physicians use ethics consultation? *Journal of Clinical Ethics* 10:122.
34. Singer, P. 2002. *One world: The ethics of globalization*, 63. New Haven, CT: Yale University Press.
35. Robinson, R. 1964. *Atheist's values*, 130. Oxford: Clarendon.
36. Dewey, J. 1934. *A common faith*, 26. New Haven, CT: Yale University Press.

37. Savulescu, J. 1998. Two worlds apart: Religion and ethics. *Journal of Medical Ethics* 24:382–384, esp. 383.
38. Singer, P. 2004. *The president of Good and Evil*, 3. New York: Penguin.
39. Bartley, W. 1973. III. *Wittgenstein*, 64. Philadelphia, PA: Lippincott.
40. Rachels, J. 2001. Killing and letting die. In *Encyclopedia of ethics*, 2nd edn, vol 2, 947–950, esp. 948. Routledge, London: Lawrence and Becker.
41. McCormick, R. 1984. *Health and medicine in the catholic tradition: Tradition in transition*. New York: Crossroad.
42. Rorty, R. 1989. *Contingency, irony and solidarity*, 8. New York: Cambridge University Press.
43. Beauchamp, T., and Walters, L. eds. 1999. Contemporary issues in bioethics, 5th edn, 11. Belmont, CA: Wadsworth.
44. McMahan, J. 2002. *The ethics of killing: Problems at the margins of life*, 190. New York: Oxford University Press.
45. Glannon, W. 2005. *Biomedical ethics*, 160. New York: Oxford University Press.
46. Krantz, S. 2002. *Refuting Peter Singer's ethical theory: The importance of human dignity*, 39. Westport, CT: Praeger.
47. Krantz, S. 2002. *Refuting Peter Singer's ethical theory: The importance of human dignity*, 38. Westport, CT: Praeger.
48. Singer, P. 2000. Extending ethics beyond our own species. In *Bioethics and the law*, vol I, 133–142, esp.140. eds. Kemp, P., Rendtorff, J., and Johansen, N. Copenhagen: Rhodes International Science and Art Publications and Centre for Ethics and Law.
49. Dworkin, R. 1977. *Taking rights seriously*. Cambridge: Harvard University Press.
50. Kant, I. 1785/1998. Grundlegung zur Metaphysik der Sitten. Weischedel, W. 14th edn. Werkausgabe vol 7, Suhrkamp Taschenbuch Wissenschaft, Frankfurt am Main; Kant, I. 1785/1998. In *Groundwork of the metaphysics of morals*. ed. Gregor, M. New York: Cambridge University Press; O'Neill, O. 1989. *Constructions of reason: Explorations of Kant's practical philosophy*, 126. Cambridge: Cambridge University Press.
51. Fletcher, J. 1979. *Humanhood: Essays in biomedical ethics*, 32–33. New York: Prometheus.
52. cf. Shibles, W. 1980. Morality in moral education. *Curriculum Review* 19:234–236; Shibles, W. 1980. Ethics as a science: Going from is to ought. *Iowa Science Teachers Journal* 3:26–32.
53. Wood, A. 1999. *Kant's ethical thought*, 3. Cambridge: Cambridge University Press; cf. Orig. Kant, I. 1785/1998. Grundlegung zur Metaphysik der Sitten. Weischedel W 14th edn. *Werkausgabe vol 7 Suhrkamp Taschenbuch Wissenschaft*, 1, 2. BA: Frankfurt am Main; Kant, I. 1785/1998. In *Groundwork of the metaphysics of morals*. ed. Gregor, M. New York: Cambridge University Press.
54. Kant, I. 1785/1998. Grundlegung zur Metaphysik der Sitten. Weischedel, W. 14th edn. Werkausgabe vol 7 Suhrkamp Taschenbuch Wissenschaft, Frankfurt am Main; G 4: 404 405; Kant, I. 1785/1998. In *Groundwork of the metaphysics of morals*. ed. Gregor, M. New York: Cambridge University Press; Wood, A. 1999. *Kant's ethical thought*, 23. Cambridge: Cambridge University Press.
55. Kant, I. 1785/1998. Grundlegung zur Metaphysik der Sitten. In *Werkausgabe vol 7 Suhrkamp Taschenbuch Wissenschaft*, 14th edn, 52. ed. Weischedel, W. BA: Frankfurt am Main; Kant, I. 1785/1998. In *Groundwork of the metaphysics of morals*. ed. Gregor, M. New York: Cambridge University Press; Kant, I. 1785/1998. Kritik der praktischen Vernunft. In *Werkausgabe vol 7, Suhrkamp Taschenbuch Wissenschaft*, 14th edn, 140. ed. Weischedel, W. BA: Frankfurt am Main; Kant, I. 1788/1997. *Critique of practical reason*. Gregor, M. tr, New York: Cambridge University Press; O'Neill, O. 1989. *Constructions of reason: Explorations of Kant's Practical Philosophy*, 126. Cambridge: Cambridge University Press.
56. O'Neill, O. 1989. *Constructions of reason: Explorations of Kant's practical philosophy*, 41. Cambridge: Cambridge University Press; Kant, I. 1785/1998. Grundlegung zur Metaphysik der Sitten. In *Werkausgabe vol 7 Suhrkamp Taschenbuch Wissenschaft*, 14th edn, 1, 2.

ed. Weischedel, W. BA: Frankfurt am Main; Kant, I. 1785/1998. In *Groundwork of the metaphysics of morals*. ed. Gregor, M. New York: Cambridge University Press.
57. Wood, A. 1999. *Kant's ethical thought*, 327. Cambridge: Cambridge University Press.
58. Wood, A. 1999. *Kant's ethical thought*, 334. Cambridge: Cambridge University Press.
59. Cassirer, E. 1953–1957. *The philosophy of symbolic forms*. 3 vols tr. Manheim, R. New Haven: Yale University Press.
60. Wittgenstein, L. 1958. *Philosophical investigations*, 3rd edn. Anscombe E tr, revised. 2002. Cambridge, MA: Blackwell Publishers.
61. Carse, T. in: Gordon, S., Benner, P., and Noddings, N. eds. 1996. *Caregiving: Readings in knowledge, practice, ethics and politics*, 88. Philadelphia, PA: University of Pennsylvania Press.
62. Pellegrino, E. 1979. *Humanism and the physician*, 89. Knoxville, TN: University of Tennessee.
63. Fletcher, J. 1979. *Humanhood: Essays in biomedical ethics*, 39. New York: Prometheus.
64. cf Arras, J. 2002. Pragmatism and bioethics: Been there, done that. *Social Philosophy and Public Policy* 19:29–58, esp. 52–53.
65. BMI. 2004. *Medical ethics today: The BMA's handbook of ethics and law*, 2nd edn, 240. London: BMI.
66. Kilner, J. 1990. *Who lives? Who dies?*, 234 New Haven, CT: Yale University Press.
67. Brown, B., and Singer, P. 1996. *The greens*, 58. Melbourne: Text Publications.
68. Wittgenstein, L. 1980. *Culture and value*, 44. Oxford: Blackwell.
69. National Commission Report on Excellence in Education. 72; April 1983.
70. Kline, M. in: Breslow, L. 2002. *Encyclopedia of health*, 4 vols, I, 305–306 New York: Macmillan.
71. Kossek, E., and Block, R. 2000. *Managing human resources in the 21st century from core concepts to strategic choice*, 276. Cincinnati, OH: South-Western College Publishers.
72. See Maier, B. 2003. Das Vertraute und das Fremde in ethischer Perspektive. Das Vertraute und das Fremde als kontextoffene Begriffe. In *Psychosomatische Gynäkologie und Geburtshilfe. Beiträge der Jahrestagung 2002 der DGPFG*, 291–296. Gießen: Psychosozial Verlag.
73. Schweitzer, A. 1923/1996. *Kultur und Ethik*, 338. München: Beck'sche Reihe.
74. cf Kopelman, L. 1997. Medicineś challenge to relativism: the case of female genital mutilation. In *Philosophy of medicine and bioethics*. eds. Carson, R., and Barnes, C. Dordrecht: Kluwer.
75. Dewey, J. 1929/1958. *Experience and nature*, 322. Dover, New York: Open Court, La Salle.
76. Sidel, V., and Levy, B. eds. 2005. *War and public health*, IV, 1301–1304. New York: University of Oxford Press.
77. Some health related organizations which stand for nonviolence are: American Public Health Association, Physicians for Social Responsibility, Physicians for Human Rights, International Physicians for the Prevention of Nuclear War, Amnesty International, etc. The Pentagon budget for 2004 was $400 billion, plus $40 billion a year to be spent on Homeland Security. The Bush administration planned to spend $2.7 trillion on the military over the next 6 years. Add to these costs for the wars in Afghanistan and Iraq, eventually over a trillion. The US spent in the region 6% of its GDP on the military in 2003. When one considers all expenses associated with the military this may possibly be doubled to 12%. It may be noted that in 2001, Jordan, Kuwait, Saudi Arabia and Oman all spent between 8 and 13% of their GDP on the military. North Korea spends 25% GDP, Democratic Republic of Congo 22%, Eritrea 16%, and Israel 10%. Nations compete to defend themselves against other nations even at the expense of the welfare, health and lives of their citizens. Hundreds of thousands of citizens are being killed. Additionally, in 2005, 45 million people in the U.S. have no health insurance. cf. U.S. Census Bureau).
78. Moser, R. 1975. An anti-intellectual movement in medicine. *Western Journal of Medicine* 122:433–449.

79. Moser, R. 1975. An anti-intellectual movement in medicine. *Western Journal of Medicine* 122:446.
80. Moser, R. 1975. An anti-intellectual movement in medicine. *Western Journal of Medicine* 122:449.
81. Singer, P. 1979/1993. *Practical ethics*, 2nd edn, 17. Cambridge: Cambridge University Press.
82. Gordon, S., Benner, P., and Noddings N eds. 1996. *Caregiving: readings in knowledge, practice, ethics and politics*. Philadelphia, PA: University of Pennsylvania Press.
83. Kuhse, H., and Singer, P. 1994. *Individuals, humans, persons: Questions of life and death*, 54. Salzburg: Sankt Augustine Academia Verlag.
84. Tranфy, K. 1999. In the absence of ethical theory. *Hastings Center Report* 29:43.
85. Rendtorff, J., and Kemp, P. 2000. *Basic ethical principles in European bioethics and biolaw. vol. I Autonomy, dignity, integrity and vulnerability*, 25. Copenhagen, Barcelona: Centre for Bioethics and Law.
86. Harris, J. 1999. The concept of the person and the value of life. *Kennedy Institute of Ethics* 9:293–308, esp. 307.
87. Kuhse, H., and Singer, P. 1994. *Individuals, humans, persons: Questions of life and death*, 64. Salzburg: Sankt Augustin Academia Verlag.
88. Easterbrook, G. 2002. Washington monthly online. November 2002.
89. Smith, W. 2000. *Culture of death: The assault on medical ethics in America*. San Francisco, CA: Encounter Books.
90. Smith, W. 2000. *Culture of death: The assault on medical ethics in America*, 22. San Francisco, CA: Encounter Books.
91. Maclean, A. 1993. *Elimination of morality: Reflections on utilitarianism and bioethics*, 3. London: Routledge.
92. Elliott, C. 1999. *Bioethics, culture and identity: A philosophical disease*, xxi. New York: Routledge.
93. Elliott, C. 1999. *Bioethics, culture and identity: A philosophical disease*, 148. New York: Routledge.
94. Elliott, C. 1999. *Bioethics, culture and identity: A philosophical disease*, xxii. New York: Routledge.
95. Schwartz, L., Preece, P., and Hendry, R. 2002. *Medical ethics: A case-based approach*, 1–5. Edinburgh: Saunders.
96. Harding, J. 2000. *Is there a duty to die? And other essays in bioethics*, 131. New York: Routledge.
97. Fletcher, J. 1979. *Humanhood: Essays in biomedical ethics*, 91. New York: Prometheus.
98. Metz, Th. 2002. Recent work on the meaning of life. *Ethics* 112:781–814, esp.798.
99. Metz, Th. 2002. Recent work on the meaning of life. *Ethics* 112: 781–814, esp.803.
100. Loewy, R. 2000. *Integrity and personhood: Looking at patients from a bio/psycho/social perspective*, 37. Dordrecht: Kluwer.
101. Nussbaum, M., and Sen, A. eds. 1993. *Quality of life*, 127. Oxford: Clarendon Press.
102. Quante, M. 2002. *Personales Leben und menschlicher Tod: Personale Identität als Prinzip der biomedizinischen Ethik*, 174. Frankfurt am Main: Suhrkamp.
103. Loewy, E. 1996. *Textbook of health care ethics*, 67. New York: Plenum.
104. cf. Shibles, W. 1998. Philosophical counseling, philosophical education and emotion. *International Journal of Applied Philosophy* 12:19–36.
105. Jonsen, A., Siegler, M., and Winsdale, W. 1998. *Clinical ethics*, 4th edn, 56, orig.1982. New York: McGraw-Hill.
106. Elliott, C. 1999. *Bioethics, culture and identity: A philosophical disease*, 147. New York: Routledge.
107. Schweitzer, A. 1923/1996. *Kultur und Ethik*, 338. München: Beck'sche Reihe.
108. Schweitzer, A. 1923/1996. *Kultur und Ethik*, 351 f. München: Beck'sche Reihe. *Erneuerung der Kultur ist nur dadurch möglich, dass die Ethik wieder die Sache der denkenden*

Menschen wird. (Renewal of culture is only possible, when ethics again becomes the concern of thinking people).
109. AMA Principles of Medical Ethics. 2001. Adopted by the AMAs' House of Delegates June 17th, 2001.
110. Hope, T., Savulescu, J., and Hendrick, J. 2003. Medical ethics and law: The core curriculum. London: Churchill Livingstone/Elsevier. See preface.
111. Kuhse, H., and Singer, P. 2006. *Bioethics: An anthology*, 2nd edn. Oxford: Blackwell.
112. Hope, T., Savulescu, J., and Hendrick, J. 2003. *Medical ethics and law: The core curriculum*, 4. London: Churchill Livingstone/Elsevier.
113. Hope, T., Savulescu, J., and Hendrick, J. 2003. *Medical Ethics and law: The core curriculum*, 6–8. London: Churchill Livingstone/Elsevier.
114. Beauchamp, T., and Childress, J. orig. 1979–1994–2001. *Principles of biomedical ethics*, 4th edn. Oxford: Oxford University Press.
115. Hope, T., Savulescu, J., and Hendrick, J. 2003. *Medical ethics and law: The core curriculum*, 10–11. London: Churchill Livingstone/Elsevier.
116. Hope, T., Savulescu, J., and Hendrick, J. 2003. *Medical ethics and law: The core curriculum*, 17. London: Churchill Livingstone/Elsevier.
117. Shibles, W. 1980. Ethics as a Science: Going from is to ought. *Iowa Science Teachers Journal* 3:26–32.
118. Hope, T., Savulescu, J., and Hendrick, J. 2003. *Medical ethics and law: The core curriculum*, 18. London: Churchill Livingstone/Elsevier.
119. Hope, T., Savulescu, J., and Hendrick, J. 2003. *Medical ethics and law: The core curriculum*, 20–28. London: Churchill Livingstone/Elsevier.
120. See Shibles, W. 1994. Cognitive-emotive theory of desire. *Journal of Indian Council of Philosophical Research* 11:25–40.
121. Hope, T., Savulescu, J., and Hendrick, J. 2003. *Medical ethics and law: The core curriculum*, 29 ff. London: Churchill Livingstone/Elsevier.
122. Hope, T., Savulescu, J., and Hendrick, J. 2003. *Medical ethics and law: The core curriculum*, 35–36. London: Churchill Livingstone/Elsevier.
123. Hope, T., Savulescu, J., and Hendrick, J. 2003. *Medical ethics and law: The core curriculum*, 34–35. London: Churchill Livingstone/Elsevier.
124. Hope, T., Savulescu, J., and Hendrick, J. 2003. *Medical ethics and law: The core curriculum*, 39–49. London: Churchill Livingstone/Elsevier.
125. Sidel, V., and Levy, B. 2003. Physician-soldier: A moral dilemma? In Pellegrino et al. *Military medical ethics*, vol I, 293–312. Pellegrino, E., Hartle, A., Howe, E., et al. Washington, DC. Walter Reed Army Medical Center and Bordon Institute.
126. Gross, M. 2004. Bioethics and armed conflict. *Hastings Center Report* 34:22–30, esp.26.
127. See Shibles, W. 2002. "Requirements before Going to War" in eBook on Humor on the web *Humor Reference Guide: A Comprehensive Classification and Analysis*. eBook on humor at http://facstaff.uww.edu/shiblesw/ or search: Shibles humor.
128. Gelfand, M. 1968. *Philosophy and ethics of medicine*, 113. Edinburgh: Livingstone.
129. Gross, M. 2004. Bioethics and armed conflict. *Hastings Center Report* 34:22–30 esp.24.
130. Gross, M. 2004. Bioethics and armed conflict. *Hastings Center Report* 34:22–30 esp.23.
131. Gross, M. 2004. Bioethics and armed conflict. *Hastings Center Report* 34:22–30 esp.27.
132. Gross, M. 2004. Bioethics and armed conflict. *Hastings Center Report* 34:22–30 esp.25.
133. Gross, M. 2004. Bioethics and armed conflict. *Hastings Center Report* 34:22–30 esp.28.
134. Sidel, V., and Levy, B. 2003. Physician-soldier: A moral dilemma? In *Military medical ethics*, vol 1, 293–312 esp.302. eds. Pellegrino, E., Hartle, A., Howe, E., et al. Washington, DC: Walter Reed Army Medical Center and Bordon Institute.
135. Sidel, V., and Levy, B. 2003. Physician-soldier: A moral dilemma? In *Military medical ethics*, vol 1, 293–312 esp. 296. eds. Pellegrino, E., Hartle, A., Howe, E., et al. Washington, DC: Walter Reed Army Medical Center and Bordon Institute.

136. Sidel, V., and Levy, B. 2003. Physician-soldier: A moral dilemma? In *Military medical ethics*, vol 1, 293–312 esp. 297. eds. Pellegrino, E., Hartle, A., Howe, E. et al. Washington, DC: Walter Reed Army Medical Center and Bordon Institute.
137. Sidel, V., and Levy, B. 2003. Physician-soldier: A moral dilemma? In *Military medical ethics*, vol 1, 293–312 esp. 309. eds. Pellegrino, E., Hartle, A., Howe, E. et al. Washington, DC: Walter Reed Army Medical Center and Bordon Institute.
138. Sidel, V., and Levy, B. 2003. Physician-soldier: A moral dilemma? In *Military medical ethics*, vol 1, 293–312 esp. 304. eds. Pellegrino, E., Hartle, A., Howe, E. et al. Washington, DC: Walter Reed Army Medical Center and Bordon Institute.
139. Sidel, V., and Levy, B. 2003. Physician-soldier: A moral dilemma? In *Military medical ethics*, vol 1, 293–312 esp. 307. eds. Pellegrino, E., Hartle, A., Howe, E., et al. Washington, DC: Walter Reed Army Medical Center and Bordon Institute.
140. Sidel, V., and Levy, B. 2003. Physician-soldier: A moral dilemma? In *Military medical ethics*, vol 1, 293–312 esp. 308. eds. Pellegrino, E., Hartle, A., Howe, E., et al. Washington, DC: Walter Reed Army Medical Center and Bordon Institute.
141. Canetti, E. 1960. *Masse und Macht*, vol 2, 64 Regensburg: Claassen/Carl Hanser Verlag.
142. cf. Shibles, W. 1999. I Ching as a metaphorical method of insight. *Journal of Chinese Philosophy* 26:343–376.
143. Ellin, J. 1995. *Morality and the meaning of life*, 322–324. Ft. Worth, Texas: Harcourt Brace.
144. Schweitzer, A. 1923/1996. *Kultur und Ethik*, 324, 339. München: Beck'sche Reihe, cf *"Nach der Verantwortung, die ich in mir erlebe, muß ich entscheiden, was ich von meinem Leben, meinem Besitze, meiner Zeit, meiner Ruhe hingeben muß und was ich davon behalten darf.*"(342) *"Gesellschaftliche Leitlinien, eine gewisse allgemeine Moral wären dazu nicht geeignet, sie blieben zu oberflächlich, nicht wirklich kontextsensitiv. (339) According to the responsibility I feel myself, I have to decide what I have to hand in of my life, my possessions, my time, and my quietness and what I may keep. (342) Societal guidelines, certain general morals would not be suitable, they would remain too superficial, not really context-sensitive (My translation).*
145. Schweitzer, A. 2003/1966. *Die Ehrfurcht vor dem Leben. Grundtexte aus fünf Jahrzehnten*, 8th edn, 107. ed. Bähr, H.W. München: CH Beck: *Meine persönliche Ethik kann nur absolut sein, sie kann sich auf nichts Weiteres zurückziehen, sie folgt keinen gesellschaftlichen Imperativen, „sie ist von Grund aus subjektiv, weil sie jedem von uns die Verantwortung zugesteht zu entscheiden, wieweit er in der Aufopferung gehen will."*
146. Chomsky, N. 2002. 9–11. New York: Seven Stories Press.
147. Der Spiegel. 2005. Ethische Empfindungen No. 1/3, Jan 1, 2005: 98.
148. Savulescu, J. 1998. Two worlds apart: Religion and ethics. *Journal of Medical Ethics* 24:382–384.
149. Dirie, W. 2002. *Desert dawn*. London: Virago Press.
150. Dirie, W. 1998. *Desert flower*. London: Virago Press.
151. Dirie, W. 2002. *Desert dawn*, 13. London: Virago Press.
152. Dirie W. 1998. *Desert flower*, 62–78. London: Virago Press.
153. Dirie, W. 1998. *Desert flower*, 357 ff. London: Virago Press.
154. Dirie, W. 1998. *Desert flower*, 351. London: Virago Press.
155. Dirie, W. 2002. *Desert dawn*, 239. London: Virago Press.
156. Dirie, W. 2002. *Desert dawn*, 236. London: Virago Press.
157. Dirie, W. 2002. *Desert dawn*, 139. London: Virago Press.
158. See UN, World Health Organizations (WHO) and circumcision web sites for up to date statistics. U.N. WHO 2002: 13.
159. See UN, World Health Organizations (WHO) and circumcision web sites for up to date statistics. U.N. WHO 1998: 358.
160. Dirie, W. 1998. *Desert flower*, 241. London: Virago Press.
161. Dirie, W. 1998. *Desert flower*, 368. London: Virago Press.
162. Dirie, W. 2002. *Desert dawn*, 52. London: Virago Press.

163. Dirie, W. 1998. *Desert flower*, 46. London: Virago Press.
164. Dirie, W. 2002. *Desert dawn*, 201. London: Virago Press.
165. Dirie, W. 1998. *Desert flower*, 60. London: Virago Press.
166. Dirie, W. 1998. *Desert flower*, 360. London: Virago Press.
167. Dirie, W. 2002. *Desert dawn*, 16; 46. London: Virago Press.
168. Dirie, W. 2002. *Desert dawn*, 191. London: Virago Press.
169. Dirie, W. 1998. *Desert flower*, 359. London: Virago Press.
170. Dirie, W. 1998. *Desert flower*, 351. London: Virago Press.
171. Dirie, W. 2002. *Desert dawn*, 205. London: Virago Press.
172. Dirie, W. 1998. *Desert flower*, 461. London: Virago Press.
173. Dirie, W. 2002. *Desert dawn*, 13. London: Virago Press.
174. Dirie, W. 2002. *Desert dawn*, 226. London: Virago Press.
175. Dirie, W. 2002. *Desert dawn*, 223. London: Virago Press.

Chapter 6
Medicotheology and Biotheology

I will not permit considerations of religion. . .to intervene between my duty and my patient. [1]

One considerable advantage that arises from philosophy consists in the sovereign antidote which it affords to superstition and false religion. [2]

Abstract Religion is the major guiding belief system of nearly every culture and almost defines culture itself. Medicine is largely influenced by such beliefs. If one follows one's religion, one is thought to be moral, if not, not. It may be noted that religion is morals, not ethics. Ethics in medicine would refer to a rational, holistic consequentialism. On this view, a religious or non-religious, supernaturalistic belief, which does not consider consequences and defeats naturalistic goals and wants, such as e.g. medical research, is unethical. It may support a religion, but does not qualify as an ethical system. This is a book on the philosophy of medicine, which therefore must involve some philosophy of religion. As such, it is a critical examination of some of the concepts and methods in each area e.g. a critique of the sanctity-of-life doctrine as well as a critique of prayer as medical treatment.

Keywords Religion · religious beliefs · philosophy of religion · humanism · consequentialism · sanctity-of-life doctrine · quality of life criteria · medicotheology · biotheology · prayer

6.1 Introduction: How Many People Have Religious Beliefs?

Religion is the most prevailing belief system in nearly every society. It therefore has more influence on medicine than any other belief-system. Ernest Hemingway once noticed that the big lie is more plausible than truth.

The world population is 6 1/2 billion. Christianity is the largest religion with 2.1 billion (33–37%; 23% of the world's population), Islam 1.3 billion (20.1%), Hinduism 851 million (13.3, 80% of India; but one can be a Hinduism follower and an atheist), Buddhism 375 million (5.9%), Jehovah's Witness 3 million. The remainder of the religions is small [3].

B. Maier, W.A. Shibles, *The Philosophy and Practice of Medicine and Bioethics*, International Library of Ethics, Law, and the New Medicine 47,
DOI 10.1007/978-90-481-8867-3_6, © Springer Science+Business Media B.V. 2011

In sum, most all people in each society have and live by supernatural belief systems. Even some Buddhists say that every belief system is an illness [4]. The philosopher and epistemologist could agree. Medicine is largely influenced by such beliefs.

6.2 The Influence of Religion on Bioethics and Medicine

Religion is one of the most significant and yet troublesome aspects of nearly every society. Religion is the major guiding belief system of nearly every culture and almost defines culture itself. It is also regarded as the most important aspect of people's lives and even defines person itself. Person and human are, however, defined differently by the different religions.

When we speak here of religion we do not refer to just Christianity, but include the various religions. Hippocrates replaced the supernatural with a natural cause of disease [5]. Yet the Hippocratic oath is dedicated to Apollo, god of medicine. What is common to all religions is a belief in supernaturalism. These religions partly or mainly govern each society and so also directly or indirectly the medical practice in that society. Christianity, for example, predominates over medicine in the Western world and keeps medical practice in line with its particular teachings. The Church versus medicine battle is a subcategory of the Church versus state battle whereby the Church often sees itself as above the state and above medicine. For example, the Catholic hospital believes that it need not comply with other community values. Religious organizations with deceptively neutral sounding names counsel nonreligious as well as religious women not to have abortions. In the U.K. a doctor may refuse to perform an abortion on religious grounds [6]. There are, for example, Islamic, Jainist, Hindu, Buddhist, Jewish, Confucian, Baptist, etc. bioethics. The largest, most active group is Catholic bioethics. But philosophy and ethics, as was shown in the Chapter 5 are often in conflict with religion. An examination of the philosophy of religion also indicates that. Religion has appropriated philosophy and science and bioethics. Christianity absorbed philosophy into theology as suggested by Shusterman's view "medieval scholasticism's subordination of philosophy to theology" [7].

Bryan Hehir is Professor of Practice in Religion and Society, Harvard Divinity School and a former Senior Research Scholar of the Kennedy Institute of Ethics. He wrote, "In the discipline of Catholic moral theology, bioethics has held a major place." He gives as the sources of bioethics a natural law position: John XXIII's encyclical *Pacem In Terris* (1963), Paul XI's encyclical *Humanae Vitae* (1968 about contraception), and the teaching of the *Second Vatican Council* (1962–1965). Hehir claims that, "The future role of natural law in Catholic moral theology, and specifically in bioethics, will. . .tilt more toward a biblical-theological analysis than toward the Catholic teaching" [8].

Churches and the presence of religion are to be found in virtually every city and town in the world. The result is that medical practice has been greatly influenced by and has even become led by religious practice. As is typically the case, the ethics in medicine institutes in Tübingen and Vienna, for example, are headed by theologians.

Hastings center was begun by, Andre Hellegers, a Roman Catholic physician, who founded the Kennedy Center for Bioethics at Georgetown University in 1971. "Biolaw and bioethics in Austria...is limited by Catholic theology" [9].

Christianity has infiltrated the U.S. Government's National Institute of Health (NIH). The National Institute for Health Science research funding was cut for fiscal year 2006. On the other hand, Complementary and Alternative Medicine (CAM) dealing with alternative spiritual and religious medicine is at a historic high at $120,379,000 [10].

Callahan who has had much influence to bring religion into bioethics, nevertheless said that the NIH has no clear view of healthcare [11]. This is especially true in 2006 under President Bush who expressed the view that God made him president and that all is theologically caused, that there is a universal divine causality. Is this to be the basis of medicine? The NIH publications can no longer speak of abortion or contraception, but must stress the government's religious view of abstinence. Alternative medicine is stressed because that allows for a religious healing perspective and hundreds of new studies have come out attempting to show that prayer heals. Some examples are reported here. See the analysis of prayer as medical treatment (at the end of this chapter, and also the Chapter 19) which tries to show that such prayer work is based on scientific statistical research. President Bush's supernaturalistic governmental Council on Bioethics also reflects the religious-political perspective of the government, and most Americans support a religious approach. NIH does CAM research for this purpose. The following gives in detail some of this tax-supported "research." It may be noted that the medical works of ancient Greece opposed healing prayers and divine explanations for disease [12].

About the approach of the new conservatives in bioethics, Macklin wrote, "The new method is mean-spirited, mystical and emotional. It claims insight into absolute truth yet disavows reason" [13].

What is said to characterize these supernaturalists is the attack on rationality, the secular, humanism, and science, and opposition to medicine and medical research, such as stem cell research. They favor religious faith and regard whatever deviates from that as "unnatural" and artificial. It goes back to the medieval slogan that whatever is against the church view is "*contra naturam*." Modern medicine is largely denied therefore as being "against nature." Macklin criticizes the supernatural approach as opposing "liberal humanitarianism" [14]. Macklin herself claims to be a "liberal, humanitarian bioethicist."

According to Leigh Turner "contributions to bioethics typically drew upon a single moral theory, religious tradition, or intuitionist model or moral deliberation" [15]. In the U.S. the bioethics debate is a battle between reasoned scientists and supernaturalists having few concepts (or universals) in common. Problems also arise because of the other different conflicting groups: religious, ethnic, economic, political, etc. There is no "common morality," "universals" or objective ethics. Rather each view is embedded with culture, history, tradition, etc [16]. This is to say that people in any culture are typically enculturated. Turner does go on to say, "To most

Jehova's Witnesses, mandatory [needed] blood transfusions for minors constitute an assault on the integrity of their religious tradition rather than a judicious exercise in ethical reasoning" [16]. To the extent that Turner supports cultural diversity and religious diversity he is advocating chaotic and irrational relativism in bioethics. However, he offers a hedge in his conclusion, The purpose of this paper is not to celebrate the existence of multiple normative traditions. . .however, I think it important to be aware of the significant obstacles to the development of a "common morality" [17]. "Early contributions to scholarship in bioethics emerged from within theological frameworks. . .However, . . . bioethics needed to be rescued from these partisan religious traditions. . .to inform secular, public, and institutional policy in a social context where no one particular religious tradition dominated public debate" [18]. The Austrian philosopher, Peter Kampits, argues clearly that religion is too dogmatic to be involved in bioethical decision-making and that philosophers should not promote religion in bioethics [19].

Three theologians presided over the beginnings and continuance of contemporary bioethics: Joseph Fletcher, a conservative Episcopalian minister; Paul Ramsey, a Methodist minister who appeals to scripture, sanctity-of-life, human obedience, and is against consequentialism and utilitarianism (also Professor of Religion, Princeton); and the Catholic theologian Warren Reich, who has a Doctor of Sacred Theology degree (S.T.D.) at the Gregorian University in Rome. He is editor of the *Encyclopedia of Bioethics* and a member of the Kennedy Institute. Other bioethics researchers are: Richard McCormick, S. J., a Jesuit moral theologian and also a member of the Kennedy Institute [20], Father Albert Jonsen, S.J., author of *Clinical Ethics* (with others) and *The Birth of Bioethics*. He was the first Catholic to obtain a Ph.D. in Religious Studies from Yale University. Baruch Brody teaches the Philosophy of Religion. Joseph Fletcher was a theologian, but left religion in favor of situational ethics, act-utilitarianism (consequentialism) and Dewey's humanistic naturalism. "The rights that Fletcher claimed for patients are less against their doctors than against the 'otiose dogmas of religious moralists.'. . .He consistently refuted their [Catholic] arguments against sterilization, contraception, artificial insemination, and euthanasia" [20]. "Biolaw and bioethics in Austria are marked by a pragmatic positivistic legal tradition that is limited by Catholic theology" [21]. "The pluralism of Belgian society is opposed to the Catholic Church" [22]. "Demand for pluralism must also be seen as a reaction to the dominance of the Church" [9]. It is a "confrontation" [23]. "The Italian debate on bioethics is marked by the opposition between Catholic and secular bioethics" [24]. The documents of the Church, *Humanae Vitae* (1968) and *Donum Vitae* (1988), give the Catholic position to be recommended for the medical profession [24]. "The Church has been very critical toward utilitarian and consequentialistic approaches to bioethics" [25]. By contrast, Norway is characterized by deontological Lutheranism [26].

President GW Bush, a committed Christian who prays and reads the bible daily, stated that he has a "'divine plan that supersedes all human plans'" [27]. Bush enthusiastically declared war in the Gulf, which resulted in 1.5 million deaths. Medical treatment was virtually ended in Iraq. His son continued his policies in these areas a decade later with Gulf War II.

The son, President G. W. Bush, has the same view. What implications does this have for bioethics? Bush appointed Leon Kass as head of the government's Council on Bioethics. While in office, Kass supported Bush's policies. Bush appointees tend to be staunch loyalists. We can see that bioethics can easily become appropriated to religion especially since the Council is only advisory anyway. It is an "elected theocracy." "President Bush is using his religious beliefs to support his public policy" [28].

Also, medical practice may therefore be seen as a form of religious practice. This is reflected in the history of medicine as well as in the writings on bioethics. The religious model of medicine has, however, come into conflict with the medical model of medicine. To what extent should medicine be determined, influenced and guided by religion?

The Church attempted to appropriate the Hippocratic Oath. Steven Miles presented some of the myths regarding the Hippocratic Oath, which follow. We do not know who wrote the Oath. Thus, it is false to say that Hippocrates did [29]. It is also false to say that the Oath is merely based on divine appeal. "I swear by Apollo, the physician and by Asclepius . . ." Apollo is the god of healing, poetry, reason and prophecy, father of *Asclepius* (meaning literally "unceasingly gentle"), who was dedicated to healing based on love. The appeal to Apollo seems to be an appeal to religion, but represents rather an appeal to reason, and natural causes, the reverse of divine ones [30]. The speculative prophecy is rather prediction and insight following the Greek aphorism, "In the case of acute disease, to predict either death or recovery is not quite safe" [30]. The National Catholic Bioethics Center rewrote the oath to render it as "I swear by the presence of the Almighty. . ." The purposeful mistranslation of the Oath by the Church also involves deceit of both patient and healthcare worker. The Oath, by contrast, states that honest prognosis and truth telling should be provided to the patient, and not invocation to a Christian god.

This is a book on the philosophy of medicine, which therefore must involve some philosophy of religion. As such, it is a critical examination of some of the concepts and methods in each area. That is, one must do both the philosophy of religion as well as the philosophy of medicine. The additional background of the discussion is the concern with the fact that the medical system is in crisis worldwide, as it cannot now adequately care for all who need medical care. The cause is not basically solely with the medical field itself, but with the faulty cultural practices and belief systems of the people and institutions, which oppose or fail to support it, the "philosophy" of culture.

6.3 Church Opposition to Medicine

> *Theology [is] a special adversary of philosophy and science. . .Theology cannot make a contribution of moral theory to the* endeavors of bioethics [31].

E.g. the opposition to contraception by the Church causes unnecessary overpopulation, costs lives and diminishes the quality of life of those who have good reasons for not having children. All artificial methods of birth control, but also medically assisted procreation is regarded as sinful [32]. The Church condemns both contraception and abortion and is anti-sexual and inhumane [33]. According to Jonsen's Clinical Ethics, physicians are not to consider even the most fantastic and dangerous belief to be incapacity regarding autonomy of the patient's choices [34]. Physicians and nurses may refuse to cooperate in actions [e.g., abortions] they judge immoral on grounds of [religious, etc.] conscience [35]. Jehovah's Witnesses can refuse life-saving transfusions except in the case of minor children or if it endangers others [36]. Organ transplantation was opposed by Jewish religious law and by the Catholic Church, as it was regarded as destruction and mutilation of the body. Many still oppose presumed consent for organ donation on religious grounds, even though the Church now allows it.

Eventually organ transplantation won at the expense of religion [37]. American Lutherans oppose a duty to donate body parts. Christian Scientists oppose transplantations [38]. The state protects religious belief over sound medical treatment. The physician may however request a legal challenge.

Singer states, "The state has a responsibility to ensure that children are not simply being indoctrinated into a narrow set of religious or political teachings" [39]. Tronto also sees caring as a non-indoctrinaire responsiveness, therefore it should not be religious or cultural [40]. Physicians need not cooperate in such indoctrination either.

The question arises as to whether the physician as a supernatural religious believer has given up rationality and the scientific approach to medicine and should therefore be disqualified as a physician except for members of his or her own church. Thus, Singer speaks of church "doctrines about immortality, original sin, and damnation...doctrines so obnoxious...that if anyone did accept them, we should be inclined to discount any other moral view he/she held" [41].

In regard to a duty to help others, religious belief cannot be given as an excuse not to help. Christian Scientists may be required to bring medical help, and Orthodox Jews may have to exert medical effort on Saturday [42].

Case example: A 60 years old psychoanalyst and physician reported that about 30 years previously when she was a young physician and when abortion already was legal (1975) she performed them in a Viennese hospital. The anesthesia was however in the hands of Catholic nuns. They disapproved of abortions but nevertheless were required to administer anesthesia for all operations performed in the hospital. Therefore, they always tried to give too little anesthesia to the woman having an abortion so the woman would experience pain for her "sins." It was regarded as God's punishment through the nun-nurses. On discovering this, the young physician from then on protested against the influence of religion on medicine (Personal communication). It is often thought by the religious that disease is a way in which God punishes one for one's alleged "sins."

Religion also undermines medicine directly by the following. According to the *Catholic Doctrine of Faith* 1987 it is illegitimate to: 1. have prenatal care to determine impaired children so abortion can be performed, 2. undergo medically assisted procreation.

6.4 Should Medicine Be Based on Supernaturalism?

Early cures were religious cures, e.g., by faith or miracle, pilgrimages, chants, rituals, charms, incense. Disease was regarded as God's punishment. The same tradition continues today.

In the philosophy of religion it is typically concluded that there are no rational arguments for the existence of god (any god). From that follows that arguments for religion are irrational.

There are no facts in religion unlike reason and science. Religion does not change its dogma on the basis of new information. We may change, "In God we trust," that is, "In human reason we do not trust," to "In human reason we trust." "Only humans can think through their moral choices [ethical] and be held morally responsible for what they do" [43].

The concepts of fate and determinism are also informal logical fallacies that take the place of intelligent action. If we are going to be supernatural there is no limit to the incredible sorts of things we might say or command. Religion and appeal to the supernatural are ways to avoid rational decision-making. John Fletcher wrote, "A high-order mistake is to choose or wrongly defend an indefensible world view that conditions one's basic perspective in ethics" [44].

6.5 Science and Metaphysical Causes?

Religion contradicts science, medicine and inquiry.

Hippocrates replaced supernatural causes with natural causes of disease [45].

The Oxford University professor, Richard Dawkins stated, "As a scientist I am hostile to fundamentalist religion because it actively debauches the scientific enterprise" [46]. He speaks against "the presumptuousness whereby religious people know, without evidence" [47].

Each religion wishes to impose its philosophy of life and "afterlife" as the only one. Society has many religions and other philosophies of life. Medicine and bioethics cannot accept relativistically all the different views as its basis, nor can it accept only one religion. It can, however, accept rational, critical, naturalistic, consequentialistic scientific thinking in general. "A rational perspective is not just one perspective among various alternative possibilities" [48].

Because Noddings is anti-principles, anti-obedience, and anti-dogma we can predict that she is anti-religion as well. She accordingly states that religion is too authoritarian, and based on obedience, accusation, and fear to be caring [49]. "Religion and ethics are different categories of human enquiry. Religion is about faith; ethics is about reason. Religion is about what biblical texts, traditions and figureheads say is right and wrong, and what some theists believe is right and wrong" [50]. Reward and punishment, e.g., in heaven or hell, does not constitute an ethical system. On the Church "morals" one does not do anything because it makes sense according to own judgment and consequences. People do not seem to be able to realize that religions are typically like oppression and require total unquestioned obedience. One must obey blindly.

Ethically nobody has the "right" to obey (H Arendt), the ethical person has the "duty" to inquire and decide and be responsible. Andrew Sullivan, himself a questioning Catholic, wrote about Pope Benedict XVI, "The Pope's dogma is a circular system that's immune to reasoned inquiry" [51]. Belief in God is rather a revelation. The Pope holds that the main values for a woman are motherhood and virginity. The Church requires us to give up that power which most characterizes humans: reason. Without rationality we are no longer human.

6.6 Case Example: Religion and Autonomy

A 28 year old pregnant Austrian woman in the 28th week of gestation is admitted to the obstetrics department. She was already the mother of a 4-year-old child. She had a twin pregnancy, one of the twins died intrauterine. This is now decomposing, but would be leading in presentation in case of delivery. A cesarean at this time would be required. In addition, the woman suffered from a lung infection, was anemic and found it difficult to breathe. I, the coauthor and the physician in charge, was told immediately that she and her husband were members of Jehova's Witness religion and would not accept any blood transfusions regardless of the danger to her life. She showed a document to verify her belief and which authorized her autonomy in favoring her belief over all medical treatment. She was unwaveringly supported by her husband in this belief. Every step of the medical treatment became suspiciously supervised by her in spite of her intense suffering. Her husband also showed similar suspiciousness. They confirmed their fixed and undeniable decisions as often as anyone of the medical staff had contact. Only gradually did the patient become better, though with proper medical treatment she could have recovered more quickly and without such extreme risk. In the 33rd week her baby was delivered vaginally, but due to the atonia of the womb she had a massive blood loss, though fortunately never lost consciousness. With her last strength left she repeated, "No blood." Her hemoglobin went down to the dangerous levels of 5.6, hematocrit to 19. Nevertheless she received all other supplies possible but no blood transfusion and eventually recovered. This is not a story of the miracle of faith, but rather of the deliberate risk to her life and loss to her family. She was an in-patient for about 2 weeks and was lucky to have survived. If a physician had withheld such treatment s/he could have been charged with malpractice. The consequence was that she was severely endangered and gave no thought to the consequences for her family, the child, or the medical workers or hospital. She obeyed her religion blindly. More services and supplies than usual were needed to care for her and she put the healthcare workers in a compromised and stressful position because the life-threatening block to treatment was counter to professional and sound medical practice. Law in the prosecution of religious belief only allows the giving of blood in such a situation if the patient loses conscience (no informed consent) and even then a governmental attorney must be called for permission. The patient must also be in ultimate danger.

This autonomous decision-making is unfair to physicians as well as to the healthcare system and unfair to others who need the usually restricted services and supplies. Should autonomy based on irrational beliefs especially if they go directly against professional medical treatment and is harmful to all involved, remain unchallenged no matter the outcome?

6.7 Religion Versus Medicine: A Common Ground?

There has been a separation of religion and bioethics over the past 20 or so years. Diverse religious views could yield no consensus in bioethics, religious language and religious "morals" are not relevant to the critical ethics, which concerns our daily lives. It is a supernatural language applying to a supernatural world, not the world we live in [52]. *What we do is to bring words back from their metaphysical to their everyday use* [53]. Albert Schweitzer said, *The embodiment of Being, the Absolute, the Worldspirit, and all expressions of this kind indicate nothing real, rather something conceived in the abstract, which therefore is absolutely unimaginable* (Author's translation) [54].

John Dewey, although an atheist and naturalist, wrote *A Common Faith* (1934) in order to attempt to interpret supernatural religious language in naturalistic terms and thereby discover some common understanding, or at least to try to make religious language intelligible for natural experience. That is, he sought a common ground. An attempt is made to see if religion is translatable into naturalism. Religion consists of supernaturalistic pre-scientific beliefs, survival from outgrown cultures. It is not *sui generis* [55].

How does "unseen" relate to the present? We must change religion from a noun to a naturalistic adjective. There is no religion in general [56].

Religion can be just a way of looking at things so it can be used as an adjective. Einstein said, "If something is in me which can be called religious then it is the unbounded admiration for the structure of the universe so far as our science can reveal it" [57].

Religion is fixed sets of beliefs. "Religious," however, is not meaning the same as church. We can instead reduce fixed belief to flexible adjectival behavior. "It denotes attitudes that may be taken toward every object and every proposed end or ideal" [58].

We give different names to the terms of religion. Such terms refer to a natural occurrence to which an emotional quality has been attached [55]. In terms of emotion, "there is such a thing as passionate intelligence" [59].

The religious quality is not a cause, but the effect produced, what it does, contentment we attain [60].

God is not an entity, but just to be reduced to human possibilities [61]. One could say the religious experience is nothing but the contentment regardless of how achieved, whether sitting by a stream or prayer. It is not fixed beliefs or practices. A philosophical insight, reading poetry, and a new perspective are such reorientations [62]. We can mean by the religious only significant moments of living [62]. It can be merely friendship, an aesthetic experience, etc [58]. Invisible powers are in rationality and our natural experiences with others. These are the amazing powers, which are creative and not fixed in stone. Our ends and goals are unifications presented imaginatively. Truth reduces to finding ends and creative problem solving. Moral conviction reduces to accepting a certain goal or end [63]. Ideals must be based on natural informed and critical experience [64]. "The objection to supernaturalism is that it stands in the way of an effective realization of the sweep and depth of the implications of natural human relations" [65]. Emancipation from religion is a form of healthful adjustment [66]. We realize that we can now use our own powers to change the world rather than waiting for divine providence to do so [67]. "The 'divine' is thus a term of human choice and aspiration" [68]. Instead of submission to the supernatural submit to nature and the environment. Submission to nature is naturalistic, submission to supernaturalistic beliefs is not. Religion isolates us from the world and prevents us from having human dignity [69]. Just as reason does not apply to religion, religion does not apply to our natural world. Similarly, we may see that the ordinary language approach to religious language is to place it in the language game in which it genuinely has its meaning [70].

Humanism is consequentialistic critical thinking (speaking) for the purpose of producing the best life possible for all human beings and their environment including animals and nature and the expansion of our knowledge and abilities as much as possible.

Religion, itself not democratic, appeals only to consensus for its support. Consensus is not an argument, but only an opinion or bias. Culture in this sense is like consensus. Culture itself is neither good nor bad, but non-ethical. Smith disapprovingly states, "The ultimate bioethics agenda is startlingly radical: dismantling traditional Western values and mores and forging a new ethical consensus based on values most people do not presently share" [71].

We may agree with this statement. Otherworldly religious beliefs of jurors and the members of the courts, and some hospital workers can predominate over medical findings and practice. Again, this is the attempt of the church to determine medical decisions, without a common ground. Religion tries to usurp medical decision-making.

6.8 Religion as Ethics

People in each culture identify morals with their religion. As was just argued, this is to bring the mystically irrelevant language into the relevant language of the world. If one follows one's religion, one is thought to be moral, if not, not. It may be noted that religion is morals, not ethics. Similarly if one follows one's cultural practices one is moral. This is true regardless of the culture or religion.

Religion is not ethics, nor is religion above ethics, a divine higher ethics, as they would have it. Religion and common normative language are uses of ethical terms, which are fraught with equivocations and fallacies. Religion is not above the philosophy of religion and the critique of ethics. It is rather the other way around.

It is not an ethical theory at all [72].

It does have commandments. Thus, we accordingly find religion being put into question as the basis of bioethics as also indicated by the following.

Peter Singer wrote, "Clergymen...do not make...satisfactory moralists [ethicists]," and "Moral [ethical] positions should be discussed and argued about not accepted on the authority of God or god-profession" [73].

Derek Parfit wrote, "Belief in God, or in many gods, prevented the free development of moral reasoning" [74]. Religion is contradictory to the very meaning of the study of ethics. Religion and ethics are different categories of human enquiry. Savulescu wrote, "Religion is as different from ethics as it is from mathematics. Religion is about faith; ethics is about reason" [75]. An extension of the attempt to overpower this view is that, "religious myths have been used to maintain the subbordination of women" [76].

Noddings states, "You are free to practice your religion as you see it, but when you enter the public arena, your commitments and recommendations must be...subject to the methods of intelligence" [77]. She suggests that if one is to teach religion one should discuss what is good about religion but also what is bad about religion. She states that religion is often seen as anti-humanistic and harmful [78]. Religion leads to war and bloodshed, fails to allow critical thinking (speaking), has an abstractionist claim to absolute truth, often as with Christianity and Islam tries to impose its beliefs on all others (Hindu-Muslim massacres, Catholic-Protestant massacres) creates inability to be critical about the tenets of religion [i.e. believers are not familiar with the philosophy of religion] even by otherwise critical people, involves and encourages violence, must assume evil god or if not a fallible one, etc [79]. Society supports the favored religion of the particular society, but religions continually oppose secularism, atheism, humanism, naturalism, inquiry and science, including medicine.

6.9 Ethics Committees

At first there was a search for a contemporary Apollo, god of medicine. Gerald Kelly, a leading Roman Catholic ethicist, stated in Medico-Moral Problems the position that bioethics should be based on Catholicism: "The Catholic moralists do have a just claim to special competence in the science of ethics.... Catholic moralists represent by far the world's largest group of specialists in the science of ethics" [80]. The use of "science of ethics" here equivocates, indicating being expert in Christian ethics, not the philosophy of ethics or a scientific basis of ethics, which they oppose.

"Hastings Center bioethicist Daniel Callahan claims (disapprovingly) that, 'the first thing' bioethicists had to do to establish itself as a profession/specialty was 'to push religion aside'" [81]. Callahan stated, "The decline of religious contributions [to bioethics is] a misfortune, leading to paucity of concepts, a thin imagination, and the ignorance of traditions, practices, and forms of moral analysis of great value" [82].

Wesley Smith, in defense of enculturation, believes that bioethics should be based on popular morality, metaphysics, and transcendent supernaturalism [83]. He advocates "sanctity-of-life" doctrine and a universal and absolute, "objective" religious basis for bioethics. Smith opposes science, secular philosophical reasoning, consequentialism, and naturalism.

Ethics committees and the philosophy of medicine must be divorced from any such attempt to indoctrinate into a particular religion. Philosophical counseling, on the other hand, is contradictory to any supernatural, fixed or dogmatic system or religion. "The secularization of Western societies delegitimated the role of religious ministries and theologians in providing. . .values" [84]. This seems, however, not to be true. Religion still dictates "morality." Engelhardt stresses the importance of philosophy in bioethics: "The philosophy of medicine in bioethics offered the needed direction for a socially central institution (i.e. medicine) that had grown secular and more democratic" [85]. Accordingly, philosophical counseling and Ethics Committees should not be for the promotion of religion. "Ethics committees and consultants can. . .be. . .producing more harm than good. . .. whenever such entities (a) see themselves constituted to enforce a particular religious point of view, (b) allow themselves to be co-opted by the institution in which they work" [86].

6.10 Humanism Versus Religion

The religious characterization of people, their nature, rights and fate is often not a kind one. The concept of hell is that because of one's sins, e.g., disobeying the rules of the Church, one is *eternally tortured* – a concept more vicious perhaps than any other imagined by humankind. Some religions are more pacifistic, e.g., Quakers, Buddhists (at least in theory), etc. The Catholic Church can be credited with at least requiring certain conditions to be met before going to war ("just war arguments"). On the other hand, if we do not think we will live forever as the Christians do, we are encouraged to regard life as more precious, encouraged to be more humanistic and care for people in this life.

We must ask if the Church kills by the following: 1. letting die by opposing stem cell research, 2. teaching that there is eternal suffering in hell, 3. supporting religious and other wars, 4. by not supporting human and bodily life, but rather soul and spiritual life, 5. by opposing science, rationality, and critical thinking (speaking). Religion in the Middle Ages prosecuted and killed those who had knowledge, midwives, researchers, and free inquirers. Medicine was thought to be interference with God's plans for humans, 6. also the Church held the view for centuries that God makes one ill either to punish one for one's sins, challenge one's faith, or to make one suffer as proof one is a good Christian (Hiob in the Bible). This is an outrageous attack on ill people and those taking care of them in a humanistic way. It makes ill people into social outcasts: "They must be sinners otherwise they would not be ill. The same is done to parents of the disabled children."

"The Italian debate on bioethics is marked by the opposition between Catholic and secular bioethics" [87]. The documents of the Church *Humanae Vitae*, a July 25 1968 encyclical letter of Pope Paul VI, which prohibits also artificial methods of contraception) and *Donum Vitae* 1987 give the Catholic position.

According to Rendtorff and Kemp [88] Norway is characterized by deontological Lutheranianism.

"The pluralism of Belgian society is opposed to the Catholic Church." "Demand for pluralism must also be seen as a reaction to the dominance of the Church" [89]. It is a move towards "confrontation."

Bioethics in Austria is intensely influenced by Catholic and religious views so that, for example, embryonic stem cell research is opposed [90].

6.11 Absolute Religious Ethics Versus Consequentialism

On a humanistic, naturalistic system of ethics, deliberate, rational, informed human wants are met and the consequences to bring about the wants of all humans are carefully considered. Such a system of ethics as that of John Dewey may be characterized as stated in the Chapter 5, as follows: Ethics is to bring about our (a) informed, (b) human wants and likes (c) deliberately (d) on the basis of inquiry (e) with as adequate and full consideration (f) as reasonably possible (g) of naturalistic and global consequences for everyone including concern for animals and nature. In a word, it is humanism or human and natural ecology on a world-wide, all-encompassing level which includes the knowledge found in the various natural and social sciences including philosophy, aesthetics, agriculture, astronomy, economics, mathematics, sociology, medicine etc. It is in this sense that we may speak of an adequate, holistic ethics. Ethics in medicine would refer to a rational, holistic consequentialism. On this view, a religious or non-religious, supernaturalistic belief, which does not consider consequences and defeats naturalistic goals and wants, such as e.g. medical research, is unethical. Religion is not holistic medicine, holistic philosophy, or holistic ethics. Religion means by "holistic," holy and spiritual medicine. The humanist means medicine, which is complete and adequate in terms of the whole person. It also stresses his/her environment and the prevention of harm and disease. Typically, religion simply appropriates all terms to their religious agenda.

On the view of the humanist-pragmatist, John Dewey, all supernaturalism is unethical. It may support a religion, but does not qualify as an ethical system.

6.12 Case Example: Deprogramming Religion in Medicine

A 41-year-old woman made an appointment with me, the author in my position as head of the fertility department. She had a desire for a child for more than 12 years. She was physically examined to determine why for many years she could not conceive. All examinations showed that there were no physical problems either with her or with her husband. No physical reason for infertility could be found. Their sexual intercourse was regular. As our department has a holistic and psychosomatic orientation we discussed other reasons why she thought she might not be able to conceive. It was soon revealed that when she was 17 years old she became pregnant. The man was too young and immature to deal with the consequences and immediately left her. Being alone with the problem she was exposed only to the views of her religious mother. The latter feared what she viewed as the shame for herself and her daughter, more than fear of the inevitable punishment of God, and thus had her daughter abort. The mother thereby not only shamed the daughter, she violated her religion. The abortion was performed with insufficient anesthesia resulting in extremely psychological and physical suffering. She received no counseling and so kept the traumatic experience within. Her only thoughts and emotions concerned her mother's rejection and threats of God's punishment, and the religious view that abortion was seen as murder and interference with God's creation. She thought and felt confused, ashamed and guilty. She thought that she should be punished. She said, "I never told that to anyone, not even my husband. You, a physician of a fertility department, are the first person that I have told. Perhaps it was because of your nonjudgmental acceptance and openness that I could talk about it." I was compassionate and told her about my medical and ethical view about abortion, about embryos as clusters of cells and about the cruelty she had to experience because of the dogmatic and anti-humanistic religious beliefs imposed on her. Having suffered with this for more than 20 years the woman began showing signs of relief. She was at last able to speak of the worst experience of her life. The counseling continued regularly with a psychologist. She made good progress in successfully dealing with her former experience. At the age of 43 she was able to get pregnant, but it resulted in a miscarriage. At this age such is not uncommon. Now the woman was able to accept it as for what it was: miscarriage because of age and genetics, and given a rational medical reason. God's punishment had now lost its influence on her as a result of the effective deprogramming and counseling by the department psychologist. The woman remained childless, but was nevertheless able to live a fulfilled life without the harmful influence of religion.

6.13 Case Example: A Real Woman

A married woman 37 years old underwent a Fallopian tube ligation a year ago. She did not wish to have more children. She is a Catholic farmer. Her gynecologist admitted her to the hospital because of depression in connection with tubal ligation. She was weeping, irritated, and experienced loss of sexual desire. She also experienced an identity crisis as a woman and increasing problems in her partnership. Her first child was born before her present marriage. The father was another man. She works hard on the farm, and takes little time for herself or the creative formation of the relationship with her partner. Her religious belief system "told her" that taking a contraceptive step such as tubal ligation interferes with God's plans. She said, "No wonder that I am depressed, having acted against God's will." She wanted now to have surgical re-fertilization or an IVF procedure to feel again like a real woman, a real wife.

The procedures she wanted were not to be paid for by insurance. The physician's advice was for her and her partner to first obtain therapy to "evaluate their thinking, emotions and partnership." This was done. After 6 months their thinking about the situation and their religious beliefs were altered and they adjusted well and happily to their present situation without having to have additional unnecessary operations and treatment.

6.14 The Person as a Soul

On one Christian religious view, to be a person is to be a soul, a non-natural "thing," a spirit. One becomes a supernatural substitute for a natural human. In this sense, we renounce mortal life, are dead as a natural person. We are God's property and so have no genuine choices or morality of our own [91].

On the ordinary language philosophy approach the language-game of soul as describing an entity leaves everything to be desired. But we can on examination of the language see that soul is an expression of a desire for "eternal" life. It is a desire, not a description of an entity. It may also be used to refer to that which is more than body. Now, what we do not know does not create an entity either. The religious landscape changes a bit [92].

If one is basically a soul, medicine is not needed or relevant. Medicine was regarded as *contra naturam* (against nature and the Church). "Nature" is used in a theological, not a scientific sense here.

6.15 Sanctity-of-Life (Human)

> The sanctity-of-life doctrine is the theoretical bedrock of medical ethics and the law [93].

She could rather have said it is the theological rather than theoretical bedrock of medical ethics. If the religious person genuinely believed in the sanctity-of-life doctrine s/he would be against all war and be a pacifist. Sanctity-of-life is not a primary principle for Christians as it is second after worship of God [94]. Thus, if God condones a religious war it overrides the sanctity principle. The sanctity-of-life principle is absolutistic rather than consequentialistic. If applied to the secular it means never abandon treatment regardless of quality of life, costs, consequences and conditions [95]. But treatment, costs, and consequences are secular, not religious terms. One may not conclude from the religious to the secular [96]. Perhaps if natural life is sacred we should create all the naturalistic life we possibly can. In regard to sanctity, sin is even inherited in the Catholic faith as original sin.

The sanctity of life doctrine is not a biological assertion. The religious concern is not for natural life, but for the soul and supernatural world. Thus, the statement should read, "The soul or spirit is sacred," or "Sanctity of soul." "Sanctity" is a religious term, so would not be meaningful to those who are not religious. In this sense, such abstract words beyond nature and intelligibility are not relevant to life or us. Similarly, if something is valuable-in-itself it has no relations or relevance to humans.

Kuhse gives the term "sanctity" an entirely different metaphorical and naturalistic meaning: "Human life has sanctity because human beings are rational, purposeful, moral beings, with hopes, ambitions, preferences, life purposes, ideals, etc" [97]. Here, like Dewey, she is giving naturalistic meanings to religious terms. It is not just the irrational and supernatural that can claim to be sanctified. Even if there were somehow metaphorical sanctity-of-life one would still have to balance it against the metaphorical sanctity of the social, mental and physical life of humans and the sanctity of moral behavior, humanism and rational, consequentialistic thinking. But, on the other hand, rational consequentialistic thinking does not need to be sanctified. It is an equivocation and mistake to try to sanctify medicine and science. The latter are instead accessible to observation and scientific examination.

6.16 General Observations Regarding the Value of Human Life

1. "Humans *should not be killed* because of sanctity of human life." "Sanctity" means inviolable, not to be treated badly. Again, this is circular. And again, it is not a *religious* contradiction to advocate killing. The deity set it up so that all people die. A deity can command anything. The religious can assert that as religion is not rational they can be as contradictory as they wish.
2. If life has sanctity, why do people kill in war? Around 90% of Americans were behind the Gulf war (and most of them being religious) killing of 1.5 million sanctified/sacred/secular people and many nations supported it or let it happen. The United States in Fall 2004 reelected a war president to continue the wars around the world.
3. Beauchamp and Childress grounded bioethics on four principles: autonomy, non-malfeasance, beneficence and justice rather than on a philosophy or ethics. Childress was a religious ethicist and deontologist so it was perhaps a way of trying to combine religion with secular thinking through principles. But autonomy is not good in-itself, and to say that "good" is beneficence (good making) or non-malfeasance (not bad making) is circular [98]. As "justice" is a value term, it is also circular to base "right on justice (right)" and in addition it is "open context".
4. According to the Christian Medical and Dental Associations Ethics Statement, "The great value of human religious life transcends that of the quality of life." (cmdahome.org) Religion always transcends the secular.

6.17 Contradictions Regarding the Sanctity-of-Life Doctrine

If people are sacred (have sanctity-of-life):

1. Why then are people sent to hell on the basis of their behavior?
2. Appeal to authority or god is not a defensible position. If it is, then one can proclaim oneself as the authority [99].
3. We cannot kill, but to let die often is of no concern also for the religious people. (This is implicitly to make a quality of life decision.) Sanctity seems to reach only a short distance.
4. Sacred (sanctity) is not natural or human, but supernatural. This life is disvalued in favor of the next one. Then the sooner one goes to the next life in heaven the more sanctified one will be. For Hamlet, even if there were an afterlife it may be much worse than anything in life.
   If the next "life" is so good why should one wait for it? There are only post-mortem rewards. To give naturalistic rational claims of the afterlife is curious. Religiously each life is equal to every other because none have naturalistic value. If one spiritually lives forever, medicine is not needed.
5. "Life is sacred" is like "life is supernatural, unintelligible, without intelligible morals, dogmatic, absolutistic." On the medical and scientific view, nothing is sacred about human life. It is more valuable than that. Rather, we can choose to practice naturalistic medicine to protect the natural human body, which is of the utmost value. On a naturalistic theory of ethics what is more important is consequences considering all of the factors of an ethical issue, not a mere dogmatic ascription of "sacred."
7. "The *sanctity* of life excludes *quality* of life criteria." This can lead to utmost cruelty in terms of consequences.
8. Innocent people should not be killed, but people are not regarded as being innocent by the Church, but rather as being guilty and sinners. What then?
9. If sanctity refers to holiness, or godliness, "sanctity-of-life" is like saying "Supernaturalness of nature," or "Something is what it is naturally not," which is a contradiction.
10. Human life is "infinitely valuable." Infinity, which is an unscientific term, does not apply. If value is undefined we cannot have a little of it or an infinite amount of it. Nor can we intelligently speak of infinite nothing or infinite something or other [100]. Similarly, "absolutely valuable" makes no sense. To merely say life or anything else has a value is not to assert anything. It may serve as a persuasive definition the purpose of which is only to encourage someone else to have an emotional attitude about something: e.g., "I oppose it, so should you," though no meaning or reason is given for doing so. The reversal of the question would make more sense: Instead of "Does life have value?" we may ask, "Does value have life?" Does value have any meaning here? Furthermore, a person or thing does not *have* meaning. Only people can give meaning. The sentence "x has meaning or value," is therefore misleading. We can rephrase it as, "It is we who give ourselves, life, etc., meaning or value." It is we, then,

who are responsible for the value of people's lives. The supernatural takes no responsibility for claiming or determining that something is valuable. Medicine cannot be left in its hands. On such a view, there is no autonomy at all for the individual, but rather choice is dictated by representatives of the various religions who would also dictate medical policy.

11. If there is naturalistic "sanctity-of-life" the irreversibly comatose would have to be kept alive as long as is possible regardless of consequences and whatever the hardship to himself or other people.
12. "Immortal life would increase boredom" [101]. If so, it perhaps also attacks the religious view of eternal life, which may also be boring.

6.18 Selected Arguments from the Philosophy of Religion

Religion is the ultimate irrationality [102].
Believing what we don't believe/ does not exhilarate [103].

The religious are not concerned about lack of rationality or rational arguments, philosophy of religion (which most have never heard of or read a book on), negative consequences of religion, scientific investigation, or evidence. There is rather pride of faith and belief without evidence. This has generated a number of arguments against religion in the philosophy of religion such as the following fallacies presented in this section.

What is meant by the philosophy of religion? The well-known philosopher-humanist, Bertrand Russell, suggested that we: "Apply solvent criticism especially to the beliefs that we find it most painful to doubt, and to those most likely to involve us in violent conflict with [people] who hold opposite...beliefs" [104].

The Oxford philosopher, John Wilson, put it this way: "Another way of defending ourselves against thinking – is to say things like 'Reason can only get you so far; after that you have to make the leap of faith,' or 'You have to rely on intuition.'...To be willing to give reasons, to have your beliefs out in public, to allow them to be inspected and challenged, is essential for all kinds of thinking" [105]. One test of rationality, then, is being open to criticism and being able to change our thinking on that basis.

Theological and metaphysical arguments are now generally regarded by philosophers as misuses of language and based on informal logical fallacies.

The informal fallacies and the Metaphorical Method have been used to classify these limitations of religion. E.g. the following:

False Reason. 1. All disease is caused by sin. 2. "'We ought to do what God wills because God will punish us if we do not obey him'...is hardly a morally good reason for doing what he commands since such consideration of self-interest cannot be an adequate basis for morality" [106].

Hypocrisy. "Acknowledgment that we do not know what we do not know is a necessity of all intellectual integrity" [107]. "A lot of words, like 'faith,' or 'revelation,' or 'intuition,' are used to cover up this idea [that believing alone makes something true], which in its naked form is obviously silly" [108]. Hypocrisy is also shown in that 75% of Catholics thought abortion should be available although it is explicitly against Catholic teaching (1998 *NY Times*/CBS News Poll).

Ignorance Humor. "Religion is regarded by the common people as true, by the wise as false, and by the rulers as useful" (Seneca).

Reversal. "Religious teachings...[are] neurotic relics: the universal obsessional neurosis of humanity" [109]. Sciences reduce the unknown to the known; religion reduces the known to the unknown. God causes disease (by opposing medical practice and research), and disease causes God (by fear). Instead of saying, "God bless you," one may say, "May reason and humanity guide you."

Simile. "Man makes religion.... it is the opium of the people" [110].

Blatant Vice "The church...not only strove against the dawning and rising science as false, but it called this science impious and anti-Christian" [111]. "'Faith' means not wanting to know what is true" [112].

"I say quite deliberately that the Christian religion, as organized in its churches, has been and still is the principle enemy of ethical progress in the world" [113].

6.19 Prayer as Medical Treatment

It has been shown above that and how religion has tried to reduce medical treatment and medical ethics to religious morals, which we have called here medicotheology and biotheology. Medical terms and methods are replaced by religious. Recently in medicine there has been stress on evidence-based medicine (EBM) as a scientific method (See critique in the Chapter 19). The U.S. Government ca. 30 billion dollar tax-supported National Library of Medicine NIH was shown to give a religious basis to medicine, for example, by stressing complementary and alternative medicine (CAM) including faith-based medicine. They have therefore recently tried to show that scientific EBM movement gives evidence for prayer as a medical treatment [114]. The prayer studies promote the idea that physicians should spread Christianity to their patients. Religion and prayer studies lead one to reject medical treatment. The prayer issue may serve as a paradigm to show how religion appropriates medicine. (See also "Evidence-Based Spirituality" in Chapter 19.)

"There is little evidence to support claims that health benefits from religious activity" [115]. From Sloan, Bagiella and Powell and Bagiella, Hong and Sloan [116] the following sorts of criticisms of evidence for faith healing are presented. The variables "religion" and "spirituality" were not defined or specified. The various studies have defined religion in operationally very different ways, e.g., as watching religious TV, prayer, etc. Positive spirituality was defined as: honesty, love, joy, peace, hope, patience, generosity, forgiveness, thankfulness, kindness, gentleness, goodness, understanding, and compassion. But not one of these would need to characterize spirituality and none of them seem to. Spirituality refers to the supernatural and mystical, certainly not to understanding. But if it is defined as stated and if there is a significant correlation it would only show that positive emotions promote health, which is unsurprising. Spiritual wellbeing is also defined as goal-oriented knowing what you want and striving to get it. This also characterizes Dewey's naturalistic, anti-supernatural ethics.

The following is a presentation and critique of some of the literature on prayer as healing.

One of the main sources for EBM Controlled Trials and Systematic Reviews is Cochrane Collaboration Library. (cochrane.org) In 2006 it listed ca. 8,000 EBM entries on prayer, 76,000 on religion. "Philosophy of Medicine" yielded less than 900 entries. Clearly, religion and prayer are regarded as being more to important scientific medicine than the whole area of the philosophy of medicine.

Prayer is usually not defined, it may be a cultural and religious myth or metaphor people are enculturated into.

Prayer may have a placebo effect in that if one believes that by praying, whatever that is, one will improve one's health it may help one to do so. (See Chapter 19 for an analysis of placebo) The cognitive can psychosomatically affect the body in positive and negative ways as illustrated by the analysis of emotions. (See Chapters 7 and 9). One problem with the placebo effect is that the belief may be irrational or false.

What is prayer and how is it supposed to heal? Prayer differs from religion to religion. It also takes many forms. One may utter different statements all of which can constitute a prayer. One can also pray for many different things. Prayer is usually regarded as a communication with a deity. It is something else if the religion has no deity, but it will involve the supernatural. To expect science to give evidence for the supernatural is a contradiction in terms. Prayer cannot by definition be evidenced-based in science. Evidence does not apply to the spiritual. This is pointed out in the literature e.g. by Cohen [117]. One main assumption is that God is all-powerful and can cause all things, including miracles, to happen. It also assumes that the deity will listen to and honor human requests. Used metaphorically, prayer can also mean a strong naturalistic wish or hope. Another view of prayer is just to use it as a way to "become closer to the deity," whatever that may mean. Now prayer is based on personification and anthropomorphism that a God can "hear" us and "talk" to us. It is a violation of ordinary language. We know what it is for a person to answer us, but what is it for God to answer prayers?

"The religious are not to suppose that God is at one's disposal." We cannot tell God what to do. This view would be one reason why requests may not be answered. The authors missed the point that the supernatural cannot relate to the natural human world of requests and answers. A supernatural request is not a real, naturalistic request. Or put differently, a supernatural request can only give a supernatural answer.

If prayer has a placebo effect, it may also have a nocebo effect as well, that is worsen the situation. Dr. Richard Sloan of Columbia Presbyterian Medical Center said, "By suggesting that religious activity promotes health, you also imply the converse, which is that bad health is associated with insufficient devotion and insufficient faith" [114].

Religion is a cause of stress: fear of hell, sin, God's wrath, fear of disobedience, guilt, blame, and inferiority. Is religion a disease needing healing? Some philosophers would certainly think so, e.g., Friedrich Nietzsche, John Dewey, etc. Koenig found that 16% of the reviews in psychiatric journals showed religion to be harmful [118].

What is left out of the authors' presentation is the most essential level of inquiry, the philosophy of religion. The critique of the concepts and methods of the various religions is totally absent from their account, which invalidates their discussion. They not only reject EBM on religious grounds, they reject the philosophy of medicine as well by omission. No criticism of religion is to be presented or allowed at all by them or to be allowed by the healthcare worker. This then is a form of indoctrination into the prevailing culture and religion.

Andrew Newberg, a neurologist, identified neurological areas of the brain that light up during prayer [119].

The following EBM trial found that remote prayer helps coronary outcomes. William Harris reported a "randomized, controlled trial of the effects of remote, intercessory prayer on outcomes in patients admitted to the coronary care unit" [120]. Nine hundred ninety consecutive patients were newly admitted to the coronary care unit (CCU) and were randomized to receive remote, intercessory prayer (prayer group daily for 4 weeks). Patients were unaware that they were being prayed for, and the intercessors did not know and never met the patients. Conclusions: Remote, intercessory prayer was associated with lower coronary care unit course scores. This result was said to suggest that prayer may be an effective adjunct to standard medical care.

Annie Bayne (2002) of the Center for Advancement for Health reported that claims that religious activity provides health benefits have virtually no grounding in the medical literature, according to an article by Richard Sloan, professor of behavioral medicine at Columbia University, and others [114]. "Belief in the health benefits of religious and spiritual activities is so widespread that many think these activities should be incorporated into clinical practice."

Charatan states that 172 children died in a 20-year period because parents did not allow medical care because it was against their faith [121]. Colorado General Assembly and Church of the First Born denied children medical treatment because illness was believed to be cured by prayer. It is based on Lord's will to "raise you up." Church of Christ Christian Scientist groups lobbied so that now 45 states have statutes allowing parents to withhold medical treatment of their children for religious reasons. Not even child abuse laws protect the child this way. In the Child Abuse Prevention and Treatment Act of 1996 no medical treatment need be given if it is against the beliefs of the parents. Thus, one can withhold treatment for religious reasons, but not for other, e.g. quality of life or medical reasons.

References

1. Hippocrates, tr. Jones, W. 1957. *Hippocrates precepts*, vol 2. Cambridge: Harvard University Press.
2. David Hume, in Benatar, D. ed. 2004. *Life, death, and meaning: key philosophical readings on the big questions*, 289. Lanham, MD: Rowman and Littlefield.
3. New York Times Almanac. 2006. Encyclopedia Britannica Almanac.
4. Maguire, D. 2001. Sacred choices: *The right to contraception and abortion in ten world religions*, 45. Mineapolis, MN: Fortress Press.
5. Brody, H. 1987. *Stories of sickness*. New Haven, CT: Yale University Press.
6. Hope, T., Savulescu, J., and Hendrick, J. 2003. *Medical ethics and law. The core curriculum*, 122. London: Churchill Livingstone/Elsevier.
7. Shusterman, R. 1997. *Practicing philosophy: Pragmatism and the philosophical life*, 198. New York: Routledge.
8. Hehir, B. 1996. Bioethics and natural law: The relationship in Catholic teaching. *Kennedy Institute of Ethics Journal* 6:333–336.
9. Rendtorff, J., and Kemp, P. 2000. *Basic ethical principles in European bioethics and biolaw*, 366. Copenhagen, Barcelona: Centre for Bioethics and Law.
10. Mitka, M. 2006 NIH Budget Slips. JAMA 295: 617.
11. Callahan, D. 2003. *What price better health? Hazards of the research imperative*, 251. Berkeley, CA: University of California Press.
12. Miles, S. 2004. *The Hippocratic oath and the ethics of medicine*. New York: Oxford University Press.
13. Macklin, R. 2006. The new conservatives in bioethics: Who are they and what do they seek? *Hastings Center Report* 36: 34–43, esp. 34.
14. Macklin, R. 2006. The new conservatives in bioethics: who are they and what do they seek? Hastings Center Report 36:34–43, esp. 42.
15. Turner, L. 2004. Bioethics in pluralistic societies. *Medicine, Health Care and Philosophy* 7:201–208, esp. 201.
16. Turner, L. 2004. Bioethics in pluralistic societies. *Medicine, Health Care and Philosophy* 7:201–208, esp. 203–204.
17. Turner, L. 2004. Bioethics in pluralistic societies. *Medicine, Health Care and Philosophy* 7:201–208, esp. 207.
18. Turner, L. 2004. Bioethics in pluralistic societies. *Medicine, Health Care and Philosophy* 7:201–208, esp. 205.
19. Kampits, P. 2004. Die Autonomie des Einzelnen. Die Presse, July 23, 2004.
20. Jonsen, A. 1998. *The birth of bioethics*, 41. New York: Oxford University Press.
21. Jonsen, A. 1998. *The birth of bioethics*, 48–55. New York: Oxford University Press.
22. Rendtorff, J., and Kemp, P. 2000. *Basic ethical principles in European bioethics and biolaw*, 154. Barcelona: Centre for Bioethics and Law, Copenhagen.
23. Rendtorff, J., and Kemp, P. 2000. *Basic ethical principles in European bioethics and biolaw*, 157. Copenhagen, Barcelona: Centre for Bioethics and Law.
24. Rendtorff, J., and Kemp, P. 2000. *Basic ethical principles in European bioethics and biolaw*, 201. Copenhagen, Barcelona: Centre for Bioethics and Law.
25. Rendtorff, J., and Kemp, P. 2000. *Basic ethical principles in European bioethics and biolaw*, 202. Copenhagen, Barcelona: Centre for Bioethics and Law.
26. Rendtorff, J., Kemp, P. 2000. *Basic ethical principles in European bioethics and biolaw*, 217. Copenhagen, Barcelona: Centre for Bioethics and Law.
27. Singer, P. 2004. *The president of good and evil*, 91. New York: Penguin.
28. Singer, P. 2004. *The president of good and evil*, 13. New York: Penguin.
29. Miles, S. 2004. *The Hippocratic oath and the ethics of medicine*, 3. New York: Oxford University Press.
30. Miles, S. 2004. *The Hippocratic oath and the ethics of medicine*, 18. New York: Oxford University Press.

31. Engelhardt, T. in Shelp, E. 1985. *Theology and bioethics*, 79–91, esp. 90. Dordrecht: Reidel.
32. Ross, J.W. 1986. *Handbook for hospital ethics committees*, 150. Chicago: American Hospital Publications.
33. Fletcher, J. 1979. *Humanhood. Essays in biomedical ethics*, 138. New York: Prometheus.
34. Jonsen, A., Siegler, M., and Winsdale, W. 1998, orig.1982. *Clinical ethics*, 4th edn, 64. New York: McGraw-Hill.
35. Jonsen, A., Siegler, M., and Winsdale, W. 1998, orig.1982. *Clinical ethics*, 4th edn, 95 ff. New York: McGraw-Hill.
36. Jonsen, A., Siegler, M., and Winsdale, W. 1998, orig.1982. *Clinical ethics*, 4th edn, 75. New York: McGraw-Hill.
37. Jonsen, A., Siegler, M., and Winsdale, W. 1998, orig.1982. *Clinical ethics*, 4th edn, 202. New York: McGraw-Hill.
38. Ross, J.W. 1986. *Handbook for hospital ethics committees*, 151. Chicago: American Hospital Publications.
39. Brown, B., and Singer, P. 1996. *The greens*, 166. Melbourne: Text Publications.
40. Tronto, J. 1993. *Moral boundaries: A political argument for an ethic of care*, 103. New York: Routledge.
41. Singer, P. 2002. *One world: The ethics of globalization*, 230. New Haven, CT: Yale University Press.
42. Ratcliffe, J. ed. 1981. *The good Samaritan and the law*, 276–277. Gloucester, MA: Peter Smith.
43. Singer, P. 2004. *The president of good and evil*, 41. New York: Penguin.
44. Fletcher, J. ed. 1995. *Introduction to clinical ethics*, 264. Frederick, MD: University Publishing Group.
45. Brody, H. 1987. *Stories of sickness*. New Haven, CT: Yale University Press.
46. Dawkins, R. 2006. *The god illusion*, 284. New York: Houghton Mifflin.
47. Dawkins, R. 2006. *The god illusion*, 314. New York: Houghton Mifflin.
48. Engelhardt HT. 1991. *Bioethics and secular humanism. The search for a common morality*, 16–17. London: SCM Press
49. Noddings, N. in Gordon, S., Benner, P., and Noddings, N. 1996. *Care-giving: Readings in knowledge, practice, ethics and politics*, 165. Philadelphia, PA: University of Pennsylvania Press.
50. Savulescu, J. 1998. Two worlds apart: Religion and ethics. *Journal of Medical Ethics* 24:382–384, esp. 383.
51. Sullivan, A. 2005. "The vicar of orthodoxy", Time 165, May 22, 2005:49.
52. Hamel, R. 1993. Speaking of god: must theology remain silent in bioethics and public debate? *Second Opinion* 18:83 ff.
53. Wittgenstein, L. 1958/1953. *Philosophical investigations*, 3rd edn. New York: Macmillan. Anscombe, E. tr. and revised 2002, #116. Cambridge, MA: Blackwell Publishers.
54. Schweitzer, A. 1990. Kultur und Ethik. Beck, München: 325 *Der Inbegriff des Seins, das Absolute, der Weltgeist, und alle derartigen Ausdrücke bezeichnen nichts Wirkliches, sondern etwas in Abstraktionen Erdachtes, das deswegen auch absolut unvorstellbar ist.* authors' translation.
55. Dewey, J. 1934. *A common faith*, 13. New Haven, CT: Yale University Press.
56. Dewey, J. 1934. *A common faith*, 9. New Haven, CT: Yale University Press.
57. Dawkins, R. 2006. *The god illusion*, 15. New York: Houghton Mifflin.
58. Dewey, J. 1934. *A common faith*, 10. New Haven, CT: Yale University Press.
59. Dewey, J. 1934. *A common faith*, 79. New Haven, CT: Yale University Press.
60. Dewey, J. 1934. *A common faith*, 16. New Haven, CT: Yale University Press.
61. Dewey, J. 1934. *A common faith*, 43. New Haven, CT: Yale University Press.
62. Dewey, J. 1934. *A common faith*, 14. New Haven, CT: Yale University Press.
63. Dewey, J. 1934. *A common faith*, 20. New Haven, CT: Yale University Press.

64. Dewey, J. 1934. *A common faith*, 21. New Haven, CT: Yale University Press.
65. Dewey, J. 1934. *A common faith*, 80. New Haven, CT: Yale University Press.
66. Dewey, J. 1934. *A common faith*, 14–16. New Haven, CT: Yale University Press.
67. Dewey, J. 1934. *A common faith*, 45. New Haven, CT: Yale University Press.
68. Dewey, J. 1934. *A common faith*, 54. New Haven, CT: Yale University Press.
69. Dewey, J. 1934. *A common faith*, 25. New Haven, CT: Yale University Press.
70. Wittgenstein, L. 1958/1953. *Philosophical investigations*, 3rd edn. New York: Macmillan. Anscombe, E. tr. and revised 2002. Cambridge, MA: Blackwell Publishers.
71. Smith, W. 2000. *Culture of death: The assault on medical ethics in America*, 7. San Francisco, CA: Encounter Books.
72. Steinbock, B., Arras, J., and London, A. 2003. *Ethical issues in modern medicine*, 6 th edn, 20. New York: McGraw-Hill.
73. Singer, P. 1994. *Rethinking life and death: The collapse of traditional values*, 62 f. New York: St. Martin's.
74. Parfit, D. 1984. *Reasons and persons*, 453–454. Oxford: Oxford University Press.
75. Savulescu, J. 1998. Two worlds apart: Religion and ethics. *Journal of Medical Ethics* 24:382–384; esp. 383.
76. Noddings, N. 2002. *Educating moral people*, 117. New York: Teachers College Press, Columbia University.
77. Noddings, N. 1993. *Educating for intelligent belief or unbelief*, 143. New York: Teachers College Press, Columbia University.
78. cf. Haught, J. 1990. *Holy horrors*. Buffalo, New York: Prometheus Press; Wilson, A.N. 1991. *Against religion*. London: Chatto and Winders.
79. Noddings, N. 1993. *Educating for intelligent belief or unbelief*, 53–61. New York: Teachers College Press, Columbia University.
80. Kelly, G. 1958. *Medico-moral problems*, 34. St Louis, MO: Catholic Hospital association.
81. Preece, G. 2002. *Rethinking Peter Singer: A Christian critique*, 15. Downers Grove, IL: Inter Varsity Press.
82. Preece, G. 2002. *Rethinking Peter Singer: A Christian critique*, 16. Downers Grove, IL: Inter Varsity Press.
83. Smith, W. 2000. *Culture of death: The assault on medical ethics in America*, 9. San Francisco, CA: Encounter Books.
84. Engelhardt, H.T. 2002. The ordination of bioethicists as secular moral experts. *Social Philosophy and Policy* 19:59–82, esp. 63, 74.
85. Engelhardt, H.T.. 2002. The ordination of bioethicists as secular moral experts. *Social Philosophy and Policy* 19:59–82, esp. 77 f.
86. Nussbaum, M., and Sen, A. eds. 1993. *Quality of life*, 51 f. Oxford: Clarendon Press.
87. Rendtorff, J., and Kemp, P. 2000. *Basic ethical principles in European bioethics and biolaw*, vol I, 201. Copenhagen, Barcelona: Centre for Bioethics and Law.
88. Rendtorff, J., and Kemp, P. 2000. *Basic ethical principles in European bioethics and biolaw*, vol I, 217; 154. Copenhagen, Barcelona: Centre for Bioethics and Law.
89. Rendtorff, J., and Kemp, P. 2000. *Basic ethical principles in European bioethics and biolaw*, vol I, 157. Copenhagen, Barcelona: Centre for Bioethics and Law.
90. Research on Human Embryonic Stem Cells. 2009. Opinion of the Austrian Bioethics Commission, March 16th, 2009 http://www.bka.gv.at/DocView.axd?Cobld=34240 accessed October 1st, 2009.
91. Kuhse, H., and Singer, P. 1994. *Individuals, humans, persons: Questions of life and death*, 54. Salzburg: Sankt Augustin Academia Verlag.
92. cf Shibles, W. 1972. Review of DZ Philip's Death and immortality. *Southern Journal of Philosophy* 10:391–394.
93. Kuhse, H., and Singer, P. 1994. *Individuals, humans, persons: Questions of life and death*, 51. Salzburg: Sankt Augustin Academia Verlag.
94. Preece, G. 2002. *Rethinking Peter Singer: A Christian critique*, 175. Downers Grove, IL: Inter Varsity Press.

95. Jonsen, A., Siegler, M., and Winsdale, W. orig.1982/1998. *Clinical ethics*, 109, 4th edn. New York: McGraw-Hill.
96. Schwartz, L., Preece, P., and Hendry, R. 2002. *Medical ethics: A case-based approach*, 115. Edinburgh: Saunders.
97. Kuhse, H., and Singer, P. 1994. *Individuals, humans, persons: Questions of life and death*, 490. Salzburg: Sankt Augustin Academia Verlag.
98. Beauchamp, T., and Childress, J. 1979–2001/1994. *Principles of biomedical ethics*, 4th edn. Oxford: Oxford University Press.
99. Kuhse, H., and Singer, P. 1994. *Individuals, humans, persons: Questions of life and death*. Salzburg: Sankt Augustin Academia Verlag.
100. Kuhse, H., and Singer, P. 1994. *Individuals, humans, persons: Questions of life and death*, 48. Salzburg: Sankt Augustin Academia Verlag.
101. Rendtorff, J., and Kemp, P. 2000. *Basic ethical principles in European bioethics and biolaw*, vol I, 50. Copenhagen, Barcelona: Centre for Bioethics and Law.
102. Whitehead, A.N. 1938. *Science and the modern world*, 208. New York: Penguin.
103. Dickinson, E. 1955/1958/1979. In *The poems of Emily Dickinson*, 3 vols, 741, L 3–4. ed. Johnson, T. Cambridge, MA: Harvard University Press.
104. Russell, B. 1957. *Why I am not a Christian*, 30. New York: Simon and Schuster.
105. Wilson, J. 1970. *Moral thinking*, 8. London: Heinemann.
106. Nielsen, K. 1973. *Ethics without god*, 4. New York: Prometheus.
107. Dewey, J. 1934. *A common faith*, 86. New Haven, CT: Yale University Press.
108. Wilson, J. 1968. *Philosophy*, 4. London: Heinemann.
109. Freud, S. 1964. *The future of an illusion*. Robson-Scott, W. tr, Chapter 8. Anchor Garden City, New York: Doubleday.
110. Kurtz, P. 1973. *A secular Humanist declaration*. Reprint from Free Inquiry 1, 1. 1980, 153.
111. Dewey, J. 1971. *The early works 1882–1898*, vol 4, 6. Carbondale, IL: Southern Illinois University.
112. Nietzsche, F. 1954. In *The portable Nietzsche*, 635. ed. Kaufmann, W. New York: Macmillan.
113. Russell, B. 1957. *Why I am not a Christian*, 21. New York: Simon and Schuster.
114. U.S. Department of Health and Health Services. nlm.nih.gov/tsd/serials/lji.html.
115. Sloan, R., and Bagiella, E. 2002. Claims about religious involvement and health outcomes. *Annals of Behavioral Medicine* 24:14–21; esp. 19.
116. Sloan, R., Bagiella, E., and Powell, T.1999. Religion, spirituality, and medicine. *Lancet* 353:664–667; Bagiella, E., Hong, V., and Sloan, R. 2005. Religious attendance as a predictor of survival in the EPESE cohorts. *International Journal of Epidemiology* 34:443–451.
117. Cohen, C. 2000. Prayer as therapy. *Hastings Center Report* 30:40–47.
118. Koenig, H. 2001. *Handbook of religion and health*. New York: Oxford University Press.
119. metanexus.net/tarp, and andrewnewberg.com/default.asp cf Newberg, A., dÀquili, E., and Rause, V. 2001. *Why god won't go away: Brain science and the biology of belief*. New York: Ballantine Publishing Group.
120. Harris, W. 1999. A randomized controlled trial of the effects of remote, intercessory prayer on outcomes in patients admitted to the Coronary care Unit. *Archives of Internal Medicine* 159:2273–2278.
121. Charatan, F. 2001. Bush's initiative could help groups that promote faith healing. *Western Journal of Medicine* 174:306.

Chapter 7
Emotion In Medicine

The dominant view in contemporary analysis of emotion is a cognitivist one. [1]

The real moral question is what kind of a self is being furthered and formed. [2]

All illnesses can be considered to involve the psychosomatic.

Abstract On the Cognitive Theory of Emotion an emotion is a cognition (assessment or evaluation), which causes bodily feeling. Emotion can be changed by changing the cognition. Negative emotions such as anger, revenge are due to faulty assessments such as failure to accept reality, failure to understand that we can only do that, which is within our power and a misuse of value terms. Emotion is not at all the sort of mentalistic thing that can be "released." It is on the basis of the cognitive theory of emotion that we may regard negative emotions as philosophy of language fallacies. Assessments are enculturated and so are emotions and in need of an ethical critique, especially in the area of medicine, for healthcare-workers as well as their patients.

Keywords Emotion · cognition · feeling · cognitive-emotive theory (CTE) · psychosomatics · mentalistic fallacy · negative emotion · apathy · anger · pity, self-pity

7.1 Introduction

To be a good physician, nurse, enlightened administrator, or healthcare worker, one must know how emotions work. Healthcare workers have also negative emotions as anyone else and there is often little staff collegiality. One cannot be an enlightened manager without a sound knowledge of emotion. Bad management is a symptom of emotional illiteracy and negative emotions (See Chapter 8). Having negative emotions is a lifestyle cause of illness and death [3].

Sotile and Sotile by their analysis about what to do with the angry physician show that they are not aware of emotion theory and philosophical clarification [4]. They say the most important thing is that emotions are contagious. What would this

B. Maier, W.A. Shibles, *The Philosophy and Practice of Medicine and Bioethics*, International Library of Ethics, Law, and the New Medicine 47,
DOI 10.1007/978-90-481-8867-3_7, © Springer Science+Business Media B.V. 2011

mean? Are they to be treated as diseases? They give no ways in which to clarify emotions or avoid or get rid of negative emotions. They speak of "Type A behavior pattern" which is aggressive coping, but the category is poorly defined. We may agree with them on the point that "addressing negative emotions is one of the most prevalent organizational and personal challenges facing physician leaders."

In the experience of one of the coauthors of this book, university students and others consistently estimate that 75% of their lives and the lives of others consist of negative emotions.

Buetow and Elwyn wrote, "Some patients have recurrently poor attitudes and behaviors" [5]. We can predict this on the basis of negative emotions reported. Lives consist of worries, fears, irritations, anger, revenge, envy, etc. We virtually never have education or training in emotions. It is too vague to use the phrase "behavior that deviates from the norm" because the norm may consist of negative emotions. To reduce "negative attitudes" in healthcare requires an analysis of emotions and emotion training. We are in any society enculturated and in this way emotionally dysfunctional. In the same way that clear, critical thinking (speaking) helps one overcome one's uncritical thinking (speaking) and enculturation, knowledge of emotion helps one overcome one's emotional dysfunction. Knowledge of emotions is empowering.

Emotion is not a feeling. It is a common belief that emotions are just feelings. It will be argued here that it is just what emotions are not.

According to the cognitive-emotive theory (CTE), emotion (E) is cognition (C), which causes bodily feelings (F). E = (C > F) Dictionary of symbols used:

= means "equals"
> = causes item on right
A = anger
C = cognition (actually an assessment)
BF = bodily feeling
NE = negative emotion
PE = positive emotion
≠ = does not does equal

In ordinary language, "emotion" and emotion words refer to both cognitions and feelings. This was pointed out earlier in a short article by Bedford arguing that emotion terms basically refer to cognitions and assessments [6]. Perkins later showed that emotions were not merely cognitive assessments, but they involve body feeling as well [7]. An analysis of emotion therefore requires an analysis of cognition, feeling (perception, sensation), and their relationships. Zoltán Kövecses also holds a cognitive theory. He states, "Anger is not just an amorphous feeling devoid of any conceptual content, but rather it has an elaborate cognitive structure" [8].

For example, anger (A) is not a mere bodily feeling: A ≠ F. Because emotions involve cognitions, it is always a mistake to say, "I feel angry." Ellis pointed out that it is more precise to say, "I think-feel angry" [9]. Bodily feeling (F) here can also refer to sensation and perception (hearing, seeing, tasting, etc.). If emotion were only a feeling it would not be intelligible. Anger is not like having pain, love not like a headache. Melden wrote: "There is no simple or single feeling one has such that feeling anger consists in having it and nothing else. Anger. . .cannot be an internal feeling or state conceptually unrelated to the functions of intelligence" [10]. It is a mistake for a therapist to just ask what one's feelings or emotions are. It is the cognitions, which are most important. Physical, neural and behavioural theories fail to tell how the emotion began. Roseman and Smith however, improperly speak of emotions as feelings, and with no explanation of feelings [11]. Feelings are thought to be irrational and the opposite of cognitions. Some people are thought to be emotional rather than rational. There is a romantic age versus age of reason. The distinction between emotion and reason breaks down. Caring and love are no more just bodily feelings than cognitive assessments. The emotive person also has cognitions.

Emotion is a Cognition Causing Bodily Feeling [$E = (C > F)$]. As far as one can see, there is, at present, no viable alternative to an appraisal. . .explanation for the general prediction of the elicitation and differentiation of emotions [12]. Scherer and coauthors have given support to the appraisal theory of emotion [12]. Events do not cause emotions, but rather only the interpretation of events. "Cognition" does not refer to mentalistic "ideas" or "thoughts." Instead, it refers here to non-mentalistic assessments. These are statements consisting of self-talk, utterances, and language-use. To a large extent our cognitions are our emotions. To see what an emotion is we need to know how one thinks and appraises. If we know that we can predict what emotions one will have [13]. Also reducing thought to language avoids the objection of making emotion too cognitive.

Thought is basically language-use. But spoken language has more potential for meaning than written language because it contains many intonations not able to be put in writing. These intonations are important in rendering connotative associations, which produce emotions. Thus emotions such as hope may be rendered by conversations, which could not be rendered by the written word. The words in novels and written narrative may be read and interpreted in diverse ways, as there is no intonation to complete the meaning. If the word "health" is used many quite different images and meanings may come up.

Roseman and Smith give no analysis of ethics or ethical terms, which form the basis of appraisals [11]. Kappas also does not present an analysis of ethical theories, but just says that appraisals are intuitive. This would mean that ethics is just intuitive. This is no rational understanding or basis of ethics [14].

Although emotions no longer exist as such, the letter E will be used to stand for $C > F$, where C is one or more specific assessment and F is a bodily feeling. The pseudo-opposition of "thought versus feeling" is often presented. "Rational aesthetic emotion" is not a contradiction. E can, however, involve rational or irrational cognition (verbal assessments).

Most think that there is a mind and there are ideas in us. This is a fallacy called mentalism. For Wittgenstein "Mental processes are just queer" [15]. For Gilbert Ryle: "The phrase 'in the mind' can and should always be dispensed with." "The mind is not a 'ghost' in the machine" [16]. W Quine repudiates mental entities as entities [17]. Mead states: "The locus of mind is not in the individual." [18] and "In immediate experience there is no mind" [19]. The bioethicist Howard Brody for example, committed a mentalistic fallacy when he wrote, "We know that we have minds" [20]. This also is rejected by recent scientific neurological findings [21]. The task of clarifying emotion becomes especially the determination of the kinds of cognitions (assessments) specifically involved in the experience. Phrased differently, it becomes an analysis of its diverse language-games.

7.2 Case Example: Non-mental Associations Provide Complexity to Cognitions

Smith and Kirby suggest that cognition does not necessarily occur only verbally and consciously, but by associative processing of various kinds, e.g., images, sensations, etc [22]. This is a useful observation also made earlier in Shibles' Association Theory of Meaning [23]. One difference is, their associative theory is mentalistic, Shibles' is not. Their terminology, for example, refers to "information stored in memory", [24] which is the classical fallacy of the refuted "container theory of mind." There is no spiritual memory container, and ideas are not things, and they are not stored in a memory warehouse. They also speak of "mental constructions", which is the view that we have a mind, another mentalistic fallacy [24]. The account of muscle-induced emotions is then explained on the association theory [25]. Muscle induced emotions is a physical and sensation reductionist fallacy, but the association theory yields a useful insight. Thus, the sight of one's dead husband's fishing gear can reduce one to tears. Associative connotations of an event can evoke a present emotion.

7.3 Emotion Is Not an Internal State

If mind, ideas, cognitions, and other alleged internal states do not exist, as was shown in the earlier Chapter 5, the status of emotion is also put in question. Perhaps emotions are not entities inside of us either? On the cognitive theory of emotion (CTE) view, they are not. Emotions are only assessments, which cause bodily feelings. Emotions do not exist as such. Having physical states, e.g., post-traumatic syndrome, compulsions, obsessions, etc., which are medical or the result of CTE is not the same as cognitions. The cause may seem to be physical but comes on gradually from assessments. The literature suggesting that there is a certain number of different emotions, four or fifty-four, is misguided. There are no emotions as such countable within us. Griffiths similarly states, "There is really no such thing as emotion." It is not a natural kind. It is not referential of a psychological state, but may refer to a practice [26]. Dewey states, *Experience is emotional but there are no separate things called emotions in it* [27]. This is especially true because there are no psychological states as such. For example, grief cannot be accurately described as "deep mental anguish." This is a mentalist fallacy. Rather, grief is often due to assessments of one's own loss, rather than concern with the death of the other. The assessment may involve death denial and so produce shock when confronted with actual death. Those who deal better with emotions (including grief) live longer. We can object to having to die, but there is no sound basis for fear of death. Anxiety and fear of death undermine happiness. Fear of death is a negative emotion, not liking life, and prevents us from enjoying life.

7.4 Emotions Can Be Changed

In opposition to the prevailing view that our emotions are inborn, are part of our fixed character, and cannot be changed, it is maintained here that the reverse is the case. Emotions are mostly nurture, not directly genetic. Emotion can be created or radically changed by changing the cognition. This is similar to a change by a different seeing-as. By changing our assessments we can prevent and eliminate negative emotions. Clinical therapeutic experience as reported in the *Journal of Rational Emotive and Behavioral Therapy* has shown this to be the case and the technique is used throughout therapy. We can create a different personality. The extent to which one may change one's emotions, then, depends on our ability to be critical and avoid faulty language-use. In regard to ethics being based on bringing about our naturalistic wants and likes, Dewey argues that even these can be changed. We can change our desires [28].

One must be able to critique one's language – admittedly few are genuinely able to do. People tend to be anti-critical and argument-illiterate and so emotionally illiterate. Thus, the title by Goleman, *Emotional Intelligence* traces that thought [29]. This intelligence we lack. People tend to generally reject learning about emotion and consider it to be of little use to their lives. For most people, including health-care workers, extensive education would be needed. It may be pointed out that in American nursing schools they used to study Freud, but this has been often replaced by the now more popular other forms of therapy, such as the cognitive theory of emotions and behavioral methods.

In order to understand emotional intelligence, however, one needs to have a sound theory of emotions. The point is also made by Zeidner and coauthors who claim that the so-called "Emotional Intelligence" is illusive. They say that Emotional Intelligence (EI) is recently claimed to be essential to improve medical and nursing education and practice, but researchers have not clearly defined what it is. EI has equivocal scientific status and is perhaps just a fantasy, a "mirage" [30]. Emotion was erroneously thought to be the opposite of cognition and intellect, a "mental" ability, or a component of a mentalistic "will" [30]. It was often reduced to facial expressions because psychologists could not otherwise explain it. Psychologists are typically shown to have presented only an "elusive status" of emotions [31].

For example, anger can be changed by changing the open-context value assessment that something is bad. Without creating negative assessments there can be no anger. Cultural encouragement and constant exposure to anger can produce the motivation to be angry. The cognitive-emotive theorists have amassed literature and clinical experience affirming the cognitive view. One may consider that anger also hurts the one who is angry, as well as others. Anger, depression and stress are characteristics of those who attempt and commit suicide (Worldwide one million a year). Similar to anger, revenge is due to the faulty assessment that it makes sense to hurt others as they hurt us. It has rather been shown that revenge and blame virtually never make sense [32]. Marcus Aurelius exhorts, "Blame no one" [33].

Several major fallacious ideas, which form the basis of negative emotions, are:

(a) Failure to accept the reality of the situation e.g. that people die or that one has a certain disease.
(b) Failure to understand the fact that we can only do what is within our power, e.g., that it not only makes no sense, but is harmful, to be torn apart when nothing further can be done to correct a disorder.
(c) Failure to see that the desired goal is within our power. We complain or worry instead of acting.
(d) Misuse of value terms (for example, think that something is bad-in-itself, or give irrational or supernatural substitutions for "bad.")

Ironically, in medicine there may be a double bind. We must accept a disease we cannot easily accept, have hope in the face of past experience, but past experience can make the situation worse. Acceptance of reality has a solid placebo effect.

7.5 The Happy Stoics: Passionate Rational Emotion

Gould states that the Stoics, Zeno of Citium, and Chrysippus, held that all emotions are bad [34]. This is incorrect. They rejected only negative emotions (NE). That NE are bad is true by definition. NE are regarded as false judgments. The position of Marcus Aurelius [35] is not that we should have no emotions and so be passionless. Rather, he opposes violent excitement (an oxymoron). He rather encourages cheerfulness and humor [36]: *Do every act of your life as if it were your last* [37]. Have *good emotions* [38], and *happiness* [39]. This would have to include all other positive emotions as well.

Against the widespread characterization, Rist argues that the Stoics did not advocate apathy [40]. On the cognitive theory of emotion (CTE), apathy is a NE. For the Stoics, it means *without disturbance*. The wise person experiences the joy, happiness, and even exhilaration, which comes from living a rational life in accordance with nature [41]. This position would argue against the Romantics' and the religious use of supernaturalism and idealism in the arts. In a healthy state, rational cognitions are identical with positive emotions [42]. Rist argues that they produce *rational feelings*, and that only the picture-book Stoic wise person is devoid of passions [43]. Gould gives us such a storybook picture in saying that for Zeno: *All emotions are bad* [34]. Anyone who seeks "apatheia" in the sense of total elimination of all feeling and emotion is asking for a state when all activities are suspended. Such a state would be equivalent to death [44].

7.6 Virtually All Judgments Involve Emotion

Any statement may produce an emotion. The formula for emotion is E = (C > F) (emotion = cognition, which causes bodily feeling). It would also follow that any assessment (belief, judgment) would be accompanied by a feeling. And this is exactly what the Stoic, Chrysippus, held. It is not strictly the case that C > F, rather

F is a part of C. For him, all judgments involve feelings or emotions. There is no such thing as emotionless thought [45]. For Collingwood also, all judgment and all of language, express emotion [46]. Emotion and cognition unite. Even mathematics is emotive. All statements in medicine are emotive. This is like placing an exclamatory mark at the end of every sentence. Collingwood states, "There is no need for two separate expressions, one of the thought, and the other of the emotion accompanying it. There is only one expression" [47]. In the healthy state, correct judgments produce positive emotions [42]. Wittgenstein [48], like Marcus Aurelius [49] also thought of rational thinking and philosophy as therapy [50].

7.7 Emotion Can Change with Bodily Feeling

Emotion can, to some degree be changed by changing the feeling (perception, sensation). Because anger is = (C > F), there is no unfeeling anger. A change in F, then, can alter the anger to some extent. Change of emotion may take place anywhere along the continuum of cognition to bodily sensation. Sometimes bodily sensations are very much involved, other times hardly at all. In regard to negative emotion, such as anger and revenge, change of the resulting bodily feeling can do little to change the emotion. It is mainly the cognition, which must be altered.

Psychosomatics refers to the influence of the cognitive (language-use) on the body. All illnesses can be considered to be psychosomatic in this sense. Pain refers to physical pain as well as to psychological pain. We look for single causes of disease when often our general overall physical and cognitive (philosophical) health is the major factor, e.g., in regard to back pain. Pain words and synonyms have a double psychological and physical meaning to them: agony, anguish, discomfort, distress, hurt, misery, shock, suffering, torment, torture. We may have painful memories. Thus, one may psychosomatically equivocate with the two meanings and assert that because physical pain is felt, also psychological pain is. This may not be the case. One may have psychological pain without physical pain and physical pain without psychological pain. One may have anger with or without noticeable bodily feelings. The term "psychogenic pain" is considered to have limited clinical or diagnostic usefulness and the preferred term "idiopathic pain syndrome" used in DSM-III-R is advocated. "The fact that hypnosis was able to induce a genuine painful experience suggests that some pain really can begin in our minds," said Dr. David Oakley [51].

A survey published last year suggested as many as one in seven Britons are in constant pain. In many cases, their condition cannot be explained by doctors who are unable to identify the cause by using conventional tests. There are still doctors writing pain off as psychological. Peveler reported that physical investigation fails often to reveal the cause of pelvic pain [52].

Still chronic pelvic pain in women is poorly understood [53]. Motion sickness can occur without any motion, e.g., by just seeing an astronaut float upside down. Hampton reports a carpenter with great pain due to a nail in his foot, but upon removing the shoe and seeing that the nail went between his toes the pain vanished [54]. On the other hand those injured in an emergency may experience no pain until the emergency is passed. Cognition is a large part of pain behavior. Pain is an emotion insofar involving cognition, and other emotions can influence pain. Fear can make pain worse and humor can lessen it. Each cognition (language-use) has a potential influence in constituting pain behavior. "Emotions and moods also modulate the perception of pain" [54]. If pain were regarded as an emotion rather than a mere sensation, it of course placed stress on the bodily feeling side of the cognitive equation.

Schechter and Smith hold that chronic, non-specific back pain can be more psychosomatic than merely physical [55]. The authors present a view combining the "physical" and "psychological" conceptions of pain. "We believe that the ultimate reason for the persistence of the pain is in the mind/brain or subconscious. This creates or perpetuates the pain in order to distract attention from emotions that are too threatening for the individual to address consciously, such as anger, rage, grief or anxiety, hence the term 'distraction pain syndrome.'" [55] Lacking however is a theory of emotion and they mentalistically assume the existence of a mind and unconscious.

Another psychosomatic influence of the cognitive (language-use) on the body is suggested by Hyman and Liponis who state that writing your thoughts and feelings has significant physiological effects [56]. If so, critical philosophy, and good creative writing and poetry would have the most beneficial effect.

7.8 Emotion Is Not Passive

Because emotions are caused by our assessments and because they are not passive bodily feelings, we are actively responsible for them. They do not just happen to us. One does not just "get into a mood" like one has a headache. Prior and present assessments cause the mood or attitude. Siemer says that mood cognitions cannot be separated from mood sensations, that moods are not just sensations. The feeling theory of moods is opposed [57]. This is not surprising because mood is a synonym of emotion. Ultimately, no one can make one happy, but oneself.

Because of the confusion between emotions and feelings, we may tend to regard emotions as being passive, as feelings are. Dewey states, "Even anger and hate are partly caused by us rather than in us" [58]. The sudden irritation that seems to be groundless is based on numerous prior assessments. We are repelled by the assessments, which do not meet our prior preferences. We can reconstruct the experiences, which lead us to have these experiences in an attempt to reevaluate them and so dispel them. In another sense, our assessments may themselves be enculturated. Our emotions are essentially derivative from the various cultures. We dislike certain people or things partly because we have learned to do so in the society in which we were raised. In this sense our anger is geographical or culture-specific. That is, our negative emotions may be based on a faulty, uncritical, morality. Emotions based on a sound, critical ethics would generate positive and eliminate negative emotions. It may be noted that a reconstruction of the reasons for anger or other negative emotions is just another language-game, not a statement about what "really" is the case. But it nevertheless has its uses as with the correction of negative emotions.

7.9 Emotions Are Unique

We can never have the exact same emotion twice. Nor can two different people have the same emotion. Mew speaks of "*that slippery phrase 'the same emotions.'*"[59] An emotion will be some combination of diverse sorts of assessments and bodily feelings to produce each specific case of emotion. The first reason for this is that there is no emotion as such. Secondly, for each emotion there is a different C and F. We have assessments constantly and cannot completely control the way in which they come to us. The same is true of feelings. For example, for each anger there is a different C and F. Anger = C_1, C_2, C_3...> F_1, F_2, F_3.... It may be noted that whereas we tend to regard them as alike, each emotion of anger is to a greater or lesser degree different than every other one. Repetition of fixed ideas can, however, produce similar angers.

Logically, no two things are the same, and it makes no sense to say the same thing is identical with itself – nothing left over. "The production of the same emotion by different contexts is impossible" [60]. But the production of the same emotion by the same context is also impossible. No two outbursts are exactly the same. Collingwood states, "The anger I feel here and now...is not quite like any anger I ever felt before and probably not quite like any anger I shall ever feel again" [61]. Our lives are written with varieties of emotion. For each emotion we must find the specific assessments and feelings actually experienced. Thus, each specific occurrence of emotion (each language-game) must be examined separately. The reasons for contextualist theories of meaning are the same as for contextualist theories of emotion.

Emotions vary continuously and are constantly changing. Emotions do not appear singly, but in combination with our other emotions, language-use and bodily feelings. These thoughts and bodily feelings may be explored by means of discussion and writing. This means a dialogue and narrative is needed to find out what can be meant. Metaphor expansion does this clarification, a view, which also is a central part of philosophical counseling (See Chapter 17). In medicine, dialogues between patient and physician are needed for comprehension and advancement. Dialogue and narrative are needed for physicians who work with teams and, for example, when healthcare workers are accused of making mistakes. Expressive writing is said to be emotionally beneficial and to improve health and the immune system and to fight infections. It supposedly also can reduce stress [62].

7.10 Rejection of the Release Theory of Emotions

It is commonly held that emotions are, and even should be released. On the cognitive theory the release theory of emotion is rejected. Because emotions are not just bodily feelings, inner states, psychic energy, or substances inside of us like steam, they can be neither discharged nor repressed. They are not entities within. It is thought that crying triggers a magical substance, which flows out in tears and helps you to get over grief. It is as if the tears are emotions falling to the ground and splashing at our very feet: grief. The funeral ceremony has been erroneously thought to be a catharsis. Rather, if emotions are to change, the cognition must be changed. This change of cognition takes the place of release. In any case, a well-adjusted person would seldom, if ever, have negative emotions requiring the therapy, which catharsis may claim to afford. What is said to be release behavior is just an intensification of anger or grief. The alleged release of anger and grief, in fact, encourages to continue being angry and to grieve. The emotion pace intensifies. Screaming, distraction, physical release, and intoxication will not change one's assessments. Such emotions will not disappear until the faulty assessments on which they are based are changed.

7.11 Case Example: Emotion Requires Assessment

Berkowitz and Harmon-Jones found that anger is caused by external "adversive conditions" and that it occurs in the absence of appraisals [63]. This statement is a fallacy. It is a misuse of language. No external condition can cause anger. It is not surprising that bad things cause anger, but it is the negativity, which we give to it. Negative emotions cannot be in others or in objects. Only we can cause our anger by our assessments. If there are no negative assessments there is no anger regardless of how disastrous the external situation in fact is. Psychologists often regard emotions in terms of the simplistic approach-avoidance behavioral model. "Strong displeasure" is thought of as sensation, rather than cognition, which is a mistake. For the authors to say that anger is due to "strong displeasure" is circular because negative assessments produce negative emotions [63]. They hold that someone must be blamed for anger to arise. This again is circular as blame includes the assessment of someone being bad. But, in any case, one can be angry without blaming. Blame is a different emotion than anger. For these authors, an "affectively determined impulse to aggression" is said to be linked to anger [63]. On their model anger can arise just from flexing muscles. In opposition to this view one may suggest that the body has no impulse to aggression. That is a cognitive thing. And to say it is affective is already to say it is cognitive and an emotion, and so it is circular. In opposition to their view, to have anger one needs a negative assessment, and no situation can produce anger without it. Clore and Cenerbar critique the above article and offer instead a recent defense of cognitive emotion theory [64]. However, their arguments were already given 30 years earlier in Shibles [13].

Typically psychological experiment involves a few abstract, undefined vague terms such as fear, aggression, frustration and then questionnaires are constructed and statistical analyses given. The result is of questionable value. For example, Kuppens gave only four appraisals as possible causes of anger: goal obstacle, other accountability (another is to blame), unfairness, control loss, and also antagonistic action tendency (or blame). They concluded that anger was to some varying degree associated with each component [65]. This, one would know without the experiment insomuch as the component involved a negative value term. Why four? We can evaluate thousands of things, which we could have negative emotions about. But a well-adjusted person may not be angry confronted with any of the components. The problem is caused by the absence of philosophical and critical thinking (speaking) about abstract terms in the experiment, by the use of unacceptable pseudo-scientific and problematic statistical methods, so that the most mundane conclusions are reached (See Chapter 19). It is stated, "Generally no specific associations between other accountability and emotions other than anger have been found" [65]. This presupposes that there are a number of fixed emotions that can be charted on a diamond figure, when there are no emotions as such and thus not at all to put on the chart. There are only value assessments, which cause bodily feelings, and the number of these is not limited. With each statement there is an emotive component.

7.12 Negative Emotions Are Philosophical Language Fallacies

Negative emotions may be thought of as forms of philosophy of language fallacies [66]. Because anger is based on fallacious cognition or statement, it may be regarded as being illogical. Negative emotions are the breakdown of our thinking and feelings. Emotions are often problematic because based on a dysfunctional culture, false beliefs. That is, if one is enculturated and has false beliefs, one's emotions will be dysfunctional. What one regards as repugnant, producing anger, or even precious will be unfounded. To have negative emotions is to harm, to destroy the body, a form of letting-die.

It is on the basis of this theory, that the therapists Paul Hauck and Albert Ellis present the following account of anger. Hauck wrote: "I cannot think of a single human emotion that is more dangerous to each of us than anger" [67]. According to Hauck, anger:

1. Puts us out of control and leads to crime and violence.
2. Leads to hatred, and such consequences as divorce and loss of friends.
3. Has the physiological consequences of stress, headache, upset stomach, heart attack, phobia, neurosis, etc. [Stress and fear cause adrenalin release, which can be harmful. Adrenal cortisol: chronic exposure ages body and depresses the immune system [68]. Positive emotions including humor strengthen the immune system and prevent illness, stress, death, old-age disability, strokes, heart attacks and angina [69]. Stress contributes to excessive cholesterol [70]. Stress contributes to inflammation. Suarez found that high levels of negative emotions as determined by psychological tests were associated with increased blood levels of CRP, a marker for inflammation of the arteries [71]. Thus, reducing anger may promote cardiovascular health. Heart attack can be triggered by emotional stress, extreme heat or cold.
4. Anger does not get at the cause of or solve the problem.
5. Always adds to existing frustration.
6. Leads to revenge.
7. Sets a poor example and is "infectious."
8. Is often based on blame and the idea that "bad people ought to be punished."
9. Is never righteous, though all think their anger is justifiable.
10. Is a form of punishment.

11. Is self-punishment if one is angry with oneself.
12. Is largely due to non-acceptance and putting a "catastrophic" negative value on a situation instead of accepting reality or doing what is within one's power to change it.
13. "Is not frustration. The latter need never lead to anger."
14. Is often due to dictatorial demands. Domineering people become angry if they do not have their way.
15. Can embitter so that you are "so disinterested in life you can't see anything beautiful anymore."

In general, anger is based on irrational thinking, is harmful, ineffective, it is non-adjustive behavior.

Albert Ellis [72] argues that anger:

1. Is due to irrational thinking and demandingness.
2. Takes the place of intelligent action.
3. Is an abusive power play of force.
4. Takes the place of joy.
5. Is aggressive behavior, violating the rights of others.
6. Is like temporary or continued insanity.
7. Is always unjustified and ineffective.
8. Destroys the needed love of others.
9. Is punitive and cruel.
10. Is a self-defeating emotion. It does more harm than good for one's own cause.

According to Herrald and Tomaka emotional experiences affect psychological and physical health [73]. Negative emotions contribute to anxiety disorders, depression, low self-esteem, cardiovascular disease, cancer, and immune system suppression whereas positive emotions contribute to subjective well-being, increased self-esteem, self-efficacy, morale, and may reduce the harmful effects of negative emotions. Anger can lead to low self-esteem, negative self-concept, interpersonal conflict, and aggression. Anger has been linked to coronary heart disease. Sadness is associated with cancer and reduced immunological functioning.

7.13 Some Traditional Examples of Philosophy of Language Fallacies

It is then on the basis of the cognitive theory of emotion that we may regard negative emotions as philosophy of language fallacies. These fallacies are also unacceptable in medical treatment and management of medicine. This may also be expressed as follows:

argument from anger: Negative emotions are virtually all fallacies.

appeal to emotions: (*argumentum ad passiones; ad populum*: appeal to the emotions of the public) The appeal to emotions is put in place of sound, or fair argument: (a) *ad amicitiam*: appeal to friendship, (b) *ad invidiam*: envy, (c) *ad mentum*: fear, (d) *ad misericordiam*: pity, (e) *ad odium*: hatred, (f) *ad superbiam*: pride. According to this theory, negative emotions, e.g., anger, revenge, blame are abusive and forms of irrational thinking [74].

fallacy of force (ad baculum): (lit. "stick" or threat; force) e.g., unkindness, pressure, threats, being political, or being angry if one does not agree with an argument (cf. *ad mentum*, fear). Virtually any use of anger is a fallacy. "Just War" arguments now state that going to war out of anger or revenge is unjustifiable.

fallacy of negative emotions: According to the cognitive-emotive theory, negative emotions are based on faulty thinking. Thus, anger, revenge, blame, worry, guilt, "feelings" of victimization, jealousy, depression, etc., are fallacies.

7.14 Pity

Pity can induce an egoism worse than the illness itself.

There can be pity for others and pity for ourselves. It may be an emotion of concern for the suffering and misfortunes of others. The German word is *Mitgefühl* (feel with). Pity is not just a bodily feeling, but an assessment of a negative state of affairs. The emotion produced may be positive or negative. Pity may lead to help and care, or to self-punishment and obliteration.

As a negative emotion synonyms of pity are: being depressed, down, sad, unhappy, *Mitleid haben* (German: suffer with). Grief may be thought of as pity for the dead or self-pity of the griever.

As a negative value term synonyms are: disadvantage, shortcoming, weakness, bad, awful, terrible, matter for regret, miserable, contemptible, e.g., "what a pity," "pitiful." Pity may be merely a term of blame and condemnation without sympathy, care or concern. Pity may be with or without mercy. Pity is caused by ourselves. Nothing is in itself pitiful. People are not in themselves pitiful. Only we can see a situation as such. "It excites pity," or "x *is* pitiful," are fallacies. No one and nothing can cause pity, but our own assessment.

As self-pity it is "feeling sorry for oneself." Instead of accepting the events of one's life, one sees them as negative and further negativizes them, e.g., "one wallows in self-pity." This is non-adjustive and makes matters worse. The fallacy of Appeal to Pity, *ad misericordiam*, refers to the misdirected sorrow for the failure of others often with blame and contempt. Commiseration is for example, weeping with another. We may out of pity help the selfish and cruel. Self-pity can be especially practiced by selfish and egoistic people. One does not bother to take care of oneself, e.g., control one's diet, or exercise, and then think-feels sorry for one's being overweight. One blames others or life for one's condition. Through pity, the physician may be doing great harm to the patient by inducing or indirectly teaching selfishness and negativity. A spoiled patient becomes demanding and untreatable. We may pity the ignorance of those who have not bothered to improve themselves. We may break a rule of fairness to others because of pity for one person.

7.15 Hope and Humor

Hope is the only liar who never loses his reputation for veracity [75].

Hope is an emotion, which may be analyzed as assessments about goals, which produce bodily feelings [75]. Hope has similarities with the placebo effect (See Chapter 19 on placebo).

The positive assessment needed for humor may take the form of hope. Synonyms for hope are trust, wish, prospect, possibility, need, and expectation. Humor is a way to reframe our lives and even a life's goal itself. Both hope and humor take and run around life's problems. A negative situation may not be within our control, but humor is. Dis-ease becomes ease. Humor is especially useful when our future looks bleek. Hopelessness has been shown to cause depression, also depression of the immune system. Snyder says that hopelessness leads to apathy, despair and rage [76]. Hopelessness can be as much of a disease as depression. Humor is a form of "hope therapy." It is a performative utterance in the sense that it is a self-fulfilling prophecy. Assuming that it is realistic, it brings about its own goal. Goallessness leads to hopelessness. Perhaps thinking of unrealistic goals is behind when Kant, Nietzsche and Kierkegaard oppose hope. The latter wrote, hope is "the passion for the impossible" [77].

Acceptance of, or adjustment to, perceived hopelessness, can bring about humor. If such is taken seriously, this leads to poor adjustment, depression, and suicide. When we are hurt, and accept that fact, we find that we sometimes laugh at our situation: the silly leg cast, bandages, which you had your friends autograph; the temporarily bounding limp foot. A broken leg is a fact. The most hopeless situation is death, which is why humor is one of the few ways in which it can be coped with or explored (cf Chapter 21).

7.16 Case Example: Patients' Negative Emotions

On Sunday night a 16 year old came into the Women's Hospital to obtain a morning-after pill after having unprotected sexual intercourse. She was required to wait as she was the 27th patient there this morning and many others had urgent problems. She was then called on, interviewed, given an ultrasound examination and provided with the required pills. She was then scheduled to return the next day during walk-in hours for further examination and anti-contraceptive counseling. This was carried out. As a precaution it is usual to have such young girls be accompanied by someone close to her. Her father did so and took her home. Several weeks later I received a phone call from her mother who quite angrily scolded me for not having taken enough time for her daughter, and accused me of being insensitive. She said that I was so incompetent that I should leave my job as a physician. As it turned out she herself had been away on her honeymoon at the time her daughter came in. It would seem that the father blamed the mother for not properly taking care of the daughter and the mother then tried to pass the blame on to me rather than accept it herself. But we did take good care of the daughter even with our busy schedule perhaps more than most hospitals would have. I asked the mother if she had a good relationship with the daughter, why she had not prepared the daughter for sexual life, and did not provide information about conceptive precautions. The health care workers took over for parental irresponsibility and then were scolded for it. Such negative emotions add to the already exhausting job of the physician, unnecessarily use up scarce resources, cause negative emotional disruption of the family, and fail to give the appropriate appreciation for helping physicians and nurses.

7.17 Can Emotions Be Reduced to Physiology?

It is a reductionist fallacy to reduce cognition or thought to mere physiology or body. Hyman and Liponis commit this error when they state, "Your brain cells fire the chemical impulses we call thought" [78]. Emotional health is dealt with almost mainly on the medical model. It is said to be, for example, due to poor nutrition. Kurzweil takes supplements and green tea for calmness and stress management and pills to help him sleep. Grossman uses an alpha-wave stimulator, massage and yoga to calm down and reduce stress [79].

Where are the neural centers of emotions located in the brain, and what do we know about them? Damasio and coauthors say the brain regions of emotions are somatosensory cortices and upper brainstem nuclei. There are dynamic neural maps of, for example, happiness, anger, sadness, and fear. The neural structures are referred to as "mental states known as feelings" [80]. This commits the mentalistic fallacy of assuming there are such things as mind, mental and thought states. They only speculate whether the neural activity is accessible to consciousness, which term is not defined. "The feeling-state of emotions might be grounded in emotion-specific neural patterns in the regions identified here" [80]. All the authors could conclude was that the neural patterns differ with each emotion. This would be true for nearly any brain activity and as no two emotions are the same each emotion would have different patterns. Emotions such as anger, etc. are too vague to find neural correlates for them, thus the experiment cannot succeed. In addition, no useful practical information regarding emotions is gained. The information may be useful in the area of brain surgery or testing drugs or nutrition. Sadness is supposedly induced by high-frequency stimulation of the substantia nigra [81]. This is interesting, but this cannot be identified with the phenomenal sadness we experience. The authors state that little is known about the neural basis of emotions and regard their work "as a first step toward a theory-driven, systematic investigation of the neurobiology of feelings" [81]. On the cognitive theory it is an error for them to call emotions "feelings" because emotions rather involve cognitions, which cause bodily feelings. The authors would have a more secure empirical basis for their investigation if they had known about and used the cognitive theory of emotion rather than common mentalistic uncritical terms.

Sirois and Burg note that depression, hostility, anger, anxiety are all shown in the literature in terms of pathophysiological mechanisms to have harmful effects in patients with coronary heart disease (CHD). Up to 65% of myocardial infarction patients supposedly have symptoms of depression, and depression is a predictor of worsening CHD, poor quality of life and high mortality risk [82]. None of the emotion terms studied was defined to any extent except circularly, e.g., anger is defined as ranging from irritation to rage [83]. Then attempts are made to count the number of times one is angry, with no definition of anger. Anxiety is associated with CHD without a definition of anxiety. In the summary they also express doubts about whether anxiety is in fact a justifiable cause [84]. In any case, whatever it is it is supposed to cause a worsening of CHD [85]. Emotions are reduced to physiology as possibly the "dysregulation of autonomic nervous system function and of hypothalamic-pituitary-adrenal axis function specifically" [85]. Anger and depression are linked to inflammation and atherosclerotic plaque rupture. In regard to treatment, in no case was education about the emotions, emotion theory, how to prevent or eliminate emotions given. Some interventions of cognitions were given in some trials. In general, it was suggested that negative interventions and treatments be given for CHD, but no concrete, practical, or useful treatments of guidelines were given. An acquaintance with the cognitive theory of emotion in philosophy and techniques of philosophical counseling would have given such concrete guidelines.

7.18 How Are Diseases and Emotions Classified?

The DSM, *Diagnostic and Statistical Manual of Mental Disorders,* is the standard reference for the classification of mental disorders (See also ICD-10-CM International Classification of Diseases). Presently there is DSM IV Multiaxial System (2000). The DSM-5th edn will be in published in 2010 [86]. Psychiatrists who use *DSM-IV-TR* are invited to submit comments and suggestions at the "*DSM-V* Prelude Project: Research and Outreach," (www.dsm5.org) as the 2007 revision begins for 2010 publication.

If there is no diagnosis, the insurance companies claim there is no disorder. There are, however, many criticisms of the DSM series [87]. The different versions sometimes differ greatly from each other. One version listed homosexuality as a disorder, but other editions did not. Wallace states that the entries in DSM are unacceptable, lack case studies, ignore sociopolitical and economic context, and have a naive view of the self [88]. The DSM lacks accuracy. Instead of merely appealing to the DSM a structured interview for DSM-III-R [and National Institute of Mental Health Diagnostic Interview Schedule] was found to improve routine diagnoses of severe mental illness by fifty percent [89]. Different therapists give widely different diagnoses from DSM IV. Reed states that DSM often has little relevance for the clinician [90]. Frances and coauthors point out, "DSM-IV has no philosophical pretensions... included no systematic philosophical scrutiny" [91]. It was collected for communication, clinical and statistical purposes. It is based on a supposedly unbiased consensus of the empirical literature, data analysis, and field trials. Realists falsely think the classifications refer to real entities independent of the knower, e.g., there is a real entity called, "schizophrenia." Organic versus nonorganic, and physical versus mental disorders are pseudo-scientific categories [92]. That is the mentalistic fallacy of assuming there is a mind and mental entities is committed. Thus, the very title, "the Diagnostic and Statistical Manual of *Mental* Disorders," is mentalistic. Unfortunately the authors prefer "psychiatric disorders" to "mental disorders." But "psychiatric" is a synonym of mental. [German: *geistig*, mental; *seelisch*, mental or psychic] All seven authors' criticisms against mentalism are themselves mentalistic. The authors correctly note that the classificatory system is abstract, obscure and unscientific. They claim to have developed a pragmatic solution, but have not mentioned pragmatism and not proposed any clear or useful model. One clear model would be to reduce each classification to a set of specific symptoms so that the classificatory term is an operational term for such symptoms. Similar symptoms may be classified together. The authors have not mentioned philosophical psychology, philosophy of action, or philosophical counseling the areas where issues relating to DSM are discussed. Nor have they mentioned theories of emotion upon which DSM should be based. Different authors may have different theories of emotion, but they at least should have some theory even if it disagrees with those of others. Also, ethical theory is relevant and the authors touched on this by suggesting that diagnoses are value-laden, but once again a theory of ethics is missing. As for emotions, basically they are not classified at all in this supposedly comprehensive work.

7.19 Case Example: Legal Recognition of Emotional Harm

Emotional abuse, e.g., anger, rejection, etc. are some of the most blatantly harmful as well as insidious forms of abuse which can ruin peoples lives, yet for 30 years only physical damage was considered for awarding legal damages. Society had not recognized that emotional injury could be even more devastating than physical injury. In 1948 the American Law Institute asserted that one is responsible for the emotional distress caused by intentional unreasonable, "extreme and outrageous" behavior even if it is regarding a third person such as a family member and even if there is no bodily harm and even if there is no prior (fiduciary) duty owed to the affected [93]. Mild insult and indignity were excluded [94]. Such irresponsibility does not seem to apply in the military situation. The law and author incorrectly speak of emotions as "mental" and refer to the "psyche" and so commit the unscientific mentalist fallacy. Without a definition, understanding or theory of emotion on the part of the law, emotional distress cannot rationally be determined. Nor can the courts determine if and what might cause an "emotion." For example, one may be entirely unjustified, thereby causing oneself to be in an emotional state and the law may support the emotional distress. A physician was successfully sued for a "brusque and unfeeling diagnosis and treatment" [95]. A physician cannot overstate a diagnosis or give an untrue or incomplete one, knowing it will likely cause emotional distress [96]. Often courts have also required that there be immediate physical harm because they cannot determine if the emotional harm is genuine. This requirement is, however, gradually being relaxed. Harsh language based on factual belief is not actionable, however. In 1959 Notre Dame Hospital kept someone's child in the hospital until the hospital bill was paid thereby causing emotional distress. It may be noted that "outrageous" and "atrocious" behavior are relativistic and arbitrary, open context terms. The courts have no ethical standard or consistent standard for their determination, so they judge merely on a case by case basis. "There is not a clear cut medical legal standard of outrageousness" [97]. It cannot be determined if the emotional harm is due to any one thing rather than another, e.g., the negligence of the physician or the fact that there is a defective birth. The jury is an uninformed collection of arbitrarily chosen individuals who have questionable therapeutic or philosophical expertise in determining emotional distress, yet they are charged to do so. The emotional damage must often be so severe as to cause neurosis, psychosis phobia or chronic depression [98]. In terms of the philosophy of science such classifications of DSM IV are in fact unscientific pseudo-categories. We do now, however, have the means by the philosophy of emotion and philosophical counseling to determine if one genuinely has an emotion and what that emotion is. This is the cognitive theory of emotion. It is a faulty assumption on the part of the court. In view of the above difficulties the recent trend is to treat emotional distress cases like the more strict malpractice cases, e.g., giving the wrong information, which require all of the elements of: duty, negligence, causation and damage [99].

The law treats grief as *suffering*, a more bodily feeling term for grief, rather than as cognition. In any case, it has been maintained by some literature that grief is an irrational emotion [100]. In general, what the courts have shown by their shifting criteria and handling of emotional distress is that they are not qualified to deal with it, but that healthcare workers and others are nevertheless subject to its anti-medical, un-philosophical and unscientific determinations. In conclusion, one may state that in the light of Coburn's analysis of the courts that, ironically, the confusion of the courts on the issue of emotional distress causes physicians to practice irrationally defensive medicine, refuse to take risky cases, and causes the physicians themselves to experience emotional distress due to the courts (See also legal issues in the Chapter 8).

7.20 Brief Summary of the Cognitive Theory of Emotion

1. Emotion is not just a bodily feeling. E ≠ BF. Instead say, "I think-feel emotion."
2. Emotion is cognition (assessment or evaluation), which causes bodily feeling. E = (C —> BF)
3. The emotive cognition (self-talk or language use) is typically a value assessment.
4. Emotion can be changed by changing the cognition. We cause our own emotions. Emotive reflexives render this well, for example, such as found in French and Gerrman (*Ich ärgere mich*, "I make myself angry.")
5. Emotion is not innate or unalterable. Personality can be changed.
6. We cannot have just the same emotion twice, because both C and F change.

7. Negative emotion such as anger, revenge are due to faulty assessments such as:

    (a) failure to accept reality.
    (b) failure to understand that we can only do that, which is within our power.
    (c) misuse of value terms, such as thinking that something is bad in itself.

8. Emotion is not at all and is not the sort of mentalistic thing that can be "released." (We are incorrectly told to "release" our anger, grief, jealousy, which cannot be done.)
9. Emotion as such is not a cause of behavior. Only (C —> BF) can be a cause.
10. Because a judgment or statement is cognition plus bodily feelings (C —> BF), any statement or judgment may be regarded as an emotion.
11. There are meta-emotions such as emotions about emotions, e.g., fear of fear, etc.

For a further application of the emotion theory to the analysis of specific medical concepts, see the Chapters 8, 9, 10.

7.21 Case Example: The Cognitions Involved in the Emotion of Interest

Silvia points out that, "Research on appraisals and emotions is relatively recent" [101]. Interest is based on, motivation, engagement, curiosity, enjoyment, etc. all of which are positive value terms and as such make the statement circular, e.g., interest is based on what I like. Silvia bases the assessments involved in interest on novelty, as well as coping and understanding potential [101]. But novelty may be and often is disliked and so would not generate interest, which, as a positive value term, by definition cannot be negative. Coping and understanding may be positive terms, but the average person puts a low value on inquiry, philosophy and understanding, but a higher value on football games and political disasters. Clearly, if understanding is the assessment needed for interest, and if one is frustrated because one does not understand, it need not be regarded as interesting. If one would wish to puzzle it out, and not just be frustrated, one may still have an interest in that sense. It is simplistic, as the author admits, to reduce interest to just a few assessments. Rather, interest is a synonym of good and whatever anyone assesses as good is at the same time of interest. Some find having babies exciting, others not. The author will not be able to make out his case because interest comes from us, not from external events. Nothing is in itself interesting, only we can make it so. We say, "X is interesting," but this is misleading. In sum, the psychological and statistical experiments conducted by the author do not clarify or give insight into the notion of interest. Such methods actually prohibit understanding rather than enhance it. The experimental methods used are unscientific. In this sense they may not be found to be interesting. The author is certainly on a potentially productive track trying to search for the assessments involved in the notion of interest on a cognitive theory of emotion, but the lack of critical analysis of ethical terms, the commitment of language fallacies and the unscientific experimental methodology prevent the specification of such assessments.

References

1. Marks, J. Ed. 1986. *The ways of desire*, 3. Chicago: Precedent.
2. Dewey, J., and Tufts, J. 1932. *Ethics*, 327. New York: Holt.
3. Gallo, L. 2004. Negative emotions and health. In *Encyclopedia of health and behavior*, ed. Anderson, N., 2 vols, 300–306. Thousand Lakes, CA I: Sage.
4. Sotile, W., and Sotile, M. 1996. The angry physician-part 1. The temper-tantruming physician. Physician exec 8 (Aug 22, 1996):30–34.
5. Buetow, S., and Elwyn, G. 2006. Are patients morally responsible for their errors? *Journal of Medical Ethics* 32:260–262, esp. 262.
6. Bedford, E. 1964. Emotions. In *Essays in philosophical psychology*, ed. Gustafson, D., 77–98. New York, Doubleday: Garden City.
7. Perkins, M. 1966. Emotion and feelings. *Philosophical Quaterly* 75:139–160.
8. Kövecses, Z. 1990. *Emotion concepts*, 68. New York: Springer. For a review of Kövecses book on emotion see Shibles, W. 1992. Book notice. Zoltan Kövecses emotion concepts. *Studies in Language* 16:255–261.
9. Ellis, A. 1994. *Reason and emotion in psychotherapy*. New York: Carol Publishing Group.
10. Melden, A. 1969. The conceptual dimension of emotions. In *Human action*, ed. Mischel, T., 205 and 216. New York: Academic Press.
11. Roseman, I., and Smith, C. 2001. Appraisal theory. In *Appraisal processes in emotion: Theory, methods, research*. eds. Scherer, K., Schorr, A., and Johnstone, T., 3–19, esp. 4. New York: Oxford University Press.
12. Scherer, K., Schorr, A., and Johnstone, T. eds. *Appraisal processes in emotion: Theory, methods, research*, 389–390. New York: Oxford University Press.
13. Shibles, W. 1974. *Emotion: The method of Philosophical therapy*. Whitewater, WI: Language Press; Shibles, W. 1995. The cognitive-emotive theory of desire. *Journal of Indian Council of Philosophical Research* 11:25–40; Shibles, W. 1995. *Emotion in aesthetics*. Dordrecht, Netherlands: Kluwer Academic Publishers.
14. Kappas, A. 2001. A metaphor is . . . a metaphor is a metaphor. In *Appraisal processes in emotion: Theory, methods, research* eds. Scherer, K., Schorr, A., and Johnstone, T., 157–172, esp.158. New York: Oxford University Press.
15. Wittgenstein, L. 1958. *Philosophical investigations*, 3rd edn, 363. New York: Macmillan.
16. Ryle, G. 1949. *The concept of mind*, 40. London: Hutchinson.
17. Quine, W. 1966. *The ways of paradox and other essays*, 208–214, esp. 213–214. New York: Random House. On mental states.
18. Mead, G.H. 1977. *On social psychology*, 100. Chicago: University of Chicago Press.
19. Mead, G.H. 1977. *On social psychology*, 89. Chicago: University of Chicago Press.
20. Brody, H. 1980. *Placebos and the Philosophy of Medicine*, 3. Chicago: University of Chicago.
21. Damasio, A.R. 1994. *Descartes' error: Emotion, reason and the human brain*. New York: GP Putnam's Sons.
22. Smith, C., and Kirby, L. 2004. Appraisal as a pervasive determinant of anger. *Emotion* 4:133–138.
23. Shibles, W. 1995. *Emotion in aesthetics*, 82–118. Dordrecht, Netherlands: Kluwer Academic Publishers.
24. Smith, C., and Kirby, L. 2004. Appraisal as a pervasive determinant of anger. *Emotion* 4:134.
25. Smith, C., and Kirby, L. 2004. Appraisal as a pervasive determinant of anger. *Emotion* 4:135.
26. Griffiths, P. 2004. Emotions as natural and normative kinds. *Philosophy of Science* 71:901–911, esp. 902.

27. Dewey, J. 1929/1958. *Experience and nature*, 42. New York: Dover.
28. Dewey, J. 1939. *Theory of valuation*, 29. Chicago: University of Chicago Press; cf. Shibles, W. 1995. The cognitive-emotive theory of desire. *Journal of Indian Council of Philosophical Research* 11:25–40.
29. Goleman, D. 1994. *Emotional intelligence*. New York: Bantam.
30. Zeidner, M. 2001. Slow down, you move too fast: emotional intelligence remains an "elusive intelligence". *Emotion* 1:265–275: esp. 266–267.
31. Zeidner, M. 2001. Slow down, you move too fast: Emotional intelligence remains an "elusive intelligence". *Emotion* 1:265–275: esp. 274.
32. Shibles, W. 1992. Verbal abuse: An analysis. *Papiere zur Linguistik* 46:30–44. See full analysis of blame in the chapter on *Care: A Critique of the Ethics and Emotion of Care.*
33. Marcus, Aurelius 1964 ed/ tr Long, G. *Meditations*, 88. New York: Doubleday/Penguin.
34. Gould, J. 1970. *The philosophy of Chrysippus*, 34. Albany, NY: SUNY Press.
35. Marcus, Aurelius 1964 ed/tr Long, G. *Meditations*, 80. New York: Doubleday/Penguin.
36. Marcus, Aurelius 1964 ed/tr Long, G. *Meditations*, 12. New York: Doubleday/Penguin.
37. Marcus, Aurelius 1964 ed/tr Long, G. *Meditations*, 20. New York: Doubleday/Penguin.
38. Marcus, Aurelius 1964 ed/tr Long, G. *Meditations*, 56. New York: Doubleday/Penguin.
39. Marcus, Aurelius 1964 ed/tr Long, G. *Meditations*, 72. New York: Doubleday/Penguin.
40. Rist, J. ed. 1978. *The stoics*, 259–272. Berkeley, CA: University of California Press.
41. Rist, J. 1969. *Stoic philosophy*, 25, 31, 37. Cambridge: Cambridge University Press.
42. Rist, J. 1969. *Stoic philosophy*, 26. Cambridge: Cambridge University Press.
43. Rist, J. 1969. *Stoic philosophy*, 259. Cambridge: Cambridge University Press.
44. Rist, J. 1969. *Stoic philosophy*, 35. Cambridge: Cambridge University Press.
45. Rist, J. 1969. *Stoic philosophy*, 35–36. Cambridge: Cambridge University Press.
46. Collingwood, R.G. 1938. *The principles of art*, 267, 296. Oxford: Clarendon Press.
47. Collingwood, R.G. 1938. *The principles of art*, 276. Oxford: Clarendon Press.
48. Wittgenstein, L. 1958/1953. *Philosophical investigations*, 3rd edn. New York: Macmillan; Anscombe, E. tr. And revised. 2002. Cambridge, MA: Blackwell Publishers.
49. Marcus, Aurelius 1964 ed/tr Long, G. *Meditations*, 103. New York: Doubleday/Penguin.
50. Shibles, W. 1974. *Emotion: The method of philosophical therapy*. Whitewater, WI: Language Press.
51. Dr. David Oakley, director of the hypnosis unit at University College London. *BBC News* 9 August, 2004. news.bbc.co.uk.
52. Peveler, R. 1996. Psychosocial factors and chronic pelvic pain: a comparison of women with endometriosis and with unexplained pain. *Journal of Psychosomatic Research* 40: 305–315.
53. Maier, B., Akmanlar-Hirscher, G., Krainz, R., Wenger, A., and Staudach, A. 1999. Der chronische Unterbauchschmerz – ein immer noch zu wenig verstandenes Krankheitsbild. *Wiener Medizinische Wochenschrift* 149:377–382.
54. Hampton, T. 2006. A world of pain: Scientists explore factors controlling pain perception. *JAMA* 296:2425–2427.
55. Schechter, D., and Smith, A. 2005. Back pain as a distraction pain syndrome: A window to a whole new dynamic in integrative medicine. *Current Opinion: Evidence-Based Integrative Medicine* 2:3–8.
56. Hyman, M., and Liponis, M. 2003. *Ultraprevention*, 326. New York: Scribner.
57. Siemer, M. 2005. Mood-congruent cognitions constitute mood experience. *Emotion* 5:296–308; esp. 305–306.
58. Dewey, J. 1958. *Experience and nature*, 251. New York: Dover.
59. Mew, P. 1985. The musical arousal of emotions. *BJA* 25:357–361; esp. 360.
60. Bosanquet, B. 1894. On the nature of aesthetic emotion. *Mind* 3:153–166; esp.160.
61. Collingwood, R.G. 1938. *The principles of art*, 112–113. Oxford: Clarendon Press.
62. Groom, C., and Pennebaker, J. 2004. Expressive writing and health. In *Encyclopedia of Health and Behavior,* ed. Anderson, N., vol I: 326. Thousand Lakes, CA: Sage.

63. Berkowitz, L., and Harmon-Jones, E. 2004. Toward an understanding of the determinants of anger. *Emotion* 4:107–130; esp. 108–109.
64. Clore, G., and Cenerbar, D. 2004. Analyzing anger: How to make people mad. *Emotion* 4:139–144.
65. Kuppens, P. 2003. The appraisal basis of anger: Specificity, necessity, and sufficiency of components. *Emotion* 3:254–269; esp. 266.
66. cf. Brinton, A. 1988. Appeal to the angry emotions. *Informal Logic* 10:77–87.
67. Hauck, P. 1974. *Overcoming frustration and anger*, 31. Philadelphia: The Westminster Press.
68. Kurzweil, R., and Grossman, T. 2004. *Fantastic voyage: Live long enough to live forever*, 282–283. New York: Plume (Rodale orig.).
69. Tugado, M., and Fredricksen, B. 2004. Emotions: positive emotions and health. Encyclopedia of Health and Behavior. In *American psychological association*, ed. Anderson, N., vol. I: 307. Thousand Oaks, CA: Sage.
70. Kurzweil, R., and Grossman, T. 2004. *Fantastic voyage: Live long enough to live forever*, 76. New York: Plume (Rodale orig).
71. Suarez, E. 2002. Psychological risk factors of cardiovascular disease in apparently healthy adults. In *Risk*. eds. Ropeik, D., and Gray, G., 349 ff. Boston: Houghton Mifflin.
72. Ellis, A. 1977. *How to live with and without anger*. New York: Crowell.
73. Herrald, M., and Tomaka, J. 2002. Patterns of emotion-specific appraisal, coping, and cardiovascular reactivity during an ongoing emotional episode. *Journal of Personality and Social Psychology* 83:434–450.
74. cf. Shibles, W. 1987. Reformatory blame. *Journal of Rational-Emotive Therapy* 5:266–281; Shibles, W. 1992. Verbal abuse: An analysis. *Papiere zur Linguistik* 46:30–44.
75. Averill, J., Catlin, G., and Chon, K. eds. 1990. *Rules of hope*, 63. New York: Springer.
76. Snyder, C.R. ed.. 2000. *Handbook of hope: Theory, measures, and applications*. San Diego, CA: Academic Press.
77. Averill, J., Catlin, G., and Chon, K. eds. 1990. *Rules of hope*, 104. New York: Springer.
78. Hyman, M., and Liponis, M. 2003. *Ultraprevention*, 188. New York: Scribner.
79. Kurzweil, R., and Grossman, T. 2004. *Fantastic voyage: Live long enough to live forever*, 259. New York: Plume (Rodale orig).
80. Damasio, A. 2000. Subcortical and cortical brain activity during the feeling of self-generated emotions. *Neuroscience* 3:1049–1056, esp.1051.
81. Damasio, A. 2000. Subcortical and cortical brain activity during the feeling of self-generated emotions. *Neuroscience* 3:1049–1056, esp.1053 Source supplied.
82. Sirois, B., and Burg, M. 2003. Negative emotion and coronary heart disease: A review. *Behavior Modification* 27:83–102; esp.84.
83. Sirois, B., and Burg, M. 2003. Negative emotion and coronary heart disease: A review. *Behavior Modification* 27:83–102; esp. 86.
84. Sirois, B., and Burg, M. 2003. Negative emotion and coronary heart disease: A review. *Behavior Modification* 27:83–102; esp.97; cf. Shibles, W. 1994. "Anxiety: A Pseudo-Concept" *International Journal of Applied Philosophy* 9:43–52.
85. Sirois, B., and Burg, M. 2003. Negative emotion and coronary heart disease: A review. *Behavior Modification* 27:83–102; esp. 89.
86. ee psycweb.com. For DSM go to psychiatryonline.com/resourceTOC.
87. See Shibles, W. 1974. *Emotion: The method of philosophical therapy*. Whitewater, WI: Language Press, for a critique of the earlier DSM III.
88. Wallace, E., Radden, J., and Sadler, J. 1997. The philosophy of psychiatry: Who needs it? *Journal of Nervous and Mental Disorders* 185:67–73.
89. Basco, M. 2000. Methods to improve diagnostic accuracy in a community health setting. *American Journal of Psychiatry* 157:1599–1605.
90. Reed, G. 2006. Clinical experience. In *Evidence-based practice in medical health: Debate and dialogue on the fundamental questions*. eds. Norcross, J., Beutler, L., and Levant, R., 13–23; esp.21 Washington DC: American Psychological association.

91. Frances, A. 1994. DSM-IV meets philosophy. *Journal of Medicine and Philosophy* 19:207–218 esp. 207–208.
92. Frances, A. 1994. DSM-IV meets philosophy. *Journal of Medicine and Philosophy* 19:207–218 esp. 212.
93. The American Law Institute. 1948. Restatement, Second of Torts, Section 46.
94. Coburn, K. 2000. The trend in claims of negligent infliction of emotional distress. *Medical Trial Technique Quaterly* 46:233–261; esp. 235.
95. Coburn, K. 2000. The trend in claims of negligent infliction of emotional distress. *Medical Trial Technique Quaterly* 46:233–261; esp. 238.
96. Coburn, K. 2000. The trend in claims of negligent infliction of emotional distress. *Medical Trial Technique Quaterly* 46:233–261; esp. 242.
97. Coburn, K. 2000. The trend in claims of negligent infliction of emotional distress. *Medical Trial Technique Quaterly* 46:233–261; esp. 247.
98. Coburn, K. 2000. The trend in claims of negligent infliction of emotional distress. *Medical Trial Technique Quaterly* 46:233–261; esp. 255.
99. Coburn, K. 2000. The trend in claims of negligent infliction of emotional distress. *Medical Trial Technique Quaterly* 46:233–261; esp. 254.
100. cf. Wilkinson, J. 2000. Is "normal grief" a mental disorder? *Philosophical Quaterly* 50: 289–304.
101. Silvia, P. 2005. What is interesting? Exploiting the appraisal structure of interest. *Emotion* 5:89–102.

Chapter 8
Enlightened Versus Normative Management: Ethics versus Morals

Without ethics, humanism and critical thinking, management is corrupt.

The prevailing picture of management is one of crisis.

Abstract We generally think of management as an independent, self-contained subject. Everybody in the working process is managing something. Responsibility for medical acting gets increasingly shifted from the patient-physician relationship so that management increasingly "performs medicine." The decisions have been made in advance, yet the responsibility for the negative result falls on the individual physician. Every time a doctor or healthcare worker is at fault, management and administration are also. The standard practice of requiring excessive overwork is bad ethics, bad medicine, bad science, and bad management. Medical professionals are among the most highly stressed occupational groups. Most stress is due to management and organizational factors. The blame for burnout is falsely ascribed to the individual burnt out physician or nurse, not to the system. Thus individualization of responsibility covers again the responsibility of perverse management.

Keywords Management · ethical management · moral management · requirements for good management · mistakes · Pragmatic Theory of Ethics · humanism · overwork · physicians on strike · understanding in management

8.1 Introduction

We generally think of management as an independent, self-contained subject. There are books and courses on management as such. It has its own terminology, methods and techniques. But management presupposes a subject to be managed, human or physical resources. Everyone is in many ways managing something. And it involves one's thinking, emotions, personality, organization and society. Management, then, involves most of our experience as humans. This means that it must be based on broad and sound principles, philosophies, goals and ethics. It is not somehow isolated from these subjects.

B. Maier, W.A. Shibles, *The Philosophy and Practice of Medicine and Bioethics*, International Library of Ethics, Law, and the New Medicine 47,
DOI 10.1007/978-90-481-8867-3_8, © Springer Science+Business Media B.V. 2011

In order to be a qualified manager one must have a sound knowledge of at least ethics, emotion, critical thinking (speaking), and preferably knowledge of critical philosophy methods, as well as having expertise in the areas to be managed. Without knowledge of such things one cannot be a good manager or administrator. The typical manager has not had a background in any of these areas except often some expertise in the area to be managed, or in economics.

Ethical versus moral management. A distinction was made to the effect that moral refers only to existing cultural, legal and normative practices, regardless of how irrational or unfair they may be. Ethics refers to the critique of morals, law, culture and normative or existing practices. Virtually all management is moral, that is, merely normative. Managers were not trained in ethics, emotion or critical thinking (speaking). Thus, management and administration are fundamentally non-ethical. We may accordingly distinguish between normative or inhumane, actual, business-as-usual "moral management" and enlightened "ethical management." Dlugacz reports that in the U.S., "Hospitals were not being managed well, nor were they managing their responsibilities adequately," thus causing the government to introduce careful regulations [1].

In 2005–2006, I was involved in a year long lasting hospital-wide strike of healthcare workers in my hospital in Austria. The central problem was and still is unenlightened management and blocked or misleading communication, but the immediate problem is excessive work imposed upon physicians, and low salaries. This is causing, inhumane and stressful working conditions, which threaten the health and lives of patients as well as of healthcare workers. Management thus often is counteracting the goals of medicine. In a strong sense, everyday normative management can be a form of letting die.

Physicians and nurses believe and are told that if they do not work the excessively long hours at the expense of their own health, the patients will be endangered. This is also the opinion of the public. In fact, the patients are endangered even more if the health care workers are burned out, depressed, psychologically ill, etc, by the excessive demands of management. Management must be held responsible for such consequences, though managers seldom are. They often have too much power. They are often not evaluated, or the evaluations are discounted.

It is the duty of healthcare workers and institutions to inform the public about harmful normative medical beliefs and practice. According to the Council on Ethics and Judicial Affairs Code of Medical Ethics, "Physicians ... shall ... strive to expose those physicians deficient in character, competence, or who engage in fraud or deception" [2]. This includes exposing failed management. Public policy and irrational or medically unsound cultural morals oppose organ donation (USA), stem cell research (Germany, Austria), etc., so that management cannot be efficient. Management therefore must be aggressive and strive to inform the public about such harmful moral beliefs for the benefit of public health.

The following are some suggestions for the sorts of things, which would be needed to be a humanistic and enlightened manager. It may be kept in mind that this applies also to all staff as everyone engages in management to some extent.

An example of the usual deceptive tricks used in unenlightened management or management as usual, is exposed by the techniques of lawyers, politicians, for example, in the work of James Hacker, *The Complete Yes Minister: the Diaries of a Cabinet Minister* 1984.

8.2 Requirements for Good Management

Altruism. The enlightened manager or administrator would be altruistic, rather than egoistic because management and administrations are by their own aims directed towards others. What or who would be managed if not others, including oneself? (For full analysis see Chapters 5 and 10).

Blame and punishment. Blame and punishment are unenlightened, problematic methods of management and administration. Dlugacz states "A good manager will. . .perhaps [encourage] a blame-free environment." "Forget about. . . blame" [3]. In place of punitive retributive blame, substitute rehabilitative blame, that is, solely attempt to correct the behavior for the future [4]. (See full analysis of blame in the Chapter 9) Punishment is never to be used regarding staff or patients. For example, Wachter and Shojania oppose shame and blame methods regarding medical mistakes [5]. They are not acceptable as a management model or metaphor. Punishment is no ethical system regarding malpractice. It debilitates the situation rather than correcting it.

Collaboration and cooperation are to be preferred to competition [6].

Communication and transparency throughout the organization is required. Communication must be open, non-threatening and adequate at every level. It is one of the main problems with organizations and managers that communication is withheld. Physicians often cause their own problems by not standing up in open communication with management and administrators. If they are new or young they do not dare risk their jobs. When they grow older they may be socialized into that pattern of not standing up; or especially in the medical profession, they cannot because of overwork, burnout, which does not allow any additional conflicts between personal and professional obligations. They are afraid to communicate with management or feel that it is too risky to challenge it, or that by so doing it will harm their career advancement. Middle management may also be afraid to question upper management. Structure tends then to be authoritarian and dictatorial and unquestioned.

Case example. "We are expected to take commands from employers, even when the orders may not be in a patient's best interest, yet it is we who are held directly accountable for patient care and outcomes" [7]. If the nurse asked questions of management they were met with job-loss threats, and Bingham, though he held an honors degree in nursing, was dismissed [8]. "In November 1997 a federal judge for the National Labor Relations Board ruled that I was illegally dismissed by the nurse executives in an attempt to silence and retaliate against me for expressing differences with management" [9]. The supervisors have since been replaced. When procedures or conditions, which healthcare workers or managers request for patient safety are denied they should be duly noted in writing and put on file so as not to be later held responsible for the harmful consequences or for lawsuits.

Cooperative rather than an adversarial relationship with other members of the organization is badly needed, but often management separates interest groups out to let them fight against each other for the smallest benefits, for example let the nurses earn some small amount of money more which can be taken from the interns who are just learners and in need of education for further careers and so can be exploited by management.

Creativity. Administration, management and staff are expected to be creative and innovative. Dlugacz and coauthors state, "Managers, along with their staff, need to embrace new ideas, to 'think outside the box,' to encourage and empower critical thinking [speaking]" [10].

Critical thinking (speaking) involves ability to present rational arguments, knowledge of the techniques of problem solving, and also knowledge of philosophy and of language fallacies, which is critical thinking (speaking). Dlugacz and co-authors state, "Quality . . .involves a kind of intellectual openness to problem solving" [11].

Cunning is a trait, which commonly is found in management, but is to be avoided. Synonyms are: devious, deceitful, guile, scheming, shrewd, sly, sneaky, tricky, underhanded, and untrustworthy features. Management often involves doing what managers think they can get away with.

Data-driven studies are to be based on holistic, humanistic, sound and enlightened management philosophies. Data collection merely for punitive or economic purposes is misdirected. There are philosophy-driven assumptions behind the data-driven studies. As in science, data are only data for a certain theory or philosophy. There are no facts or data as such. Data are not to be used except in connection with the mission of the organization and tied specifically to the particular context. It is often used without a transparent or proper basis or to introduce a prejudicial policy. Data may be collected also to evaluate the management and administration as well as the staff. Not even in management, should one think one could reduce quality to quantity, to mere formal mathematical systems. Numbers can cover all of one's faults. Rather, human understanding, reason and humanism are always involved, and if these are left out of such systems, inhumanity and chaos take place. Total quality management systems tend to minimize the human quality and stress quantity. It is more like maximization of profits or total monetary management causing mistrust and hardships for the people concerned.

Case example. "Their resignation to the unsafe conditions to which our once proud and caring unit had sunk – for the sake of a few dollars on staffing – alarmed me. Powerless to protect my patients, I began to wonder what I was doing as a nurse" [12].

43% of nurses in the U.S. score in the burned-out range in stress levels, and 23% planned to leave their jobs within a year [12].

"The lack of sufficient nursing attention likely prolonged [baby] Jessie's hospital stay by months and added tens of thousands of dollars to her care" [13].

Should a woman have a cesarean section or a vaginal birth? In the U.S. the percentage of cesarean births has risen for the last 7 years (See CDC 2005). Cesarean is major surgery. The U.S. has about 30% cesareans, whereas 15% may be regarded as the more acceptable number based on indications and necessity (the author of this book is an obstetrician). Why might cesarean sections be increasing? A cesarean takes about 30–45 min and can be scheduled for the physician's convenience, whereas a vaginal birth may take place in the middle of the night and last for many hours. Physicians tend not to be sued for cesarean sections as much as for vaginal deliveries. Insurance companies sometimes do not insure physicians if they perform special vaginal births. In most hospitals, vaginal birth for breech presentation is not even an option and many physicians are not qualified. Midwives are losing their jobs. Also, cesareans can cost two and a half times more than a vaginal birth, and so physicians and hospital benefit financially. However, in Austria the charge for each is the same and the cost is minimal because just part of the physician's hourly rate. But also the patient may demand a cesarean. Cesarean sections are often performed for financial reasons and to avoid lawsuits. But they are often unnecessary and can be harmful in the short run or fatal to the woman in the long run. Compared to vaginal birth, cesareans may involve more thrombotic and embolic complications, more risks of infection, respiratory complications, and a longer hospital stay. This means the hospital earns additional income as well. The medical risk of cesarean is 1 in 2.500 the risk of vaginal birth is less than 1 in 10,000. For cesarean scar pregnancies (scar from previous cesarean) early cesarean can possibly prevent uterine rupture, but adds another operation to the previous. Vaginal birth allows more often the mother to have shorter recovery times [cf. International Cesarean Network]. Women are often not given the choice to have natural childbirth, especially in special cases. Nobody then is available who has the special experience to assist in more challenging vaginal deliveries.

Decision-making. The manager must have the proven ability to make good and fair decisions (See Chapter 3). Economic reasons are not acceptable reasons for inadequate management, or risk producing, or endangering practices (cf. Quality requirement). The World Medical Association (wma.net/e/) Code of Ethics states as being un-ethical: Paying or receiving any fee or any other consideration solely to procure the referral of a patient or for prescribing or referring a patient to any source. This code is often violated. The goal of administration and management is to produce qualitative, safer practices within their budget. Not to do so is a clear sign of failed management. If a hospital has insufficient budget, it must, for example, restrict treatment, make treatment more effective, reduce over-large administrative, management and non-essential salaries, but not endanger the lives of patients and healthcare workers. Unfortunately, from the author's many years of personal experience, endangering is what is being done. In the author's Salzburg hospital, experienced physicians are leaving due to antagonistic management and are being willingly replaced by inexperienced, lower-paid ones. There is far too little or no investment in the professional staff. There will be few qualified people left to be managed. Management should not be just dehumanized economics anymore than language should be dehumanized mathematical calculus (formal logic), or the person in medicine should be dehumanized body. Anti-humanistic management is self-contradictory and self-defeating, not just unreasonable.

Case example. According to hospital management consultant, Karl Wehkamp, in a lecture at the Paracelsus Medical University in Salzburg as well as in his books, responsibility for medical acting gets increasingly shifted from the patient-physician relationship so that management increasingly "performs medicine" [14].

In the practice of many medical institutions these developments led to a battle between cultures, in which the medical values sharply confront the economic/business values. Management without ethical consciousness or concern for healthcare becomes itself a danger for medicine. Management, in short, now endangers medicine. The dean, a physician, and other high ranking medical university staff attended Wehkamp's lecture and almost 50 out of a total of 80 students voluntarily attended. By contrast, though all were invited, not a single one of the managers attended. This is the level of immediate decision-making, taking over responsibilities, planning, cooperation, etc. Physicians are given responsibilities for which they cannot be responsible, because decisions have been already made on those other levels of administration, hospital, society, culture, etc., which restrict opinions, complicate procedures, and make it often impossible to make good, balanced, positive decisions for the health of the patient. These other levels imprison the area of discretion of the individual physician, nurse, or healthcare worker. The decisions have been made in advance, yet the responsibility for the negative result falls on the individual physician. The physician must expose and reject this faulty system of decision-making. As a physician, one has to have expertise, but if, for example, the nurse calls too late in an emergency, expertise is useless. There is also the inadequate framework in which reasonable and balanced acting is impossible.

Emotions. Managers must have knowledge of emotion theory and practice to deal with the personalities of the staff, manage crisis, enhance employees, and maintain their own stability. No one should be victimized by the emotional abuse of others. For good management there must rather be *emotional intelligence* and emotional consistency rather than the prevailing enmeshment and dysfunctional emotional involvement. Emotional integrity is needed in all aspects of one's life, at home as well as work. One affects the other. One may support an unfair practice or selfish employee, manager, institution or family member because one is accustomed to do so (enmeshed) without realizing the harm done to oneself and others. It may be stressed that institutions can also be regarded as selfish, authoritarian, and abusive. It is also a fallacy to think that intellectual integrity can compensate for emotional integrity and emotional consistency. The question is whether one can exist without the other.

Emotion literacy must also be encouraged in the staff. Managers, administrators and other members of the system must have regular in-service programs on positive emotions especially as negative emotions constantly prevail in everyday cultural life. *Negative emotions are always unprofessional and unacceptable at every level.* Failed management is a symptom of emotional illiteracy, and negative emotions and emotional illiteracy are symptoms of failed management. Emotional stability of all of the staff must be encouraged. Especially recommended is the philosophical cognitive theory of emotion. Without knowledge about how to handle emotions one cannot be a good manager or staff member. It is also necessary for adequate care and management of patients. Emotional esprit de corps and collegiality are necessary and built within good management. Envy, irritation, jealousy, revenge, and other negative emotions are prevalent in the work environment. They are harmful to all and should be corrected whenever they appear. As management books virtually never give detailed information about emotion, a specific example will here be included (See also the Chapter 7).

Anger. In-service or courses specifically on management of anger may also be required as well as on the more available burnout courses. Anger (irritation, annoyance, rage, etc) is virtually always unacceptable in management as well as in interpersonal relations. But without knowledge about emotion one will not be able to create such an atmosphere (See Chapter 7).

Negative emotions are virtually all fallacies, for example, the fallacy, "argument from anger." It may be noted that there is some precedent for this as some emotions have traditionally been regarded as informal logical fallacies such as appeal to friendship, envy, fear, pity, hatred, pride, fallacy of force, unkindness, pressure, threats, being political, being emotional or being angry if one does not agree with an argument. Anger and negative emotions have no place in enlightened management, and it is vital that their appearance be immediately noticed and corrected (See Chapter 7).

Enhancement. Employees must be enhanced in their work instead of opposed and dehumanized, which is the usual case. This is essential to enhance the work of the team and gives each one pride in belonging to a group.

Errors. Physicians are dedicated to restore health and save lives, yet medical errors are the fourth or fifth cause of death. Of course, errors might be similarly expected in any profession (See also analysis of mistake in Chapter 3). Because the consequences are more disastrous in medicine, physicians should be much better compensated than they often are and supported with adequate salaries, resources and research, and free time to recover from stress. This is not being done in most countries. Error and mistake are value terms, synonyms of "bad" or "wrong." Nothing is a mistake or error as such. A standard, rule, protocol, or viewpoint must be held according to which something is a mistake. It can be seen that we often equivocate with the word mistake, because it is a chameleon word, which can take on all sorts of meanings. One can say that a mistake (sense 1) is not really a mistake (sense 2). It is false to say, "There should never be mistakes," because this statement is circular and because of equivocation of mistake. If one should never use an open context value term, and because mistake is such a value term, one should never say that there is a mistake as such. A mistake or error may produce beneficial as well as harmful results. Many discoveries in science result from errors or mistakes. Mistakes and errors are contextual. "The definition of what constitutes a medical error is so culture-bound that it changes as attitudes change" [15]. Error is often due to lack of time, due to overload, due to not allowing one to further think about a case. Reconsideration often allows one to come to a better diagnosis or solution, recheck and reevaluate cases and so present or correct errors. Error-free practice is often the assumed standard of medical practice. Such perfectionism is a delusion of adequacy. It is not to be had. Spath argues, "Quality experts agree that the most common cause of performance problems is the system itself, not the individual functioning within the system" [16]. Admittedly "system" is an abstract term needing clarification.

Every time a doctor or healthcare worker is at fault, management and administration are also. It is not necessary that errors and incidence reports be met with blame and punishment, but rather with prompt action to correct the errors. This would apply to managers themselves. Errors are basically due to lack of rational, critical thinking (speaking), and bad communication. Time and motion requirements (TQM) can themselves cause errors. Errors can be of various sorts: individual judgment, policy, protocol, memory, administrative, mismanagement, patient error, incorrect dosages, etc. Errors may ultimately be traced back to administration and management. In short, administration and management, in this way, let people die. Root causes often end with recommendations for improvement especially in management areas, which include system improvement. Unenlightened management and administration violate the Hippocratic rule, "Do no harm." Harvard Malpractice Study (1993) found that one million potentially preventable medical errors result in 120,000 deaths per year. There must be many more than those reported. In another report, there are 380,000–800,000 preventable errors in the U.S. each year [17].

Physicians and scientists have only a very limited knowledge and so operate in a world of high risk and many unknown factors. We do not possess complete medical knowledge, so must treat far from anything like complete knowledge. Medicine is a practice of probability and uncertainty.

Ethics literacy. To be acceptable management and administration needs besides rationality, critical thinking (speaking), knowledge about emotions (how to practice and encourage positive emotions and eliminate negative ones), also knowledge of ethics. Managers as well as medical staff might at least be familiar with a naturalistic theory of ethics such as that of John Dewey. Ethical codes may be established for administrators and managers to adhere to. A brief description of Dewey's pragmatism and naturalistic ethics follows: All knowledge is practical. He rejects intuition, formal logic, abstraction, metaphysics, supernaturalism; fixed ideas, principles, rules, knowledge; commandments, fixed universals (e.g., Kantian or religious), fixed duties, indoctrination, dogma, absolute truth, absolute certainty, mind or spirit-body dualism, mentalism (thinking there is mind and ideas as such; mind is only acting and speaking, inner and outer conversation, not a spiritual entity), non-participatory education, appeal to authority, objective ethics or right-in-itself or wrong-in-itself, intrinsic values, mysticism, unscientific or impractical medicine, atomistic rather than dynamic thinking, a priori reasoning rather than a posteriori reasoning, deductive logic. One cannot be an enlightened manager without knowledge of such ethical and pragmatic theories.

Evaluation of administration and management is regularly needed. Management and employee attitude surveys at each level can be given regularly and used to identify and correct problems.

Honesty and candidness is expected of all members of the organization. Violations may be reported and corrected. Integrity means honesty but it can also mean to integrate one's enculturated, inconsistent moral behavior, into an ethical system. In the United States, both hospitals and physicians are rated and given a "report card" by several independent agencies. Another aspect of overwork is excessive compulsion on the healthcare worker to perform to the point of exhaustion. This is sometimes referred to as "Performance Addiction" [18]. When work related activities are included many physicians, such as the author of this book, work all days, 7 days a week sometime with little sleep.

Hiring, unethical. If hiring of a family member is nepotism, so also is the hiring of friends or the hiring on the basis of gender or personal interest. Hiring should be based on clear and fair principles, such as that of humanism and on a naturalistic ethics, which would promote competence and fairness.

Humanism. The enlightened manager would be humanistic, but should also know what the view of philosophy humanism is. Virtually few do, which means that management and institutional goals are not based on such a philosophy. All of the positive and reasonable basic bioethics principles mentioned earlier are already in humanism which is much more clear than other views and extensive and is grounded on a naturalistic philosophy, e.g., the pragmatism of John Dewey [19]. It is presented more fully in the Chapter 17, but this entire book is centered around this approach. Some of the things it involves are: inquiry, critical thinking (speaking), altruism, naturalistic ethics, rationality, pragmatism, humanistic concern for all people, anti-dogmatism, anti-supernaturalism, naturalistic ethics, consequentialism. Its agenda is concrete, contextual problem solving (cf. clinical case method). As stated earlier, humanism is more concerned to care for all of the world's desperate people than virtually any other ethical, moral or normative, or bioethical view. It is the ethics for a globalizing world.

Knowledge and responsibility sharing. Medical techniques, skills, and information might best be shared and not used to empower oneself or to prevent another from advancing in their profession. In-service education can be provided. In this respect, there is the responsibility of management to educate and share knowledge and skills.

Modeling behavior. Administrators and managers are to act as models of thinking and behavior ("Follow me" or "Act as I do" management model). This applies to the modeling lifestyle of all of the staff as well.

Models of management. There is no model of management as such or best method, but rather various models or metaphors for management. Some models are: humanistic management, profit-motive management, top-down management, team and system management, participatory management, chain-of-command management, command-obedience management as in the military, management as guidance, etc. A minimal choice of a model would be one that does not dehumanize or exploit employees, but one that is humanistic and enhances employees toward best accomplishing institutional goals.

It was stated in the Chapter 3 that models of decision-making may be based on such models as the following: profit, religion or supernatural, politics, culture, science, family, obedience, punishment, selfishness, altruism, exploitation, legality, rules, or human and natural ecology and humanism, etc. Such models may apply to hospital, management, workplace, home, family, community, political structure, relationships, communication, goals, etc. It also suggests that institutions as well as other areas of one's life may be best conceived of on the basis of well-thought-out metaphors. All of them need to be harmonized. A dictatorial workplace and democratic home life, or vice versa is contradictory. On the first, obedience to superiors is the most important quality of a manager as indicated by, "Loyalty is job number one"; "You are either with us or against us"; "Like it or leave it." A supernatural belief system clashes with science-based medicine. An unhappy home life is a negative influence on a successful career. Religion clashes with humanism and concern for research and consequences. The profit motive clashes with the humanistic treatment of staff and employees. A punitive environment clashes with efficiency and humanism. Law often clashes with ethics. Politics often clashes with honesty and humanistic concern. Democracy clashes with authoritarianism.

Mutual trust and respect is required. This is built by cohering to the other recommendations.

Outcomes. "Outcomes" is an open-context term as is "consequences." We need to know what kind of outcomes or consequences are acceptable. Good outcomes are not necessarily an indication that the administration or management was good. It is my experience as a senior gynecologist and obstetrician and also it was worked out by an evaluation team that in my hospital department that the healthcare workers overcame the inadequate management by covering for it and by excessively working to the point of exhaustion. We are extremely understaffed – almost to the extent of dangerous practice, especially in night attendances. Outcomes and consequences must be specified in terms of the goals of the institution in concrete terms. The US ranks 30th in infant mortality rate lower than nearly all countries of Western Europe. There were 7 deaths per 1,000 in 2002, 29 per 1,000 in 1958. England has only 5 per 1,000 (CDC). We must determine why this is the case especially since the U.S. spends more on healthcare than any other nation.

Overwork. One of the most blatant examples of adverse and anti-humanistic management on the individual and institutional level is the requirement that healthcare workers may have to work 80–115 h a week. This is kafkaesque. The official workweek in Germany, France, and Belgium is 35–40 h a week. Denmark has a 34–37 h workweek. By contrast, physicians work up to three times as long and often without equivalent rest breaks or without having time to eat. If one cannot understand that this is harmful to staff and patients there is little else that management can be trusted to understand. This is an example of the failure of understanding, which is discussed at the end of this chapter. Credibility as a healthcare provider and medical professional is lost. Healthcare overwork is a paradigm example of the captivation by a historical and irrational model. The widespread practice is so significant that it is dealt with extensively at the end of this account of management.

Patient management. The hospital or healthcare unit might consider admitting no more patients than the staff and hospital can reasonably handle, and unnecessary patient requests could be carefully screened. The number of patients seen (or operations performed) per time unit must be limited. Adequate screening of patients is needed to prevent trivial and untimely requests for treatment. A patient's oath should be required of each patient, just as physicians have an oath and ethical rules to follow (For such an oath see Chapter 12: Patients' Duties and Patient Code of Ethics).

Philosophy and goals of the organization. The organization including administration and management must have a clear, written philosophy and ethics, which is formulated and carried out in participation with the staff. The philosophy would advisedly be more than merely an economic or profit-motive one.

Political power. Political pressure and power might well be avoided as much as possible. One central meaning of political tactics is to use crafty and unprincipled methods and influence. "Political decisions" is another term for a fallacy, e.g., the fallacy of force. The use of political power should be transparent and serve as a regulation for fair management.

Power. Power, as such, is often wielded over others. The manager is often bossy, controlling, calculating, scheming, domineering, officious, authoritarian, overbearing, imperious, tyrannical, haughty, arrogant, superior, and/or egotistical. It is often said that power corrupts, and absolute power corrupts absolutely. One speaks of the arrogance of power. One's position is too often used to accumulate power for oneself or be self-serving, instead of using power as a force for being creative, helpful, effective, and achieving goals, which could be wished for by the whole staff.

Senior surgeons and heads of medical departments have been observed by the authors to withhold medical instruction of the staff so as to maintain and enhance their power and prevent others from having the needed skills, they in this way violate their professional duties as physicians and should be corrected or replaced. Healthcare managers and department heads are often more interested in personal power than in patients. For example, instead of teaching others their surgical and other skills they prevent others from learning them or advancing. They violate their code of medical ethics to "teach and mentor those who follow us for they are the future of our caring profession" [20].

The manager must be psychologically sound and free from dysfunctions. Narcissistic personality disorder is an example of the misuse of power due to a dysfunctional personality disorder. But often this sort of disorder is a prerequisite to becoming head.

Publicity. The public would advisedly be regularly informed about healthcare services regarding hours, treatments, financial conditions and whether the hospital needs more funds, limitations of resources and ability to treat, rules regarding treatment, etc. The public has a right to know that medical services will be curtailed due to lack of funding, to also be informed about what is required of them as patients including filling out a patient oath regarding their duties as a patient. In the U.S. there are Hospital Report Cards and reports of institutions and healthcare workers, which are available to the public. If hospital funds are too low it is the public, which must know so they can seek sufficient funds. Lack and misuse of financing puts lives and health of patients at risk and it may be stressed this is also true of the lives and health of healthcare workers whose health is often ruined, burnt out by the present healthcare system [21].

Qualifications for management. Knowledge of ethics, emotion and critical thinking (speaking) as well as expertise in the area managed is all required.

Quality Management Committee may be established consisting of administration, physicians, nurses, staff, and others to identify problems, formulate plans to improve, change policies and procedures. This may also be called a Total Quality Management (TQM) Committee. Criticism of the communication process and quality of communication itself must also be presented. Meetings would preferably be held regularly involving communication with each level of the system. Dlugacz challenges responsibilities: "From the top leadership to the hourly workers, everyone involved in healthcare should feel responsible. . .for the entire process of care" [22]. In one sense everyone is a manager and responsible to critique bad policies and actions including one's own concrete activities. Dlugacz adds, "A sophisticated and elaborate committee structure monitors quality care throughout the institution and allows each caregiver a platform to express concerns and priorities" [23].

If quality management (TQM) is to be used, it should be used for checking management as well.

Quality often is sacrificed to quantity. There may be mainly economic requirements, or unneeded operations to produce income. The medical staff may have the goal of producing the best therapeutic result, but administration may instead have the goal of producing the minimum care for the lowest price. In the medical experience observed by the author, patients treated in the cheapest way often do not get better. Administrators who are only economists or business oriented would be in need also of the expertise of a medical supervisor. The medically empty economic model is unacceptable as a way of letting people die. A correction is to produce as much safe quality care for the lowest price if resources are scarce. Dlugacz puts it in clear terms, "The budget can set limits on what the organization can offer to patients, but not on the quality of the care that is offered" [24].

Case example: health versus financial motivation. At a meeting of physicians, gynecologists and obstetricians in practice in Salzburg in 2004, a presentation was made showing the benefits of breech presentation vaginal delivery over elective cesarean section and that the rate of more invasive cesarean sections could be reduced by more than 50% with more favorable outcomes for child and mother. One obstetrician in private practice then raised the following complaint: "Is it then the case that I have to inform the pregnant woman with breech presentation about that option – sending her to Women's General Hospital instead of delivering her by cesarean in a private hospital which means that I would lose income?" Other irrational objections were motivated by the fear of losing income, without regard for the health and wellbeing of the delivering woman. Therefore, some did not even want to inform the woman about the vaginal delivery option. Some physicians in various countries automatically perform cesareans simply because of the financial motive. Some also do it because their own skills are not good enough to perform the vaginal deliveries in difficult cases, which is reasonable and legitimate. Others do so because they do not want to be disturbed by a possible night delivery.

Respect. It is important to respect all staff at all times. Violations may be reported immediately and promptly corrected.

Risk. To what extent and in which ways do the policies of management and administration endanger the health and life of healthcare workers and patients? This has to be part of risk evaluation and risk management.

Salary schedules and salary transparency. Salary schedules are sometimes made open for all to see. In fairness, salaries of management would be consistent with other members of the organization with similar qualifications. Salary discrepancies and differences could be made more fair. Differences between professional physicians or nurses and administrators could be kept at a minimum and the latter paid less if the experience and qualifications are less. In the U.S., the CEO salary was in some cases 728 times that of the ordinary minimum wage earner [25].

If there is not enough money for salaries the hospital might consider limiting services or close. It is not a reason to say, "There is not enough money to pay staff adequately." The faculty in the author's teaching hospital is not paid to teach there but only to do this during their working time in the hospital. The salary equivalents are not even remotely comparable to those of teaching members in other medical schools. In short, they do two jobs and the work of up to three and a half people. The hospital salary for me, a physician is roughly one-seventh of what I would be paid in the U.S. excluding the lectures in the teaching hospital. It would be roughly one-tenth, were the latter included. Additionally, Austrian physicians' salaries are typically so low that in order to survive financially they are required to work extra hours, e.g., take on too many night attendances which leads them into burnout.

Hospital physician yearly salaries are (in thousands): U.S. $165–268, France 102–116, Spain 42–67, Denmark 50–73, Germany 35–56. Austria is equally low (Source: NERA). Net salaries can be half of this amount [26]. 44% of U.S. nurses average between $25–35,000 and start at $15 hour. The salaries of vice presidents of healthcare are around $240,000 per year, but they are for a 40 h week, so twice to two and one half times as much as long working physicians' salaries.

2005–2007, in Germany, 10,000 doctors were on strike for free time and financial compensation for time worked. The strike was to show that medicine is not merely a societal obligation, but a profession. 76% of Germans were reported to be for the strike. Desired was a 10 1/2 h working day maximum. What is the senior physicians' salary per hour in Austria? 14.5 Euros, the average for night and day shifts in Salzburg (SALK) hospital. And for the residents the salary is less than 10 Euros per hour. In the U.S. hours for residents were cut to a maximum of 80 per week yielding net about 20 dollars (or 15 Euros) per hour, however residents receive only token salaries. Suppose the minimum wage is $5. If the resident works 80 h per week the gross hourly rate is then $2.5 per hour. Doctors in Bratislava earn an average of $555 monthly and work twice as much for it as those do in other fields. They also threatened to strike as the Germans did.

Physicians resist strikes because it is against their professional code to endanger patients. This allows administration to take advantage of them. Administration knows this and so has pushed physicians to seriously endangering their health in order to earn profits from the advantage. However, the situation is so bad in Germany, and some other countries, that the physicians across Germany went on strike in 2005–2007. A 2004 study by UK economic research group NERA showed German hospital physicians at the bottom of a list of 11 western countries in terms of compensation and a similar OECD study showed German physicians earning 15% less than their counterparts in the UK, and 40% less than US doctors. Nightshifts often follow dayshifts without breaks. Mandatory overtime is common, and often doctors receive neither extra pay nor compensatory free time. Physicians in Germany worked 50 million hours overtime annually without pay with a loss of 1 billion Euros. The average hourly wage of an assistant physician is currently €14,5 or in dollars $17, $7.50 after taxes. This is nearly half the rate earned by a skilled worker in any other profession. Cuts have been also made in Christmas and holiday pay. The TdL demanded a working week of 42 h instead of the current official 38.5 h, although the actual workweek is over 70 h. The main negotiator, Hartmut Möllring denounced the walkout and asked, "Are the physicians less tired after 60 h work, when they receive 30% more money?" It did not occur to Möllring who is finance minister for Lower Saxony from the Christian Democratic Union – CDU, that working 70–100 h a week is inhumane and life threatening to both healthcare workers as well as patients. It may be noted that doctors' demands would cost the state up to 10 billion Euros, indicating what their loss in salary has been. 6,300 German hospital physicians had already left their jobs to work abroad.

Also a gender issue is relevant here. Women are often underrepresented in the management workforce. In many cases this is because they lack the work-time needed to arrive at such a position. Men often block the advancement of women in the workplace. The reverse can also be the case. Men are underrepresented in nursing and teaching in the early grades. Men and women block the advancement of other men and women as well. Management often hires on the basis of networks, friendships, religion, gender, political and social connections. This is done in practice in spite of civil laws against doing so. It is not objective and fair to men or to women. It is not professional. 50% of U.S. physicians are women [27]. In Jena, Germany, more women (83%) study medicine than men [26]. Men and women, with the same years of experience and qualifications are often earning different salaries. Salaries are often negotiated. Merit and other factors also make for salary discrepancies. Nevertheless, a humanistic approach to transparent fairness in promotion advancement and salary distribution for men and women is a basic requirement for enlightened management.

Case example: hospitals sued for mismanagement. W. Schmidbauer argued that management in Europe exploits patients and healthcare workers [28]. It happens in the U.S. as well. A number of tax-exempt U.S. hospitals pay management excessive salaries, excessive travel expenses to resorts for meetings in tropical settings. Unnecessary private jets are used for management travel, aggressive legal suits are engaged in to collect from the uninsured for excessive charges. The American Hospital Association encouraged such aggressive billing practices. They also may have extremely large financial reserves. Mississippi lawyer, Richard Scruggs, charged 300 hospitals in 26 states for aggressively overcharging the uninsured instead of giving charity care which failure to do so is in violation of their tax-exempt status. As a result, $100,000–200,000 refunds were given by St. Dominic Health Center in Jackson Mississippi, alone as reported the *Health Law Reporter* Aug. 12, 2004.

Size of management. Administration and management would preferably be no larger or more costly than necessary. Salaries might be kept very close to, or lower than, other professionals depending on qualifications. Managers and administrators may have lower salaries than those with more expertise and stronger backgrounds and performance. Oversized management may be due to power games rather than due to workloads.

Staffing would be adequate for the work to be done, allowing for sick leave, vacations, and emergencies, etc. without requiring overwork or stress on certain individuals. An efficient management would replace as promptly as possible staff, which leaves, which in my experience is almost never the case. For example, in the author's hospital some physicians have been methodically not replaced for at least 3 months, thereby causing the staff overwork and stress. It is the physicians and nurses who primarily constitute the hospital and need to understand the needs of the patients. The managers may serve merely in an organizing administrative and supportive role. Administration and managers may be conceived not as superiors, but rather as assistants to the physicians and staff for the purposes and goals of the hospital, for the purpose of treating patients adequately, as physicians and nurses may be conceived as serving the patients.

In 2004 in the U.S. 126,000 more nurses were needed. 1 in 7 hospitals was understaffed. It is expected that more nurses will be leaving than entering nursing by 2020. There is a projected one million shortage of nurses [29]. Thus more stress and burnout is expected. 95% of the nurses are women [30]. Roughly half of the nurses reported dissatisfaction with their jobs [31]. Surgical patients have 31% greater chance of dying if the hospital nurse-patient ratio is less than 1:7. Nurse-patient ratio in Intensive Care Unit should be 2:1; in the ward 6:1 [32].

Stress management seminars, which identify and attempt to correct the sources of stress are highly recommended. A well-qualified senior physician or surgeon should be more valuable and may deserve to be paid a higher salary than an administrator. An administrator may be easier to replace than a skilled physician or surgeon. Patients are in need of good physicians and nurses. Administration is just to help provide such conditions. It should be to prevent "fatigue of staff" mentioned above and "burnout."

Suicide. U.S. male physicians have a 70% higher chance of suicide than men in general, including other professionals. Women physicians attempt suicide three times more often than men, but succeed four times less [33]. Management must be enlightened so as to prevent rather than contribute to the suicide rate. The staff is often used or misused until worn out, looking for another job, ill or working just at the minimum necessary level.

Transparency is honesty and openness regarding communication and revealing what is actually going on in the hospital. Hospitals and healthcare units often are more concerned with appearances and how they look on paper than with patients, healthcare workers or the actual mission. Healthcare units may have the highest ratings or public image though the management and policies are at their most unhumanistic, inefficient, and dangerous levels. Politics often prevails over patients and healthcare workers.

Understanding in management. We may take a look at the notion of understanding to see why in management people do not understand one another and why they cannot communicate very well. Can people understand one another at all? To this end the following is a characterization of understanding in its various dimensions.

Understanding may mean to become clear about something. It is to have an emotion of success. "Oh, now I understand."

Understanding can mean to obey the rules and follow them regardless of their merit. It is to know one's boundaries, or a statement of being overpowered.

To understand may require that one cares or is involved. To just follow rules or repeat policy does not entail understanding.

One understands a view, but objects to it. We may accept a view, but not believe it, or try to understand the other's opinion.

There are degrees of understanding of the various sorts mentioned here: total commitment, perfunctory compliance, disbelief, uncommitted working relationship, conviction, officiousness, etc.

The understanding of one who cannot organize and present or express arguments is significantly different from one who can. Understanding presupposes rationality, the ability to follow rational arguments. Without good communication and sound critical thinking (speaking) there is no understanding. Discussion illiteracy undermines communication. One could say that those who cannot critically discuss, also cannot understand. Discussion illiteracy is widespread.

"I understand you" can suggest, "I like you," showing emotional involvement. Understanding, then, can be an emotion or an appeal to emotions. Understanding can mean an appreciation or positive regard of someone. Thus, one may not understand someone whom we do not like, or we have little respect for. Adversarial or cold management can expect little understanding from the staff in this sense.

Ethical versus moral understanding is also at stake here. Moral understanding is merely uncritical enculturated normative views as opposed to informed critical ethics. Disagreement often arises because of the two types of understanding. It is like the difference between critical and uncritical thinking (speaking).

Humanistic versus anti-humanistic understanding may involve seeing understanding in terms of altruism vs. egoism. Communication often fails because neither holds the other's position, e.g., the egoist cannot understand needs of others.

To understand another's view may only mean that one agrees with it. It is a form of intellectual agreement. "They just cannot understand" is equivalent to "They just cannot agree." One can also agree without understanding and understand without agreeing.

We usually use terms abstractly, a form of language insensitivity, without clearly knowing at all what is meant. This does not provide real understanding.

Normative understanding versus critical understanding. People use the usual senses of the term, but did not develop philosophical critical understanding. People have common, normative, prefixed, not critical understanding. If they do not have critical thinking (speaking), they cannot understand philosophically (a logical point). A physician as a physician, though an expert in surgery or a technical medical specialty cannot automatically be expected to be knowledgeable about philosophy, ethics, or critical thinking (speaking).

Understanding, like education, can involve the ability and/or willingness to change one's views. Thus, dogmatic views and the lack of philosophical and interdisciplinary discourse block understanding. Openness opens it.

Understanding involves the ability to comprehend theories as well as to apply them in practice in diverse circumstances. If we cannot apply theory we do not understand what it is about.

Understanding is not fixed knowledge, but a developing process of inquiry. Conflict arises when one sees understanding as a closed, fixed position. Understanding does not involve certainty, but on-going inquiry. Understanding is not merely a matter of persuasion, but one of ongoing investigation and interpretation (Compare fixed versus pragmatic law). In addition to the schools of natural law, British and Austrian analytical positivism and historical jurisprudence may consider humanistic pragmatic (instrumentalism) views of law. This stresses reason and consequences and abandons fixed law, deduction, authority, and precedent. Reason is needed for judges as well as physicians [34]. Decision-making must be contextual and personal, sensitive.

One can be acquainted with, conversant with, merely familiar with, guess, or claim certitude. There is quantity versus quality of understanding. One of the major errors in medicine is claiming certainty when there is only probability, statistical and partial knowledge ignoring clinical practice.

The evidence of understanding in one's own case is different than the evidence for understanding in another's. The latter we have indirect evidence for. "I understand" is not the same understanding as in, "You (they, we, he/she) understand(s)."

Different types of understanding as philosophical understanding, legal understanding, medical understanding, scientific understanding, supernatural understanding (a contradiction to philosophical and scientific understanding), client or patient versus professional understanding, etc. challenge contexts. Each person has his/her own understanding in terms of what they know and feel. Understanding a sentence or word is not like understanding a picture. We do not have understanding as such. As there is no set meaning of understanding, no understanding as such, no two people can understand one another as such. That there are these different types can be barriers to the understanding of each other. Understanding can be putting in perspective the physician's, managers, etc. perspective.

Understanding is not a mental process. Understanding is not a mental process such as thinking or having internal ideas in one's mind. Thinking, ideas, and mind are pseudo-psychological concepts with no corresponding reality. Rather thinking is basically language use, talk to ourselves or others. Without language there would be no thinking. It is not clear what there would be (For a full analysis see Chapter 18).

Value understanding is orientation towards goals. One sense of understanding is to find meaning or enjoyment in something. It is to appreciate something. We would ask trying to understand, "What is the meaning of this disease for my quality of life?"

Because of the many meanings of understanding, one can easily equivocate. For example, one can think someone understands in one sense when they only understand in a different sense. Good management communication involves an awareness of the nature of and problems with understanding.

Violations of any points mentioned here might be brought up in meetings for correction.

Case report: understanding and autonomy. A woman 31 years old was diagnosed with extra-uterine pregnancy, pregnancy in the Fallopian tube. She presumably was in the 8th week of gestation. She already had free fluid, probably blood, in her Douglas excavation, which is not uncommon in such pregnancies. The fear is that the pregnancy will rupture through the thin walls of the Fallopian tube causing possible life-threatening blood loss. The patient was told, "Please remain in the hospital because of your Fallopian tube pregnancy. We may have to perform pelviscopic surgery at any time if the extra-uterine pregnancy ruptures. In order to be prepared for the operation, you should also not eat or drink. In any case, surgery will be performed tomorrow". The patient replied, "I cannot stay as an inpatient. I have to take care of my two cats at home." She asked, "What is the worst scenario, if I leave and wait until tomorrow?" I (the author and the patients ́physician) answered with as much clarity and emphasis as possible, "It might cost your life, especially if no one is at home with you and you were to lose consciousness due to blood loss into the abdomen and no one could bring you to the hospital. She nevertheless signed a paper indicating that she left the hospital fully on her own responsibility and that she was informed by the physician beforehand about the risks of doing so. That same night she came back as an emergency patient. It was fortunate that she was even able to find a way to return. I operated (the author) on her and the outcome turned out to be favourable. However, because of the emergency conditions she did require stored blood, which she most likely would not have needed if she had stayed in the hospital, thus costing extra danger, services, expenses, and medical resources. As is usual with operations I visited her the following day and she said to me, "I would not have left the hospital if I had known more clearly what the consequences would have been." In what clearer way could I have explained to her what was at stake? Information often does not seem to clarify for a patient clearly lacking the capacity to understand. She just thought of her immediate concerns without consideration of the significant longer-range consequences although they clearly were pointed out.

Case example: economizing by management. It is almost a universal that health resources are limited and that it is, in any case, desirable to economize. The problem is where we may economize without harming the patient. The walk-in at the Salzburg Women's Hospital Gynecological Department is not used very often at night by patients, especially after 10 p.m. Before that, patients come in regularly. During the night a nurse is on duty to answer the many telephone calls, prepare tools for the next day, give information to those seeking the obstetrics department or help with emergency cases. In order to save money it was proposed that the nurse attendance shift after 10 p.m. be omitted. The operating theatre nurses and night-attendance physicians were to somehow take over these additional tasks although it was no plan as to how this would be done. This would mean firing two and one-half nurses. The head hospital nurse manager called together those involved, explained the situation and asked them to designate which team members would be let go or work half time. By having to engage for weeks in these discussions the team was demoralized as they were participating in the firing of their own team members. There was a confusion of kindness and desperation, trust and mistrust, positive and negative emotions. The goal of economizing was left aside in all of the confusion and the former trust and team spirit was torn apart.

Case Example: Hospital strike: physicians on strike. In 2004 there began a protest and strike by the physicians of the nonprofit Salzburg General Hospital (Salzburger Landeskliniken SALK), including the author, as was reported by Der Salzburger Arzt [35]. In 2005 this strike ended. On an absolute scale and as compared to other professionals of similar responsibilities and educational level, physicians of the hospital are paid low basic salaries, which serve to provide only low retirement incomes. Their salary is among the lowest in Austria. In order to make a sufficient living to support themselves and their families they are forced into taking excessive night attendances and additional excessive responsibilities such as being on call. A Senior Physician, age 35, has a basic salary of 3,061 Euros gross per month and after taxes up to half of that, or 1,530 (18,360 yearly) Euros. With four night attendances this can rise to 2,740 monthly after taxes (roughly 33,000 per year), but this may require more than an additional 30 h a week. For comparison, cleaning women receive about 10 Euro per hour and they tend to take more sick leave than physicians. During a physician's strike for better wages, the Business Director of the Salzburg hospital Landesklinik, released inflated salary figures to the newspaper, Salzburger Nachrichten [36], giving only the before taxes salary. It was an attempt to whitewash the issue. The base salary of an Assistant Physician is 2,335 (slightly more than a privately working cleaning person, but less per hour), of a Specialist is 3,061, and Senior Physician is 3,676 Euros per month, but the take-home pay after taxes and regular dues, etc. is about less than half of this. A physician responding to this announcement in a subsequent issue of the Salzburger Nachrichten stated, "Gross Salary is irrelevant" [37]. The physician calculated that the basic salary for even a Senior Surgeon is only about 2,000 Euro net monthly. If one worked additional hours (four night attendances and 1 weekend a month) totaling 273 h a month the actual income of the physician is 14.5 Euros per hour. People normally work 160 h a month, or 40 h a week, whereas physicians work over 68 h a week just to earn a basic living. Retirement is only calculated on the basis of the lowest basic salary. Because the hospital is kept seriously understaffed, some physicians have little choice but to work around 60–80 h a week and sometimes, while totally exhausted, operate on or treat patients. Me as well as my collegues just after a day and night attendance were on several occasions required to fill in for a sick physician because no other physicians were available to do so. Furthermore, in Austria, physicians can only retire at age 65, whereas nurses, railway workers, and others can retire much earlier. In the U.S. women are allowed to retire with full benefits earlier than are men. If I use the hospital for a private patient, I do not receive any of the amount charged by my department. Physicians who write or speak to newspapers to discuss hospital mismanagement may expect never to be promoted and to endanger their careers.

Additionally, the Head Physicians receive up to ten times more salary than the regular physician. The administration keeps their salaries hidden from the other healthcare workers. As a result, we can even project the worst unfairness. In companies in the U.S., Chief Executive Officers have even been known to have salaries 720 times that of the average or beginning workers. The salary gap is also unfair because the hospital is in financial trouble. How did the administration try to resolve the strike in Austria? They decided to cut supplies and non-medical services, economize, hire too few physicians, give the physicians only 15% salary increase in 10 years, but not reduce or touch the salaries of the heads and administrators who are so grossly overpaid in comparison to the staff. Presumably the heads would then side with administration, but many were often put in such jobs most likely because of loyalty and political or social reasons in the first place. The managers have a hostile adversarial relationship with the healthcare workers below them. This is an example of extremely arrogant and non-communicative management. If management had been good it would have already economized, and communicated with the staff to avert a strike.

Another lesson is that failed management is expensive, not just in terms of money, but in terms of the health and lives of patients and physicians. The rational thing to do would have been to reduce the head's and administrator's salaries to the levels of senior physicians, but lower if their experience and background is less. The behavior of the management is contrary to the Hippocratic oath and medical ethics.

For comparison, I am a senior physician and with my advanced degrees and experience would in America command a salary of $350,000 dollars per year (Source: internet average physician's salaries. High malpractice insurance might reduce this, however). My income in Austria, for example, doing a cesarean section with total responsibility is only fifteen Euros. That is the physician payment which is per hour not per task. Also, the number of night attendances is over the legal allowable limit especially because of severe understaffing causes. Some work sometimes around 100 h per week and sometimes average 80 h per week, up to two to three times the normal workweek. *Der Salzburger Arzt* reported an overall average of 60 h per week, sometimes 72 h, and some attendances last 49 h at a time, six times the normal work day. Furthermore, the number of patients since 1996–2005 increased 20%. This overwork has caused physical and psychological burnout and illnesses such as depression, tinnitus, sleep deprivation. Family life is also disrupted. Many physicians, including some of the most experienced, left the hospital as a result. The hospital was generally regarded as one of the worst places a physician could work which is evidenced by the strike and by management consultant results. The result was that patient care was also put in danger. *Der Salzburger Arzt* reported "Ongoing pressure to save money expresses itself in the quality of work of physicians. It endangers patient's care to the utmost" [38].

The work of the physicians and surgeons demands high alertness and skill, a positive emotional balance and strength in treating patients. At present an exhausted physician or surgeon who has not slept for 30 h may be asked to treat a patient or perform a long and difficult operation. Would you wish to have such a surgeon operate on you? These negotiations for salaries had been going on for 10 years with at most only the verbal promises of the administration, but no action. As explained by Dr. Terje Hovdar, this resulted in a strike concerning such low wages, duties, goals, flexible schedules and inferior management of the hospital [39]. He added that the politicians lack the required medical knowledge, and do not have adequate communication with the physicians, and as a result the hospital itself lacks an adequate healthcare plan and has been in financial crisis for 10 years. This shows the lack of concern about health. The hospital business manager made the outrageous proposal to cut the salaries from younger physicians to compensate for the salaries of the seniors. A May 2004 institutional study showed that only 15% of the people in Austria trust their politicians, but 91% trust their physicians. Vice President Hovdar of the Physicians Medical Council states that it is clear why this is so [39].

The Salzburg government had decided for a reform of physician's salaries in February 2004 to raise the basic salaries, pay lump sums for the extra night and weekend attendance hours, provide more flexible and better patient-oriented work and office hours. Work for the Private Medical School was to be voluntary, but such work and teaching was to be minimally paid. However, it would not be paid as if one were a professor at a medical school, though physicians often had qualifications exceeding such professors. There is no payment at all or practically nothing for such teaching.

In terms of reform no fixed date was given. Once again, as during the 10 years delay, the physicians began to realize that the proposed reform was an empty promise so they took labor-unionist measures to force the reform to begin in 2005. Several hundred physicians of the approximately 550 affected met on Sept. 2, 2004 to protest by deciding to block administrative work and thereby block accounting and the ability of the hospital to receive any funding for the work being done there. As of summer 2005, there was no success. It is an extremely rare case when hospitals go on strike, but the management was so corrupt that this extreme measure was taken by even well qualified and caring physicians. Emergency care for patients would always be available, if needed. Christine Rühle, the vice-CEO of the hospital said that although she is sympathetic with the physician's requests they cannot be realized in times of scarce resources. The recently elected governor of Salzburg, Gabi Burgstaller (SPÖ), also referred to the present financial crisis of the country and hospital, and she tried to discuss the issue with all parties to try to reverse the crisis. To date this was apparently without success. The questions then became: why did they for 10 years offer so many services which they did not have staff or funds to support, why do they have so many highly paid bad managers, and why did they open a new medical school without proper funding in a hospital so badly managed, as they are also an indirect part of management, why did not the physicians strike earlier?

Case example: analysis of the above strike: As a senior physician in the hospital and author of this book I will give my views regarding the above crisis. I am engaged in the procedures, discussions and decision-making of the above discussed and am fully for the protest of the hospital physicians. Where management mismanages and physicians have had for many years to cover for that at the expense of their psychological and physical health, there should be a reaction. Physicians must demand and initiate full communication with management and be themselves participants in management. It should be realized that they are the essential "objects" of management and should be fully consulted on all important issues. Management was so unresponsive that nothing else was left to do but strike. The physicians also had to strike because they could be individually blamed for failings due to the structure of their work instead of the management. It is management's job to communicate well and fully with physicians, to not offer services which cannot be funded, to try to obtain sufficient funding for the services performed, to let the public know if funding is insufficient, to manage humanistically (though they show not having any knowledge of what humanism is), to raise morale, to encourage physicians in their jobs and careers, to see that the patients have the best care possible, to provide a good environment which will attract and keep the best doctors. Management has failed on all counts. The other requirements for humanistic management mentioned in this chapter may also be used to assess what has happened in this Salzburg hospital. We can learn much about actual management from such actual examples.

8.3 Special Section on Overwork: A Failed Metaphor of the Medical System

A physician shall. . .recognize a responsibility to seek changes in those requirements which are contrary to the best interests of the patient [40].

8.3.1 Introduction

One of the greatest malpractices in medicine is management's requirement of the amount of hours that healthcare workers must work. Most hospitals management is built upon overexploitation of staff, especially in the education process. This way healthcare workers get used to exploitation and loose sensitivity for themselves as well as for patients. It is one of the major indications that organized healthcare is not capable of responsible medical care. Any medical organization which not only allows, but forces its employees to work under the present 80–100+ h workweek and 12, 24, 36 and 48 h shifts should be disqualified as a medical unit. The standard practice of requiring excessive overwork is bad ethics, bad medicine, bad science, and bad management. Some argued that these excessive hours are needed for education or as initiation practices.

The overwork issue shows that the prevailing normative morality and practice is not an ethical practice. It is founded on cultural irrationality and indoctrination. It also shows the insensitivity of the public as well as of the medical profession.

One of the problems of speaking of overwork is that the term "work" remains a vague abstraction. In the present context it may refer to paid employment to perform one's occupation. It need not have the value connotation that it is desirable or undesirable. One may like one's work or not. One may by choice enjoy working 14 h a day or 100 h a week at one's profession, but even then this can be unhealthful and even stressful. The term "work" can cover many entirely different kinds of things. It can cover exhausting physical labor as well as writing poetry or baking a cake. Accordingly, work can mean very different kinds of things to different healthcare workers. We cannot then speak of overwork as such. We must consider what each job requires. Some require night attendance, intense emergency room and surgical operations, others do not.

8.3.2 How Many Hours Do Physicians, Nurses and Healthcare Workers Work?

Hours worked vary from country to country, hospital versus private practice, and depend upon the specialty in which one works. Administrators and administrative staff may work a normal 40 h workweek. Nurses are often in short supply and may have extremely long workweeks.

In contrast, physicians work hours are 60–100 h per 5 or 7 day week. This may include or exclude on-call hours. Thus, some physicians and nurses have only several hours left for sleep and all else. If a healthcare worker is on a 24–48 h shift, all of the hours may be worked without sleep for up to 2 days. In addition, work is brought home and extra hours are spent at home on the phone with staff or patients, and one is often additionally on-call. A good physician makes un-required follow-up calls when off duty out of concern for patients. Also healthcare workers are often expected to give lectures, publish papers, take advanced courses and keep up with the medical literature. Physicians are in addition required to stop for accidents outside of their practice, meaning that they are liable to spend much more time than regular working hours.

The hours of healthcare workers may be contrasted with normal work hours in other professions. A number of European countries have reduced the time to 38 and 35 h per week. The average U.S. worker works 34.5 h per week [41].

8.3.3 Attempts to Limit the Number of Work Hours

New York law allowed for residents an 80 h week maximum, less than 24 h shifts followed by an 8 h rest period. The Accreditation Council for Graduate Medical Education (ACGME) has limited work hours for U.S. medical residents to less than 320 h in a four-week period, but this means one could work more than 80 h a week. Instead of reducing hours organizations just shift the hours around.

In July 2003, the Accreditation Council for Graduate Medical Education (ACGME), as the body responsible for the accreditation of more than 8,000 programs that collectively provide for the education of 100,000 residents, made recommendations for resident physicians that would allow residents to work an average of 104 h a week, with provisions permitting 125-h weeks once a month (acgme.org). The ACGME's common duty hour standards acknowledge scientific evidence that long hours and sleep loss have a negative effect on resident performance, learning and well-being. The standards include: An 80-h weekly limit, averaged over 4 weeks; an adequate rest period, which should consist of 10 h of rest between duty periods; a 24-h limit on continuous duty and up to six added hours for continuity of care and didactics; 1 day in seven free from patient care and educational obligations, averaged over 4 weeks. But such rules are often not complied with.

"58% of long-haul truck accidents investigated were fatigue-related" [42]. Thus legislation has been passed to limit their hours of work in the U.S. New Zealand restricts emergency room residents to 10 consecutive hours and a maximum of 50 h a week. Other residents are restricted to 16 consecutive and 72 total hours a week. This is still excessive. Denmark, Norway, and Sweden, having often the most humanistic and rational policies, restrict medical residents to 37–45 h per week, less than half as many hours as the larger industrial nations [43]. The authors believe one reason for the long hours is financial gain for the institutions, though not for the physicians [44]. If so, this would indeed be perverse. It may be a reason why the practice is not abandoned. Rather, the main reason for working long hours, marathon shifts, is an uncritical medical tradition. The American Medical Association, for example, has resisted reform [45]. When reforms were made in New York State (12 consecutive hours for emergency room residents, 24 h for others, and an 80 h week limitation), the hospitals violated the rules and were fined. It was found that the new U.S. resident hour restriction was not well complied with anyway, and more hours were worked than allowed [46].

U.S. bus and cab drivers can only work 12 h in a 24-h span. Truck drivers are forbidden to drive more than 10 consecutive hours before taking an 8-h break and cannot work more than 70 h over any 8-day period. Rules proposed in 1999 would create a further limit of 12 h of total driving within any 24-h period. Railway workers cannot work a shift longer than 12 h, and any 12-h shift must be followed by at least 10 h of rest. "Aviation safety management...stands in stark contrast to safety and error reduction in hospitals" [47]. Pilots, under U.S. Federal Aviation Administration rules, cannot fly more than 8 h in any 24-h period or more than 30 h in 1 week. Flight attendants are governed by FAA regulations that include a requirement for increased staffing on shifts of more than 16 h and that specify minimum hours of rest after each overtime shift. Pharmacists in the U.S. are not allowed to fill over 10–20 prescriptions per hour.

The Japanese word *karoshi* refers to a worker who has suddenly died from overwork. "Official statistics show that Japanese put in the longest working hours in the industrialized world: a year long total of 2,088 h on average versus 1,500 for Europeans and 1,800 for Americans" [48]. In Japan, as many as 10,000 people may die from *karoshi* each year. However, the average physician and nurse work much more than the Japanese. Excessive work of the Japanese people causes health problems, such as depression, burnout syndrome, and chronic fatigue. Overwork can cause mental disorder and sudden death of employees especially where there is high demand, minimal control, and poor social support. There is also increasing "*karo jisatsu*." (suicide from overwork) [49]. Michie and Cockcroft also report Japanese suicide due to overwork [50].

They recommended that the working week be severely reduced [51].

Similarly in the U.S., regarding having too little control of their working conditions, physicians rate healthcare organizations as having lower standards than private practice regarding the quality of their work lives. The major issue was that physicians had little control of policies in the organizations [52].

According to the EU time directive (hmso.gov.uk/so/si2003/20031684.htm), from August 2004 to July 2007, the weekly working time limit for doctors in training will be 58 h, to be reduced to 56 h actual working time, excluding time which might involve sleeping at the hospital "on call," from August 2007 to July 2009. It is an absurd reduction of 2 h when the hours are still excessive. In 2009, the limit on hours falls to 48 a week.

The American restriction is only to 30 consecutive hours and 80 h a week averaged over 4 weeks. This means one can work more than 80 h during a particular week, and this only applies to residents not to physicians and nurses [53].

In Austria 60–80 h hospital workweek is standard except for certain specialties. Mistakes and an increase of worker auto accidents result from overwork. There was also an increase in percutaneous injuries during night work and extended work duration [54].

In regard to "on-call" time in Germany, on-call service and stand-by are categorized as rest time, apart from the periods during which work is actually performed. Should not also the on-call duty performed be deemed work time in its entirety? A regional labor court agreed – even where the doctor in question is permitted, as occasion allows, to rest and sleep during periods of inactivity. The court said that a doctor being required to be available at the place determined by the employer cannot be regarded as being at rest during the periods of on-call duty when he or she is not actually carrying out any professional activity [55].

8.3.4 Do Physicians and Nurses Also Cause the Problems of Overwork?

Members of the medical profession have been enculturated into accepting working up to two and one half times as long as the normal workweek. By not exercising critically thinking over tradition and by not protesting the long working hours they also cause their own overwork. They are captivated by a metaphor, taking the prevailing practice as literally correct. There are some legal movements to reduce hours of residents. Healthcare workers sign permissions, which allow them to overwork, but the employees are usually forced to work over 40 h a week.

Physicians or nurses may overwork to cover for inadequate management. This is the same as encouraging such management to continue. They may overwork so their colleagues will not have to overwork. In all cases they might ask themselves if it is fair to the health of patients and if it is their professional duty to do so. In addition, Pfaff states that surgeons are working long hours for the economic benefit of the medical unit [56].

If one overworks, the rest of the time must be saved for sleep and enhancement of one's well-being to prepare for emergencies and long operations, etc. Healthcare workers are, however, not likely to spend such time in this way, preferring to do other things. They often do not take care of their own health and wellbeing. Physicians cannot help others if they cannot attend to their own health. "We [healthcare workers] need to. . .demand a whole life" [57].

AFSCME, together with other nurse organizations, health care unions and the AFL-CIO, 287 created a set of "principles" for federal legislation to ensure that the laws have a real impact. All health care facilities are included. Overtime is defined as any time exceeding a pre-determined shift (the pre-determined shift should never exceed 12 h in a 24-h period or 80 h in a 2-week period). On-call is included in the 80 h. Mandatory overtime may be used only in emergencies. Mandatory overtime shifts of three or more hours must be paid at double time for the entire mandated shift.

"Physicians tend to neglect their own need for psychiatric, emotional, or medical help" [58]. Suicide in Japan due to overwork is said to occur partly because of the stigma of going to therapy even when it is available [49]. According to the Canadian Medical Association (CMA) Code of Ethics, the medical system is fundamentally to blame for the low level of physician's wellbeing, but physicians have to some lesser extent been to blame because of their failure to obtain enough rest, exercise, pleasurable activities, and healthy nutrition. "The policy makes it clear that physicians may need to be taught the importance of taking care of themselves and each other." Once again, the burden is placed upon the healthcare worker. Physicians are asked to themselves take continuing medical education courses in stress management, communication, safety, and health and wellbeing of physicians, instead of advocating fewer work hours. The CMA Code of Ethics requires that physicians seek help from colleagues and appropriately qualified professionals for personal problems that adversely affect their service to patients, society, or the profession. Physicians are reluctant to do so especially as it shows their incapability.

A standard on-going team group discussion of each physician's recent errors and successes helps overcome such threats as all are involved. If one is prone to mistakes, proposals of help may be offered. Insurance in the U.S. does not pay for physician psychiatric illnesses. The proposed wellness policy calls for more research into this complex area. This is a senseless tactic because it should be obvious to all that working over 40 h per week is harmful to all and allows little remaining private and personal life for healthcare workers. It may be rather suggested that they work no more than 35 h per week especially in such areas as emergency rooms and surgery, obstetrics, anesthesiology and pediatrics because of the challenge, which comes with such areas [59].

Physicians should supposedly look after their "well-being," be "altruistic" by means of the "ritual" of working long hours, should have a "spiritual life," spend time with their "family," as though "family" were a guarantee for rest, whereas it is often like a full-time obligatory job. The literature on the wellbeing of healthcare workers is here typically devoid of critical thinking (speaking) and philosophical clarification, knowledge of ethics or emotion theory. The medical profession is disciplined in working up to 100 h, but healthcare workers are in general not at all disciplined about their own lifestyle. Overwork is often just another aspect of numerous physicians who practice an unhealthful lifestyle. Physicians have the responsibility to keep themselves healthy, guard their wellbeing, and to live an unstressful lifestyle. They should attempt to limit themselves to a normal workweek rather than one of 60–100 h a week [60].

8.3.5 What Is the Effect of Overwork?

a. Fatigue of staff may be avoided simply by limitation of hours worked, e.g., have an 8-h day including on call contacts and work taken home. It is contradictory for medical institutions to permit staff to drink alcohol, but require excessive fatigue by overwork, which is equivalent to intoxication. Dawson and Reid showed that after 24 h without sleep, psychomotor impairment can be so severe that it is equivalent to or greater than alcohol intoxication [61] (Nature.com). Cognitive function deteriorates to a level equivalent to having a 0.10% blood alcohol level, 30% higher than the legal limit for driving in Michigan.
   After a month of 80–90 h per week and shifts of up to 36 consecutive hours of challenging work, fatigued physicians showed impairments in driving and other tasks, which required constant attention and quick reactions, comparable to having consumed three or four alcoholic drinks. Residents in a simulator attempted to drive at 60 miles per hour and drove off the road at least once [62]. 36% more serious errors were found in longer work hours (85 h with two 30 h shifts, as opposed to two 20 h shifts) [63].
   In Europe doctors must be given 11 h off every day, only 13 h can be worked [64]. This rule is typically violated. One can still work 91 h a week on this system.
b. Studies on sleep deprivation of physicians performing clinically relevant tasks have not been conclusive and the literature is unsatisfactory. Most studies of recurrent partial sleep deprivation have suggested that sleeping only 5–6 h a night can lead to impairment. These decrements in performance accumulate with continued partial sleep deprivation.
c. Medical errors and accidents endanger patient's lives. "Medical errors" cannot technically be called errors, because they are deliberately caused by a system, which requires overwork and thus the mistakes are inevitable. Mistakes are thus often due to the failed management of medical systems.
d. Excessive work demands increase mortality in intensive care units [65].
e. Errors increased significantly for nurses who worked over 12 h shifts and significant errors occurred if they worked over 40 h a week and even more for over 50 h workweeks. Nursing shortage forces nurses to work additional overtime and on days off, which they sometimes euphemistically call "voluntary overtime." Some would lose their jobs if they did not comply. Twenty-four hour shifts have become common [66]. If 50 h a week is dangerous to patients and staff we can only imagine what 100 h a week would be [67]. Having less than 5 h sleep a night endangers the lives of patients and healthcare workers [68].

Errors are increased even when fewer than 24, 36 or 48 shifts are worked. 38.7% of the nursing shifts examined exceeded 12.5 h. The risks of making an error are significantly elevated when nurses worked more than 12.5 consecutive hours or worked longer than scheduled [66].

f. Thirty percent of residents reported clinical depression during their residencies. Emergency room residents are seven times more likely to have a motor vehicle accident due to falling asleep at the wheel during their residency than before it, according to a 2,000 article in *Academic Emergency Medicine* [69].
g. Overworked workers are less caring. "Today's medical-surgical nurses who work in hospitals often feel they do not have time to listen or show caring toward their clients" [70].
h. In one U.S. study, three-quarters of the residents had burnout with negative effect on patient care [57].
i. General Motors (Michigan) found that compulsory overtime up to 60 h per week led to increased accidents, injuries and illnesses as well as long term detrimental effects of stress [71].
   It was found that 84% of managers in various organizations worked an average of 50 h a week, but their productivity dropped and injuries, stress and illness increased. Overwork was found to be cost ineffective [72].
   Nevertheless, 50–60 h a week does not compare with the physicians' 80–115 h a week.
j. In terms of quality of life there is in general little time remaining for life, especially not for quality.

8.3.6 Overwork Harms Health of Staff

Sleep deprivation causes mood change, memory loss, slow reactions, decreased alertness and concentration, and promotion of errors of judgment. Circadian rhythm is disrupted. There are psychosomatic complaints [73].

Ying Liu of the National Cancer Center in Tokyo and associates found that men whose average workweek was greater than 60 h for the past year were twice as likely to have heart attacks as those who averaged 40 or fewer hours. Men who slept 5 or fewer hours for each working day had double the heart attack risk compared to those sleeping for more than 5 h [74].

Scott found that overtime work hours increased the risk of errors and decreased nurses' vigilance. They support the Institute of Medicine recommendations to minimize the use of 12-h shifts and to limit nurses' work hours to no more than 12 consecutive hours during a 24-h period [75].

8.3.7 Overwork Increases Sick Leave

The reasons for this are clear. It harms one's health.

Psychosomatic disorders are prevalent, addiction is often the consequence when other coping strategies have broken down.

8.3.8 Overwork Causes Stress and Burnout and Addictions

"Medical professionals are among the most highly stressed occupational groups" [76]. Most stress is due to management and organizational factors [76]. How long will burned-out physicians last to continue to be able to effectively treat patients? Extremely challenging work situations and stressful personal experiences can greatly increase the chance of a mistake. This is hardly surprising, but needs to be said because administrative practices often typically contribute to stress. In many hospitals, this is extreme to the point of widespread depression and psychosomatic disorders, demoralization and lack of joy of life. The best physicians and surgeons often leave the hospital because of unenlightened management. Kossek and Block speak of the necessity for "satisfied, non-stressed workers who are able to perform at the best of their abilities" [77]. This is stated because the opposite situation is often the case. An administration, which allows or encourages frustration and stress in employees is one, which needs to be replaced. Clever argues that burnout is a totally preventable disease [57].

In medicine there is also the stress of having to diagnose on the basis of incomplete evidence and with unknown factors, and so encounter death and suffering. Additionally, to be exposed to unenlightened management, unfair and inhumane working conditions, threat of disease, excessive workload, night work, disturbed sleep, etc. aggravates the situation.

Spickard found that two-thirds of Canadian physicians in 1998 considered their workload too heavy [78]. Burnout is poorly defined, even as "erosion of the soul" [79]. Two factors are an exaggerated sense of responsibility and compulsiveness [80]. The situation may be extremely conducive to burnout but is ultimately often perpetuated by oneself. One accepts abusive conditions because one does not wish to protest or change one's lifestyle or does not wish to take a different job. The physician may wish to work as a physician. One does not wish to strongly confront an abusive and non-communicative manager or department chair although burnout can often be due to such management and systematic mismanagement such as requiring healthcare workers to work long hours. One does not understand that one can serve others better if one takes better care of oneself. A burned out physician is no longer of use. Female physicians are 60% more likely to be burned out. "The odds of burnout in women increased 12–15% for each additional 5 h worked per week of more than 40 h" [80]. Imagine what the situation must be for those many thousands of physicians and other healthcare workers who are required to work 80–100 h per week.

One suggestion given is "adopting a healthy philosophical lifestyle" [81]. Unfortunately this cannot be done if one does not know what it means. We may suggest that it means to be a critical thinker, rational, know about emotions and ethics, know about what humanism is, have healthful nutritional and physical exercise lifestyles and apply these to one's practical life. Spickard [81] has rather suggested that one develop a religious and spiritual life, that is, undermine reason. Holistic management includes the whole life of the worker as it does with care of the patient. Management decisions need to involve the philosophical as well as all of the other levels of management decisions. The literature, however, only gives a few arbitrary and narrow factors involved in burnout. What is said to be needed by the healthcare worker to counteract burnout: autonomy, self-acceptance, environmental control, purpose of life, personal growth. These are empty terms. They also speak of unscientific "energy renewal" and each physician as a "locus of energy exchange" [82]. This reminds one of Freud's notion of "psychic energy." It is said that one should have "mindfulness" [82]. What could that be? The account given of burnout is uncritical and almost pure metaphysics under the guise of basing information on controlled EBM studies. Burnout is referred to as "emotional exhaustion" but no account is given of any emotion theory. The article, however, is important for bringing to our attention that medical management and organization has been so unhealthful and shockingly dreadful that healthcare workers literally work themselves to illness and death.

The blame for burnout is falsely ascribed to the individual burnt out physician or nurse, not to the system. Thus individualization of responsibility covers again the responsibility of perverse management.

Chopra stated, "Burnout is a syndrome defined by the three components of emotional exhaustion, depersonalization, and diminished feelings of personal accomplishment" [83]. This is a simplistic view of burnout. One may have burnout without having any of these factors and one can have all of these factors and have no burnout. Furthermore, "feelings" are here as elsewhere confused with emotions. Emotions are not bodily feelings (See Chapter 7).

8.3.9 Overwork and Suicide

"The suicide rate among male doctors is 40% higher than that among men in general, whereas the rate among female doctors is 130% higher than that among women in general" [84].

Physicians have a higher rate of mental illness, anxiety, alcoholism, drug use, and depression than the general population. This is especially true of anesthesiologists, pharmacists and psychiatrists. For female physicians suicide is three to four times as great as in the general population [85].

8.3.10 Overwork Causes Loss of Quality of Life

There is virtually no time left for any other activities. Fatigue and illness also take the quality out of work hours. Interest in life is lost. A qualitative relationship with a partner or friends becomes impossible to obtain or maintain.

8.3.11 Overwork Is a Cause of Negative Emotions

Sleep deprivation results in mood swings [86]. In a survey of 3,604 resident physicians, residents who worked more than 80 h per week reported experiencing more irritability and making more fatigue-related errors than did residents who worked fewer than 80 h per week [87].

The negative emotional and psychological consequence of sleep-deprivation is also intense. Physicians otherwise well-tempered develop negative attitudes toward patients as well as toward each other. One skilled, senior physician reported wishing to machine-gun the patients flooding the hospital waiting room. Patton quoted a physician reporting that he wished the patient would die so he could just get some sleep [88]. One overworked physician is reported as having stated, "I don't want the SOB to make it. . .to keep me up two more hours." "I don't want the asthmatic SOB to live if it means I don't sleep" [89]. However, many statements are made in medical, blatant vice humor to stress a point, here about overwork. It need not be taken seriously but is a metaphorical way of expressing oneself.

8.3.12 Overwork Causes Loss of Interest in Medical Practice

Too much is too much. Loss of interest is a defense-mechanism. One needs to get out of continuous involvement with work.

8.3.13 What Is the Legal Result of Overwork?

Kowalenko found that emergency room residents are 6.7 times more likely than others to have a motor vehicle crash due to falling asleep at the wheel during their residency [90].

According to Patton, overworked healthcare workers place both the workers and the institutions directly and indirectly in legal jeopardy [91]. An automobile accident of an overtired employee may bring a suit against a hospital. Of course, the hospitals and HMOs could do a cost-benefit analysis and in some cases conclude that it would be cheaper to violate the law and pay the fines rather than limit the excessive hours worked. Physicians are required to tell the patient if they are sleep deprived before treatment or a surgical operation. The authors conclude, "It is beyond reasonable debate that chronic sleep deprivation negatively impacts one's physical and mental abilities. . .[and] is particularly alarming when it occurs. . .in the life and death realm" [92]. Charles Czeisler, Professor of Sleep Medicine, says that doctors should report to patients if they slept less than 2 h in the last 24 h [93]. Because the medical institutions and professionals resist reform, Patton recommends effective legal and union reforms to address the problem.

Nationwide strike of physicians in Germany is concerned with the excessive work and too little pay. Jens Flintrop, wrote about "*The Revolt of the Physicians and How it Came to That*" [94]. Signs read (translated from German): "85000 unpaid overtime hours." "Beware: Murder-Doctors," "How many hours has your doctor slept?" "Charité Hospital Shame Management." "Hospital stinginess," "Doctors are not piggybanks." "Scarce resources." Physicians work 70–100 h a week, equivalent to 2 and 1/2 jobs worked by one person. In any other usual profession this would be called slave labor. On a per hour basis, the salary would come to around or less than minimum wage. Physicians were not able to support a family. 90,000 countrywide clinic doctors quickly joined the strike even with the threat of being fired or not having their contracts renewed. Other healthcare workers also joined them in the strike. In Germany, twenty university clinics issued warning strikes. The physicians sought 30% increase in salary, paid overtime work, and reduction of work hours to 48 h weekly. Even 48 h in other professions would be considered to be outrageous. The recent Greens Party proposals were rather to reduce working hours for all workers to 38 h per week to allow people some little life in addition to work. The hospitals and other employers claimed that there was no money to pay for such requests. 10% of the hospitals were closed, and there was a 20% increase in patients. The physicians on strike replied that the health care units would then all have to be closed for they would leave their jobs.

Overwork leads to strikes as a last chance to react. This threat does not seem to bother the administrators. The political and economic administrators took advantage of the physicians' humanistic devotion to patient care to reduce the physicians literally to slavery-like conditions. There had to be the most extreme insult to the lives, health and working conditions of the physicians to move them to join in a strike. The Hippocratic Oath says the physician should do no harm, but it also says that physicians are obligated to engage in politics to advocate for better health policies. It is a contradiction for hospitals and other healthcare units to ruin the health and lives of the healthcare workers themselves. Humanistic management must replace the typical healthcare administration and management.

8.3.14 Denial that Healthcare Workers Overwork and/or that It Is Harmful

It is interesting to note that no thinking regarding the harm of overwork is allowed. The simplest thing must be proven by EBM test results, but no common reasoning is allowed. So it is initially denied that anything is wrong with working 100+ h a week. Hopefully, the physicians do not give this advice to patients as well. In fact, death and hospitalization in those with higher workloads is more than twice that in groups with low workloads. Job strain predicts mortality.

Fischer supports 92–95 h a week and argues that 80 h a week is not enough a time for surgery residents to provide consistent care. He opposes the 80 h proposal and the other new work rules of the Accreditation Council for Graduate Medical Education (ACGME) as being a loss for surgery. "The 80-h workweek is...the denial of the foundation of one of the most closely guarded and almost religiously regarded axioms of surgical care: the concept of continuity of care" [95]. There is an unscientific indoctrination and captivation by a model here. The author also disparages research and medical education, proposing that long hours of patient care is basically all that is needed. "In the real world, the 80-h workweek represents a complete denial of the value system of U.S. surgery" [95].

William Halstead in 1904 adopted the German-Austrian strict discipline of long hours for physicians for Johns Hopkins (Halstead Method). With the July 1, 2003 resident rules the workweek decreased from 100.7 to 82.6 h, excluding on-call hours [96]. The online Maslach Burnout Inventory-Human Services Survey was used in the study reported. Burnout was not significantly affected with the 80 h week. But it meant formal in-hospital education was reduced although it is one of the most important aspects of residency training. Long workweeks also would mean physicians would have little or no time to keep up with the medical literature or do evidence-based medicine which requires reading the literature. Residents reported depersonalization and exhaustion. Insofar from the study little difference was found from before and after reducing hours worked one may conclude that working 100 h a week is acceptable, this is totally misleading. Rather, one may conclude the obvious, that the hours worked have not been reduced enough.

Stryjewski and Slonim [97] state that studies fail to show adverse effects of working longer hours because of the poor design of such studies and the inability to adequately measure all of the tasks performed. To a large extent the studies show that the studies themselves are unacceptable. One specific study does show that there were 50% more errors by the sleep-deprived in evaluating electrocardiograms [97]. It is revealingly pointed out that it would be unethical to do a "randomized, prospective, double blind controlled trial of sleep deprivation versus rested surgeons" [98].

Pfaff states, "There is little empirical evidence on the effects of working hours on relevant clinical outcomes" [99]. Instead of restricting the excessive number of hours worked, the author recommends that surgeons learn to better handle the stress and to compensate for the loss of sleep by means of prevention strategies. Economists claim that it would be too expensive not to have healthcare workers overwork. One system works as follows: High managing managers are provided with salary increase if they save costs for the hospital. They often do that at the expense of staff by having overworked staff and/or too few employees, but not by cutting the services, although this can also occur. Supposedly the "same" service can be taken care of by those who are less professional healthcare workers. Older experts are not desirable to maintain because their salaries are higher and beginners can supposedly provide the "same" service. But because of their long clinical experience such experts are not replaceable. Medicine often advances by way of such experience, rather than by fixed formulae and "objective" evidence-based medicine (See Chapter 19).

8.3.15 Is There Evidence for the Harm of Overwork?

Kuflik states, "The traditional system has proved resistant to change" [100]. In other words medicine has been unscientifically and unprofessionally captivated by a dysfunctional model or metaphor. It is a good example of the irrationality sometimes found in science. "How could an unwarranted policy persist so long and be so resistant to serious efforts that have been made to reform it?" [101]. The practice is "medically unsound and morally unjustifiable" [102]. The requirement that it be proven that overwork does harm is even less plausible than the cigarette companies originally claiming that it has not yet been proven that cigarettes cause lung cancer. The burden of proof should be on those who require longer than 35–40 h a week [103].

Continuity in patient care is the argument used for defense of long schedules. Continuity is harmed, not helped, by long schedules [104]. The author argues from studies that long hours decrease learning [105]. Training to work long hours is training to be sleep deprived and to ruin one's health. Also constant loss of sleep does not prepare one for having to work extra emergency hours [106]. "Sleep depriving training has been, for many, an embittering experience, often bordering on the traumatic.... with post-traumatic stress." Tired physicians tend to take shortcuts, which cause them less effort and allow them to leave shifts earlier. "A humane shift schedule may not suffice unless we also have a genuinely humanistic medical curriculum as well" [107]. Kuflik likens the long hours to a form of fraternity-like "hazing" [108]. In sum, the traditional practice of working long hours results in inferior patient care, inferior medical education, stress and burnout, and shows total lack of concern with the health and wellbeing of physicians.

Weinger and Ancoli-Israel noted that only about 12 studies on sleep deprivation were published in the last 10 years and that previous work was methodologically unacceptable [109]. They summarized recent findings of trials on the effect and results of sleep deprivation on physicians as follows: 1. cardiac arrhythmias were detected, 2. saddness, fatigue and unsureness, 3. surgical residents made more errors and performance was slower (e.g., with less than 3 h sleep in 17 h, or less than 4.5 h sleep in a 32 h shift), 4. diminished creative thinking, 5. mood impairment, 6. increased errors, 7. falling asleep during treatment of patients. They recommend that physicians should warn the patients when they are fatigued so as to protect them from maltreatment [110]. Virtually all of the studies gave concern only to resident physicians as if regular or experienced physicians were not affected by lack of sleep. The authors note that only one study was done on sleep deprivation of experienced physicians: [111] In that study mention of chronic insomnia, stress, burnout and a number of other effects were not considered.

In aviation they found the causes of their errors, where hospitals did not in regard to excessive hours worked. This is again not evidence-based medicine. If medical management cannot use science and EBM to conclude the most obvious and anti-medical practice, it must be given up as a dangerous myth.

The requirement of overwork shows that it is not "evidence-based," the industry has accepted an irrational historical paradigm and the practice is almost universal in the Western world [112].

An Institute of Medicine report estimated that 44,000–98,000 Americans die each year as a result of medical errors [113]. Scientific research shows that sleep-deprivation produces profound motor and cognitive deficiencies [114]. The effects of sleep-deprivation were found to be like those of alcohol intoxication [115]. To allow such a practice is to violate the basic professional purpose and ethics of medical practice. One who supports or engages in such practice cannot call himself a professional healthcare administrator, manager or worker. The administrators, however, often just themselves have 9–5 h. In spite of this the American medical educators steadfastly oppose reform and there is "cultural resistance to reform." It is referred to as "entrenchment" [116]. Rationalizations given for the long-hours requirement are: to teach self-sacrifice, to indoctrinate, teach physicians to be humble, to perform a cultural rite of passage, to promote bonding, for financial reasons to have cheap labor, etc. New York State saved around $300 million dollars yearly on cheap resident's labor before their hours were reduced [117]. Now hospitals can be sued for harm and death results from their policies, which involve unhealthful, long working hours. A lawyer can claim the patient was treated by sleep deprived healthcare workers and so quite possibly prevail against the hospital's defense. The hospital has no defense if it maintains such policies [118]. The healthcare worker must inform the patient about the full nature of risks and outcomes so as to make an informed decision. It is suggested that the healthcare worker inform the patient also how long s/he has not had any sleep. In court, the issue may arise that the physician neglects to give this relevant information to the patient. Without giving the patient this information, the right to be fully informed can be violated as may be established in a court of law. Hospitals risk lawsuits also because they often require workers to work over legal limits, and firing those refusing to do so.

Whereas physicians may have no sleep at all at night, truck drivers are thought to be a serious risk on the highway if they had as little as 5 h sleep a night. In one study 58% of long-haul accidents were fatigue related [116]. Ironically the American Medical Association recommended sufficient sleep before driving, but opposed it for physicians. Airplane pilots cannot work more than 8 h in a 24 h period. "The aviation industry may serve as a model for medical education reform" [119]. It may be noted that the healthcare worker's work is far more exhausting than that of the pilot. In Denmark, Norway and Sweden, residents work only 37–45 h a week [120].

"Surgery accounts for about half of the adverse events in chart-review studies of errors in medicine." Half of these were preventable and the major cause is excessive workload, sleep deprivation and its mood and stress impairment, and additional on-call duties [121].

Trivial and expensive EBM studies are conducted to show the obvious instead of shortening the work hours. Dula sought to determine whether working 5 serial night shifts (11 pm–7 am) in the emergency department results in a decline in physician performance as measured with an intelligence test. Night-shift workers average approximately 25–33% less sleep than do day workers and 75% of night workers experience sleepiness on their shifts with 20% falling asleep while at work. Emergency physicians at Stanford University Hospital were found to have a decline in decision-making and problem-solving after working 3 consecutive night shifts [122].

The results of this study should come as no surprise.

Tempelaar states that research on sleep deprivation is too poor to prove anything [123]. If the implication is that because we do not know much about sleep deprivation healthcare workers should work 80–110 h a week. The tobacco industry for years said that it has not been proven that smoking is harmful. The reverse conclusion should be drawn that people should work no longer than 35–40 h a week until evidence is given that no harm is done by working longer hours.

References

1. Dlugacz, Y., Restifo, A., and Greenwood, A. 2004. *The quality handbook for health care organizations: A manager's guide to tools and programs*, 8. San Francisco, CA: Jossey-Bass.
2. Council on Ethics and Judicial Affairs Code of Medical Ethics: Current Opinions with Annotations. American Medical Association. (2002–2003):280.
3. Dlugacz, Y., Restifo, A., and Greenwood, A. 2004. *The quality handbook for health care organizations: A manager's guide to tools and programs*, 33, 155. San Francisco, CA: Jossey-Bass.
4. Shibles, W. 1987. Reformatory blame. *Journal of Rational-Emotive Therapy* 5:266–281.
5. Wachter, R., and Shojania, K. 2004. *Internal bleeding: The truth behind America's terrifying epidemic of medical mistakes*, 380. New York: Rugged Land.
6. Kossek, E. and Block, R. 2000. *Managing human resources in the 21st century: From core concepts to strategic choice*, 21.17. Cincinnati, OH: South-Western College Publishers.
7. Adams, B. 2002. Accountable but powerless. *Health Affairs* 21:218–223.
8. Adams, B. 2002. Accountable but powerless. *Health Affairs* 21:221.
9. Adams, B. 2002. Accountable but powerless. *Health Affairs* 21:222.
10. Dlugacz, Y., Restifo, A., and Greenwood, A. 2004. *The quality handbook for health care organizations: A manager's guide to tools and programs*, 27. San Francisco, CA: Jossey-Bass.
11. Dlugacz, Y., Restifo, A., and Greenwood, A. 2004. *The quality handbook for health care organizations: A manager's guide to tools and programs*, 24. San Francisco, CA: Jossey-Bass.
12. Bingham, R. 2002. Leaving nursing. *Health Affairs* 21:211–217; esp. 214.
13. Bingham, R. 2002. Leaving nursing. *Health Affairs* 21:211–217; esp. 216.
14. Bircher, J., and Wehkamp, K.H. 2006. *Das ungenutzte Potential der Medizin. Analyse von Gesundheit und Krankheit zu Beginn des 21.* Jahrhunderts. Zürich: rüffer & rub Sachbuch.
15. Dlugacz, Y., Restifo, A., and Greenwood, A. 2004. *The quality handbook for health care organizations: A manager's guide to tools and programs*, 198. San Francisco, CA: Jossey-Bass.
16. Spath, P., ed. 1999. *Error reduction in health care: A systems approach to improving patient safety*, xxvi. Chicago, IL: American Hospital Association Press.
17. Preventing Medical Errors Report. July 2006. nap.edu.
18. Ciaramicoli, A. 2004. *Performance addiction*. New York: Wiley.
19. Shibles, W. 1995. Analysis of the definitions of Humanism. *Scottish Journal of Religious Studies* 16:51–61.
20. Council on Ethics and Judicial Affairs. Current Opinions with Annotations. *AMA* 2002–2003:292.
21. Bircher, J., and Wehkamp, K.H. 2006. *Das ungenutzte Potential der Medizin. Analyse von Gesundheit und Krankheit zu Beginn des 21*, Jahrhunderts. 3. Zürich: rüffer & rub Sachbuch.
22. Dlugacz, Y., Restifo, A., and Greenwood, A. 2004. *The quality handbook for health care organizations: A manager's guide to tools and programs*, 45, 53. San Francisco, CA: Jossey-Bass.
23. Dlugacz, Y., Restifo, A., and Greenwood, A. 2004. *The quality handbook for health care organizations: A manager's guide to tools and programs*, 179. San Francisco, CA: Jossey-Bass.
24. Dlugacz, Y., Restifo, A., and Greenwood, A. 2004. *The quality handbook for health care organizations: A manager's guide to tools and programs*, 23. San Francisco, CA: Jossey-Bass.
25. Kossek, E., and Block, R. 2000. *Managing human resources in the 21st century: From core concepts to strategic choice*, 27.12. Cincinnati, OH: South-Western College Publishers.
26. Harro, A. 2006. *Halbgötterdämmerung*. Die Zeit, June 8, 2006.

27. Myers, M. 2001. The well-being of physician relationships. *Western Journal of Medicine* 174:30–33.
28. Schmidbauer, W. 1997. *Helfen als Beruf. Die Ware Nächstenliebe*. Reinbek: Rowohlt.
29. Wachter, R., and Shojania, K. 2004. *Internal bleeding: The truth behind America's terrifying epidemic of medical mistakes*, 216. New York: Rugged Land.
30. Wachter, R., and Shojania, K. 2004. *Internal bleeding: The truth behind America's terrifying epidemic of medical mistakes*, 217. New York: Rugged Land.
31. Wachter, R., and Shojania, K. 2004. *Internal bleeding: The truth behind America's terrifying epidemic of medical mistakes*, 218. New York: Rugged Land.
32. Wachter, R., and Shojania, K. 2004. *Internal bleeding: The truth behind America's terrifying epidemic of medical mistakes*, 368. New York: Rugged Land.
33. Weissman, J.L. 1999. Prevalence of suicide ideation and suicide attempts in nine countries. *Psychological Medicine* 29:9–17.
34. Summers, R.S. 1982. *Instrumentalism and American legal theory*, 172. Ithaca, New York: Cornell University Press.
35. "Protest! Teilbetriebsversammlung der SALK Ärzte beschließt Protestmaßnahmen." *Der Salzburger Arzt* Sept, 18th 2004. 12–13. See also *Salzburger Nachrichten* Sept. 3rd, 2004:3, *Salzburger Krone* Sept. 3rd, 2004:3.
36. *Salzburger Nachrichten* Dec. 3rd, 2004:3.
37. *Salzburger Nachrichten* Dec. 3rd, 2004:13.
38. *Der Salzburger Arzt* 2004. 18, 9:12–13.
39. Dr. Terje H., Vice President of the Chamber of Physicians Medical Council, Salzburg in *Der Salzburger Arzt* Sept. 9th 2004:13.
40. AMA Principles of Medical Ethics III. Adopted by the American Medical Association's (AMA'S) House of Delegates, June 17, 2001.
41. Time Almanac. 2006 for year 2005:120.
42. Patton, D. 2001. Legal considerations of sleep deprivations among resident physicians. *Journal of Health Law* 34:377–417; esp.385.
43. Patton, D. 2001. Legal considerations of sleep deprivations among resident physicians. *Journal of Health Law* 34:377–417; esp.387.
44. Patton, D. 2001. Legal considerations of sleep deprivations among resident physicians. *Journal of Health Law* 34:377–417; esp.402.
45. Patton, D. 2001. Legal considerations of sleep deprivations among resident physicians. *Journal of Health Law* 34:377–417; esp.383.
46. Landrigan, C. 2006. Intern's compliance with accreditation council for graduate medical education work-hour limits. *JAMA* 296:1063–1070; esp. 1063; 1069.
47. Patton, D. 2001. Legal considerations of sleep deprivations among resident physicians. *Journal of Health Law* 34:377–417; esp.394.
48. Do Rosario, L. 1991. Dropping in harness: Salary-men stalked by "unmentionable" killer. *Far Eastern Economic Review* 151:30–31.
49. Inoue, K., Matsumoto, M. 2000. Karo Jisatsu (Suicide from overwork): A spreading occupational threat. *Occupational and Environmental Medicine* 57:284–285.
50. Michie, S., and Cockcroft, A. 1996. Overwork can kill. *BMJ* 312:921–922.
51. Karoshi, H. karoshi.jp/english/overwork.html.
52. Beasley, J. 2005. Quality of work life of independent versus employed family physicians in Wisconsin: A WreN study. *Annals of Family Medicine* 3:501–506.
53. Landrigan, C. 2006. Intern's compliance with accreditation council for graduate medical education work-hour limits. *JAMA* 296:1063–1070: esp.1063, 1069.
54. Ayas, N 2006. Extended work duration and the risk of self-reported percutaneous injuries in interns. *JAMA* 296:1055–1062; esp.1061.
55. European Industrial Relations Observatory. eiro.eurofound.ie (Accessed Dec. 5, 2006).
56. Pfaff, H. 2004. Surgical safety and overwork. *British Journal of Surgery* 91:1533–1535.
57. Clever, L. 2002. Who is sicker: Patients – or residents? Residents' distress and the care of patients. *Annals of American Medicine* 136:391–393; esp.392–393.

58. Schernhammer, E. 2005. Taking their own lives – the high rate of physician suicide. *NEJM* 352:2473–2476.
59. Puddester, D. 2001. The canadian medical association's policy on physician health and well-being. *Western Journal of Medicine* 174:5–7
60. *Medical Ethics Manual.* World Medical Association. 2005. Ferney-Voltaire Cedex, France: 117.
61. Dawson, D., and Reid, K. 1997. Fatigue, alcohol and performance impairment. *Nature* 388:235 ff.
62. Harder, B. 2005. Dead tired. *Science News* 168:163–164.
63. Cromie, W. 2004. Overworked interns prone to medical error. *Harvard University Gazette* Oct. 28, 2004:1.
64. Czeisler, C. 2006. Doctor fatigue hurting patients. Harvard University Gazette Dec 10th, 2006:1.
65. Tarnow-Mordi, W. et al. 2000. Hospital mortality in relation to staff workload: A 4-year study in an adult intensive care unit. *Lancet* 356:185–189.
66. Rogers, A. 2004. The working hours of hospital staff and patient safety. *Health Affairs* 23:202–212.
67. Greene, J. 1999. Residents say long hours hurt patient care. *American Medical News* 42:30–31.
68. Patton, D. 2001. Legal considerations of sleep deprivations among resident physicians. *Journal of Health Law* 34:377–417; esp.378.
69. Ardati, A. 2002. Don't overwork physicians, imperil public. *Detroit News*, February 13th, 2002.
70. Ignatavicius, D., and Workman, M.L. 2002. *Medical-surgical nursing care: Critical thinking for collaborative care*, 4th edn, 2 vols, 10. PA: WB Saunders, Philadelphia.
71. General Motors Michigan. 2000. *Leadership and Organization Development Journal* 21:307–310.
72. Simpson, R. 1998. Presenteism, power and organizational change: Long hours as a career barrier and the impact on the working lives of women managers. *British Journal of Management* 9:37–52; esp. 42.
73. Tempelaar, A. 1997. The problem doctor. In *Problem Doctors: A conspiracy of silence*. eds. Lens, P., and Van der Wal G., 31–55; esp. 46–47. Amsterdam: IOS Press.
74. *News Report.* New York, Reuters Health, 7/10/02.
75. Scott, L. et al. 2006. Effects of Critical Care Nurses' work hours on vigilance and patients' safety. *American Journal of Critical Care* 15:30–37.
76. Power, K., and Swanson, V. 2000. Medical profession and stress. In *Encyclopedia of stress*. ed. Fink, G., vol 2, 708–713; esp.708. San Diego, CA: Academic Press.
77. Kossek, E., and Block, R. 2000. *Managing human resources in the 21st century: From core concepts to strategic choice*, 11.7. Cincinnati, OH: South-Western College Publishers.
78. Spickard, A., Gabbe, S., and Christensen, J. 2002. Mid-career burnout in generalist and specialist physicians. *JAMA* 288:1447–1450.
79. Spickard, A., Gabbe, S., and Christensen, J. 2002. Mid-career burnout in generalist and specialist physicians. *JAMA* 288:1447–1450; esp.1447.
80. Spickard, A., Gabbe, S., and Christensen, J. 2002. Mid-career burnout in generalist and specialist physicians. *JAMA* 288:1447–1450; esp.1448.
81. Spickard, A., Gabbe, S., and Christensen, J. 2002. Mid-career burnout in generalist and specialist physicians. *JAMA* 288:1447–1450; esp.1449.
82. Spickard, A., Gabbe, S., and Christensen, J. 2002. Mid-career burnout in generalist and specialist physicians. *JAMA* 288:1447–1450; esp.1450.
83. Chopra, S. 2004. Physician burnout. *JAMA* 291:633.
84. Schernhammer, E. 2005. Taking their own lives – the high rate of physician suicide. *NEJM* 352:2473–2476.

85. Power, K., and Swanson, V. 2000. Medical profession and stress. In *Encyclopedia of stress*. ed. Fink, G., vol 2, 708–713; esp.710. San Diego, CA: Academic Press.
86. Martini, F. 2006. *Fundamentals of anatomy and physiology*. San Francisco, CA: Pearson.
87. Baldwin, D., Daugherty, S., Tsai, R., and Scotti, M. 2003. A national survey of resident's self-reported work hours: Thinking beyond speciality. *Academic Medicine* 78:1154–1164.
88. Patton, D. 2001. Legal considerations of sleep deprivations among resident physicians. *Journal of Health Law* 34:377–417; esp.377.
89. Patton, D. 2001. Legal considerations of sleep deprivations among resident physicians. *Journal of Health Law* 34:377–417; esp.378.
90. Kowalenko, T. 2000. Emergency medicine residency related MVCs – Is sleep deprivation a risk factor? *Academic Emergency Medicine* 7:451.
91. Patton, D. 2001. Legal considerations of sleep deprivations among resident physicians. *Journal of Health Law* 34:377–417; esp.395–397.
92. Patton, D. 2001. Legal considerations of sleep deprivations among resident physicians. *Journal of Health Law* 34:377–417; esp.402.
93. Czeisler, C. 2006. Doctor Fatigue Hurting Patients. *Harvard University Gazette* Dec. 10, 2006:1.
94. Flintrop, J. 2005. "Der Aufstand der Ärzte—und wie es dazu kam." ("The Revolt of the Physicians and How it Came to That.") *Deutsches Ärzteblatt* 51–52:2518–2523.
95. Fischer, J. 2004. Continuity of care: A casuality of the 80-hour work week. *Academic Medicine* 79:381–383.
96. Gelfand, D. 2004. Effect of the 80-hour workweek on resident burnout. *Archives of Surgery* 139:933–940.
97. Stryjewski, G., and Slonim, A. 2002. Who says You're too tired? *Critical Care Medicine* 30:2396–2397.
98. Stryjewski, G., and Slonim, A. 2002. Who says You're too tired? *Critical Care Medicine* 30:2396–2397; esp. 2397.
99. Pfaff, H. 2004. Surgical safety and overwork. *British Journal of Surgery* 91:1533–1535.
100. Kuflik, A. 2001. Traditionally long shifts for medical residents: Medically unsound, morally unjustified. *Australian Journal of Professional and Applied Ethics* 3:2–21; esp.3.
101. Kuflik, A. 2001. Traditionally long shifts for medical residents: Medically unsound, morally unjustified. *Australian Journal of Professional and Applied Ethics* 3:2–21; esp.3–4.
102. Kuflik, A. 2001. Traditionally long shifts for medical residents: Medically unsound, morally unjustified. *Australian Journal of Professional and Applied Ethics* 3:2–21; esp.4.
103. Kuflik, A. 2001. Traditionally long shifts for medical residents: Medically unsound, morally unjustified. *Australian Journal of Professional and Applied Ethics* 3:2–21; esp.7.
104. Kuflik, A. 2001. Traditionally long shifts for medical residents: Medically unsound, morally unjustified. *Australian Journal of Professional and Applied Ethics* 3:2–21; esp.10.
105. Kuflik, A. 2001. Traditionally long shifts for medical residents: Medically unsound, morally unjustified. *Australian Journal of Professional and Applied Ethics* 3:2–21; esp.11.
106. Kuflik, A. 2001. Traditionally long shifts for medical residents: Medically unsound, morally unjustified. *Australian Journal of Professional and Applied Ethics* 3:2–21; esp.12.
107. Kuflik, A. 2001. Traditionally long shifts for medical residents: Medically unsound, morally unjustified. *Australian Journal of Professional and Applied Ethics* 3:2–21; esp.13.
108. Kuflik, A. 2001. Traditionally long shifts for medical residents: Medically unsound, morally unjustified. *Australian Journal of Professional and Applied Ethics* 3:2–21; esp.17.
109. Weinger, M., and Ancoli-Israel, S. 2002. Sleep deprivation and clinical performance. *JAMA* 287:955–957; esp.956.
110. Weinger, M., and Ancoli-Israel, S. 2002. Sleep deprivation and clinical performance. *JAMA* 287:955–957; esp.957.
111. Smith-Coggins, R. 1997. Rotating shiftwork schedules: Can we enhance physician adaptation to night shifts? *Academic Medicine* 4:951–961; Weinger, M., and Ancoli-Israel, S. 2002. Sleep deprivation and clinical performance. *JAMA* 287:955–957; esp. 956.

112. Patton, D. 2001. Legal considerations of sleep deprivations among resident physicians. *Journal of Health Law* 34:377–417; esp.393.
113. Patton, D. 2001. Legal considerations of sleep deprivations among resident physicians. *Journal of Health Law* 34:377–417; esp.378.
114. Patton, D. 2001. Legal considerations of sleep deprivations among resident physicians. *Journal of Health Law* 34:377–417; esp.381.
115. Patton, D. 2001. Legal considerations of sleep deprivations among resident physicians. *Journal of Health Law* 34:377–417; esp.380.
116. Patton, D. 2001. Legal considerations of sleep deprivations among resident physicians. *Journal of Health Law* 34:377–417; esp.383.
117. Patton, D. 2001. Legal considerations of sleep deprivations among resident physicians. *Journal of Health Law* 34:377–417; esp.389.
118. Patton, D. 2001. Legal considerations of sleep deprivations among resident physicians. *Journal of Health Law* 34:377–417; esp.397.
119. Patton, D. 2001. Legal considerations of sleep deprivations among resident physicians. *Journal of Health Law* 34:377–417; esp.385.
120. Patton, D. 2001. Legal considerations of sleep deprivations among resident physicians. *Journal of Health Law* 34:377–417; esp.387.
121. Pfaff, H. 2004. Surgical safety and overwork. *British Journal of Surgery* 91:1533–1535; esp.1533.
122. Dula, D. 2001. The effect of working serial night shifts on the cognitive functioning of emergency physicians. *Annals of Emergency Medicine* 31:152–155.
123. Tempelaar, A. 1997. The problem doctor. In *Problem doctors: A conspiracy of silence*. Lens, P., and Van der Wal, G., 31–55; esp. 50. Amsterdam: IOS Press.

Chapter 9
Care: A Critique of the Ethics and Emotion of Care

Abstract This chapter aims at the clarification of the notion of care on the basis of the cognitive-emotive theory of caring. Care theories are presented and critiqued. Bonding is analyzed. Empathy, sympathy and helper's syndrome are examined. A philosophy of caring is presented rather than only morals of caring. An analysis of caring is seen to require an analysis of ethics, the self, causes of action, motivation, and emotion. It is also shown how caring may be redefined and based on a naturalistic, humanistic theory of ethics. Philosophy and ethics of personality involve the emotion of care towards the humanistic concept of rational care.

Keywords Care · caring · care theories · bonding · care as emotion · rational caring · The Patient's Hippocratic Oath · empathy · humanism · Philosophy and Ethics of Personality

9.1 Introduction

This chapter aims at a clarification of the notion of care and care theory that has been discussed recently in the area of bioethics and the philosophy of medicine. This means that it is a criticism of the language and methods involved in care.

Even if we theoretically know we should help others who are in desperate need, why do we not actually do it? We often basically care only when it is our professional duty to do so and delegate care to those on duty alone even when we could share care. We may do things to or at people rather than for or with them. If we regard the other as part of oneself (identification) it can be just a form of egoistic caring or self-caring. We can identify with a patient or physician. We may also care because by so doing we have a good view of ourselves which generates a positive emotion. By contrast, a rational naturalistic humanist is universally caring – not because of ego or direct self-gain, but because it makes sense to do so in terms of consequences just as it makes sense to be a critical world citizen. Civil or social caring is concerned with others in society both local and international. The world is my family. Universal caring could put care for the individual in the context and perspective of caring on a larger scale. We may compare the Greens' slogan: Think globally, act locally. The naturalistic ethical theory does not support

B. Maier, W.A. Shibles, *The Philosophy and Practice of Medicine and Bioethics*, International Library of Ethics, Law, and the New Medicine 47,
DOI 10.1007/978-90-481-8867-3_9, © Springer Science+Business Media B.V. 2011

selfishness because humanistic ethics is based on inquiry and consequences leading to the understanding that one wants to live harmoniously and ecologically in the world and society at large. Bringing about one's own wants as well is not selfishness. Selfishness rather is defined as doing good for oneself while not caring about or doing bad for another. However, by definition, bad is nothing one would want to do. Naturalistic ethics, more than any other ethics, has led to humanism and altruism, both of which may be recommended as goals in the helping professions. Basically, the view of John Dewey, humanists and pragmatists is that ethics is consequentialistic involving enhancing human wants and abilities to the maximum in harmony with others and with nature now and in the future. The basis of naturalistic ethics is the bringing about of one's likes, wants, enjoyments and desires. Caring would have the same basis.

The question of what causes us to act in general is raised. The caring aspect with stress on individual wants and needs is especially contained in a naturalistic ethics. Thus, care may be regarded as balanced, humanistic treatment thereby producing care. Thus it cannot be negative or interfering. In caring, on this view, one should not be controlling, but humanistic, one should not be dogmatic, but open to specific needs. There might be problems of balancing of caring for one person against all others, and of having only inadequate resources to provide proper care. There might be unenlightened management, which prevents adequate caring, etc. It was found that one-third of 40 doctors showed "almost total lack of awareness how to interact in a professional manner" [1].

In the following, the theories and views, which will be presented, are contrasted with such a humanistic, naturalistic ethics.

9.2 Care Theories

The ethics of care for Noddings view is basically a deduction from anti-patriarchal feminist philosophy [2].

Accordingly, Noddings states that caring is feminine, cared-for is masculine. Caring comes from women [3]. "Logos, is the masculine spirit, whereas the more natural and, perhaps, stronger approach would be through Eros, the feminine spirit." This view commits the fallacy that emotions (female) are different from cognitions (male), whereas emotions involve cognitions. Her view also falsely regards emotion as a female quality, and reason as a male quality. Women and men both have emotions and reason. "Ethics has been discussed largely in the language of the father: in principles. . .the mother's voice has been silent" [4]. She thinks caring is based on the maternal instinct, is "maternal caring," caring as mothering, a certain kind of bodily feeling without language. This would exclude males and male language. Thus it sounds contradictory when Groenhout states that we should all engage in good mothering practices [5]. The move for Noddings is to only allow feminine knowledge and emotions, which only females can have. Caring comes from women. Her view, then, is that of gender caring, specifically, matriarchal caring instead of patriarchal caring. Groenhout, however, critiques Noddings view about mothering and suggests that there can be better and worse mothering, and that mothering can often be oppressive exploitation [6]. On her view, care theory is built from the "familial relationship," not just any mothering.

Noddings states, "Women have actually done most of the care-giving work for centuries" [7]. This is a questionable statement because men have been and still are the primary family providers. However, Noddings has recently reversed her thinking that women are more caring: "I have no idea whether women are, by nature, more caring than men. I doubt it" [8]. How unclear this theory remains is indicated by the title: "The Chaos of Care and Care Theory", which was the topic of the 23rd issue of the *Journal of Medicine and Philosophy* [9].

In regard to Noddings anti-patriarchal feminist view Jecker and Reich state, "The close association of care with gender and the feminine voice may hinder efforts to develop a broader understanding of care" [10]. Perhaps there are gender or role differences in regard to the kind of care provided.

Caring is also further restricted. We need care only for the immediate circle, family and close friends. It is not general or universal caring, but limited caring. The question also arises as to what extent, if any, it is available to men. Noddings is opposed to universals, but the one exception is for the bodily feeling of caring [11].

To care for friends or for one's family only is inconsistent with care for all other desperate people in the world. Ethics is not limited to one's own patients, friends and exclusive cliques. Our actions and policies affect people we do not know. We may tend to withhold caring from those we do not know. This is the old morals of helping no one except some family members and friends who help you. It is egoism because the self is just extended to the people one identifies with. Perhaps for these reasons Noddings in her most recent work has extended her notion of caring for someone to "caring about" people in general, e,g., the starving in the world [11].

Noddings caring is connected to the mother's (or father's) bonding of which there are various types. Bonding is made to seem a mystical physical connection. It has a physical basis as everything else we experience and do. But such a reduction is not a good model for rational caring.

Bonding and care are confused. Is the capacity for care based on bonding experience, on mother-child-bonding and the development of bonding-styles? Bonding-theories have an influence on the psychological interpretation of the physical process of bonding, the answers of the bonding experience allowing or avoiding mother [12].

D. Eyer's PhD dissertation on bonding is on the surface interesting, but could have benefited by adding the depth contained in philosophical analysis, philosophical psychology and the philosophy of science. Only a few philosophy texts were briefly touched on. However, her questioning of medical research on bonding was interesting [13].

Part of the evidence was from observations of animal behavior, e.g. female goats separated from their infants for a few minutes will supposedly reject them. The gosling after the third day "learns" to follow anything it takes to be is its "mother," even if it is a sailboat. To think the newborn knows it has a mother is, in any case, a personification. The infant does not "know." Bonding is confused also because it is one thing for the mother to "bond" to the child and another for the infant to "bond" to the mother. It is a big jump from saying the teen is difficult to manage or there is poor child development because the child was not held as an infant. Parenting requires much more than mere touching at a "magical" or alleged "sensitive" period of life. These are called confounding variables and were failed to be considered as such in the research designs [14].

In her book, *Mother-infant bonding: a scientific fiction*, Eyer fails to deal with emotion or to present a theory of emotions regarding caring. The study gives no analysis of the reasons why women behave as they do toward their offspring. Bonding was never clearly defined and varied with each study [15].

On the other hand, it makes good common sense to take gentle, protective and constant care of the infant regardless of the research or alleged consequences [16]. Reason here again can trump mechanical research. But as Eyers points out bonding should not be based on a mythical unscientific and fixed doctrinaire principle. As a result of the bonding research we have no knowledge about what it means for the mother to love or not love her child. We are given no knowledge about what love or any other bonding underlying emotion is at all. "I would like to urge the impossible – that we discard the word [bonding] entirely. Doing so would force us to recognize that strong relationships require many ingredients...love, understanding, trust, money, sharing, giving, stimulating, and inspiring"[17]. Bonding is a form of simplistic reductionism [18]. We cannot get around that entirely by a mere mechanical touching as bonding which is, of course the initial incentive, a neural paradigm induction. Bonding is a form of the physical medical model, as we know that emotion is embodied, as well as reason. But this all does not let us escape the question of rational bonding. What makes bonding successful in terms of consequences, in terms of the development of bonding structures which allow for autonomous development as well as for safely being bound to other people.

Tronto's view of caring is closer to a naturalistic, humanistic theory: we should do all we can to make this a better world. Unlike Noddings view that caring only applies to one's immediate circle, Tronto's care is a practice in the overall social context, something one needs to do for a social context to be a social context [19]. Noddings stresses caring as a relationship, what Veatch criticizes as being unintelligibly vague [20]. Caring as "relationship" is an abstract term without analysis and not specific enough to generate a theory as caring theorists require. There is no "relationship ethics" because ethics includes relationship already. Sociability is almost a synonym of caring. To care is to be sociable. To be in a relationship requires one to participate in it. To love or care is created by one's conscious deliberate cognition to do so if it is to be a social context. There is no relationship if one does not relate, no communication if one does not communicate.

Noddings is anti-principles, anti-rules, anti-obedience, anti-dogma, anti-reason, anti-empirical, anti-universals, anti-natural rights, anti-definitions, anti-detachment, anti-negative emotion, anti-consequential, anti-systematic [believes it is a mistake to be systematic], anti-rational (not empirical or logical or with proofs or knowledge, but rather on bodily feelings) [21]. "Aesthetic care" is care about things, ideas or money [22]. She also states that her care theory is not consequentialistic [23]. Because she is anti-principles, anti-obedience, and anti-dogma we can predict that she is anti-religion as well. She accordingly states that religion is too authoritarian, and based on obedience, accusation, and fear to be caring. Tronto also sees caring as a non-indoctrinaire responsiveness therefore it is not religious, cultural, or indoctrinative caring [24]. Accordingly, one might suggest "humanistic care".

Barnes defines caring in practice as good communication, understanding the patient, as action, and in terms of a therapeutic relationship (trust, respect, honesty, frankness, empathy), a most beneficial outcome. He criticizes the feminist account of Noddings [2–4], [25], and [26], who put caring on a maternal relationship only because this may lead to burn-out, favoritism, lack of objectivity, empty or vague emotional involvement at the expense of effective cure or treatment [27].

According to Tronto care is an emotion and therefore one must first understand how emotions work to then be able to discuss care. Noddings does not say how one can become caring [28]. *See in this book the* Chapter 7.

Groenhout [5] and Noddings hold that Care Theory is a genuine ethical theory. Veatch, however, gives significant criticism of care as a theory. He asks if it is a virtue theory or a theory of right action? [20] The argumentation here is that care theory is not an ethical theory at all. Care is an emotion. Groenhout's final statement that caring needs standards of good and bad shows that caring is not an ethics, but rather requires one [29]. "Care Theory must be grounded in a substantive theory of the good" [30].

The distinction made in the earlier chapters of this book between morals as only the normative, uncritical beliefs and practices of the culture, and ethics as the critique of morals, applies here. Enlightened caring cannot be based on morals, but only on critical ethics. One can then speak of mere "moral caring" as opposed to philosophically enlightened "ethical caring." This also implies that caretakers should be as caring to each other as to the patient.

Caring must be reciprocal. "A relation cannot properly be labeled caring in their [reciprocal or mutual responses] absence" [31] (See critique of mutuality later). Noddings recent formulation of caring shows the absolute necessity of reciprocity: "(A, B) is a caring relation...if and only if i. A cares for B – that is, A's consciousness is characterized by attention and motivational displacement – and ii. A performs some act in accordance with i), and iii. B recognizes that A cares for B" [32]. If this were required it would possibly eliminate caring in most healthcare situations. This would also be true of many or most parent child caring relationships. Healthcare workers may often be defined as people who help other people who do not help them in return. Where do they then fill their resources again? This is one of care-taker burn-out reasons.

There is supposedly intrinsic giving or caring for its own sake. But this contradicts her view that we care for others because we want to be cared for ourselves [33]. Furthermore, if caring is based on mothering of a child we possibly identify ourselves with caring is based then on egoism. Can one be caring to oneself? A mother may care for a child because she wanted a child, and identifies with a child as her's, as part of herself. Women may also have children for security, self-esteem, avoidance of work outside the house, for some imagined "fulfillment" of themselves, entertainment ("Children are so cute."), etc. Then in effect she cares for herself. Groenhout speaks of "identity egoism." These commitments "constitute a part of my identity" [34].

Care is seen as being non-rational [35]. Care is supposedly a pre-act feeling or attitude. Noddings uses the term "engrossment" to describe caring [36]. It means to wholly occupy or take one's entire attention for the other not the self. This is by definition an excess and so not a rational policy. It is like the term "enmeshment", which is also negative involvement and dysfunction. In caring one is "engrossed," "feeling with." "Mothers quite naturally feel with their infants" [37]. Noddings seems to think we can feel without language. Language is regarded as a male, patriarchal institution and so to be demoted. If to be ethical requires one to be deliberate, then Noddings' care theory is non-ethical. Caring may be unintended, habitual, automatic, or for no reasons. One may not know why one is caring. Her theory comes close to that view. Caring is a mystical, spontaneous "natural" inclination or activity [38]. From this supposedly arises an "ethical theory," though she opposes theory.

Noddings views would seem to support the case method in bioethics. However, as it is anti-reason and anti-empirical, it would also be anti-clinical. Casuists are often relativistic, but naturalistic humanists base their views on adequate reason involving concrete situations, which would be a better basis for a case method. On Nodding's view, we should consider only the others' wants we should act for persons, not for principles.[36] We would be concerned for the other, not oneself. This is the traditional definition of altruism so the critical literature on altruism should be brought in. Rational caring is where one also considers oneself as Tronto also believes [39]. Care only for the patient with no consideration for the physician or nurse generates burnout. It also generates irresponsibility in the patient or other.

9.3 The Word-Field Meanings of Caring

9.3.1 Introduction

According to the American Medical Association Principles of Medical Ethics [38], a physician shall be dedicated to providing competent medical care, with compassion [40]. But what is "compassion" and what is "care"? Caring is an abstract term. Care is basically an empty, open-context, value term. Context is needed to determine its meaning. Care may mean just "good" treatment where "good" is open-context also. What one thinks is care another may think is harm or a smothering relationship. How does one care? Could one follow a rule to do so? To answer this question an interpretation of caring is needed.

There are as many uses or meanings of caring as indicated by a world-field analysis of the term, language-game, or family of related terms and synonyms. Thus, equivocation arises when caring is used in the different senses. For example, that one is caring professionally does not mean that one is caring personally, or that one is caring in the sense of being understanding or unselfish. The pragmatics of the language-games of caring shows that caring has numerous different uses in language. The word "caring" may involve a claim, an exaggeration, a description, a lie, a joke, a way to keep the conversation going, a rationalization, an excuse, an apology, a question, an exclamation, etc. Each kind of caring may also be qualified, e.g., medical caring, therapeutic caring, holistic caring, ethical caring, humanistic caring. Caring may also be used metaphorically and applied to the entire environment. It is the extension of human caring to the harmony of all of nature in which humans live. Personified caring is to ascribe caring to animals, medical institutions (with a motto of "We care"), of governments, etc. There is also a complex pragmatics of language. Care may be an emotive expression. "He really cares!" can mean "Wow!" or "Amazing!" or "I adore him!" Care may also be used as a performative utterance: "To say one is caring is to be caring." We may compare, "To say, 'I love you' is to love." The saying is the doing.

Each meaning may be combined with other meanings to yield what caring means in any particular situation. Caring is not pure. There is always a mix of different types of caring with other cognitions and emotions. Also, at least the minimum claim may be made here that "caring" is a vague abstraction with many meanings and many fallacious uses.

One may be caring about a belief or an idea, e.g., one's discipline, for example, medicine. One may have a passion for critical thinking (speaking) as well as humanism. Caring can apply to any subject or activity, e.g., one may take care of the medicine, children, books, hospital finances, etc. Caring in medicine done for money is not personal caring, although it can involve other types of caring. There are also as many types of carelessness as there are types or meanings of care. One can be caring in one sense, but careless in another, e.g., if one is too caring one may tend to be careless. Also, "healthcare" has several meanings. Is the care just professional healthcare or is the healthcare caring? Healthcare can have as many meanings as care has.

One method of investigation is to analyze through concept. Any word can be elevated to a theory. Thus, we may gain insight by investigations through the concept of care. "Situational ethics," or "love theory" was also thought to be a new ethical theory by Joseph Fletcher [41]. One could speak of "x theory" where x is any ethical, emotion or cognitive word, e.g., client theory, patient theory, clinical theory, dependency theory, etc. That is, there is a philosophy in every word, which may be expanded.

Caring in healthcare may be given insight by comparing it to a love relationship. If so we would not wish a caring partner to be just an economic relationship as it is in some healthcare contexts, or to be treated merely as an independent, partial relationship relating only to the body. We would often wish a full, holistic, passionate love relationship within the bounds of appropriateness and discretion.

9.3.2 The Synonyms and Word-Field of the Term "Caring."

Caring may be good or bad, positive or negative.

Positive meanings of "caring" are: accepting, accommodating, affectionate, altruistic, attentive, benevolent, careful, charitable, circumspect, concerned (emotionally and cognitively), considerate, cooperative, courteous, dutiful, enthusiastic, erotic, farsighted, friendly, giving, forgiving, generous, gentle, helpful (stresses autonomy), humane, kind, indulgent, intimate, involved, judicious, loving, patient (compare with the word the "patient"), personal, prudent, reciprocal, respectful, responsible (for oneself or another), sensitive, understanding, unselfish.

Negative meanings of "caring" are: controlling, enduring, indoctrinative, intrusive, long-suffering, mothering, over-protective (unnecessary suffocating, or smothering caring), over-sympathetic, paternalistic, perfunctory (caring only out of duty), pitying, selfish, egoistic, self praising ("I am caring."), supernaturally caring (agape), worrying. Care as worry/anxiety (*Besorgnis*) is dysfunctional caring. Any negative emotion such as sorrow, worry, pity are dysfunctional forms of caring.

9.4 Irrational Forms of Caring (See Also Empathy and Sympathy)

Defense mechanisms. Caring may be used as a defense mechanism, e.g., a rationalization. One may claim to care as an excuse, lie, apology, etc. One may verbally claim to care, but not actually do so. One may give the appearance of caring to cover one's genuine attitude. Care may be a fixation, transference (e.g., if one transfers one's love of another person to the one cared for), symbolization, introjection (internalize the fears or grief of the cared-for), identification of the caretaker for the cared-for, projection (e.g., attribute one's own illness to others), conversion (physical manifestation of a psychological problem, e.g., over–nervousness, stress), denial (e.g., false assumption that the cared-for can care for themselves), repression, intellectualization, acting-out, regression, sublimation, compensation. One may help others to compensate for one's own guilt. One may care for someone in order that one might oneself be taken care of. This is caring as projection of one's own desires. Dewey and Tufts state that it is possible that by false sympathy and compassion one "is weakening the character of others, and, while helping them superficially, is harming them fundamentally" [42].

Dependency-producing care. This involves making another obligated. Calculated care may be given so as to control another by making them dependent. This often occurs with mothers who thereby prevent children from becoming independent or self-capable. This may also occur, for example, with healthcare workers who do not allow their interns to gain the experience needed to be on their own. In the medical profession, the patient sometimes is made over-dependent on the physician or health care system. Care can promote dependency, overprotection, and imprisonment. The caregiver thus becomes controlling or domineering.

Detachment. Taking care of someone may mean to look after someone, but for this no emotional attachment is needed. There can be caring with a psychological distancing. It allows one to observe the most tragic situations in an accepting way so as to appreciate it in whatever ways we can. Constant negativity and tragedy are unhealthful. Humor and psychical distance have significant places in caring.

Identity egoism. Identity egoism is care as self-identification with another, a group, a clique, or with a nation or belief system. It is a form of egoism, e.g., to identify with a race, a country (nationalism), or one's family, without care or concern for others. Other is already defined in terms of the self – as other than self. When one thinks of a person as other, it almost precludes the obligation to care. One may not even care for oneself. The identification egoism of the mother with the child was critiqued in the section on Noddings' care theory.

Insufficient or excessive amount of caring. One can care too much or too little. The adequate amount of caring must be specified. It is often stated that one, in general, need not legally help others in distress, but may consider helping others if it is convenient. On the other hand, one may give as much money, time, effort as reasonably possible even at great expense to oneself. From this perspective, anyone who buys luxury goods may be seen to be letting people die. In terms of sufficiency, care must often be constant, consistent and dependable. Also, we may care in one sense of care, but not in ten other senses of care. We do not or cannot care in every sense of caring. One may, for example, care emotionally, but not cognitively or efficiently, show concern, but not do what is necessary for the patient. Hopefully, one would care in as many of the positive senses as contextually possible.

Intrinsic caring. Some hold that caring is intrinsically good or good-in-itself. Metz, for example, wrote, "It is intrinsically *valuable* to *help* others" [43]. This is a fallacious use of ethical terms because good-in-itself is unintelligible. "Intrinsic" commits the fallacy of absoluteness and the statement is also circular.

Irrationality of perfection in care: There are mistakes made by physicians and patients. One of the leading causes of death and negative outcomes of treatment is mistakes. One must accept that if one works, one will make mistakes. Physicians, like others, are not all-knowing. For financial reasons there exists lack of medical funding to increase medical knowledge, and there is strong religious opposition to needed research. People in societies usually contribute to medical mistakes by lack of support of the medical profession. After the healthcare workers do what is reasonably possible to prevent and eliminate mistakes, we must accept that mistakes will nevertheless happen. This should be understood by all concerned with medical treatment. There is always cause for correction and improvement on the part of the public, patients, physicians, and administrators, but never cause for punishment.

Patients make the greatest mistakes of all if they are not as informed and careful with their health than as they could be. Need for medical treatment may well be due to mistakes people make in not taking care of their health and their health care systems. People are overweight, smoke, drink alcohol, have unhealthful lifestyles and belief systems, fail to exercise, do not follow medical orders, etc. In this case, negligent patients seem to rarely think about their responsibility for their health, though the negative impact of not taking care of themselves sufficiently might be greater than that brought on by healthcare workers' inevitable errors. Here one may speak of the irrationality of believing that medicine and the physician can perfectly correct all of the disorders caused by patients' own poor lifestyle. In this sense, this is also a form of the fallacy of perfectionism on the part of the patient.

In all of the cases above, fallacies in regard to caring and not caring are the result of uncritical thinking. What is also revealed is the importance for the patient to be responsible to himself or herself (self-care) as well as caring toward the healthcare givers.

It is this that would help to generate positive mutual caring in the context of medical care instead of the one-sided view that the medical profession must alone be caring to the patient. For the best emotional and medical outcome the patient must be reciprocally caring to the healthcare workers. People often, or typically, do not show appreciation even if a healthcare worker bears extreme effort and sacrifice in saving their health or lives. When the cared-for do not show appreciation for the one who cares, it also undermines the mutuality of caring that, while not necessary, is nevertheless important for personal caring. Otherwise it puts caring on an impersonal duty level.

Case example: IVF Society of Austria Survey, Sept. 30, 2004 Meeting.

Patients in reproductive medicine reported the priority patients give for care in the following order:

1. Caring on the part of healthcare workers.
2. Professional competence.
3. Success regarding outcome.

This means that it is more important to be caring than competent or even to have a successful outcome! This is surprising but reveals patient's priorities.

Supernatural caring (Metaphysical caring, agape). Supernaturalism means beyond science, beyond knowledge, beyond evidence or proof, in short, beyond what we do know. Basically, a category-mistake, misuse of language, e.g., personification such as: a caring Buddha, an angel, god, or a "caring" universe is involved.

The demanding versus the caring attitude. It is a logical point that one cannot give to one who will not receive. One cannot demand that which can only be given by choice, otherwise it is not choice. In this sense, one cannot give to a demanding person. What is taken cannot be given. The caring aspect is removed by turning the need or request into a demand. A demand allows no alternative for the caregiver – and shows lack of concern for the caregiver. The demand for a personal or loving gift is a contradiction. The demand for personal medical care, by definition, cannot be given. It can only be complied with, forced or extorted. One may request one's right to medical treatment, etc., but demandingness is always an abuse, especially when one has no right to what is demanded. It reflects an uncaring attitude. One may have a right to treatment, but not a right to abuse to obtain it. A right to the sort of medical care that one assumes one has cannot therefore be given to one. Ungratefulness and lack of appreciation undermine the mutuality of a caring relationship. To be caring to the ungrateful can be misdirected by spoiling the cared-for or making them more ungrateful or dependent. The ungrateful patient expects and demands undeserved care and takes little responsibility for his/her own health.

Uncaring: takes the form of dislike, hate, apathy, indifference, ungraciousness, cruelty, unfairness, bad attitude. Thus one is not prepared to help. Not caring is not merely a neutral stance, but rather a negative emotion. The dehumanization fallacy is the fallacy of treating humans as if they were inanimate "Another cancer patient" is to treat one as a classification and statistic. It is a form of not caring. The personification fallacy is to treat things or animals as if they were human and in the context of care, one may regard things as more important than people.

Unrealistic caring. The physician or healthcare worker may be personally and emotionally devastated if over-involved with the patient in regard to every treatment or when involved with a death. This may cause the healthcare worker to distance himself or herself from the patient so as not to be so negatively affected. One need not do so. One can still be caring if one understands that one must accept reality that people do die including oneself and that not all treatment and operations are successful. But this can only be successfully done if one has knowledge of a sound theory of emotion. If one has done all one can to prevent disease and death, one need not have negative emotions. The physician as physician especially does not need to feel guilty. Those who have done nothing to prevent illness, or do not agree to organ donation at death, support military and war instead, are those who may rather think-feel guilty. Sympathy or grief on their part is merely hypocrisy.

Burnout Syndrome. Helpers cared too much and so were over involved. Often people who are over-concerned were influenced by humanistic and altruistic ideas or organizations, pacifists, or sensitive thinkers. They often receive too little in return, are over-criticized and unappreciated by those helped or their relatives. They become "helpless helpers" [44]. Without knowledge of emotions and care one becomes a helpless helper, burns out, or becomes ill.

Pity. Pity can be seen in various ways. It may be seen as a thought-feeling: for the suffering and misfortunes of others. *Mitleid* (German, literally sym-pathy), pity is not just a bodily feeling, but also an assessment of a negative state of affairs. The emotion produced may be positive or negative. Pity may lead to help and care, or to grief and desperation.

As a negative value term: it expresses disadvantage, shortcoming, or weakness. It characterizes bad, awful, terrible, matter for regret, miserable, contemptible, e.g., "what a pity," "pitiful." Pity may be merely a term of blame and condemnation without sympathy, care or concern. Pity may be with or without mercy.

As self-pity one is "feeling sorry for oneself." Instead of accepting the events of one's life, one sees them as negative and further negativizes them, e.g., "wallows in self-pity." This is non-adjustive and makes matters worse. It is also an egoistic attempt to extort the sympathy and care of others.

9.5 The Cognitive-Emotive Theory of Caring

In philosophical literature the cognitive-emotive theory is the prevailing one, and its counterpart in therapy is the Rational-Emotive and Behavioral Theory (REBT). There is a fortunate collaboration here because philosophers can work out the theoretical exploration while the therapists provide the actual clinical experience (See *Journal of Rational Emotive and Behavioral Therapy*).

According to the Cognitive Theory of Emotions as we interpret it, emotion (E) is a cognition (C) that causes a bodily feeling (F). E = (Cognition > Bodily Feeling). In ordinary language, "emotion" and emotion words refer to both cognitions and feelings. An analysis of emotion therefore requires an analysis of bodily feeling, emotion, cognition, and their relationships. A number of characteristics of aesthetic emotion may be deduced from the cognitive-emotive theory.

9.5.1 Feeling

Caring is not a mere bodily feeling. Because emotions involve cognitions, it is always a mistake to say, "I feel caring." Ellis pointed out that it is more precise to say, I "think-feel" caring. F here can also refer to sensation and perception (hearing, seeing, tasting, etc.). If emotion were only a feeling it would not be intelligible [45]. Melden wrote: "There is no simple or single feeling one has, such that feeling anger [caring] consists in having it and nothing else. Anger [caring]. . .cannot be an internal feeling or state conceptually unrelated to the functions of intelligence" [46]. We may accordingly answer Wittgenstein's question: "What similarity has my admiring this person with my eating vanilla ice and liking it?" [47]. We may note that "I like ice-cream," is a taste (F), "I admire her," is an emotion (E). We do not ask why we like ice cream, but we do ask why we admire someone.

The prevailing polarized view of "reason versus emotion" or "caring versus reason" is rejected. The dichotomy reason (cognition) versus emotion dissolves or is restructured once it is seen that emotion is cognitive, that cognition produces feeling: (Cognition > F). The dichotomy is based on a faulty view of emotion. "Cognitive caring" is not a contradiction. Noddings view is also based on an opposition between emotion and cognition. She speaks of caring as an emotion, affect and attitude, but it can be none of these as she rather regards them as mere bodily feelings [48]. Nevertheless, both feeling and reason are embodied. Without a brain, there is no reason, and no feeling.

In contrast to the cognitive theory, care for Noddings is just a primitive bodily feeling. It does not involve knowledge, but is a non-verbal impulse [49]. It is like Hume's sentiments of sympathy. However, if caring can just come, it can just go, like a headache or pain [50]. It is said to be a natural innateness like animals, such as cats, have. But cats cannot have human emotions as they lack our language and way of thinking. There is no emotion or feeling without some kind of cognition. Noddings fails to give a theory of emotion as a basis of her theory. If caring were innate so would love, hatred, and anger be and out of our control, and fatal. . . what they often are – without reason.

9.5.2 Caring Is a Value Cognition Causing Feeling [Caring = (Cognition > Feeling)]

"Cognition" does not refer to mentalistic "ideas" or "thoughts." Instead, it refers here to non-mentalistic assessments involving value terms. These cognitions, then, are statements consisting of self-talk, utterances, and language use. As care, anger and other emotions no longer exist as such, the letter E will be used to stand for (C > F), where C is one or more specific assessments.

Caring can involve rational or irrational cognition. The cognitive theory of emotion thus gives a therapeutically useful, as well as philosophically clear, picture of emotion. No essentialistic claims are made. The analysis of caring becomes an analysis of its diverse language-games. For one to be consistently caring one must deliberately and consistently assess that one will be the most caring (kind, attentive, concerned, etc.) possible. What caring will mean is determined by the definition one gives to it. A more adequate definition might include: (a.) communication in order to solve problems and conflicts when they arise, (b.) communication to enrich the relationship, (c.) acceptance of the other person in various ways, (d.) ability to encourage positive emotions, (e.) mutual trust, (f.) the humanistic use of humor that can help one adjust to one's situation and, as humor involves acceptance, create friendship, (g.) care for everyone if one sees that people are not to be blamed or that they only do harmful things out of ignorance [51].

9.5.3 Caring Is Based on Positive Cognitions

The emotive cognition is typically a value assessment [52]. The cognition in caring must involve a positive value term or assessment, e.g., "I think you are a valuable person," Basically caring is caused by our assessment that doing so is good. We may think of caring as helping. "Good," however, is a notoriously vague term. As it can mean almost anything, it may be seen as an empty or open-context term – like a blank check. Consider the statement by a physician, "I will give the best [caring] treatment to my patients." What will be done? No specific commitment is actually made. "Good" is meaningless until something is substituted for it. The philosopher, Derik Parfit, wrote, "We benefit someone only if we do what will be better for him" [53]. This is circular. It does not specify the value terms used.

By definition, caring expresses a positive value, which is already contained in the very word "caring." In expression caring becomes clarified. We can analyze the emotion in practice. The assessments involved in caring are diverse, but may be something like, "You are in need and so I will help you." Caring is made possible by the acceptance of reality and the motivation to improve the given situation best we can.

9.5.4 Emotions Can Be Changed

Emotion can be created or changed by creating or changing cognition. This is similar to a change by a different sensing-as, for example, hearing-as. Psychogenic pain is cognized pain. Change of emotion and caring may take place anywhere along the continuum of cognition to sensation. Caring can be changed by changing the value assessment. To care one must oneself have the assessment(s) to care on some definition of caring. Cultural encouragement and constant exposure to uncaring behavior can produce the influence to be uncaring and the enculturated assessment that one need not care. Without the ability to create positive emotion there can be no caring. The ethical challenge is to gain insight into the consequences of being or not being caring.

9.5.5 We Cause Our Own Emotions: Caring Is Caused by Ourselves

One's caring is caused by oneself, not by objects or others. We cause our emotions by our value assessments of certain objects. No one can make one personally care or love, but oneself. Dewey states, "Even anger and hate are partly caused by us rather than in us" [54]. Caring must be done in the first person case: "I will care for others." One may always ask oneself, "Could I have been more sensitive and caring?"

On the cognitive emotion theory, one can deliberately commit oneself to love or care for others. This is necessary for a deep and ongoing caring relationship. One can count on the other as a partner, friend or caretaker. One can speak of uncommitted caring as one can speak of loose, uncommitted friendship. It is unreliable. Healthcare as impersonal law-bound care may be unreliable and limited in terms of time, amount and quality. And be based solely on financial negotiation.

One might only take on the role or appearance of caring because it is one's job as a healthcare worker, but this is role or pretense caring. It is not personal or first person case caring, but second and third person caring: "You require or they require me to be caring." This generates a different or more impersonal emotion of caring. It may be noted here that impersonal caring is still an emotion and is still caring. There can be negative emotions of mere "professionalism" or cold "objectivity." The "objective" sounding monotone lecture is similar to the emotion of boredom. Such a professor is not enthusiastic about his students.

If one helps others merely because of the money one thereby earns or because it is one's job or obligation to do so it will not generate genuine caring or enthusiasm. One form of caring involves being excited about helping another. Because of the many meanings of caring there are many emotions of caring. There is no caring as such, just as there are no emotions as such, but only "caring" assessments, which cause bodily feelings. What caring is for any individual depends on their known or implied definition of caring. The various definitions were given earlier.

Without an analysis of emotion few would have a clear understanding of what caring is. One may call this "blind caring." If asked, one cannot answer why one cares. One may have an attitude of personal caring without professional caring or identity caring, etc. Similarly, the notion of "helping" may be a personal, subjective helping or objective, impersonal helping. It can be a perfunctory act. It is like mere social politeness versus being genuinely concerned. It is hypocritical caring if personal caring is actually selfishness (self-caring). It is the unhumanistic smile.

Just as emotion is not innate or unalterable, neither is caring. Caring is not an innate quality of humans, but can be learned or unlearned by change of assessments and understanding. We can radically change our caring behavior. Being uncaring often shows a faulty sense of goals and philosophy of life. It is therefore interesting that the literature on caring stresses the social aspect of caring and caring as a relationship. It also means that the investigation of caring belongs on the level of philosophical counseling.

9.5.6 The Passionate Stoics: Rational Emotion, Rational Caring

The Stoics did not reject emotions they only rejected negative emotions such as uncaring. Uncaring behavior would be "false judgment." The position of Marcus Aurelius is not that we should have no emotions and so be without passion. Rather, he opposes violent excitement (an oxymoron) [55]. Have "good emotions" (58), and "happiness" (74). We could include in this the emotion of caring. Rational judgments produce positive emotions, for example, happiness (*eudaimonia*). Against the widespread characterization, Rist argues that the Stoics did not advocate apathy. On the cognitive-emotive theory, apathy is a negative emotion [56]. The wise person experiences the joy, happiness, and even exhilaration that come from living a rational life in accordance with nature [57]. Rist argues that they produce "rational feelings" [58], and that only "the picture-book Stoic wise person is devoid of passions" [59]. Concerning this Rist states, "Anyone who seeks 'apatheia' in the sense of total elimination of all feeling and emotion, is asking for a state when all activities, even mental activities, are suspended. Such a state would be equivalent to death" [60]. If one is not caring, one is not neutral, but uncaring.

9.5.7 Negative Emotion Changes with Feeling

Emotion can, to some degree, be changed by changing the bodily feeling (perception, sensation). Because caring is cognition, which causes bodily feeling, there is no unfeeling caring. A change in bodily feeling, then, can alter the caring. In regard to negative emotions, such as being uncaring and revengeful, change of the resulting feeling can do little to change the emotion. It is mainly the cognition that must be altered. If it were only the feeling no physician would get up at 3 o'clock in the morning to take care of a patient in emergency.

9.5.8 Negative Emotion Is Not Passive

Being uncaring is not passive like bodily sensations or bodily feelings. Because feelings are used metaphorically to refer to emotions, we may tend to regard emotions as being just bodily feelings and so as being passive. They are not. Even the sudden caring concern that seems to be groundless is based on numerous prior assessments. We are attracted by those who meet our prior preferences. We can reconstruct the experiences that lead us to have these experiences. In another sense, our assessments may themselves be enculturated. We are uncaring partly because we have learned to be so in the society in which we were raised. A reconstruction of the reasons for being caring or uncaring is just another language-game, not a statement about what "really" is the case. But it nevertheless has its uses as with the correction of failure to care.

9.5.9 Each Emotion of Caring Is Unique

We can never have the same emotion twice. Nor can two different people have the same emotion twice or at all. The first reason for this is that there is no emotion as such. Secondly, for each emotion there is a different cognition (C) and a different bodily feeling (F). We have "ideas" constantly and cannot completely control the way in which they come to us. The same is true of feelings. Thus, again, for each emotion of caring there are different C and F. Caring $= C_1, C_2, C_3 \ldots > F_1, F_2, F_3 \ldots$.. Each instance of caring is to a greater or lesser degree different than every other one. Our cognitions do not stand still, and nor do our bodily feelings. And two people can only have a *roughly* similar emotion of caring.

Caring_1, Caring_2, $\text{Caring}_3 \ldots$ can be distinguished by their different cognitions. "The production of the same emotion by different contexts is impossible" [61]. For each act of caring we must find the specific assessments and feelings actually had. A physician exhausted from night attendance and/or from irrational demands of patients will have different thought-feelings of caring. A specific act of caring will be some combination of diverse sorts of assessments and feelings. Thus, each specific occurrence of emotion (each language-game) must be examined separately. The above analysis hopefully provides a reply to Hoaglund's complaint, "Philosophers have not really come to grips with the wide disparity between our simple emotion terms and our complex emotional life" [62].

9.5.10 Rejection of the Release Theory of Caring

Emotion is not a sort of mentalistic thing that can be "released." We are wrongly told to "release" our anger, grief, jealousy, etc. Dewey states, "Experience is emotional but there are no separate things called emotions in it" [63]. And experience is embodied, so are cognitions, emotions, and feelings. The change of cognition takes the place of release. In any case, a well-adjusted person would care because it makes sense to do so in terms of consequences for the cared for as for oneself.

9.5.11 Judgments Generally Involve Emotion

Virtually all judgments involve emotion. Belief (judgment, cognition, scientific statement, etc.) itself may be regarded as an emotion. The formula for emotions is: emotions are cognitions, which cause bodily feelings. It would follow that any cognition would be accompanied by a feeling. And this is exactly what the Stoic, Chrysippus, held. It is not strictly the case that cognitions cause bodily feelings rather bodily feelings are part of cognitions. For him, all judgments involve feelings or emotions. There is no such thing as emotionless thought [64]. For Collingwood also, all judgment and all of language, express emotion. Emotion and cognition unite [65]. "There is no need for two separate expressions, one of the thought, and the other of the emotion accompanying it. There is only one expression" [66]. We can now see that there are caring judgments. As the word-field of care shows, it may be involved in many judgments and emotions, e.g., love, help, etc.

9.5.12 Metaemotion

Altruism, caring and egoism are not causes. Only cognitions-causing-bodily feelings can be causes. Morality has largely to do with feelings. If benevolence is regarded as an emotion, it is not a cause of altruism, but a meta-emotion, an emotion causing another emotion, or more strictly speaking, one cognition-causing-a-feeling that causes another cognition-causing-a-feeling.

9.6 Caring and Negative Emotions

It has been argued above that negative emotions are due to irrational thinking. As a person has mostly negative emotions s/he cannot to this extent be caring. Physicians or nurses, who are overworked, frustrated with problems and demands, stressed, or burned-out cannot adequately provide needed positive caring emotions. A simple test of one's professional emotional competence is the following: If one does not have positive emotions with colleagues, it is certain that one will also not have them with patients or others, and even with oneself.

Total quality management (TQM) is meaningless if one does not know what quality means. To know about quality, an open-context value term, one must know about both ethics and emotions.

The negative emotion of blame often blocks caring. With anger and revenge, blame tops the list of the foremost negative emotions. With such an emotion one cannot easily have, for example, personal caring, though one can have perfunctory tolerant professional caring. It has been argued that blame does not make sense because: 1. We cannot change the past or what is, and 2. People do harmful things out of ignorance (e.g., because indoctrinated or enculturated or uneducated). No one could have done otherwise than they did or they would have done so. To claim otherwise is to think one always has all knowledge and control of oneself. We do not have such knowledge, and especially not of emotion. If we cannot be certain what we will do in the future, how can we claim to know what we would have done in the past? The claim that anyone at any age and stage of development could have done otherwise is to judge from an unrealistic, ideal, after the fact perspective. We do the best we can from what we know regardless of how adequate. Although it therefore makes no sense to blame, it does make sense to correct behavior. We thereby substitute rehabilitative blame for retributive blame. We treat the worst patients, but we educate them to get out of their destructive behavior as well. Without blame or unfounded negative emotions one is free to love or be caring to everyone. Understanding is a form of caring. The more we understand others the less we can blame them. This gives support to the physician's Hippocratic Oath requiring the physician to do no harm, and to treat people equally. However, this does not mean that the equality should be blind and ignore other relevant factors. One may not blame, but unconditional caring is self-defeating. It does not make sense to give care to all without regard to the consequences (see "Nighttrain to Lisbon" by Pascal Mercier). We may also have to psychologically and physically protect ourselves from them. One may nevertheless have a positive attitude without anger or bitterness toward everyone if one does not blame anyone. However, that we do not blame also means that those more capable have a responsibility to educate those who do not. The healthcare worker may serve as a model of caring.

9.7 Mutuality of Caring

It was earlier mentioned that for Noddings, caring must be reciprocal. "A relation cannot properly be labeled caring in their [reciprocal or mutual responses] absence" [67]. Groenhout agrees that reciprocal caring is required, even eventually of a child. If this were true it would possibly eliminate caring in most healthcare situations [68].

Caring relationships need not be mutual. One may be caring toward (for, with, of) another without reciprocation. The healthcare worker-patient, parent-child, teacher-student relationship may involve non-reciprocal caring. In love this would be called "unrequited love" or "pure love," a caring when nothing is expected in return. Otherwise love and caring are like barter involving egoistic benefit. Nor can one expect care to be recognized by the cared-for. One cares because it makes sense, not because one is thanked for it. Although it would be humanistic and supportive, patients often do not show appreciation being cared for. They may merely expect it, take it for granted, or demand it. They are far from appreciative caring. The cared-for may even be noncompliant or betray the caregiver. Noncompliance undermines reciprocity. Good Samaritan laws even had to be enacted to protect the rescuer from lawsuits by the rescued.

One can assess to love/care for someone though s/he does not or cannot reciprocate. This is especially true if we do not blame people, but if we rather seek to help better them. If people, including health care workers, do not know about how emotions work, they are to this extent not capable of fully caring.

9.8 The Patient's Hippocratic Oath

To avoid the shortcomings of caring mentioned above we may propose the following as the Patient's Hippocratic Oath.

1. Do no harm to yourself, the healthcare provider, or others.
2. Live a healthy lifestyle. Make intelligent decisions regarding risks and diseases.
3. Develop your critical abilities and other abilities and potentials to achieve the maximum quality of life.

4. Cooperate and participate in your own cure and recovery by inquiry and obtaining information to become as healthy as possible.
5. Find out what ethics is about and create a sound ethical system.
6. Learn about emotions so as to encourage positive emotions and eliminate negative ones.
7. Do what you can to promote medical treatment and research. Offer to donate organs upon death, do not oppose medical research for supernatural reasons, donate to research and other health organizations, etc.
8. Make sure you have good reasons for having a child and whatever is necessary to care for the child.
9. Contribute whatever is reasonably possible to the people of the world who are in desperate need of food and medicine or are otherwise desperate.
10. Honestly reveal to the healthcare worker all information required for treatment.
11. Cooperate with the healthcare worker regarding treatment including the following of instructions given.
12. Resolve differences by means of good communication, ethics counselors, or philosophical counselors rather than by appeal to punitive laws.

9.9 Empathy and Caring

Empathy can involve pity, to think-feel sorrow for someone or something because of the poor or depraved state they are in. One may or may not act to help "out of pity." It is an emotion that may motivate, and is partly a negative emotion. It is a failure to accept reality and the fact that we can only do what is within our power.

Webster's Dictionary defines empathy as: "Imaginative projections of one's own consciousness into another being." It is, however, often said that philosophical counseling is not to involve projecting one's views, emotions or consciousness into another. It may be dangerous and indoctrinative to try to project or impose one's views in this way.

Is empathy then having a similar emotion (pathos) to that of another? If so, can one genuinely do that without being the other and in the same situation? One gets the impression that "sympathy" may mean to have the same negative emotion as the other. If they are depressed, you are depressed. Caring as sharing another's grief can also be debilitating (The German word *Sorge*, can mean both, care or worry in a fearful way; *sich sorgen um* means worry about, *sorgen für* means to take care of). If "sympathy" means *sym-pathy* (German *Mitleid*, "suffer with"), having the same negative emotions or feelings someone else has, Nietzsche is right in seeing it as life-denying. "Commiserate" means, literally, to be miserable with. Eisenberg even calls empathy, "emotional contagion" [69]. One should instead help without negative emotions. If one were to have sympathy it would preferably be holistic and humanistic sympathy, not just a feeling.

Some of our acts are sympathetic, others not. We have selective altruism and egoism. Should we have empathy for military or religious fanatics, for those who are anti-humanistic, anti-inquiry and hold dogmatic positions regardless of consequences? Should we be empathetic with those who continually frustrate, oppose and undermine us? Whatever is the case, if philosophical counselors are to speak of empathy they need to define it and offer a clear and tested theory of empathy. The same would be true of vague value concepts such as "caring." Like caring, "empathy" is an "emotion" and has here been similarly analyzed on the cognitive theory of emotion. There are, of course, a few theories and criticisms of empathy. In social work E Munro regards empathy as being unreliable [70]. In the context of care theory Tong wrote, "Empathy. . .can be useful for good or ill. The more I am able imaginatively to project into your psyche, the more I can help or harm you" [71]. With empathy one may supposedly come to a more complete understanding of another. If so, it appears to be a cognitive concept. But a full and deep understanding of another (e.g. *verständnisvoll* – full of understanding) involves much more than empathy and it may not involve empathy at all.

Another danger with the emphasis on empathy is that it seems to be regarded as a special non-cognitive, feeling form of intuitive knowledge [72]. Two objections to this are: 1. Empathy as a mere feeling may produce irrational self-justification: Something is true because we somehow feel that it is true; 2. According to the cognitive theory of emotion, emotions are not just bodily feelings. It would be a fallacy to regard empathy as just feelings, or bodily feelings. The widely accepted cognitive theory of emotion comes against the typical definitions of empathy. It may be noted that the German word, *Einfühlungsvermögen* refers to both feeling + cognitive ability to understand. We acquire a sensitive insight into someone or something and thus get into the very nature of the experience. It comes close to phenomenological philosophy.

Bohart and Greenberg edited a book on empathy in psychotherapy in which they conclude, "Research in the area has generally been plagued by confusion over operational definitions and lack of adequate measures" [73]. Different authors meant different things by "empathy" [74]. The definitions and mechanisms of empathy are thought to be unclear [75]. The question was raised as to whether or not empathy is even a distinct quality, and if not, it would not be measurable [76]. The general conclusion was that one does not know what empathy exactly is and that attempts to measure it have failed [77]. Four types of empathy were given: cognitive, affective, shared, and nurturant (supportive) [78]. Empathy was further defined variously as: person-centered promotion of safety and trust in the self, an intervention, bonding, restoring the relational interaction, "Gestalt", reframing, group therapy, dialectical synthesis, narrative organization of another's experience, existential concept, understanding or knowing of the other, unconditional positive regard for the other, sympathy, shared feelings with each other, same feelings as the other, having cognitive understanding of the other, non-judgmental acceptance of others as they are, immersion of the self in the other, co-construction of symbols to clarify and understand another's experience, feeling and thinking as if one were the other, gain of access to the unconscious (Freud), "reflection of feelings" (Carl Rogers), concentration on what the other is feeling, saying, meaning in the moment, communication of another's views back to them, communicative attunement, kindliness, encouragement and justification of the other's negative emotions. The sado-masochist can use empathy to take pleasure in the pain of another and do so without sympathy [79]. The therapist can have empathy for another, yet not care about them at all. The main problem with the psychotherapeutic approaches is that they do not have a sound theory of emotions and lack critical analysis as would be had with knowledge of philosophical psychology. The closest account to this is the work of the cognitive psychologists [80].

In sum, empathy, like caring, is too vague, unscientific and unsatisfactory as a concept to base philosophical counseling or medical action on it. Barnbaum stated, "Teaching empathy should be part of the medical ethics curriculum" [81]. She assigned by lottery diseases for students to imagine they had and required them deal empathetically with such diseases. From the above we can see that this would be an impossible task. Better would be to learn about and apply a sound theory of emotion.

9.10 Summary

On the cognitive theory the question, "Why do we help others?" becomes, "Which cognitions are involved in altruism?" We help others because of the following: force, threat, guilt, duty, cultural expectation, actual reward, power, financial gain, recognition, pleasure, pity or sympathy, trade-offs, influence, etc. We may do unselfish things for selfish reasons, e.g., for affection.

What is called by Wolfgang Schmidbauer the "Helper's Syndrome" is self-sacrifice to hide a lack of goals or an emptiness in one's life. People with helper's syndrome are not in touch with their true emotions. One may find oneself in helping an ill patient or one's needy child in order to get in touch with one's inner self. In helping, the physician may have lost the ability to be in a genuine mutual exchange relationship and interdependence, not an adult to adult relationship. Thus, resentment may develop [82]. On the other hand, it is also pointed out how altruism may be "negotiated" to the disadvantage of those who try to help [44].

It makes sense to care for others because caring and love are needed in the immediate and larger world. People wish to satisfy their basic needs, desires and wants as human beings and it is rational in terms of our goals and consequences to help them do that. As people, we act, have cognition, emotion and feeling. We would not wish to live in a world of either negative altruism or negative egoism, a world where no consideration is given to anyone, no flowers are grown or given. Perhaps this is what Hume partly meant by saying that sympathy cannot be further defined. Hume and others based altruism on emotions and psychological, factors such as sympathy. Altruism versus egoism is, then, part of larger questions such as, "What is the basis of ethics?" Altruism versus egoism, or caring is a fragmentary, narrowly focused issue.

Why should we help others? On the view of psychological egoism it is part of our character to do so and so has no rational basis. According to Dewey's ethics, value terms reduce to naturalistic terms. "Good," for example, refers to deliberately bringing about our informed wants and likes on the basis of adequate inquiry and knowledge of consequences. On such a view, we would be altruistic because, in terms of inquiry and an adequate knowledge of the consequences, it makes sense to do so. It is a "rational altruism or egoism." On this view we would be egoistic or altruistic, assuming the terms are clearly defined, whenever it is rational to do so in terms of the consequences. Similarly, JS Mill held that altruism is supported by our desire for harmony, a "feeling of unity" with others; because one seeks pleasure for oneself, by analogy, one ought to seek the pleasure of all [83].

Rational caring may be contrasted with irrational caring. Rational caring would include rational thinking, critical thinking (speaking), holistic philosophical thinking, humanistic thinking (ethical concern and care for humans, animals and the environment), knowledge of ethics and emotion. Without these abilities one could only have irrational caring, inconsistent or contradictory caring. One cannot be pro-killing or pro-war, or anti-medical research, anti-science and be a rationally caring person. *The creation of rational caring generates a new emotion.* Not caring, like apathy and depersonalization, are fallacies of negative emotions.

We live in a society in which we must interrelate with each other and so are concerned with each other. We are world-citizens. The problem of altruism and egoism is a holistic problem. It is a high-level problem of value and fairness. It is a problem of how to make the most adequate decisions possible for all involved. We would have enlightened management not because it makes a self-interested profit for the employer, but because it makes sense for all concerned. The assessments involved being based on reason and adequate inquiry produce an emotion of being civilized, of satisfaction in the knowledge that one has produced an intelligent social act of caring. To express this in another way, a rational or positive caring reduces to humanism. Humanism appears to be a useful guide upon which to base caring because its basic guidelines involve altruism, love of people, open education, free inquiry, free choice, freedom of speech, an ethics based on reason, consequentialism, enlightened democratic methods, anti-sexism, anti-racism, etc. It opposes enculturation, vague abstraction, indoctrination, censorship, morals based on fear and punishment, supernaturalism, appeal to authority, absolute values, etc. On this view, we are altruistic or egoistic not out of a feeling, or because of duty, but because it makes sense in terms of the consequences. This does not mean that ethics is a matter of achieving consensus. Consensus is the appeal to majority fallacy. That most people believe something does not make it true or fair. Consensus is a way of *not caring* about reasoned argument. It is also a way to retain the prejudices or supernatural beliefs of the populace.

Rational reasons for caring are fundamentally simple. People have wants, likes and enjoyments and these may be enhanced by our efforts. It makes sense to do so as we have interrelationships with everyone in the world directly or indirectly. We help others as world citizens, as humans. The negative side is to help others because they are in desperate need. The positive side is to enhance others to improve their knowledge, capabilities, aesthetic lives and living situations. To care about people is to help them be all that they can be. In the medical profession it translates into helping to rebuild the patient's potentialities and trying to give the patient new goals or insights in life appreciation, even art. A Viennese surgeon of our acquaintance does exactly that and has authored a book on the art of physicians [84]. Rebuilding potentialities is also one of the fundamental goals of philosophy and ethics.

Human ecology, humanistic medicine, and bioethics are forms of this caring. In this respect, one could say that caring is a passion for critical thinking (speaking) and practical, problem solving. In a naturalistic theory of ethics, the more we know the more we can bring about our informed needs and desires. To care is also to see what caring means in detail. This is a philosophy of caring, rather than only morals of caring.

An analysis of caring was seen to require an analysis of ethics, the self, causes of action, motivation, and emotion. To analyze caring is to analyze how people think and behave generally. It is not just an isolated quality. It was also shown how caring may be redefined and based on a naturalistic, humanistic theory of ethics.

Philosophy and Ethics of Personality involve the emotion of care. From the above analysis of emotion and care theory we may create an area of investigation called the Philosophy and Ethics of Personality or Character. It includes the area of philosophical psychology, philosophical counseling, analysis of character and the traditional virtues. We have seen that a positive personality is needed for the emotion of caring. As the personality of the physician is like the so much needed drug for the patient... it is of eminent importance to work on it professionally and privately.

This relates to the question sometimes raised as to what kind of people should we be? It involves also the question of the use of medicine and sperm banks, etc. and the possibility of having designer children. What personality qualities and goals ought an individual seek to develop and for what purposes? How can one think more rationally? To what extent can one eliminate negative emotions and develop positive ones? How can one become more ethical? How can one live a more aesthetic life? Philosophical Practice (philosophical counseling) is an attempt to guide one regarding such questions. Rational caring is one of the personality characteristics that may be considered for adoption.

To determine what caring is presupposes knowledge of ethics, emotion and philosophical critical thinking (speaking). This is the area of Philosophical Practice. We saw that caring is not itself a theory of ethics. It is an emotion. Caring leads to a wider and deeper philosophical investigation.

References

1. Fallberg, L., and Borgenhammer, E. 1997. In *Problem doctors: A conspiracy of silence*. eds. Lens, P., and Van der Wal, G., 31–143 esp. 135. Amsterdam: 1 IOS Press.
2. Noddings, N. 1984. *Caring. A feminine approach to ethics and moral education*. Berkeley, CA: University of California Press.
3. Noddings, N. 1984. *Caring. A feminine approach to ethics and moral education*, 8. Berkeley, CA: University of California Press.
4. Noddings, N. 1984. *Caring. A feminine approach to ethics and moral education*, 1. Berkeley, CA: University of California Press.
5. Groenhout, R. 1998. Care theory and the ideal of neutrality in public discourse. *Journal of Medicine and Philosophy* 23:170–189: esp. 174.

6. Groenhout, R. 1998. Care theory and the ideal of neutrality in public discourse. *Journal of Medicine and Philosophy* 23:170–189: esp. 174; 180.
7. Noddings, N. 2002. *Starting at home: Caring and social policy*, 28. Berkeley, CA: University of California Press.
8. Noddings, N. 2002. *Educating moral people*. Columbia University, New York: Teachers College Press.
9. The Chaos of Care and Care Theory. 1998. *Journal of Medicine and Philosophy* 23.
10. Jecker, N., and Reich, W. 1994. Care. In *Encyclopedia of Bioethics*. ed. Reich, W., 5 vols, 2nd edn, 336–344; esp. 342. New York: Macmillan.
11. Noddings, N. 1984. *Caring. A feminine approach to ethics and moral education*, 27–28; 79; 25. Berkeley, CA: University of California Press.
12. Maier, B., Braun, C., Schuster, D., Wenger, A., and Rauchfuß, M. eds. 2010. *Ver-Bindung. Bindung, Trennung, Verlust in der Frauenheilkunde und Geburtshilfe*. Frankfurt am Main: Mabuse-Verlag.
13. Eyer, D. 1988. *Mother-Infant-Bonding: portrait of a paradigm*. PhD dissertation, University of Pennsylvania.
14. Eyer, D. 1992. *Mother-infant bonding: A scientific fiction*, 19. New Haven, CT: Yale University Press.
15. Eyer, D. 1992. *Mother-infant bonding: A scientific fiction*, 15. New Haven, CT: Yale University Press.
16. Cierpka, M. 2010. There is no such thing as a baby – das Eltern-Kind-System. In *Ver-Bindung. Bindung, Trennung, Verlust in der Frauenheilkunde und Geburtshilfe*. eds. Maier, B., Braun, C., Schuster, D., Wenger, A., and Rauchfuß, M., 49–58. Frankfurt am Main: Mabuse-Verlag; Vavrik, K. 2010. Frühe Bindung – späte Folgen. Zur Bindungsentwicklung bei peripartalen Risiken und Krisen. In *Ver-Bindung. Bindung, Trennung, Verlust in der Frauenheilkunde und Geburtshilfe*. eds. Maier, B., Braun, C., Schuster, D., Wenger, A., and Rauchfuß, M., 71–92. Frankfurt am Main: Mabuse-Verlag.
17. Eyer, D. 1992. *Mother-infant bonding: A scientific fiction*, 42; 199. New Haven, CT: Yale University Press.
18. Eyer, D. 1992. *Mother-infant bonding: A scientific fiction*, 200. New Haven, CT: Yale University Press.
19. Tronto, J. 1993. *Moral boundaries: A political argument for an ethic of care*. New York: Routledge.
20. Veatch, R. 1998. The place of care in ethical theory. *Journal of Medicine and Philosophy* 23:210–224 esp. 221.
21. Noddings, N. 1984. Caring. *A feminine approach to ethics and moral education*, 1–11; 5; 190. Berkeley, CA: University of California Press.
22. Noddings, N. 1984. *Caring. A feminine approach to ethics and moral education*, 24. Berkeley, CA: University of California Press.
23. Noddings, N. 2002. *Starting at home: Caring and social policy*, 30. Berkeley, CA: University of California Press.
24. Tronto, J. 1993. *Moral boundaries: A political argument for an ethic of care*, 103. New York: Routledge.
25. Gilligan, C. 1995. Moralische Orientiering and moralische Endwhichlung. In Weibliche Moral. Die Kondroverse unline gescheechlsspezifische Ethik. edn. Nunner - Winkler G., 79–100. Minchen: dtv
26. Held, V. 1995. Justice and care. Boulder, Co: Westview Press.
27. Barnes, A. 2004. Am I a carer and do I care? *Medicine, Health Care and Philosophy* 7:153–161.
28. Tronto, J. 1993. *Moral boundaries: A political argument for an ethic of care*, 103. New York: Routledge.
29. Groenhout, R. 1998. Care theory and the ideal of neutrality in public discourse. *Journal of Medicine and Philosophy* 23:170–189 esp. 105.

30. Groenhout, R. 1998. Care theory and the ideal of neutrality in public discourse. *Journal of Medicine and Philosophy* 23:170–189 esp. 185.
31. Noddings, N. 1984. *Caring. A feminine approach to ethics and moral education*, 19, 41, 61, 69. Berkeley, CA: University of California Press.
32. Noddings, N. 2002. *Starting at home: Caring and social policy.* Berkeley, CA: University of California Press.
33. Noddings, N. 1984. *Caring. A feminine approach to ethics and moral education*, 5. Berkeley, CA: University of California Press.
34. Groenhout, R. 1998. Care theory and the ideal of neutrality in public discourse. *Journal of Medicine and Philosophy* 23:170–189 esp. 178.
35. Gordon, S., Benner, P., and Noddings, N. 1996. *Caregiving: Reading in knowledge, practice, ethics and politics*, 61. Philadelphia, PA: University of Pennsylvania Press.
36. Gordon, S., Benner, P., and Noddings, N. 1996. *Caregiving: reading in knowledge, practice, ethics and politics*, 161. Philadelphia, PA: University of Pennsylvania Press.
37. Gordon, S., Benner, P., and Noddings, N. 1996. *Caregiving: Reading in knowledge, practice, ethics and politics*, 31. Philadelphia, PA: University of Pennsylvania Press.
38. Noddings, N. 2002. *Starting at home: Caring and social policy*, 29. Berkeley, CA: University of California Press.
39. Tronto, J. 1993. *Moral boundaries: A political argument for an ethic of care*, 103. New York: Routledge.
40. American Medical Association AMA Principles of Medical Ethics. Adopted by the AMA's House of Delegates, June 17, 2001.
41. Fletcher, J. 1966/1997. *Situational Ethics: The new morality* Louisville, KY: . Westminster Knox Press.
42. Dewey, J., and Tufts, J. 1932. *Ethics*, 330. New York: Holt.
43. Metz, T. 2002. *Recent Work on the Meaning of Life. Ethics* 112:781–814; esp. 803 ff.
44. Schmidbauer, W. 1993. *Hilflose Helfer. Über die seelische Problematik der helfenden Berufe.* Reinbek/Hamburg: Rowohlt.
45. Ellis, A. 1994. *Reason and emotion in psychotherapy*. New York: Carol Publishing Group.
46. Melden, A. 1969. The conceptual dimension of emotions. In *Human action*. ed. Mischel, T., 199–221; esp. 205; 216. New York: Academic Press
47. Wittgenstein, L. 1966. *Lectures and conversations on aesthetics, psychology and religious belief.* ed. Barrett, C., 12 Oxford: Blackwell.
48. Noddings, N. 1984. *Caring. A feminine approach to ethics and moral education*. Berkeley, CA: University of California Press.
49. Noddings, N. 2002. *Starting at home: Caring and social policy*, 44. Berkeley, CA: University of California Press.
50. Noddings, N. 1984. *Caring. A feminine approach to ethics and moral education*, 33. Berkeley, CA: University of California Press.
51. Shibles, W. 1987. Reformatory blame. *Journal of Rational-Emotive Therapy* 5:266–281.
52. Bedford, E. 1964. Emotions. In *Essays in philosophical psychology*. ed. Gustafson, D. Garden City, NY: Doubleday; Shibles, W. 1974. *Emotion: The method of philosophical therapy*. Whitewater, WI: Language Press; Shibles, W. 1995. *Emotion in aesthetics*. Dordrecht, Netherlands: Kluwer Academic Publishers.
53. Parfit, D. 2004. In *Life, death, and meaning: Key philosophical readings on the big questions*. ed. Benatar, D., 119. Lanham, MD: Rowman and Littlefield.
54. Dewey, J. (1958/1929) *Experience and nature*, 251. La Salle, IL: Open Court, New York: Dover.
55. Marcus Aurelius. 1964 tr. Long, G. *Meditations*, 80. Doubleday: Penguin.
56. Rist, J. Ed. 1978. *The stoics*, 259–272. Berkeley, CA: University of California Press.
57. Rist, J. 1969. *Stoic philosophy*, 25, 31, 37. Cambridge: Cambridge University Press.
58. Rist, J. Ed. 1978. *The stoics*, 260. Berkeley, CA: University of California Press.
59. Rist, J. Ed. 1978. *The stoics*, 259. Berkeley, CA: University of California Press.

60. Rist, J. 1969. *Stoic philosophy*, 35. Cambridge: Cambridge University Press.
61. Bosanquet, B. 1894. On the nature of aesthetic emotion. *Mind* 3:153–166 esp.160.
62. Hoaglund, J. 1980. Music as expressive. *BJA* 20(4):340–348; esp. 344.
63. Dewey, J. 1958/1929. *Experience and nature*, 42. La Salle, IL: Open Court, New York: Dover.
64. Rist, C. 1969. *Stoic philosophy*, 35–36. Cambridge: Cambridge University Press.
65. Collingwood, R.G. 1938. *The principles of art*, 267, 296. Oxford: Clarendon Press.
66. Collingwood, R.G. 1938. *The principles of art*, 276. Oxford: Clarendon Press.
67. Noddings, N. 1984. *Caring. A feminine approach to ethics and moral education*, 19, 41, 61, 69. Berkeley: University of California Press.
68. Groenhout, R. 1998. Care theory and the ideal of neutrality in public discourse. *Journal of Medicine and Philosophy* 23:170–189
69. Eisenberg, N. 1986. *Altruistic emotion, cognition, and behavior*, 31. Hillsdale, NJ: Erlbaum.
70. Munro, E. 1998. *Understanding social work: An empirical approach*. London: Athlone Press.
71. Tong, R. 1998. The ethics of care: A feminist virtue ethics of care for healthcare practitioners. *Journal of Medicine and Philosophy* 23:131–152; esp.134.
72. Lipps, T. 1906. *Die ästhetische Betrachtung und die bildende Kunst, vol 2*, Hamburg, Leipzig: Leopold Voss Verlag; Noddings, N. 1984. *Caring. A feminine approach to ethics and moral education*. Berkeley, CA: University of California Press.
73. Bohart, A., and Greenberg, L. eds. 1997. *Empathy reconsidered: new directions in psychotherapy*, 15. Washington, DC: American Psychological Association.
74. Bohart, A., and Greenberg, L. eds. 1997. *Empathy reconsidered: New directions in psychotherapy*, 13. Washington, DC: American Psychological Association.
75. Bohart, A., and Greenberg, L. eds. 1997. *Empathy reconsidered: New directions in psychotherapy*, 5. Washington, DC: American Psychological Association.
76. Bohart, A., and Greenberg, L. eds. 1997. *Empathy reconsidered: new directions in psychotherapy*, 19. Washington, DC: American Psychological Association.
77. Bohart, A., and Greenberg, L. eds. 1997. *Empathy reconsidered: new directions in psychotherapy*, 23. Washington, DC: American Psychological Association.
78. Bohart, A., and Greenberg, L. eds. 1997. *Empathy reconsidered: new directions in psychotherapy*, 194. Washington, DC: American Psychological Association.
79. Bohart, A., and Greenberg, L. eds. 1997. *Empathy reconsidered: New directions in psychotherapy*, 63. Washington, DC: American Psychological Association.
80. cf. Safran, J., and Segal, Z. 1990. *Interpersonal process in cognitive therapy*. New York: Basic Books.
81. Barnbaum, D. 2001. Teaching empathy in medical ethics: the use of "lottery assignments. *Teaching Philosophy* 24:63–75; esp. 64.
82. Schmidbauer, W. 1993. *Hilflose Helfer. Über die seelische Problematik der helfenden Berufe*, 13. Reinbek/Hamburg: Rowohlt.
83. Mill, J.S. 1971. *Utilitarianism, liberty, and representative government*. London: J. Dent.
84. Ferraz-Leite, H. 2000. *Malende Ärzte Österreichs (Painting physicians in Austria)*. Amstetten, Austria: Selva Verlag.

Chapter 10
Egoism and Altruism in Medicine

Abstract The definitions of altruism and egoism are shown to be vague and ambiguous and so is our language. A clarification of altruism and egoism requires an analysis of ethics, the self, causes of action, motivation, and emotion. The problem of altruism versus egoism is seen to be a pseudo problem. So altruism and egoism have to be redefined and based on a naturalistic, humanistic theory of ethics in order to make sense in contexts of medicine as well as in our whole lives. A rational, humanistic altruism based on a naturalistic theory of ethics welcomes positive altruism and positive egoism in terms of positive consequences. Schweitzer wrote, *"According to the responsibility in me, I have to decide what I have to give away from my life, my possessions, my quietness, and what I may keep."* A physician must decide that.

Keywords Altruism · egoism · self · other · emotion · selfishness · sympathy · rational altruism · rational egoism · humanism

10.1 Introduction

In the profession of medical care issues of altruism/egoism play a crucial role. In many hospitals, especially in Europe, the staff is thought of as demanded to work altruistically, even to the extent of burnout. Demands of patients, society, and administration are put into question. Also working hours are discussed or more adequately expressed forced upon on the basis of focusing morals. Physicians, nurses, etc. are required to be altruistic even to the extent of the defilement and destruction of their lives. What is altruism here? How is the word (mis-)used? Which kind of critical evaluation of these questionable value terms is needed to put into perspective what they mean, and for whom and with which consequences applied? [1].

10.2 Common Definition of Altruism and Egoism

Altruism (A) and egoism (EG) are abstract terms used both in commonplace as well as in philosophical, scientific and other contexts. The terms as mostly used are so

B. Maier, W.A. Shibles, *The Philosophy and Practice of Medicine and Bioethics*, International Library of Ethics, Law, and the New Medicine 47,
DOI 10.1007/978-90-481-8867-3_10, © Springer Science+Business Media B.V. 2011

vague and ambiguous as to generate numerous language fallacies even to the point of equivocation. Auguste Comte first coined the term "altruism" [2].

10.3 Definitions of Altruism

Some immediate typical definitions of altruism are the following:

Concern for others as often contrasted with concern for oneself and one's own interests, the view that everything relates to other people, taking others or society as a starting point of knowledge and action, ethical concern for others (e.g., humanism) as the motive of all of one's actions, and as a valid end of all action, the view that one ought always to seek concern for others, and prefer their interests over one's own where they conflict (universal altruism) even at the expense of oneself, the view that I ought to promote the interests or goals of others, and that others should seek the interests of those other than themselves as well.

Moral or cultural altruism is uncritical, leading to normative practices and customs, and is not critiqued by one's own critical evaluation and concern for consequences.

10.4 Definitions of Egoism

Some immediate typical definitions of egoism are the following:

concern only for one's self, only for one's own activities as contrasted with another and their activities, the view that everything relates to oneself and even to taking the self as a starting point of knowledge and action, ethical self-interest as the motive of all of one's actions and end of all action. The view that one ought always to seek one's own self-interest and prefer one's own goals where it conflicts with that of other's (universal egoism), others should seek my interests as well. Others should be altruistic while I am egoistic. Concern for the self is to be put through even at the expense of other people.

Uncritical moral or cultural egoism leads to questionable normative practices and customs. Because the terms "altruism" and "egoism" are typically inadequately defined and ambiguous, the numerous questions and problems, which arise from them cannot be properly addressed without further detailed analysis. For example, the question of whether or not one should be altruistic or egoistic cannot be answered without a clarification of which meanings and definitions of ethics, self, altruism, egoism, etc., are being referred to. Without such definitions, altruism and egoism remain pseudo-scientific terms. Krebs states, "The concept [altruism]. . .is still unclear and no way has been found to measure its unmotivational base" [3].

10.5 An Analysis of the Word-Fields of Altruism and Egoism

The various definitions given in the literature and reference books are summarized to create a "word-field" of the usage of the terms.

10.5.1 The Word-Field of Altruism

An examination of the word-field of altruism reveals the following formula of meanings: self (one or more units such as I, we, nation, etc.) (in an altruistic *relation* with) other, for example, such as "I will help you" [4].

Altruistic positive relations may come up: do good for, be kind to, have empathy for, be concerned about, give to, promote, favor, be generous to, love, be interested in, value, please, be benevolent towards, be selfless in regard to, be polite to, sympathize with, support, understand, share interests and views of, be friendly to, be unselfish in regard to, be humane toward, be benevolent toward, etc. The relation may be rational or irrational. It is the context to be taken into consideration to determine the relation.

Altruistic negative relations may be: pity, be sentimental toward, sympathize with, be infatuated with, grovel before, be devoted to, submit to, worship, acquiesce to, yield to, capitulate to, succumb to, be a martyr to, sacrifice for. Relation negative must be distinguished from relation positive in certain concrete contexts and in terms of consequences. Some may sometimes refer to nature or any living thing. Some say that plants are altruistic if they suppress their own growth to favor that of others. In this case, altruism and egoism are used metaphorically and in such a way as not to require consciousness or deliberate thought. This may also be the case on the level of cultural practice, morality, and normative behavior. For example, one may be altruistic without knowing why or even that one is altruistic. Numerous other metaphorical uses of altruism may be generated from the word-field above. In addition, the antonyms of the definitions of egoism may be added.

10.5.2 The Word-Field of Egoism

An examination of the word-field of egoism reveals the following formula of meanings: self (one or more units such as I, we, nation, etc.) [is in a (positive) *relation* with] self, for example, such as: "I will only help myself" [5].

Egoistic positive relations may be: proud of, enhances, is good to, is concerned about, values, promotes, pleases, and loves. It would be possible for the relation to be how one is to oneself: polite, kind to, generous to, selfless, interested in, egoistic, kind, empathetic with, concerned about, generous, benevolent towards, sympathetic, supporting, encouraging, understanding, sharing, friendly, selfish, humane toward, benevolent toward, etc. The relation may be rational or irrational, certainly contextual.

Egoistic negative relations may come up with words used such as: preoccupied with (e.g., I am "preoccupied with" or a martyr to myself), centered on, excessively guarding toward, sentimental toward, sympathetic with, infatuated with; or that one grovels before, is devoted to, submits to, and worships, acquiesces to, yields to, capitulates to, succumbs to, or sacrifices for (oneself).

Making the self the center and excluding others except insofar as others support the self is self-destructive in the long run. All that is an obstacle to one's self must be destroyed at any cost... is a destructive attitude destroying oneself as well (Nationalism is similar as is going to war for one's goals without concern for those killed). There is almost a daily crisis because there will always be some threat, however, trivial to the egoistic self. The metaphorical device of synecdoche is involved, that is, taking the part for the whole.

10.6 Altruism Versus Egoism

Altruism and egoism are usually presented as being polar opposites. Questions are formulated such as "Should one become an altruist or egoist?" and "Which is better, altruism or egoism?" Can altruism and egoism be reconciled? [6].

To structure the matter in this narrow way is to commit the either-or fallacy. Why cannot one be somewhat altruistic or egoistic without the either-or fallacy? It is to exclude gradations, other possibilities and to falsely represent them as being opposites. It is thought that one must have either altruism or egoism, whereas to have egoistic egoism and egoistic altruism, together is not only possible, but reasonable. Altruism and egoism can be dialectically intertwined. This is a form of its metaphorical possibilities. Even the egoist considers others in one's self. It is not clear what sense "opposite" would have here, considering that both altruism and egoism have such a wide range of meanings. To approach these issues is to metaphorize creatively [7].

For example, some philosophers hold that altruism is independent of egoism, and others hold that an informed egoism requires altruism in some sense [8].

Hobbes holds that altruism is self-seeking in disguise [9]. In place of altruism or egoism, there may be cooperation, mutualism, or compromise. Some object to strict or one way altruism defined as excluding oneself is martyrdom. One may try to deal with the problem to help all people, yet still enjoy life. One can still be caring, but not at the expense of one's health. Altruism, though concerned with the interests of others, need not exclude concern for oneself. The altruism of a healthcare worker to the extent of giving up him-/herself and running into burnout is unacceptable in terms of consequences for physician and patient. Egoism is needed for the physician to keep healthy and there must be a balanced system between altruism and egoism in some senses of the two words. The method does not mean that patients are fully cared for while the healthcare workers struggle for time and support and are not sufficiently compensated. A socialistic system should not exclude the healthcare workers' benefits for the benefits only of patients or other forces.

It comes to an understanding that altruism versus egoism is a false and misleading dichotomy [10].

One may also have mutual altruism. That is one may not choose to be altruistic to egoists, but only to altruists. One may only be altruistic if it is mutual. One could speak of unrequited altruism. The altruist may find that no one is grateful for his or her help as is the case. The patient may merely demand or expect treatment. Altruism even in regard to one's children can be undermined by egoism.

10.7 The Problem of the Self

Whether or not one places emphasis just on oneself or other selves, a clear concept of self is required. The concepts of altruism and egoism are determined by, and can be no clearer than, with the clarity of definition of what is meant by "self." The self is a linguistic story and so is altruism and egoism. "The human self is created by the use of a vocabulary" [11]. We can speak of the script of one's life.

The problem of self in philosophy also puts altruism and egoism into question. What are we to mean by the "self"? It may refer to one's body, thinking, or actions, or how others see us. If thinking is reduced to one's language-use, as numerous philosophers have suggested, then "self" refers to us as language users. The self may be viewed through almost every word in the language. This is also part of the insight of the Metaphorical Method. It is a form of viewing-as or presentation as. Entities are created by what one might say. It may refer to any context in which it is used. There are different selves. We may, for example, be egoists behaviorally, but altruists intellectually. We may have intellectual altruism, but behaviorally not put it into action; one may hold a child in one arm, and throw a grenade with the other. There is, then, no self as such. In this sense, we are all selfless.

The self may become the other in several ways. "Self" sometimes refers to several people as in: "We won the game." "Our nation has reached the moon." "My family is my life." A family, nation, democracy, group, or business may be selfish. In addition, one may identify oneself with others or with a concept: duty, hero, culture. Harris opposes speciesism or the view that humans are naturally superior and "the definition of humans as Race, gender, nationality, religion, or any other non-moral characteristic is, and has always been, disreputable" [12]. This view also undermines the prejudice of nationalism over a world perspective. If one thinks, idealizes, or imagines oneself as others, then altruism becomes egoism and egoism becomes altruism. I become you. We role-play and put ourselves in another's shoes. Enemies are friends and friends are enemies. One may love one identity (e.g., one's own country idealized) more than another identity (physical self) such that one would gladly give up one's life for one's country. One political party, *Die Grünen,* the Greens Party, in its global platform, holds that there are no enemies, that participatory democracy does not stop at the border and that war is suicide (*Programs of the German Green Party,* 1983). The world-citizen can identify in some ways with all people. The therapist can replace condemnation with understanding. What is claimed to be altruistic toward one's children can rather be egoistic if the children are thought of as one's own "flesh and blood," or if one identifies with one's own family. One may identify with one's profession so that promoting one's profession is an act of egoism. "We fight to defend foreigner's rights," can be egoism if one does not fight personally, but one's fellow citizens instead risk their lives. In these senses, "To help others is to help yourself" is not only causal, but also circular. The either-or fallacy of altruism versus egoism dissolves.

As a quantitative matter, the self versus other creates a paradoxical mathematics. The self is, for example, two selves or many selves: We criticize, blame and flatter ourselves. We say, for example, "I was angry with myself." Self-reference generates paradoxes similar to that of the "liar paradox." How can one be deceptive (or good, polite, giving, loving, selfless, sympathetic, unselfish, etc.) in regard to oneself? For example, "I am selfless to myself." We may find such reflexive constructions paradoxical.

Even the body changes so that it is not the same body. As the diagnoses, examinations and tests are entered into the medical chart the body takes on new meaning, becomes a different bodily self (See Chapter 13).

As several selves, one may be either altruistic or egoistic in regard to oneself and different in different situations. Self-altruism may produce pride, and self-egoism may produce remorse. It becomes possible to sacrifice oneself for oneself. "Be good to yourself" is a common expression. Thus, one form of the altruism versus egoism question becomes, "Are we or should we be good to ourselves?"

10.8 The Ethical Basis of Altruism and Egoism

10.8.1 General Remarks

Both altruism and egoism claim to produce some "good" whether for oneself or for others, but without an ethical system or theory, it is not clear what could be meant by such ethical terms as "good," altruism and egoism. A Kantian ethical theory would result in a different view of altruism than would a naturalistic or utilitarian theory of ethics. Whatever theory of ethics (morals or values) one takes, it is first necessary to determine what characterizes ethical statements. How may one identify a statement as a moral statement as opposed to a descriptive statement?

By an examination of our everyday language, we find that statements which are said to be ethical, moral, or value statements, contain moralized and moralizing words: "good-bad, right-wrong, ought or ought-not," their synonyms or behavioral counterparts. On this view, there are no moral issues as such. An action or object whatsoever may be said to be good or bad. However, this merely shows the difference between moral and descriptive statements, but we still need to know what such terms mean. This is the task of ethics. There are, of course, as many meanings as there are ethical theories and language usages. On one theory, such terms as "good" and "bad" are meaningless terms. To say something is "good" is to say nothing at all. For example, if we say, "Be good to others," it is not clear what this will involve. "The war was bad" may mean that we should have won it, or on the other hand, that there should have been no war at all. "Be responsible!" does not tell us what we should or should not do. Thus, what characterizes ethical terms is that they are open-context, empty terms. They only make sense if we substitute some concrete meaning or special contextual reason for them. They are like blank checks which have no value until they are filled out, addressed and signed.

"Good" from the point of view of the altruist may even be interpreted by others as harm. Giving to the poor in some situations may make people dependent rather than teach them to be self-sufficient. Giving to the ungrateful and demanding is counterproductive. Because value terms are subjective as well as open-context, the receivers may actually regard the "good" one gives as "bad" for them. Egoists may actually be doing harm to themselves, and altruists may be covert egoists. When one wishes to do good, it is wise to reassure oneself if it will be interpreted as being good by the recipients. The United States gave tractors to India, which were worshipped, rather than used for agriculture.

The following illustrates the implications of several prominent ethical theories for altruism and egoism:

10.8.2 Utilitarian Altruism

On this view, one should produce the greatest good for the greatest number of people. It is, on the face of it, closely tied to altruism, which is interpreted as doing "good" for the greatest number of others. One problem with this is that we must first inquire as to what this open-context "good" means. The second problem is that it seems to be a formula, which can never be successfully carried out. We would not know by the formula alone whether or not to stress the greatest "good" or the greatest number. Suppose that by "good" is meant to provide medical education. It is not clear whether we should have an elitist system and admit in universities those in the top ten percent (the greatest good), or develop a comprehensive educational system allowing nearly everyone (the greatest number) to be admitted. The utilitarian formula allows for both. One could, on this theory, even claim that only one brilliant student should be admitted because that would produce the greatest good of all. In order to determine whether or not to stress the greatest number or the greatest good (quality), criteria outside the utilitarian formula would have to be found. The traditional criticisms of utilitarianism, therefore, raise difficulties for its being considered as a basis for altruism.

10.8.3 Ayn Rand's Objectivist Egoism

Ayn Rand describes in her book, *The Virtue of Selfishness,* "selfishness" as being concerned with one's own interests [13]. One's interests may include most anything. For Rand they should be rational and consequentialistic interests and thus include consequences for others. This is not selfishness, however. It is self-enhancement. It is "rational selfishness" [14]. Selfishness is doing good for oneself at the expense of others, or minimally, with no concern for others. By definition, selfishness is unacceptable because doing good (to oneself), but bad (to others), is focused on one's own definition of "bad." On this view, selfishness is self-contradictory. Rand should not have used the term "selfishness." Thus, already the title of her book is misleading, self-contradictory. She also exposes the contradictoriness of sacrificing the self to preserve the self. "No society can be of *value* to man's life if the price is the *surrender of his right* to his life" [15 Italics added].

Her view of altruism is narrow and does not take cognizance of the various meanings of the terms explicated by the word-field analysis given earlier. She writes that "altruism" is excluding benefit for oneself: She states, "Altruism *permits no concept* of a self-respecting, self-supporting" person and is merely self-sacrifice [16]. On her view, egoism involves:

(a) Concern with one's own interests as the essence of existence.
(b) Gain from any action performed.
(c) Self-interest rationally based on principles. It is "applicable *only* in the context of a rational, objectively demonstrated and validated code of moral principles, which define and determine his or her active self-interest" [17].
(d) The self-interest cannot be motivated merely by "irrational emotions, bodily feelings, urges, wishes, whims," or "blind desires" [17]. Emotions are rather to be based on one's ethical cognitions. Although she does not have a theory of emotion, her view would cohere with the cognitive theory of emotion presented in this book.
(e) Substitution of "rational self-interest" (or rational egoism) and "rational selfishness" (a contradiction) for mere selfishness or mere egoism, whatever "mere" means [18].
(f) So-called "objectivism, a theory to be built on an objective basis because the ethical basis of action must be inter-subjectively rational" [13].
(g) The motivation of all actions by a single goal: to maintain one's life. What furthers life is good, what threatens it is bad. This is a Darwinian or evolutionary view according to which "good" means to survive. It is also a teleological fallacy, that is, it purports to tell us what the single purpose of organisms is, whereas there rather are diverse purposes – those, which humans choose to have. By productive work she actually means the use of one's abilities both physical and psychological to the greatest extent possible [19].
    This coheres with a naturalist and humanist theory of ethics (See also the discussion of motivation and desires presented later). By analogy, the main purpose of medicine is not just to have people survive, but to survive humanistically.
(h) No sacrifice to anyone. This may be true by definition, because "sacrifice" means to harm oneself. Her view is actually humanistic in that it would not allow war, force, or violence as means of obtaining advantages. War may only be used in self-defense. Hobbes had also argued that altruism is in our self-interest to protect ourselves against war.
(i) Rejection of supernatural or religious basis of ethics in any form of mysticism, i.e., any claim to some non-sensory, non-rational, non-definable, supernatural source of knowledge [20]. This "individualism" or "objectivism" turns into a kind of utilitarianism because of Rand's view that what is good for me is also

good for the society. In summary, by means of redefinition, her egoism turns into altruism, and a utilitarianism, and, to some extent even humanism.

10.8.4 The Ordinary Language Basis of Altruism and Egoism

One approach to the analysis of altruism and egoism is the ordinary-language approach developed by Wittgenstein in his *Philosophical Investigations* to reduce abstract and vague terms to specific examples and usages of them in everyday language [21]. He called such usages "language-games." For example, it is misleading to ask what "hello" really means. To find its meaning we must look at its usage in the everyday language of greeting people. In doing this we find that many of the words we use seem to make sense, but are in fact misuses, or have no ordinary uses. This is especially true with abstract and value terms. A language-game is the specific sentences and contexts of usage. The way in which we pronounce a greeting such as "Hello" can be said to mean different things depending upon intonation. Accordingly, Alasdair MacIntyre states, "In most of my dealings with others of a cooperative kind, questions of benevolence and altruism simply do not arise, any more than questions of self-interest do"[22]. This is why the altruism-egoism question may be seen as a pseudo-problem.

The terms used are too theoretical and too ambiguous to know really what is meant in any particular case. Suppose one says, "Seek your own good," or "I only act out of the desire for pleasure." What could be meant and in what circumstance? What especially does "good" mean here? Is it pleasure, fulfillment of a desire, long-term goals, etc.?

Altruism and egoism also have different meanings in the first, second and third person cases. I am altruistic/egoistic, is not the same meaning as "You are altruistic/egoistic." We may not agree with each other. The Church may view abortion as being selfish; others may see it as being unselfish. Therefore the altruism of one person may not mean the same as the altruism of another person. Similarly, the meaning of egoism of one may not be the same as for another. Also, the altruism of one person may mean the same as the egoism of another person. The analysis of altruism becomes an analysis of its diverse language-games. On this view, "altruism" and "egoism" and their related expressions must be further analyzed in terms of their specific rhetoric and language-games. Rorty writes, "The self is created by the use of a vocabulary" [11]. The self is a linguistic story.

10.9 Altruism and Egoism as Emotions

Altruism and egoism have been regarded as emotions. There are emotions of benevolence, fellow-feeling, sympathy, pity, compassion, selfishness, etc. relevant here. Hume and others based altruism on "feelings" such as sympathy. It was thought that only emotion not reason can move us to do anything. Therefore especially, we must first ask what an emotion is. In philosophical literature, the cognitive theory of emotion is prevailing. Rational-emotive therapists base their work on this theory [23].

According to the cognitive-emotive theory, emotion is a cognition, which causes bodily feelings [24]. In ordinary language, "emotion" and emotion words refer to both cognitions and bodily feelings. Many characteristics of altruism and egoism and fallacies with the use of language may be deduced from viewing them as emotions.

10.9.1 Altruistic and Egoistic Emotions Are Not Mere Bodily Feelings

On the cognitive theory, emotions are not feelings. Because emotions involve cognitions, it is always a mistake to say, "I feel altruistic/egoistic." The rational-emotive therapist, Albert Ellis, pointed out that it is more precise to say, "I think-feel" altruistic. For Thomas Nagel, "Altruism is not a [bodily] feeling" [25]. Rather emotions relate to the functions of intelligence.

Sympathy, fellow-feeling, benevolence, and compassion are not just bodily feelings and cannot serve as the basis of altruism. The traditional dichotomy that has been established as a result of the above-mentioned confusions between cognition versus emotion (bodily feeling) is a false one. If the emotions of altruism and egoism are thought to be bodily feelings, it may be also held that they are irrational, subjective and not conscious or deliberate. On the cognitive-emotive theory, the reverse is the case. There the dichotomies dissolve or are restructured once it is seen that emotion is cognitive, that cognition causes bodily feeling.

10.9.2 Altruism and Egoism Are Cognitions Causing Bodily Feelings [26]

Rational emotion, rational altruism and rational egoism are not contradictions. Each emotion, altruism, egoism can be rational or irrational. They are neither cognitions nor bodily-feelings, but both. Altruism/egoism are not just amorphous feelings devoid of any conceptual content, but rather have an elaborate cognitive structure which can be analyzed especially also in the context of healthcare, for professionals as well as patients, and is needed to be analyzed because of the misuse of language and action in this field. Egoism affects and determines other emotions causing self-pity, anger and blame of others for any unfulfilled desire, lack of gratitude, demandingness, uncaring attitudes, etc.

10.9.3 The Emotions of Altruism and Egoism Can Be Changed

One may change from an altruist to an egoist or vice versa depending on contexts and consequences. Emotions are not necessary, innate, or fixed parts of our character. They can be created or eliminated or changed to certain metaphorical extents. This contrasts with Blum's view that emotions are entirely distinct from reason and rationality and we are passive in respect to our feelings and emotions. He says they are not in our control and thus we are not responsible for them [27]. On that view, we could be overcome by altruism or egoism. Pity, sentimentality, and other negative emotions connected with altruism and egoism are based on faulty thinking. They may be changed by changing such thinking. My conceit is irrational because it is excessive regard for myself. We are not altruistic and egoistic by nature. We develop altruism/egoism in developing ourselves. We do not have fixed emotions and characters that cannot be changed. This argues against psychological altruism/egoism, the view that it is in our psychological nature to be altruistic or egoistic. We can change or prevent negative emotions by changing our assessments.

10.9.4 Altruism and Egoism Are Based on Value Cognitions

Ethical (E) terms were seen to be open-context value terms. Therefore, altruism (A) and egoism are also open-context terms, not meaningful in themselves. Altruism in itself and egoism in itself are empty, open context phrases. Altruism, egoism and other emotions based upon such value cognitions are, in this respect, faulty. It is meaningless to ask if altruism is in general good or intrinsically good. The specific meaning of good (or other ethical terms) must, in each case, be explicitely given. Altruism connotes the value "good," so it is always by definition good. The question "Is altruism good?" is therefore circular.

On a naturalistic ethics, "good" may reduce to the deliberate bringing about of one's wants (including consideration of the wants of others) on the basis of knowledge and inquiry regarding consequences. This would exclude an altruism based on duty-in-itself or on anything spiritual. One could not be ethically altruistic if it is merely a duty, or on the basis of authority, especially if that authority be irrational or supernatural. Normative or moral A/EG is mere custom, as opposed to critical, ethical A/EG. That is, ethical terms are ambiguous because they may be used to refer to normative moral use, philosophical use, aesthetic use, or religious use, or legal use (often involving rights), etc. and the meaning is different for each. Such a situation invites equivocation and confusion as they are used to mean different things. However, the cognitive theory can give the correct usage formula for all of them.

The following examples illustrate typical circularities and misuses of value terms (Value terms are italicized):

Blum defines altruism in an open-context way: "By 'altruism' I will mean a regard for the *good* of another person for his *own sake*" [28].

For Rescher, "The welfare of others is the *intrinsic value* to him, *prized in its own right*" [29].

"From the *moral* point of view, we *ought* to strive individually to realize a *morally* well-ordered society" [30]. "To *disregard* the *interests* of others is to be not merely *immoral*, but *inhumane*" [31]. To say something is immoral is neither informative, nor intelligible without instantiation, that is, without telling us what is meant by the ethical terms used.

Also sympathy is under certain circumstances morally *obligatory* [31]. "Morally obligatory" is redundant and the phrase suggests the misuse, "Duty-in-itself." We have no obligation to support altruism or egoism. One does not have to be good to others or to oneself. This is *a fortiori* true because altruism and egoism are definitionally ambiguous. In this sense, one cannot be obliged to support them even if one wished to. Without giving the actual meaning of value terms, neither the altruist nor egoist can know what they are about.

Because they are cognitions, which cause bodily feelings (perception, sensation), there is no unfeeling altruism and egoism. A change in bodily feeling, then, can alter them somewhat, e.g., by a physical feeling of well-being, or after a good meal one may be in a more altruistic, giving mood. However, it is mainly the cognition, which must be altered.

There is no altruism or egoism, as such. For each altruistic and egoistic act there is a different cognition and bodily feeling. The emotion of altruism = cognition1, cognition2, cognition3, etc. plus feeling1, feeling2, feeling3, etc. Each altruism or egoism is thus to a greater or lesser degree different than every other altruism or egoism, respectively, in terms of quality, quantity, perspectives. Repetition of fixed ideas can, however, produce a similar form of altruism or egoism though in changing situations. Each altruism or egoism can be distinguished especially by its different cognitions.

Our assessments may themselves be enculturated. In this sense our altruism and egoism are geographical or culture-specific, not ethical. A reconstruction of the reasons for altruism and egoism is just another language-game, not a statement about what "really" is the case. In medicine, for example, the physician, nurse has to be altruistic, whereas the patient can be egoistic, in some senses of these words. We cannot get outside of our language-games.

Meta-emotions, meta-altruism/egoism are emotions about altruism/-egoism for example, happiness, which comes from being altruistic, or being think-feeling about being selfish. A/EG, like value statements, are statements about statements, meta-statements.

10.10 Sympathy

Hume treated sympathy (fellow-feeling, sentiment) as a "feeling" that motivates us to altruism [32]. Here a feeling is said to cause cognition – a view opposite of the cognitive-emotive theory. Sympathy is seen as being a basic, unexplained, empirical and psychologically given, like a bodily feeling. However, it is argued that Hume built into this feeling an a priori cognition (feeling = cognition), which would then conform to the cognitive theory [33]. The argument is that sympathy presupposes an inter-subjective language and society. If one relates to others, relation can mean: communicates with, accepts the meanings of, is sympathetic with, is altruistic toward.

Also, sympathy may be a poor basis of altruism because it may move us to harm rather than good, as Hume assumes. Feeling of unity and harmony in society are not feelings as Hume claims. Unity and harmony are value assessments. They can however be good reasons why one should be altruistic and so form a basis for altruism.

If communication presupposes otherness, support is given to the view that all people are sympathetic (altruistic) [34]. "We can form no wish which has not reference to society," according to Hume [32]. This is the reverse of the egoistic view that people do not act except to promote their own good. Egoism is almost defined here to mean "antisocial." Positive correlations are found between social extroversion and altruism. Negative correlations are found between altruism and antisocial tendencies "Egoism" may even mean "unfriendly."

It is thought to be self-contradictory to have a society consisting of one person, although Oscar Wilde wrote in *An Ideal Husband*, "Other people are quite dreadful. The only possible society is oneself" [35]. Egoism could never be the basis of the social community of values.

Whether or not the overall analysis is clear or correct, sympathy is sometimes treated as cognition, rather than as feeling. Sympathy is defined circularly in various ways: "Sympathy motivates one to judge that altruistic behavior is good." But, "sympathy" is one synonym of "altruism" [33]. One could equally argue that altruism is the basis of sympathy. Hume's view of sympathy as a cause of altruism may be seen as being true in the sense of circularity. And to say that sympathy presupposes value is circular because sympathy is a value assessment. It is circular to argue as Lipkin does, "Why we *approve* of what is useful. . .is just that it *pleases* us" [36] (authors *underlined*).

Some of our acts are sympathetic, others not. We have selective altruism and egoism. Blum states, "The capacity for one altruistic attitude is no assurance of the capacity for others" [37]. Like altruism, sympathy may be positive or negative.

If one relates to oneself, where the relation is negative sympathy, one may find oneself in self-pity, regret, etc. On its positive side, sympathy becomes self-confidence, self-acceptance, etc.

10.11 Selfishness

Glover speaks of "our limited capacity for altruism" and of a history of cruelty and killing [38]. Spinoza wrote, "No man [or woman] neglects to seek his [or her] own advantage," and "The more each one strives, and is able, to seek his [or her] own advantage. . .the more he [or she] is endowed with virtue" [39]. Hobbes stated, "No one gives but with intention of good to oneself" [40].

It is oversimplifying to say that one always acts for one's own good. We act for diverse reasons, some of which are against our own survival.

To say that altruism, where one is related to another, always reduces to egoism, where one is related to oneself, is to equivocate and deny others. That altruism equals egoism is in some sense virtually always true, for example. "You were happy to starve, so your children could live a better life." But the argument may be reversed. Egoism equals altruism: Virtually all egoism may be seen in some interpreted or rationalized sense as altruism, for example, Ayn Rand's wide-ranging view that what is good for General Motors is good for the nation [41]. Jean Ziegler in his book *The Empire of Shame* holds the opposite view on an ethical base of considering context and consequences [42].

The claim that we always act for our own self may be circular and trivially true. I notice that my self is always involved in whatever I do. I notice that I have my own thoughts and goals. My body is mine and my actions belong to no one else. If egoism means only that one's actions are an implicate of oneself, there is no objection to egoism. One could never be selfless in this respect. "We only do what we prefer doing. . ..My preference must be my own" [43]. We may say there is no beneficence in us independent of a relation to oneself. But that the self is involved does not mean that one is selfish.

Suppose that by being altruistic, one thereby experiences a good emotion. When a cognition is positive, rational and fair, a pleasurable bodily-feeling naturally follows, and that it does should not fault the cognition. Altruism need not come at the expense of either egoism or pleasurable feelings. Rather, the ideal state is where altruism equals egoism. Rescher holds, "An unselfish act does not become less selfish because one gets satisfaction from its performance" [44]. Dewey and Tufts state, "An act is not wrong because it advances the well-being of the self, but because it is unfair, inconsiderate, in disrespect to the just claim of others" [45].

In the above, good feeling accompanies altruism. It may, however, be irrational to be altruistic in order to achieve a bodily feeling. If one is altruistic to achieve a thought or bodily feeling, it contradicts the meaning of "altruism" in the sense of "helping others." It is then, by definition, not altruism, but egoism. One uses altruism out of self-interest. One appears to help the poor, whereas one's genuine objective is to obtain votes. Corporations or individuals may engage in philanthropic activity to obtain tax breaks and compensate for a bad reputation.

If egoism means that one is enhancing oneself, there is no objection to egoism. If it means selfishness, there is a problem. Selfishness is self-contradictory. It is to do good for oneself while doing bad for another (even additionally gaining *Schadenfreude* (German) = "malicious joy"). Selfishness should not be confused with self-enhancement. The advice often given to "be more selfish," is to endorse self-contradiction. The definition of altruism as "unselfish regard for others," is always correct in the sense that one should, by definition, always be unselfish, that is, do what is good and not do what one thinks is bad. The problem is with one`s thinking again.

Obligatory egoism involves the view that we *should* help only ourselves. Consider the possibilities:

1. I help myself and help others (A = EG).
2. I help myself and hurt others (Selfishness).
3. I hurt myself and help others (Martyrdom).
4. I hurt myself and hurt others (Senseless act).

Only row 1 clearly makes sense. Row 4 would refer to an unintelligent act. Rows 2 and 3 might be acceptable if there are degrees of harm/help: One may need the money one gives as gift much less than one who is starving. Physicians in Europe often sacrifice time, health, and money for the good of the patient. But to starve to give to a frivolous or false cause is martyrdom. Schweitzer wrote, "According to the responsibility in me, I have to decide what I have to give away from my life, my possessions, my quietness, and what I may keep" [46 Author's translation]. A physician must decide that. To be selfless is to fail to consider oneself at all. In addition, "self" in "selfless" is as ambiguous here as elsewhere. The same would apply to: a body cannot be bodyless.

Would or should one help another if it is known that the act in no way helps oneself? In one respect, this is a contradiction because "help" means "do good." I cannot help unless I am doing something I deem good. Positive altruism and positive egoism are always desirable. They presuppose needs. If there are no needs, there is no altruism or egoism. One cannot have altruism and egoism where nothing is expected in return. To do so is to act unethically. It is to act for no reason whatsoever, contribute to an unknown cause, desire more of anything at all, risk one's life for nothing. Such behavior is neither altruistic nor egoistic, but nihilistic. It is only a manner of speaking to say that love is where nothing is expected in return. We may experience joy or love in helping others.

10.12 Rational Altruism and Egoism

Why should we help others? According to Dewey's ethics, *we would be altruistic because, in terms of inquiry and an adequate knowledge of the consequences, it makes sense to do so.* It is a "rational altruism or egoism." We live in societies in which we must interrelate with each other and so are concerned with each other. Community consists of a unity or harmony of individuals. The problem of altruism and egoism is a holistic problem. It is a problem of how to make the most adequate decisions possible. We would have enlightened management not because it makes a self-interested profit for the employer, but because it makes sense for all concerned. From "rational altruism and egoism" new emotions are created. The assessments involved being based on reason and adequate inquiry produce an emotion of being civilized, satisfied that one has produced an intelligent social act. To express this in another way, a rational or positive altruism and egoism reduce to humanism. Humanism appears to be a useful guide upon which to base altruism and egoism because its basic guidelines involve love of people, open education, free inquiry, freedom of speech, an ethics based on reason, consequentialism, enlightened democratic methods, anti-sexism, anti-racism, etc. It opposes indoctrination, censorship, morals based on fear and punishment, the supernatural, appeal to authority, absolute values, etc. On this view, we are altruistic or egoistic (rational and positive egoism) not out of a bodily feeling, or because of duty, but because it makes sense to be so.

10.13 Summary

The definitions of altruism and egoism were shown to be vague and ambiguous and so is our language. A presentation of the word-fields of the terms was given and critiqued. A clarification of altruism and egoism requires an analysis of ethics, the self, causes of action, motivation, and emotion. The problem of altruism versus egoism was then seen to be a pseudo problem. So altruism and egoism have to be redefined and based on a naturalistic, humanistic theory of ethics in order to make sense in contexts of medicine as well as in our whole lives. A rational, humanistic altruism based on a naturalistic theory of ethics serves as a useful model for altruism in the medical field.

References

1. Extensively revised from Shibles, W. 1992. Altruism versus egoism: A pseudo-problem. A cognitive- emotive analysis. *International Journal of Applied Philosophy* 7:21–29.
2. Comte, A. 2006. *Encyclopedia of Philosophy* I:136.
3. Krebs, D. 1970. Altruism: An examination of the concept and a review of the literature. *Psychological Bulletin* 73:258–302.
4. For concise symbolization see: An examination of the word-field of altruism reveals the following formula of meanings: xRy, where y = other(s), x = one or more unit(s) such as: (I, we, nation, etc.); and R = relation, e.g., "I help you").
5. For concise symbolization see: An examination of the word-field of egoism reveals the following formula of meanings: xRx, where x = one or more units(s) such as: I, we, nation, etc.; and R = relation (positive), e.g., "I will help myself."
6. cf. Paul, E., Miller, F., and Jeffrey, P. eds. 1993. *Altruism*. Cambridge: Cambridge University Press.
7. An analysis of altruism (A) and egoism (EG) and their accompanying problems may be broadened to include possible formulations such as: A or EG, A vs. EG, A not EG, EG not A, neither A nor EG, both A and EG, A = EG, A independent of EG, A as bound up with EG.
8. Rand, A. ed. 1963. *The virtue of selfishness: A new concept of egoism*. New York: New American Library; Dewey, J., and Tufts, J. 1932. *Ethics*. New York: Holt.
9. Hobbes, T. 1958. *Leviathan*. Indianapolis, IN: Bobbs-Merrill.
10. Badhwar, N. 1993. Altruism versus self-interest: Sometimes a false dichotomy. *Social Philosophy and Policy* 10:90–117; Shibles, W. 1992. Altruism versus Egoism: A Pseudo-Problem. A Cognitive- Emotive Analysis. *International Journal of Applied Philosophy* 7:21–29.
11. Rorty, R. 1989. *Contingency, irony and solidarity*, 7. New York: Cambridge University Press.
12. Harris, J. 1999. The concept of the person and the value of life. *Kennedy Institute of Ethics* 9:293–308; esp. 296.
13. Rand, A. ed. 1963. *The virtue of selfishness: A new concept of egoism*, X. New York: New American Library.
14. Rand, A. ed. 1963. *The virtue of selfishness: A new concept of egoism*, XV. New York: New American Library.
15. Rand, A. ed. 1963. *The virtue of selfishness: A new concept of egoism*, 31. New York: New American Library.
16. Rand, A. ed. 1963. *The virtue of selfishness: A new concept of egoism*, xii. New York: New American Library.
17. Rand, A. ed. 1963. *The virtue of selfishness: A new concept of egoism*, xiv. New York: New American Library.

18. Rand, A. ed. 1963. *The virtue of selfishness: A new concept of egoism*, xv, 28. New York: New American Library.
19. Rand, A. ed. 1963. *The virtue of selfishness: A new concept of egoism*, 22. New York: New American Library.
20. Rand, A. ed. 1963. *The virtue of selfishness: A new concept of egoism*, 21. New York: New American Library.
21. Wittgenstein, L. 1958. *Philosophical investigations*, 3rd edn. New York: Macmillan.
22. MacIntyre, A. 1968. *Egoism and altruism. Encyclopedia of philosophy*, 462–466. New York: Macmillan.
23. cf. Ellis, A., and Grieger, R. 1977. *Handbook of rational-emotive therapy*. New York: Springer.
24. According to the cognitive-emotive theory, emotion (E) is a cognition (C) which causes a bodily feeling (F). E = (C > F).
25. Nagel, T. 1970. *The possibility of altruism*, 3. Oxford: Clarendon Press.
26. [A/EG = (C > BF)]
27. Blum, L. 1980. Compassion. In *Explaining emotions*. ed. Rorty, A., 507–516; esp. 2 Berkeley, CA: University of California.
28. Blum, L. 1980. *Friendship, altruism, and morality*, 9–10. London: Routledge and K Paul.
29. Rescher, N. 1975. *Unselfishness*, 69. Pittsburgh, PA: University of Pittsburgh Press.
30. Rescher, N. 1975. *Unselfishness*, 103. Pittsburgh, PA: University of Pittsburgh Press.
31. Rescher, N. 1975. *Unselfishness*, 15. Pittsburgh, PA: University of Pittsburgh Press.
32. Hume, D. 1888/1978. *A treatise of human nature*, 2nd rev edn. ed. Selby-Bigge, L., Book III: I, 173 Oxford: Clarendon Press.
33. Lipkin, R. 1987. Altruism and sympathy in Hume`s ethics. *Australasian Journal of Philosophy* 65:18–22; esp.18.
34. Krebs, D. 1970. Altruism: An examination of the concept and a review of the literature. *Psychological Bulletin* 73:258–302; esp. 282.
35. Wilde, O. 1932. A ideal husband. In *The plays of Oscar Wilde*, Act III. New York: Modern Library.
36. Lipkin, R. 1987. Altruism and sympathy in Hume`s ethics. *Australasian Journal of Philosophy* 65:18–22; esp. 21.
37. Blum, L. 1980. Compassion. In *Explaining emotions*. ed. Rorty, A., 507–516; esp. 509 Berkeley, CA: University of California.
38. Glover, J. 1984. *What sort of people should there be?*, 181 New York: Penguin.
39. Spinoza, B. de 1985. *The collected works of Spinoza*. ed. Tr. Curley, E. vol. II, 224, 557. New York: Princeton University Press.
40. Hobbes, T. 1958. *Leviathan. Bobbs-Merrill*, 53. Indianapolis, IN: Bobbs-Merrill.
41. Rand, A. ed. 1963. *The virtue of selfishness: A new concept of egoism*. New York: New American Library.
42. Ziegler, J. 2007. *Das Imperium der Schande. Bertelmann, München (the empire of shame)*. Random House Verlagsgruppe: Pantheon.
43. Nielsen, K. 1959. Egoism in ethics. *Philosophy and Phenomenological Research* 19:502–510; esp. 505.
44. Rescher, N. 1975. *Unselfishness*, 13. Pittsburgh, PA: University of Pittsburgh Press.
45. Dewey, J., and Tufts, J. 1932. *Ethics*, 324. New York: Holt.
46. Eisenberg, N. 1986. *Altruistic emotion, cognition, and behavior*, 1. Hillsdale, New York: Erlbaum.
47. Schweitzer, A. 1923/1996. *Kultur und Ethik*, 351, 342. München: Beck'sche Reihe.

Chapter 11
Letting Die

One life is of so much consequence. [1]
Everyone deserves to have needed medical treatment.

Abstract Allowing death = killing = murder. The reasons typically given for killing are fallacious. If one reason is given to promote it, the door is open for any reason whatsoever. Killing is a self-defeating position because if we justify killing others, others can justify killing us. It is similarly self-defeating to block medical research and then expect to benefit from such research or to thereby allow others to die. One cannot decide the issue of justifying killing x number to save y number on the basis of numbers alone. The utilitarian theory is not a genuine ethical theory, but a mechanical formula which cannot be applied unless a more adequate ethical theory is employed to do so. A holistic naturalistic humanistic ethics was suggested for this. The problem is more how to save lives without killing.

Keywords Killing · (in)direct killing · letting die · allowing death · withholding treatment · utilitarian formula · Samaritan help · reverence for life · (assisted) suicide · euthanasia

11.1 Introduction

The question of killing or of letting die has remained fundamentally untouched in philosophy and bioethics. There are various opinions: one typically ends up saying that it is acceptable in war. Or that it is intrinsically wrong, that it violates the sanctity-of-life, that one has no right to kill, or that it violates some commandment or universal principle, etc. Meanwhile in direct and indirect ways people continue to kill or let people die. They are influenced and determined by the culture to do so.

Glover speaks of "our limited capacity for altruism" and our history of cruelty and killing and our almost complete failure to provide food for the starving and those who need medical care [2]. War technology, he believes, has grown faster than our humanity. "We have a tribal psychology well adapted to survival in the stone-age" [3]. His point, however, is that the picture of humanity is perverse and needs to be

B. Maier, W.A. Shibles, *The Philosophy and Practice of Medicine and Bioethics*,
International Library of Ethics, Law, and the New Medicine 47,
DOI 10.1007/978-90-481-8867-3_11, © Springer Science+Business Media B.V. 2011

changed. His point is in fact understated. It is astonishing that one even has to argue against killing, but one does. It is a clear sign of the intellectual and philosophical poverty and depravity of even the most technologically advanced nations and its individual citizens. If this cannot be understood, what else can?

On the naturalistic theory of ethics outlined in the Chapter 5, a basis is given upon which one can argue that killing is wrong. It takes away one's desires and enjoyments and one's very self. Such desires are the most important qualities a person has and they even define the self. Ethics is bringing about our informed wants and likes. Killing is contradictory to our desires and goals, personhood and humanity. When one's life goes, one's world goes. In the context of medicine, as well as world citizenship, it makes sense to preserve the health and quality of life of as many people as possible, in the ideal case all people. The world thereby becomes a better place. Another argument against any killing is that once one gives one reason to kill, the door is open for any reason whatsoever, as is presently the case in international politics. It is a Fallacy of Rationalization to say we cannot deal with so many people in the world. A situation is genuinely self-defense only if there are no genuine alternatives available. It is not just defense of one's *way of life*. It may be noted that lack of preventative measures (bad policies and bad management, voting against medical resources and funding, voting for military interventions, and letting die by not helping those in the world who are in desperation) are also equivalent to killing. Letting-die is a failure to provide medical treatment. Letting preventable disease and disability happen is the same as letting-die (See Chapter 16).

Blame is one cause of killing. People often kill because they blame people and so wish them punished or eliminated. Blame is an irrational emotion and cannot ethically be the basis of killing. Blame is due to the faulty assessment that one can change the past by punishing, or that one could have done otherwise than one did. Neither is the case. We cannot change the past, and people could not have done otherwise than they did, or they would have. All we can do is try to change ourselves or others in the future. We should, then, substitute rehabilitative blame (education, correction and therapy) for retributive blame (revenge and retaliation) [4].

Blame, then, is no excuse for killing. It is only an indication of the need for rehabilitation. What follows from this is that we should find out as much as possible how and why we knowingly or unknowingly let people die, and attempt to correct such behavior in the future. We cannot undo or re-choose the past, but we can learn to make better choices in the future to the extent that education and going beyond culture allows. Most people cannot change because they are blocked by indoctrination, tradition, culture, anti-inquiry attitudes, and dogmatic beliefs. It would come as news with outraged denial that we are each in many ways and every day causing deaths. We would not be honest enough or have the ability to face up to this.

The question, for example, as to whether or not one should kill one or a few to save many, but not even lives, but only life-styles, wealth, oil,. . . exists against a background of publicly created scarce resources, widespread insensitivity to killing, contradictory beliefs and practices, and a populace opposed to critical inquiry and argument. People do not take the responsibility for others whom it is easily within their power to help. It is a blatant contradiction and hypocrisy for one to oppose abortion, but not oppose killing in war, and not be concerned with the billions dying of starvation and in need of medical care. People are quite prepared to ignore or rationalize such contradictions. In terms of consequences, letting die is basically the same as killing, and both involve intentions. This is a metaphorical reversal of what is usually thought. One cannot hide behind the fact that one did not personally kill. To have others slaughter and kill (e.g. by your vote) is the same as if you had done so yourself. People put killing under a different more favorable definition, or euphemisms. Nations claim not to be killing, but "defending themselves." But there are few cases of genuine self-defense. The National Guard claims in the U.S. to defend their way of life, instead of their lives. This means to use war to achieve whatever goods we wish. Also, value-laden terms are avoided, e.g. the U.S. supposedly did not slaughter one and one-half million in Iraq, but rather "saved lives" [5].

Where were these lives, the population of Iraq? It is not clear how they were saved. To kill one and one-half million to save hypothetically some thousands should be something to be ashamed of and to justify being tried for war crimes. We are supposedly not killing, but making the world safer. We are also supposedly not killing if we just are "withholding treatment." (cf. "just creating an embargo", which ends up killing civilians and others in Iraq) People kill more by inaction, uncritical beliefs, poor decision-making, bad management and un-humanistic policies than by direct personal killing. It is neither ethical nor unethical to treat or not treat. These are neutral acts. It is hypocrisy and a rationalization to think that letting-die is moral, whereas killing is not. It is only the consequences, which make either behavior ethical or unethical. The failure or opposition to rationally inquire about life and death consequences is unethical.

The healthcare system is in crisis worldwide as it cannot now adequately care for all who need medical care. As a result not everyone can survive. "Because there are not enough medical resources to meet the medical needs of all people, we need a system of rationing our resources to meet these needs as best we can" [6].

Faulty cultural practices and belief systems of people and institutions fail to sup port adequate medical care and medical research. "Governments that finance access to medical care for their citizens can never afford all that could be consumed and so rationing becomes inevitable" [7]. The result is loss of life. Life and death choices are built into medical and societal policy and decision-making (See Chapter 3). It is a false assumption that not all can have access to medical care. It is only because people let-die that this is the case. In medical-surgical situations, catastrophies and accidents triage is used to determine which patients are to be operated on first or at all. We use inoculation, medicine, surgery, anesthesia, knowing that in attempting to save many people some may die. The question of risk management becomes relevant. How many lives should we risk in trying to save the lives or quality of life of many? From the point of view of a narrow utilitarianism one could try to justify the killing of one person to save the lives of more than one. The humanist-pragmatist position would be aggressively and actively pacifistic and against any killing or letting die for any reason. However, treatment in medical practice is often invasive and entails that in attempting to heal some, life risks are involved. Ideally, any life risk would first have to be fully and critically evaluated and agreed to by the informed patient. A humanist-pragmatist position would include the holistic view that it is best to produce the greatest adequate benefit considering all rational arguments, not just satisfy the greatest number or greatest pleasure factor. That is, the decision is a rational and ethical one, not a matter of numbers or fixed principles such as utilitarianism's greatest good for greatest number.

The healthcare worker in practice often heals without needing reasons, justifications or euphemisms. We will see, however, that there are personal and cultural beliefs, inconsistencies, and practices also in medicine, which do lead to harm, killing, and letting die.

In triage there is medical prioritization of patients according at least to those who will: 1. die anyway. 2. live anyway. 3. live only if treated. It may be noted that even such classifications as "serious need," "critical condition," or "emergency treatment," are open-context ethical terms and, as such, unspecific and resting on a questionable basis. They may also be ethically manipulated. The criteria are too narrow because other criteria may be crucial, for example, a physician may be treated first so as to save others, or someone having information which can save many lives might be given priority for treatment. That is, one could determine priorities based on treating the patients in such a way as to save the greatest number with consideration of as many other relevant factors and consequences as possible.

Value of life considerations and societal consequences in triage decisions are typically covered up, but are really relevant there. How can one, e.g. a physician or healthcare worker determine such? But they often or always in some way do. Who will be treated first? This goes without question. The claim not to use quality considerations even if based on first appearance is hypocrisy. The healthcare worker cannot hide behind the rationalization that one has incomplete knowledge of the patient. One must judge as best one can and there are many clues to go by. Often these are the same ones they use but claim not to use (In normal hospital admissions we can have a whole history of the patient).

To determine what is to be regarded as relevant would require a sound ethical system. At present, when not all can be treated, there may be a tendency to treat first a family member, famous entertainer, well-known politician; a member of one's own race, gender, religion, society, army, group of friends; someone one has excessive sympathy for, etc. This kind of decision-making is based on biases, favoritism, false sentimentality, prejudices and irrationality.

In *Night-Train to Lisbon*, Pascal Mercier, a philosopher, dealt with the issue of the responsibility of a physician towards the individual patient and consequences, which might derive from doing so. What does it mean to be a physician? The protagonist, a physician, saved Mendes', a mass-murderer's life, who was called the "Slaughterer of Lisbon." When Mendes lay before him, he had seen him only as a special, individual human whose life was at stake. He had not been able to see this life as something one had to evaluate in regard to other lives in terms of the total situation. And it was exactly this, which was held against him: that he had not considered the consequences, which as well concern other individual lives, many individual lives, which would be endangered by this murderer's survival in the future. That he had not been prepared, to sacrifice this individual not to harm the many other individuals made him severely suffer. The retreat to the position of being a physician and doing only what a physician's work is, help patients without any other considerations did not really appease his conscience. His later view was to have not helped him [8].

11.2 Misuse of Ethical Terms

Physicians, like others, are typically bound to the cultural and religious beliefs of the society. Therefore, they would be qualified to make normative or cultural decisions, but not ethical ones. Culture itself is neither good nor bad, but non-ethical. Culture is common practice, not an ethical system. Culture in this sense is like a mere consensus. However, the existence of many beliefs in culture undermines genuine consensus. Ethics is not the expression of one's opinion. One's opinion is irrelevant to ethics [9]. Supernatural reasons or appeal to one or more of the diverse belief systems cannot form an adequate ethical basis for decision-making.

We speak of killing being "justifiable" or "unjustifiable." If one has not studied ethics and the uses and misuses of ethical terms, the question of anything being justifiable or not cannot be answered. One would have to ask, justifiable in what sense, according to which considerations?

Supporting the point that people at all levels are non-ethical and misuse ethical terms the following serve as a few examples: [Circular terms in italics].

"Killing is *wrong* because it is *immoral*." This is to define a term by a synonym (Fallacy of circularity).

"Killing is *wrong* because it involves a failure of *respect* for the *worth* of the victim" [10].

"Taking life is *intrinsically wrong*." [To say something is intrinsically wrong is not a reason. It is a bias and misuse of ethical terms. It is, in fact, to state that there can be no reason, otherwise it would not be intrinsically wrong].

"Killing is *wrong* because we have *no right* to take another person's life." There is no ultimate sanction for this. If the right is given by law it is not a reason, and furthermore it can be overridden or taken away (e.g. death penalty). It is taken away if one is drafted into the military, convicted of a crime, unable to afford medical treatment. The fact is that in our society there is a right to take another's life in the sense that we can do so, and we sometimes do in practice. That is, "right" is an open-context term meaningless unless some meaning is substituted for it.

Glannon states, *we do not have a right to health care* [11]. He is right in the sense that there is no absolute reason or meaning of "right" given here. But certainly, we could give reasons and arguments supporting the view that all should have health care. In Austria, unlike the U.S., virtually everyone does have a legal right to health care.

"We *should not* kill because it involves failure of *respect*." This is just to say that killing is bad because it is bad, and so is circular and is not an argument.

"We *should* not kill because people have human *dignity*." "Dignity" is a value term and so the statement is circular. But in war and in indirect ways people kill needlessly anyway.

"*Mature* people have *value*." *Mature* means *value* and so the statement is circular.

"*Moral worth*." Redundant circularity.

"The *value* of life includes *quality* of life criteria." *Value* and *quality* are synonyms.

"We *should* not kill the *innocent*." *Innocent* is a value term so the statement is circular. Also, according to whose ethical standard or which ethical theory is one innocent? One cannot be innocent as such.

11.3 Criteria for Preferential Treatment: Non-contradiction

Although we would ideally wish to treat all who need treatment this is not always possible and so choices have to be made. The following are sample cases when life and death choices must be made: The blood supply is limited. The food and water supply is limited – for billions of people in the world. Funds for medical research and care are limited. The supply of medical drugs is limited permanently or at the moment. Healthcare workers cannot give treatment to all who need it in the situation. Two or more patients are in need of immediate care at the same time and no one else is available. There are not enough organs available for transplant. The hospital or state has limited funds and cannot treat all patients for all desired treatments. Patients are admitted to the hospital only if they have the ability to pay. Permission to treat is sometimes not given by the patient or relatives. The patient would require extraordinary and heroic treatment. The overall consequences of treating would cause greater harm than not treating. Often the principle, "First come, first served" prevails, which completely excludes intelligent assessment regarding priority. The only rule for the physician to follow is simply: "Next."

Criteria given for preference for medical treatment on the egalitarian view are: 1. save no one as we cannot take one life to save another (Would this mean that we should not give medicine because one in 100,000 will react negatively to it?). 2. medical necessity. 3. first come, first served basis. 4. random selection [12].

If one were to be fair, patients would be given equal treatment only if they share the burdens equally, that is, care would be distributed according to merit and fairness as determined, for example, by a naturalistic and humanistic ethical system, but even then equality never stands as an ethical principle. This principle could generate a universal: Whatever opposes or undermines sound medical practice or research is self-contradictory, and immoral. It is self-contradictory, for example, to oppose embryonic stem cell research or oppose organ donations for normative or supernatural reasons. To oppose presumed consent to organ donation is to let die.

11.4 Case Example: Oregon Healthcare Prioritizing

Can healthcare be restricted? The Oregon (U.S.) Health Services Commission with limited health resources prioritized health care services by determining the high versus low priority disease-plus-treatment pairs based on cost effectiveness, including longevity and quality of life considerations (mainly physical qualities). The intention was partly to find some way to expand Medicaid coverage to the 16% of Oregonians who were uninsured. The Health Survey Commission ranked medical condition/treatment pairs from effective medical benefit at a reasonable cost to minimum effectiveness at high cost. Top priority was given to acute, fatal conditions the treatment of which could restore one to normal health. Priority was also given to the prevention of disease. Unfortunately some choices were questionable due to inadequate ethical criteria, for example, vasectomies ranked higher than hip replacements [13]. Of over 700 possible treatments, which already excluded about 1,400 diseases and diagnoses, the cut-off for treatment was placed at the 587th pair after actuaries determined the cost for each treatment. No coverage for Medicaid was to be given beyond this point. The proposal was rejected in 1992 by the U.S. Government Secretary of Health and Human Services. Issues had been raised such as bias against the elderly and disabled, and the principle of quality adjusted life-years (QALYs), that is the value of additional years produced by treatment, adjusted for quality. The elderly may not have many quality-life-years, but a young person in pain may have more negative value years. A fertilized egg may be unfairly given preference over people because it may have a longer lifespan. A person who sacrificed and worked hard to support a family might unfairly have less priority than members of his family who have never contributed to the welfare of the family or society. Glannon wrote, "Despite its shortcomings, something along the lines of the Oregon plan still seems to be the most promising way of realizing these goals [of best allocating medical resources]" [14]. One major problem with the quality of life criteria given is that they are too narrow.

11.5 What About Self-Caused Illness and How to Determine?

We let others die, but we also let ourselves die. People directly and indirectly by their beliefs and actions can cause their own illnesses. We all more or less do so in many different ways. See also Chapter 16) Basically, however, medical policy generally assumes that patients are hardly responsible for anything. Nobody should be blamed, but treated, and educated to live a better life-style. Hope, Savulescu, and Hendrick state, that communitarianism is our responsibility as part of a community for the common good, rather than merely egoistic preferences [15]. The common good, however, presupposes a well-founded theory of ethics. If one, including many physicians, does not take care of one's health, it is like a firehouse burning down. Communitarian theory also may simply refer to the mere uncritical normative and traditional practices of a society, in which case it does not serve as an ethical standard [16].

Those who cause their own injuries by risky behavior may be candidates for lower priority for scarce medical treatment and organ donations. What care is deserved and owed if you purposely put yourself at needless or careless risk and expect others to save you even at their risk and expense? According to the American Medical Association, "a physician shall, in the provision of appropriate patient care, except in emergencies, be free to choose whom to serve, with whom to associate, and the environment in which to provide medical care" [17]. Bahro and co-authors maintain, "Distributive justice in [rationing] medical care can only be accomplished if the reimbursement system is based on. . .a stronger emphasis on individual responsibility for one's own life and health and the insight that health is an individual rather than a collective good" [18] .This would encourage preventative medicine, which the authors suggest is lacking. Health is a collective good. Patients may oppose medicine and not support medical funding. One may cause one's own illness by failing to cooperate, by refusal to follow doctor's orders or to take prescribed medicine. Disorders are often directly or indirectly caused by the patient, when they could have been avoided or prevented (See patient responsibility in the Chapter 12).

Glannon argues that priority can be given to prevention, which can ensure better health and save more lives and resources than treatment after one has the disease. "Preventing disease promotes equal opportunity at the same time that it maximizes benefits and minimizes harms" [19]. That is, we each have equally, in terms of our individual genetic and physical condition, the opportunity to try to keep ourselves healthy, but the natural as well as social conditions of different people may be quite different and unequal. Nature does not provide equal chances, therefore those not so well equipped, have to be especially supported. We each have equal opportunity to live a better lifestyle in terms of what one can be in control of. Decisions for healthcare for some take resources from others and place an unnecessary burden on society and all those involved. Glannon states, that if one is responsible for one's condition because of poor, unhealthy lifestyle, etc. then one should be given lower priority for medical treatment [20]. We should consider the medical treatment fairness in terms of our whole lifespan and total situation, rather than just at an isolated point in time ignoring all of the contexts and causes. Thus, Glannon argues for the "prudent lifespan account of health" [20].

One problem is that we may harm when we mean to help. Should we risk our life to save the lives of ten? But 10 people minus 1 person are not nine persons. If this were the case they would not be persons at all, but indistinguishable things, depersonalized. Mathematics assumes total equality of items calculated, so no one has any value or quality. Helpers and destroyers are the same. We cannot judge if killing or saving is right on the basis of number alone. It is hypocrisy to think that it is acceptable to kill mathematically one to save many unless that one person is you.

Priority should not be based on mere ability to pay or mere arbitrary lottery or "first come, first served." In Canada rationing is often done on the basis of queuing. In the U.S. it is often rationed on the basis of the ability to pay. Those who cannot pay for healthcare are not treated, although some hospitals have the mandate to serve a certain number of poor patients. Health care is determined by market and economic principles [21]. Myers' proposal to let healthcare be determined by the marketplace is like letting it be determined by lottery [22]. And basically there is the lottery of genes as well as of the society born into, both of which are unfair and unequally provided. Not everyone starts with the same chance to health and knowledge. Glannon states that it is not unfair to treat irresponsible people unequally, "If people can make autonomous choices that adversely affect their health, then. . .they can be responsible for their health" by getting lower priority for health care [23]. He continues, "We need to use factors such as control and responsibility in devising a fair system of rationing. . .. Autonomy cannot be separated from responsibility" [23]

(See also Chapter 12). One should also suggest that if autonomy is to be responsible, sufficient education must be provided in order to make this possible. Bircher and Wehkamp [24] suggest the introduction of potential in regard to the concept of health and disease. We do not have equal chances in terms of our genetic make-up, family background, social status etc., but we do have the challenge and the responsibility to meet the challenge in order to develop our potential in the framework of the conditions given [24]. This is the task for us as individuals as well as societies which have the task to provide the developmental structures including a healthcare system based on fairness and solidarity.

In 2007 approximately 45 million people in the U.S. have no health insurance. From 1999 to 2003, 5.7 million people in the U.S were homeless some part of the time [25]. One solution to this is to require each citizen to buy insurance or be a member of a healthcare system just as they are required to have auto insurance if they are to drive. One Massachusetts bill required residents to obtain health coverage by July 1, 2007. Individuals who can afford private insurance will be penalized on their state income taxes if they do not purchase it. Government subsidies to private insurance plans will allow more of the working poor to buy insurance and will expand the number of children who are eligible for free coverage. The plan is expected to cover 515,000 uninsured people within 3 years, about 95% of the state's uninsured population.

Insurance companies are beginning to significantly increase premiums for those who do not take care of their health. Some health insurance companies charge overweight people or smokers up to 70% more for health insurance. This would be unfair for those who are overweight due to genetic rather than any lifestyle factors. It should be noted that secondary smoke can be as dangerous as smoking, e.g. if one lives or works in a smoke filled environment. A family member or employee who smokes is threatening the lives of one's family members or other employees. On a merit-based distributive system, their health costs would be paid for by the individual rather than by the insurance company. After a number of accidents one's health and auto insurance can increase to the point where even it is unaffordable. The question arises, "Can we give care only to people who have taken the necessary precautions not to become ill?" [26]. But, when resources are scarce we must make choices. One cannot distribute equally when there is not enough to distribute. Preferably all will be treated, but as indicated earlier, people themselves play a large part in the services not being there and in creating unnecessary needs for services.

Billions of dollars are spent on dangerous sports and sport medicine when millions of people are dying each year for lack of needed basic health care. Such sports increase health costs and overburden the healthcare system. In the U.S. there are about 100,000 concussions a year from football. According to Beauchamp and Walters people should be responsible to live a healthy lifestyle and avoid risks [27].

Many illnesses in the U.S. are caused by taking illegal drugs [28]. In terms of legal drugs, in 2000 the U.S. consumed 7 billion gallons of alcohol, 25 gallons for each person. According to Roepeik and Gray, "No amount of alcohol has been found safe for a pregnant woman" [29]. Marlboro prints on its cigarette carton: "Smokers die younger." "Smoking seriously harms you and others around you." Lucky Strike advertises, "Smoking kills." People buy these products anyway, and because of the blatant honesty, it may even increase sales.

One of five Americans dies of cancer. One of eight women will have breast cancer. In 2004, only 6% of women in the U.S. obtain mammogram screening as recommended by the American Cancer Association [30]. Billions are spent on cosmetics, which are poisonous, e.g. hair dye poisons the body. Most industrial and household chemicals are poisonous to touch, breathe, ingest, or have in the house. Car "accidents" are now called "crashes" because they could have been prevented. Exposure to noise over 85 decibels can lead to permanent hearing loss, yet people listen to music and expose themselves to lawn mowers and other machines which are significantly louder.

In the United States there is no national health care, and insurance companies basically exclude coverage for mental disorders. President Obama, nowadays, is oriented towards a reform of the healthcare system. U.S. is like a third world healthcare system compared with, for example, Austria which is much further advanced in health care services generally and offers not only comprehensive universal national healthcare and presumed donation of organs, but does cover mental disorders. In Austria, the state even reimburses patients for fertility clinic treatment.

One may argue that consumers should be examined for and disclose their genetic status for health before being insured. There should be no disadvantage for the consumer to do so as genetic structure is not a choice. We can now detect some inherited risks and so sometimes provide preventions. Those making bad health and reproduction choices should be encouraged not to do so. It is in the interest of the insurance company as well as the general public to have policies, which produce a healthy lifestyle. Many people do not even try to obtain insurance and so let the burden fall on others. "Presently health care spending constitutes roughly 15% of the U.S. GDP, which is well ahead of the percentage in other developed countries. Paradoxically more than 15% of the population of America has no health care" [31]. By comparison, Swiss health insurance costs 30% less and they have universal coverage [32]. The United States spends over $2.1 trillion in 2006 on healthcare, about 5% of the world's GNP. In spite of this, at least 77 million people in the U.S., two out of every five adults, cannot pay for medical bills, even if they have health insurance.

11.6 The Hippocratic Oath: Pacifism in Medicine?

> I conclude that war cannot be justified as an ethical activity. Therefore. . .pacifism becomes the only ethical position [33].

> The WMA believes that the development, manufacture and sale of weapons for use against human beings are abhorrent [34].

Medicine endangers people to help people. The surgeon takes a chance on a patient's life by operating. Drugs are always an on-going experiment and adverse effects are continually reported. Sometimes the drugs have to be removed from the market as a result. These are risks to be taken because of the advantages involved. But the military kills for no sound reason at all. They have neither full nor adequate requirements for going to war for one simple reason. In short, in terms of sound rational philosophical arguments, they do not know why they go to war. They do not know the just war arguments and many other arguments against war.

An enlightened pacifism is based on reasons, the harmful consequences of killing. We may first note that if killing is allowed for any reason, regardless of how convincing, it opens the door to allowing individuals and cultures to find all kinds of reasons to kill. It is often done by simple institutional policy or lack of policy. Killing for whatever reason encourages the institution of killing generally. Pacifism allows for not killing people except when personhood is lost or not yet attained. The just war arguments and requirements before going to war are virtually never considered and virtually never satisfied [35]. Gulf War II was originally supported by perhaps 80–90% of Americans (fewer in other countries), but from 2005 on over 50% of Americans thought it was a mistake. The Vietnam War was eventually almost universally thought to be a mistake. Trillions have needlessly died as a result of faulty thinking.

Over one hundred American court-convicted, death-row criminals were recently found to be innocent of their crimes because of the evidence of genetic testing. One can project that many thousands of prisoners of all types would be released from prison if we had such corroborating evidence. U.S. is one of the few countries having the death penalty. Only China and Iran execute more people [36]. Other countries might also be added. The European Union (EU) does not admit membership if the country has the death penalty. Thus, the U.S. would not be admitted into the EU even if it were located in Europe.

The Hippocratic Oath (fourth century B.C.) is still adhered to today by the medical profession. "Do no harm," means that the cure should not be worse than the disease [37]. It is an oath, which supports pacifism in medicine. It is directly opposed to the methods of war. In medicine, virtually no reason is allowed to justify killing a patient. And if it is wrong to kill in medicine we have a prime paradigm for not killing at all, even in war. If it is wrong to kill, we must accordingly ensure the quality of life for all humans and enhance the capabilities of all.

In war an unlimited number of people, soldiers and civilians, are brutally slaughtered in ways, which can only affirm the complete lack of intelligence and humanity on the part of individuals and nations [38]. It is one of the clearest paradigms of human cruelty and is a definition of individual and cultural psychopathy. This means that physicians cannot follow the masses in culture and society for guidance in regard to medical treatment. In contrast, one of the duties of the physician is to advocate for social, economic, educational and political changes that ameliorate suffering and contribute to human well-being [39]. Medical schools should include humanitarian and global issues [40]. Declaration of Geneva 1948 states, "I will not use my medical knowledge contrary to the laws of humanity" [41]. The physician is under oath to not harm or kill. The basis of the professional duty of the physician is to be active as a pacifist.

For Miles who wrote on the history of the Hippocratic oath, it is false to say that the job of a physician is merely to passively and neutrally serve the society. He notes that, "Institutionally, the AMA has strongly worked against universal healthcare"

[42]. "U.S. physicians have generally organized *against* efforts to address the lack of affordable healthcare" [43]. "Relatively few physicians work in the politicized area of healthcare reform" [44]. The oath recommends rather that physicians become active in healthcare reform. "The American Medical Association (AMA) has opposed virtually every governmental effort of this century to provide greater access to medical care, including Medicaid and Medicine" [45].

The Oath also states, "I will not give a drug that is deadly to anyone." Part of the reason for this is that physicians were asked to collaborate in giving poisons to political enemies. The exact same problem arises today. Has there been progress? Should physicians collaborate in helping to develop bacterial warfare, and treating soldiers whose main job is to kill? The main job of physicians is to save lives and improve health, thus it is against professional medical principles to be asked to serve political and military practices, which involve killing and wounding people [46]. The issue is falsely put in terms of rights. "War fundamentally abridges an individual's right to life." and "Combatants lose their right to life as they gain the right to kill" [47].

The issue is not only that physicians and healthcare workers give up their humanity when they support the military, but that the individual is of no concern regarding military goals. Human life is not important. Furthermore, the interests of the state may not even be supported by the majority of citizens [48]. "Utility allows military necessity to trump other military constraints on military action" [47]. "Officers rule each soldier for the good of the state and its armed forces" [47]. Ultimately no medical care need be given. "War. . .undermines each actor's right to medical care" [47]. Soldiers have no individual right to refuse medical treatment or refuse to die. They can, however, daily be commanded to die. Combatant risks of 50–100% casualties are usual. Soldiers can be required to take experimental drugs [49]. Care cannot be given for soldiers [50].

Some physicians have objected to helping any war effort as being contradictory to the medical profession and so advocate withholding their services in order to oppose war [51]. Physicians for Social Responsibility and the International Physicians for the Prevention of Nuclear War only oppose nuclear war. On the level of normative morality such contradictions can be expected, but not on the level of critical ethics. Clearly, in medicine it is a duty to actively save lives, but in the military it is a duty to kill.

Koch recently observed that President George Bush on January 18, 2002 decided that the Geneva Prisoner of War Convention does not apply to prisoners [52]. "For the first time since the Nuremberg trials. . .a major political power consciously weaponized medicine as a tool for the progress of military goals" [53]. Dr. David Tornberg, Deputy Assistant Secretary of Defense for Health Affairs, wrote: "Physicians assigned to US intelligence. . .have no doctor patient relationship with detainees. . .and no obligation to offer medical aide" [54]. Dr. Tornberg. . .denied any special ethics to medical practitioners [53]. Thus, ethics does not apply to medical practice in the military. Healthcare workers can be completely subject to the demands of the military including cooperating in various forms of torture of prisoners [50]. In addition, the orders of a superior are presumed to be legal [53]. Koch concluded, "The U.S. has militarized medicine" [52].

Kottow similarly holds that the military requires physicians to give up medical ethics for military purposes [55].

Wounded soldiers need to be treated only if they have salvage value [56]. The reason and ethics of healthcare workers is taken away and they are made subservient to military judgments.

Federal research and development spending by the U.S. Government National Institute of Health in 2006 increased by $2.2 billion to about $135 billion, but of the increase, 97% will go to weapons development and spacecraft programs, said the American Association for the Advancement of Science [57]. The emphasis is on lethal weaponry rather than non-harmful methods, which preserve health. Considering all relevant expenditures together the military is spending amounts to arguably half of the U.S. budget when all is considered. In one of the most uncaring political acts, the U.S. Social Security funds were robbed to pay for other things such as the military so that the Social Security system is headed toward bankruptcy, which puts healthcare and survival for the elderly and others in jeopardy for the foreseeable future. The killing of others jeopardizes the lives of one's own citizens and exhausts the budget.

11.7 Should We Kill X to Save Y? The Numbers Game

> Ethics goes only as far as humanity goes, that is, the consideration of the existence and happiness of the individual human being. Where humanity stops pseudoethics begins [58].

> [In Somalia] *A man's life is measured by camels, with one hundred camels being the price for a man who has been killed* [59].

With every act we benefit some while depriving others. If a physician treats one person others may be at the same time directly or indirectly thereby disadvantaged. If money is spent on weapons it cannot be spent on welfare. Everyone is in a position to help others in some way or other. In addition, sometimes we may be forced to act to prevent immediate harm where the number of people harmed is at issue, e.g. which the trolley argument represents. In philosophy, the issue was represented by the "trolley" example: Which is more ethical, as a passenger to let the brakeless trolley run over six people on its present course, or switch it to run over only one? Thomson's conclusion is roughly that one may divert a harm that already threatens people so that fewer will be harmed assuming no overriding rights of the fewer will be violated [60]. The problem is not solved because "rights" is an open context term that still needs to be made clear. We believe an examination of the literature shows that philosophers have not been able to clarify or resolve the ethical issue satisfactorily [61]. Instead of the trolley example, one could use other examples. Should a limited supply of medicine be used to keep one person alive longer, or rather be given to save five people for a shorter period? In any case, on the global level we could be absolutely sure that we could save millions of lives by our help at little expense and no deaths [62].

Should we help one person or many? This is the utilitarian question of producing the greatest good for the greatest number. But we may note that we could produce more good for one person rather than less good for many. When the issue involves one's life it would seem to be that all goods (lives) may be treated as of equal value and so one should save the greatest number of lives. The utilitarian issue is unacceptable because it is too simplistic. An ethical decision on a naturalistic theory would have to consider all of the reasonably involved contextual factors and consequences, not just numbers, which never represent lives. They can only mathematically be of equal value. People do not regard lives as equal. If that were the case, billions would not be allowed to starve and there would be universal world health care. In addition, some people have been responsible and contributed to desired goals and others have not or have even undermined them. No two people are equal in this sense. Should we save someone who would not save anyone but destroy lives, or instead save someone who would go to great risks to save others? If we had the choice of saving the Red Cross or the National Rifle Association the choice would be clear for the humanist. In actual everyday life, one life is given value over others even if the value given is misguided. Usually one would save a friend or family member regardless of merit. One cannot ethically choose merely on the basis of number.

Similar to the trolley example, should a physician give preference to his or her patient even when by instead giving the patient's medicine to others it could save five lives instead of just one? Taurek refers to the trolley case as a trade-off situation, but every ethical decision is a trade-off in one way or another [63]. It is a trade-off to buy an expensive auto when the extra money could be used to keep someone else alive. All decisions are trolley decisions. On an ethical pacifist position one would try to save all who can be saved. Also, to treat all equally regardless of merit is to be unethical by definition. This is because the criteria of action are mechanical, not critical. Numbers are only one factor in decision-making. One may add particular circumstances and contexts so that all is not simplistically equalized. By speaking of other things being equal we unfairly exclude the very things, which should be considered in order to come to an ethical decision.

Suppose the trolley example involved giving up one's own life to save five people. There is in itself no reason why one should give up one's life to save infinitely many other lives because one will not be around to experience the benefit. The question arises how many people one would let die so that one could live? For some people the number could have no limit. Numbers of lives, just as numbers of objects in themselves have no value and are ethically irrelevant [63]. But the relative value we give lives, people and objects is what is of ethical importance. We may regard our own life as more important than the lives of others. Lives and objects become valuable only to the extent that we give them value.

The extent of selfish preferences for family, friends and country form the basis for allowing an infinite number of others to be killed elsewhere. At least the question of whether or not it is acceptable to kill x number of people to save one is answered by the reality of everyday life. It is done regularly and without being challenged.

We have considered some of the values of those to be given priority for life. Now we may consider how many lives it is justified to risk. How many is too many?

If a country is thought to have weapons of mass destruction (WMD) and biological warfare, other countries will nevertheless send soldiers into battle with the clear likelihood that *all* of the soldiers and *all* of the inhabitants of the country *and beyond* will be killed by such WMD as was the case in Iraq. In the military, armies often "fight to the last soldier," regardless of the consequences and failure of doing so. If killing some is allowed for some purpose, killing all is allowed for some purpose as the contemporary world situation shows.

What is the magic number in the killing game? Some say the death of one is too many (See Schweitzer's quotation at the beginning of this section). One killing is wrong enough. With one killing we have already established the proof of the killer's insensitivity and lack of understanding the value of one's life. Our life is a whole world. Even if it may prove lifesaving for many, if one life is lost, a whole world is. There is no mathematical formula. The use of a naturalist-humanist theory of ethics may be used to sort out the holistic consequences in each case.

Contradictions expose the failure of our cultural intuitions: 1. We do not condemn a rich person (or institution) who gives nothing to help the dying people of the world, but do condemn someone who does not feed his child because he is poor. 2. We praise those who killed a million and a half in Iraq, but condemn a physician who performs one requested abortion or makes a mistake while trying to help. 3. We condemn killing of one person who asks for death to relieve suffering, but let thousands die by fatal healthcare policies, scarce resources, and not having either national or international health care protection. 4. It is not illegal not to help someone drowning whom one could easily save, but it is illegal to directly drown someone (See Samaritan discussion later). 5. One would condone killing untold numbers of foreigners to take revenge for just one killed family member. Strangers are often much more needy and deserving of care than friends or family, yet we do not help them. 6. We see no connection between having an expensive house and auto and those in the world who are dying due to lack of food and medical care. 7. Doctors are sentenced for "wrong" treatment even if intending the best.

Why not kill? The academicians have not been able to answer the question. "No one, to my knowledge, has ever offered an account of why killing is wrong that even begins to do justice to the full range of commonsense beliefs about the morality of killing" [64]. "Conventional moral views about killing are often intellectually unsatisfactory" [65].

Utilitarianism is problematic. It is defined as bringing about the greatest good for the greatest number. It does not define good adequately. The utilitarian formula is not an ethical system, but presupposes an ethical system it does not present. In this sense, it is more a way of avoiding making ethical decisions. One could, then, justify killing many people to achieve a greater "quality" of life for a few, or kill no one thereby stressing the greater number, yet have a lesser "quality" of life. The greatest number in a universal utilitarian formula is everyone. Utilitarianism is often identified with a consequentialistic view, but this is not correct. A naturalistic position has always to consider consequences, but it does not entail that a utilitarian formula be used. Consequentialism is basically opposed to absolutism. Glannon is incorrect in writing, "A consequentialist theory. . .is more concerned with the total amount of good like health than with the rights and interests of the people" [66]. Utilitarianism might be, but not necessarily is a consequentialist view.

Consider medical risk. A physician immunizes patients knowing that one in y number (e.g. ten, hundred, thousand, million) will thereby become seriously ill or die. Should we experiment on humans directly instead of animals, which yield only irrelevant or remotely suggestive results, in order to save as many humans as quickly as possible? We find out more about drugs and treatment from the medical treatment of patients. People being treated are at the same time being used to gain experience. Adverse and negative effects on patients are then noted and reported to the Federal Drug Administration as experimental feedback. They are reported to "NHS Direct" (National Patient Safety Agency). Unfortunately the NHS has not the staff or inclination to properly use such information although it is one of the best sources available to increase medical knowledge.

The utilitarian formula does not help us decide between the greater "good" or greater number, so the principle breaks down. Furthermore, it would be desirable to produce the greatest good for all, which goes even beyond universal utilitarianism.

A surgeon performs an operation knowing that the survival rate of such an operation is x%, e.g. 30%. Should a commitment to life require that operations not be performed unless the risks are below 50%? A 99% risk operation may be performed if death is the alternative. But a critical, contextual and patient-specific analysis also needs to be done for all such statistics. In surgical risk, the general percentage of risk may be set too high for a healthy person and too low for an unhealthy one. Statistics often do not work so well for the individual.

The utilitarian formula is merely a mechanical formula where good is left unspecified which avoids one having to make ethical decisions. Majority rule and consensus (democracy) are forms of utilitarianism by means of which the majority is satisfied. However, it may be at the expense of the minority and may involve 49% being dissatisfied. The majority rule principle can exercise tyranny over a minority. Also "rule" can mean brutal force. In philosophy, the majority rule principle and appeal to consensus are informal logical fallacies. What is rather needed is a sound ethical theory.

To correct the shortcomings of utilitarianism we could substitute a naturalistic theory of ethics, which would allow us to determine how to decide between the greater number and the greater quality. On such a system, introduced already in several chapters of this book, by ethical terms, e.g. good, right, etc. we could mean the following: "Good" means: bringing about our (a) informed, (b) wants and likes (c) deliberately (d) on the basis of inquiry (e) with as adequate and full consideration (f) as reasonably possible (g) of the naturalistic and global consequences including the bringing about the informed wants and likes of everyone including concern for animals and nature. That is, naturalistic ethics may be seen as humanism or human and natural ecology on a world-wide, all-encompassing level, which includes the knowledge found in the various natural and social sciences including philosophy and aesthetics, mathematics, and astronomy, etc. It is in this sense that we may speak of an adequate, holistic ethics. Ethics in medicine would refer to a holistic, qualitative consequentialism.

Some criteria given on the utilitarian view for determining who should be favored, when resources are scarce, are: 1. medical success, 2. immediate usefulness, 3. ease of treatment, 4. patient with largest responsibility, 5. social worth [67].

Schweitzer thinks differently: *Humanity means that no human is ever sacrificed for any purpose or reason. The ethics of ethical personality wants to preserve humanity. The ethics set up by society is incapable of it* [68].

On this view, we sacrifice no one. We sacrifice zero people to save x number of people. This is close to the Hippocratic Oath. Unfortunately, if this were the case we would eliminate many medicines, medical procedures, and inoculations where there is a risk in taking such medicine, but the medical risks are to avoid bodily harm and death. Where risk is concerned we may obtain consent, but we need always try to determine how the risk can be prevented. Patient consent may also be a way to avoid making an ethically and medically sound decision. The Rule of Double Effect refers to a foreseen wrong event, which is not intended, e.g. administering analgesic (pain killer) to a terminally ill patient even if it hastens the patient's death. It is foreseen, but death is not intended. Nevertheless, this is a form of allowing to die.

Returning to the "trolley" type of example, suppose we are driving and there is a sudden accident forcing us to choose between hitting a car with one person in it or hitting a car with six people in it. Which should we do? Clearly, hit the car with one person in it. But, what if the six are known criminals and the one is a physician? Would we refuse to sacrifice one person's life to save many people also having a holistic quality of life, because even one such life is of infinite value to that one person? One answer is that we should not directly take any lives in medicine, or in war, because it leads to allowing killing for any reason.

We speak of numbers: We kill x *number* to save y *number*. This treats people equally, ignoring all distinctions between them. Decisions must be shown in each case by means of, for example, a sound naturalistic ethical system and by an informed and detailed consideration of consequences. Are we only numerically different? Who is it that is being killed? "Killing" is a strong word implying that it is always wrong to kill, however, our political, social, and often lacking medical practices all generate policies, which negatively affect people's health and lives.

If we are to wish fairness and equality of treatment, then along with this must be considered fair and equal responsibility of the patient to prevent disease and maintain a good health lifestyle. With benefit and priority comes responsibility. Priority should be given which promotes prevention of disease in the society, rather than encourages it as it now does. Blind equality is an empty concept and not a sound public policy as it undermines distributive justice, and people thus will not be encouraged to practice preventative medicine.

With a naturalistic theory of ethics the answer to why killing is unacceptable is simple: We enjoy life. We wish to live. It is the greatest wish and enjoyment we have. Enjoyment is what we largely mean by the personality and self. It is contradictory to wish for the cessation of enjoyment and therefore of the cessation of the self.

We do not want to lose ourselves. We want to have a future, the opportunity to materialize our goals in life. We do not need to be lawgivers or to have abstract universal principles to understand this. We do not need proof for the beauty of a flower or proof for the pleasure of an embrace or loving touch. This is rendered by naturalists as ethics based on wants, likes and desires and avoidance of the opposites. If one wished a further analysis one could inquire into what a desire is [69]. Singer's position of "preference utilitarianism," for example, comes close to this view, but preferences are values [70]. It is to say, "This is better than that," or "I favor this over that." "Good is based on preferences" is a circular statement. A naturalistic reason for not killing anyone is that it permanently destroys their most desired wants and enjoyments, what they value most and it destroys their selves. If one's life is not able to be enjoyed one may wish to end it, e.g. to end intolerable pain. If one's life is threatened by another after trying all the alternatives to protect oneself, there might be some case when one may have to risk killing the aggressor, rather than just wounding them. But, it is an extremely rare case when this would be necessary. War is virtually never necessary. It can only be waged in conformity with all of the "just war" and other requirements satisfied before using force. If this were done it would never be waged [35].

Those who attempt to kill others, in war or elsewhere, thereby subject themselves to being killed and make it justifiable to being killed.

11.8 Allowing Death = Killing = Murder

> Killing is no different morally than allowing to die [71].
>
> The distinction between killing and allowing to die is less clear-cut than we commonly think [72].
>
> The hand that signed the paper felled a city [73].

The arguments presented here have been ways to try to prevent direct killing. One can even prevent indirect killing. One form of indirect killing is allowing to die. We kill by omission as well as by commission every day. Whether we kill someone or do not help them in their desperation, we cause their death [74]. You are to blame for all the deaths and harm you could have prevented. To not help save people's lives is the same in terms of the result as killing them.

The following is a brief examination of the equivalent outcomes: allowing to die = killing = murder.

Definitions of "let die": 1. passively allow to die (e.g. by doing nothing) to avoid it. 2. actively let die (by an action one performs) "I let her leave the hospital too soon." 3. ethically let die. 4. one or many may let die. 5. we may stipulate that people let die whether they do or not. 6. encourage to die. 7. persuade to die. 8. "let die" as killing. 9. legal definitions, e.g. let die by negligence, failure to act, omission, violation of rights, etc [75]. 10. pragmatically not save. 11. consciously or unconsciously let die. As with negligence, we are typically obliged to know what we are doing and to take the necessary care to protect life. 12. prevent from saving. 13.

risking one's life to save others (letting oneself die). 14. fail to provide preventative measures or education. 15. withholding treatment in the United States is not regarded as euthanasia. But it is letting die.

To omit action is an alternate action. One cannot omit as such or withdraw or withhold as such. Death caused by omissions of action may be of different types. One can only omit, withdraw, or withhold a specific action. To omit is to actively leave out or exclude. You did and caused something by not doing something. Negation is always a positive configuration of events. To do nothing is to act in a certain way. One cannot just do nothing exempt of any action. There is no pure negation. To not feed a child is to starve a child. To not help is to harm. To not exercise care is to be negligent. By not giving advice you are giving advice (Cf. metaphor of reversal). Not doing something is aggressive action, just as is negligence. Inaction is not contradictory to action. John Harris regarding killing versus letting-die said that it is the consequences, which are important not whether the act is active or passive [76].

A positive is equivalent to a double negative. To say that killing = not not killing. Let die = kill. I let someone die. "Let" means that I allowed, caused, did nothing when I could have kept someone alive, but chose not to. The choice could have been deliberate as when watching someone drown whom you could have saved, or helping the desperate you chose not to help. Ordinarily, if one were well informed and then still could not have done otherwise it is not letting die. Letting die could be unintentional if you were uncritical or unaware that help was needed. In this sense we have an obligation to be aware critical thinkers in order to help those in need. Otherwise we kill by remaining ignorant. Unintentional letting die is thus still killing. Letting die may be intentional, unintentional or accidental if we could have prevented the dying. We kill by not preventing accidents (by defective auto safety measures or by not developing Tsunami warning systems, etc.) (See Chapter 16) In the recent Tsunami 280,000 were killed (*New York Times Almanac 2006).* We cannot say that tragedies were an act of nature if we could have prevented them.

Kuhse argues that it is a myth to hold that letting die is not equivalent to killing [77]. Letting die is still murder even if it is a choice, obliquely intended, an act of omission or non-action, indirect action, withholding treatment, or merely foreseen. She wrote, "From the moral and legal point of view, intentional killings and intentional letting die are, other things being equal, the same" [78]. Rachels also holds that killing is no worse than letting die. According to the "equivalence thesis," killing is the same as letting die [79]. Feinberg in his discussion of the legal aspects of the failure to prevent harm, argues that "the distinction between harming and not preventing is insignificant" [80]. The distinction between killing and letting die breaks down. People are just as unnecessarily dead. Gratton, a philosopher and member of Toronto General Hospital Ethics Committee stated, "There is no morally relevant distinction between letting die and making death happen, and between withholding and withdrawing life-support" [81]. On the other hand, one could argue that letting die is often worse than killing because it shows that one does not care, that one only acts to help because forced to by law, religion or culture. People now have the inalienable right to allow others to starve and die. They exercise that right.

11.9 Letting-Die and the Samaritan

Rescue is everyone's obligation [82].

Why is withholding treating a patient different from killing? We feel we have no obligation to help others, or strangers, or even friends. People assert with legal right, "I am not required to offer anyone Samaritan help." They protest, "Why should I help others?" The U.S. does not compel active benevolence toward others. There is no duty to be a good Samaritan. But U.S. citizens do have an obligation to kill or support killing for their country. However, one cannot leave the scene of an accident where someone is hurt. We are responsible for those we let allow on our property. Also we are responsible for passengers in our car if we drive recklessly. Thus, sometimes we have responsibility and sometimes, not.

English speaking countries have basically not regarded it as wrong in tort or criminal law to fail to rescue [83].

In the U.S. "There is no general duty to assist a stranger in distress" [84]. There is in general no legal duty to help a sick, helpless, or desperate person [85]. On the contrary, if one does volunteer to help a desperate person one must do no harm and make the situation no worse. One does not need throw a rope to a drowning person, save a child lost in the forest, warn a blind person of an open manhole, lift a drowning person's head from the water. There is simply no general liability for failing to act. Beneficence is superogatory, i.e., not required, but voluntary.

Only if assigned a duty or entrusted with someone, who needs care, it is required to rescue him. There is a duty to the state, one may be drafted into the military [which may be legal, but is unethical], be required to help in flood control, must give time and property to the state if it requires; and doctors may be mobilized to provide services in emergencies. The police may ask one to offer help even if it is at the risk of one's life. Religious belief is not an excuse not to help. Although they might oppose medicine, Christian Scientists may have to bring medical help, and Orthodox Jews may have to exert effort on Saturday [86]. To be a tort a prior duty must have existed. In this sense we need not call it "help" or "rescuing" at all, but rather a necessary part of work or services rendered. It is not altruism, but job performance. Nonfeasance is failure to do what duty requires. Malfeasance is wrongdoing or an unlawful act. Parents have a duty to save their child, but only if with little risk to personal safety. A parent may refuse to give his/her life preserver to his/her drowning child if it would increase the survival risk of the parent [87]. In the U.K. physicians outside the hospital are not legally required to act as good Samaritans. Private physicians within their geographical area may be so required [88].

One must also help another if one is oneself involved in an auto accident, or if one is a parent or spouse, etc. On the other hand, over fifteen European countries do require punishment if one fails to rescue [89]. Feinberg concludes, "My own intuition is that 'bad Samaritan' statutes [requiring one to help or rescue] are morally legitimate" [90].

One problem is, ironically, that the person helped out of distress sometimes sues the rescuer. The Good Samaritan laws protect the rescuer against such lawsuits, but the U.S. laws on this vary from state to state. Thus, people may decide not to help others in emergencies, because they may later be sued for doing so. Relatives may sue if resuscitation (CPR) is performed too late so that a seriously brain-damaged or defective person is kept alive. CPR workers therefore sometimes fear doing, or refuse to perform CPR to try to save a person. On the other hand, a rescuer can sometimes claim damages, if needed, which are caused by the one in need.

A comparison was made before 1981 indicating public views about Good Samaritan helping. Should helping those in distress be only a matter of conscience without legal consequences? In favor of this were: Germany 62%, Austria 42%, U.S. 75%. Favoring jail for not helping: Germany 22%, Austria 15%, U.S. 2%. Favoring legalizing the duty to help: Germany 86%, Austria 26%, U.S. 19% [91]. The French criminal code in 2006 states that it is a crime not to help someone needing assistance when it can be provided at no or limited risk to oneself.

"Negligence," by definition means something wrong, or a "failure to act." "Carelessness" is failure to be careful. It is like neglecting one's duty. The "omission" of an act is to neglect to do the act, or not to do an act. However, in some cases of death, the Supreme Court does recognize omission as killing if a death is thereby caused. One need not do what one ought to do. If one omits doing something because it helps not to do it, it is not wrong. A healthcare worker may conclude that no treatment will succeed better than invasive treatment. If one omits by mistake, however, it is wrong. Legal negligence can be limited in scope, and so allow actions which lead to harm to others. If one thinks one is an island unto oneself and has no duties to anyone else, one may see no reason to help another. But, if so, then s/he should not expect others to help her or him. This is the ultimate egoistic position. If, on the other hand, one sees that one is dependent on others for medical care, food, jobs, education, and in an infinite number of ways, like love, respect, exchange of ideas, one can recognize that we have more than mere legal duties to one another. We are, in a strong sense, interdependent world citizens. This means that we have good reasons to help people everywhere (See also Chapter 10).

Medicine is ideally a reciprocal relationship involving acts of humanism on the part of both patient and physician. The physician is obliged to treat disorders, but patients have obligations to live according to instructions. To what extent, if any, should a Samaritan or physician help people who do not help others or, in fact, harm others? Within the resources available treatment can be provided but also education as would be in preventative medicine. In short, some need help in the form of therapy or education in addition to medical treatment. The physician and medical profession have central roles in providing such education.

We are also letting die by having more children without first caring for the needy children in the world who are already here. Suppose a person had several children and could not feed them. Should they then have another? "We should not bring them [unwanted births] forward on the grounds of appeal to some abstract right to life that somehow lets us tell ourselves that we are good people who respect human life when, in fact, human life is neglected all around us" [92].

Although we need not do so according to the U.S. law why should not one have an elliptical obligation to help all people? Singer states, "We have an obligation to help those in absolute poverty, which is no less strong than our obligation to rescue a drowning child from a pond" [93]. In one sense, Singer is arguing for a universal utilitarianism: "Helping is. . .something everyone ought to do" [93]. His utilitarianism is, however, limited for he says that it should only be done "without sacrificing anything of comparable significance" [94]. On his view, for example, we need only give, for example, 10% of our income to help others [95]. Singer himself gives 20%. Feinberg also stresses that the obligation to help should involve little or no harm or inconvenience to oneself [96]. Again, it would seem that saving lives is so insignificant that it should involve hardly any inconvenience to the rescuer. It is like saying, "I would save you from death, but I am busy just now." The minimalist effort criterion reveals a low level of morality. This would produce the least good for the least number. It is like a token, rather than genuine, contribution by the rich to alleviate criticism, or (if any felt) guilt.

We usually let die indiscriminately without knowing whom we let die. People rationalize their letting die by imagining such people do not exist or that they have no responsibility for them. Is killing a specific person we know and sympathize with worse than killing those who are unknown? But the needy are not invisible, we just have to look.

Allowing to die is also hypocritical because people who are dying are often not within sight. We would let thousands of people in Ethiopia starve, but rarely someone in our own household. If one has an obligation to one's family, why would not one have a similar obligation to all others?

Schweitzer pointed out that the usual taking over of morals only creates compromise, but that you personally have to decide. "According to the responsibility, which I experience I must decide what I must give up in my life and what I may keep: my possessions, my rights, my happiness, my time, my peace" [97]. The question nevertheless is raised as to what the limits of obligation are. Should one go into irretrievable debt and destitution to pay the medical bills of a friend or relative? To say that consequentialism or altruism demand too much is to say that one does not understand ethics. If something demands *too much* it is, by definition, *unethical*. A humanistic, naturalistic theory of ethics uses reason to balance out what is enough, reasonable or too much. Again, an adequate ethical theory or altruism includes also the wants and needs of the individual giver or healthcare worker. Masochism is self-defeating. The above decision about what one must give up applies to all areas as a physician, a researcher, a teacher, a mother, a father, a (humane) human being. I have to make my very personal decision here in terms of my specific situation and sensibility, not merely according to vague and superficial, contextless, societal guidelines, but according to an individual lived-through (*erlebte, durchlebte*) ethical decision [98]. Ethics is, then, radically individualistic [99]. It must be determined by our own informed sensibilities. The self is now defined as its relationship with the world. We cannot retreat into an isolated, unrelated self as is done in egoism and nationalism. The unrelated self is a non-self [100]. The Greens Party thinks globally, acts locally, here and now you do what you can do in your particular situation.

Physicians, fire fighters, police and others are often required by professional ethics to take great risks. Also in non-professional capacities it may, in certain situations, be quite reasonable to take great risks or endure inconveniences to help others, e.g. by emergency rescues, or taking financial risks. However, an adequate and rational altruism should not completely exclude one's own interests and one's safety. Voluntary international emergency health workers often put themselves at the greatest risks.

Uncritical thinking (speaking), indolence and selfishness are several of the most central reasons for allowing others to die. Not helping also promotes the practice of selfishness. Indirect killing has a triple harm: One not only kills, one fails to help survive, and one makes life worse when one could rather have made it better. If we begin to not let die, the model and practice of helping more people may increase.

Death is not just a natural cause. It is unacceptable to give the excuse that death is just nature's way. We are all responsible for extending the healthy life of all people. It is certainly possible.

One does not let die if one has done all one reasonably can to prevent or treat the desperate and dying. One does let die if one has not. The public need not complain about a physician's limited knowledge, equipment, time or research limitations unless it has promoted medical research, cared for the dying person to prevent the death, promoted organ donation; did what they could in order not to cause their own health problems, give to medical research rather than to the Church, the rich, the military, etc. In an inescapable sense, one is the cause of one's own and one's friend's death. The rationalization commonly given is that it is too demanding to always be doing something useful or to help others.

The central way in which people kill people is because of their faulty thinking. It is because they in fact oppose genuinely critical philosophy, which is the clarification of concepts and methods in the various disciplines. They, for example, go to war not knowing the many arguments against war [101]. Their failure and strong opposition to the questioning of our enculturated beliefs in each culture causes people to be killed and allowed to die.

11.10 Albert Schweitzer on *Reverence for Life*

Albert Schweitzer offers some reasons why we should not let others die, and why we should act as beings intertwined with life [102]. "Respect" refers to: esteem, high regard, wonder, amazement, to consider with deference or dutiful support, avoid interference or intrusion on something or someone, or to relate to something. *Ehrfurcht* means reverence for, have awe of, have respect for. However, the term may be used metaphorically and is secularized to mean concern for or sensitivity in reference to life. This is respect for life. Respect for life is the ruling principle in Schweitzer's ethics and actions of a physician. But what is the meaning of life and reverence for it? It is not so clear as it might seem at first glance. *Ehrfurcht vor dem Leben* is usually translated as "reverence for life," especially because of

Schweitzer's theological background. If respect and awe were regarded as emotions, the ethics would be based on the emotion of awe. But there is more to it. The feeling of respect can mean both positive and negative: sensitivity for the vulnerability to life, and concern for life. We can be induced to preserve life because we are so vulnerable. We arrive at this view only as sensitive and thinking individuals.

For Schweitzer, reverence is a practice-oriented concept, the realization, "I am life, in the middle of life, which wants to live" [103].

It engages one, makes one yearn for the promotion and care of life – one's own as well as for that of other people, animals, plants, etc. Nevertheless, the statement is circular – life is not defined, but used in an abstract way in spite of Schweitzer's saying that abstractionism is the death of any ethics [104]. His concern is rather with sensitive, humanistic action. He says that we surrender to life, which means active engagement into life [105]. Life can be only understood in the concrete case, something, which lives, as somebody who lives, not an abstract principle. I do not live in the abstract. In this sense, I am not life nor do I live life. It is our concrete understanding and knowledge in the world and in activities, which move us to action. In a sense, we are the world and the world is us. "We live in the world and the world lives in us" [100]. His view is like the position of the pragmatists that we only live in interactive action and in doing our tasks, which we determine by our critical intelligence as such making sense to do. It is revealed by experience. Knowing is something which we do based on our experiences with the world and with others. "Live" is a verb, not a noun. We are what we do. The evaluation is formed in the process, not something taken from the outside. Similar to the view of pragmatism, what we are compelled to do is determined by our own critical assessment of the full concrete situation in terms of all we know and feel in the situation and in terms of future consequences.

We may take a closer look at the term "life" [102]. Life as noun is an abstraction out of many concrete lives and out of the experience of living, in different conditions and stages. The verb is more concrete, dynamic: I live, my cat lives, the roses in my garden live, but very different lives of course. So we have to ask of whose life we speak. The respect for life is a very basic principle, but needs to be filled in with concrete content, needs to get in touch with reality again. Linguistic constructions such as, "I am living my life," are circular. In one sense, what it means to live is to experience in language, to express experiences. We may ask who is living, how, where, under what conditions. Life and living are to do, to experience and act, to understand and create, to get involved and improve or destroy. "Life" as a term in bioethics and medicine is used differently. The term is taken from the Greek "bios", which meant life as in one's biography, a "whole life" concept, but today it is only used in terms of biological life. It is reduced to biology [102].

It is questionable if one could genuinely sort out different kinds of life: the biological, the biographical, etc. How could we do this? A more holistic concept of life would be one as is expressed in German, "*leibhaftig leben*" (to live embodied, personified here and now). Interestingly, Schweitzer never speaks about the value of life neither as subjective nor objective values [106]. There is a desire to live that goes beyond any evaluation of lives. It is a basic characteristic of humans. It opposes the threat and fear of destruction or impairment by pain and disease. This opposition shows how fiercely people long for life. We may, however, sacrifice our lives for an ideal. Life is what we think-feel. We ascribe it also to plants by analogy. This is surrender and action in regard to respect for life. Schweitzer has no contemplative view of life, but an active one, which engages oneself as a person who has to take responsibility for [107]. In another sense, it is the basis of ethics to bring about our informed wants and likes. Pragmatism has elaborated on such action and responsibility. This insight has radical personal consequences. For Schweitzer personally it meant to become totally engaged in a helping profession, e.g. by becoming a physician.

The tragedy in our lives in spite of aiming at the respect for life involves self-contradiction. Schweitzer left here his profession as a theologian. Living, we destroy lives – intentionally or unintentionally. We eat red meat or make fatal mistakes. We cannot deal with lives carelessly. Only if it is necessary, not avoidable, can some forms of life be destroyed. We may set priorities. Vegetarianism and pacifism are such sensitive and respectful priorities. We surrender to the next person in a personal and caring relationship [108]. Engagement for life is especially being sensitive to not letting die whenever we could do something to prevent it. The physician's Hippocratic oath becomes applied to everyone, rather than remain exclusively for physicians. The concept of life has to differentiate between "having a life" and merely "being alive," the personal versus just the biological dimension.

Biological life is only one aspect of life. Individual people, and not mere biological life, are those who have morally relevant human qualities [109]. What remains for us as complete humans is to bring out the best of our thinking, sensibilities and actions, to be the best that we can be, for ourselves as well as for others.

Mary Warren offers the following criticisms: [110]

1. A universal "will to live" (*Wille zum Leben*) is a supernatural view. Schweitzer personifies animals, plants, etc. as having such will to live.
2. Schweitzer states, "Ethics... [consists in] responsibility without limit toward all that lives" [111]. But this sometimes is impossible. Even Schweitzer did not: he killed microbes through antibiotics as a physician. Humanity is something one must earn. Only humans can make ethical judgments and base these judgments on reason.
3. Schweitzer's theory is guilt producing because some bacteria and cells must be killed in order for humans to live. He describes himself as a "mass murderer of...bacteria" [112]. We supposedly cannot escape guilt in our lives, involved in life [113]. It is, however, non-adjustive to encourage guilt (See Chapter 7 for full discussion of negative emotions).

In terms of the above analysis, however, Schweitzer is basically a humanistic pragmatist who is coerced to action and involvement to preserve life and health on the basis of his critical and rational understanding and emotions in his concrete situation.

11.11 Negative Emotions Kill and Let Die

Expressing negative emotions is a way of letting die. Negative emotions also take the place of much needed positive emotions, which make life worth living. One of the greatest threats to the quality of life as well as life is negative emotions (See also the Chapter 7). For example, anger is a notorious killer because it can cause heart attack, and violent behavior. Anger makes life not worth living for both the angry person and the one abused by the anger. In short, people kill or are killed by their anger, revenge, hatred, etc. Negative emotions also have a significant negative effect in the medical context. Without knowledge of emotion one cannot be an enlightened, successful physician or healthcare manager (See also Chapter 8).

11.12 Lack of Organs for Transplantation as a Form of Letting Die

In 2003, 85,000 people were on the U.S. national organ waiting lists. 6,500 died because no organ was available. In the U.S. roughly 17 die each day due to lack of organ donations. From 1996 to 2005 people on kidney waiting list increased from 20,000 to 30,000. In 2001 roughly 100,000 were on the waiting list for bone marrow transplants. Roughly, half of all the children on U.S. organ transplant lists die from not receiving a transplant. The U.S. has a strictly volunteer organ donation system. It is a system in crisis with little hope for improvement and which costs thousands of lives each year. The basic reason for this situation is that the people in the U.S. will not consent to presumed organ donation unlike over 22 countries in Europe. They would rather let these people die even if it means they themselves must also die because of lack of organs (See Chapter 14).

11.13 Suicide and Euthanasia

> There is but one truly serious philosophical problem and that is suicide. Judging whether life is or is not worth living amounts to answering the fundamental questions of philosophy.
> Albert Camus

Physician assisted suicide is generally not allowed, but was legal in Oregon since 1997 [114]. Physician (or otherwise) assisted suicide was not illegal in the Netherlands, Luxemburg, Uruguay. In 2001, suicide was the 11th cause of death in the U.S. and homicide was 13th. To a large extent, suicide is letting ourselves die. The cause of death is ourselves, one must be aware that one is committing suicide (cf. Chapter 16). Only a rational person can make a rational choice to commit suicide or not.

There are many forms of committing suicide:

The decision to join the military is similar to committing suicide.

There is emotional suicide. We cause suicide due to negative emotions, and by not learning about emotions.

There is slow suicide by our lifestyle and beliefs. For example, lung cancer is the most frequent type of cancer and is often due to smoking.

There are many forms of committing suicide:

The right not to use life support machines or heroic measures is a form of euthanasia.

Do not resuscitate order is a form of euthanasia.

We allow ourselves to be killed because of vague or false beliefs, e.g. freedom, religion, etc. We have no duty to a supernatural being to stay alive or to die. The Church is against suicide but supports martyrdom.

Refusal to contribute organs, and blocking stem cell research after death are not just forms of euthanasia, but forms of killing.

It is contradictory that one can legally refuse medical treatment, but not consent to euthanasia. If the physician must honor refusal of treatment then why not also be able to honor assistance in dying which is a form of refusal to treat? What is the difference? There is a difference in the law between acting and omitting to act. However, a pain-stopping drug may be given even if it hastens the death of the patient [115]. Ferguson argues that the distinction between letting-die and killing is minimal or irrelevant and that the law should allow the physician to administer requested euthanasia if the patient's condition and quality of life is sufficiently deteriorated as opposed to withholding nourishment [116].

The following models given for suicide treat the suicidal person as a victim.

1. Medical model: suicide is a disease or product of mental illness.
2. Cry for help model. Suicide in order to blame others, or as a manipulation strategy [117].
3. Sociogenic model. Social forces cause suicide [118].

An opposing model is that individuals, to a large extent, cause their own suicide by their assessments and emotional dysfunctions.

Kant says that suicide is self-contradictory: "To use the power of the free will for its own destruction is self-contradictory" [119]. It is freedom, abolishing freedom, life, abolishing life, oneself, abolishing oneself. It is contradictory for the lawgiver to eliminate the lawgiver. Kant thinks one has a duty to oneself, but duty here is open context, and, in fact, Kant proceeds to say there are higher duties, e.g. to avoid disgraceful conduct, one may sacrifice one's life. He allows suicide if one does not live ethically or honorably. His point seems to be that it is contradictory to use our freedom to destroy our freedom. We cannot use ethics to destroy ethics. To kill oneself (anti-life) from self love (pro-life) is self-contradictory. It is certainly true that one cannot get to war out of love.

Each year in the U.S. around 800,000 people attempt suicide and 32,000 succeed in committing suicide; worldwide one million a year, 16 deaths per 100,000 people (Source: WHO). In the U.S., four times more men than women commit suicide, but three times more women attempt it. It is the third leading cause of death in the age group 15–24, and eighth among ill people. The highest number is among the elderly. Once again, as with most disorders, suicide is thought to be preventable, e.g.

according to the American Association of Suicidology and most other sources [120]. Some of the reasons given for attempting suicide are: thoughts of hopelessness, having no reason for living, escape from problems, stress, and depression.

Suicide may be seen as an irrational killing of oneself. One "commits" suicide. If so, it should be prevented. Euthanasia, on the other hand, may be a rational choice to end one's life. It is a choice. Strictly speaking, both suicide and euthanasia may be rational or irrational. A consistent political policy would treat rational suicide and rational euthanasia in the same way. That they do not shows how semantics can mislead sound policy. The problem is not with suicide or euthanasia, but with whether or not they are based on reason. Most countries prohibit suicide, such as Ireland and Italy. Denmark, England, France, Germany, Norway, Sweden, do not ban suicide. But such countries do ban *assisted* suicide. Oregon, Switzerland, and the Netherlands allow assisted suicide. One would think that (irrational) suicide would be banned, but that rational assisted euthanasia would be allowed. Polls in the U.S. indicate that between 50 and 75%, depending on the year, are in favor of allowing physician assisted suicide or the physician to comply with the wishes of a dying patient (AARP and Harris Polls). In Dutch law, for assisted suicide, the suffering may be psychological, not physical, and no consideration of how long the person is expected to live need be considered, whereas Oregon has many restrictions including a less than 6 month expected survival period.

Hume argues that we must rather use reason to balance the interests of all involved in the specific context and use that as a basis for considering suicide [121]. Suicide is also allowed by utilitarians such as J Bentham and JS Mill. In some situations it may be rational for one individual to choose one's death to prevent disasters from happening to others. War, however, is still to be regarded as unacceptable. There is no duty as such to live or to die.

In addition to the alleged duty of a citizen to die for the state, certain professions carry a high risk, such as firefighting, police work, certain healthcare work, etc. Harding argues that there may be practical and consequentialistic reasons for not prolonging one's life (depending on one's condition, age, and illness) such as the following.

Harding, however, commits the fallacy of ageism in stating, "The duty to die becomes greater as you grow older" [122]. Most societies discriminate between citizens on the basis both of age and life expectancy. Elderly people or those whose life expectancy is short should not necessarily have commensurately reduced claims on their fellows for priority in health care where resources are scarce [123]. The *Journal of the American Medical Association* reported in a clinical trials study of approximately 6,500 women with early-stage breast cancer that older women receiving aggressive chemotherapy care showed the same survival rate as women who were much younger [124]. The point was that we would be mistaken if we routinely thought that those who are older necessarily are worse than younger patients. Harding seems to be using "duty" in "duty to die" in an open-context way. One has no duty as such, and therefore no duty to die as such. For that matter, one has no duty as such to stay alive. One can, however, conclude that one has no duty to live regardless of the circumstances, or at all costs. Harding concludes that one should be able to choose to die if one regards it as best [125].

There is a higher prevalence of psychiatric disorders among physicians than in the general population. We as physicians must care not only for our patients but also for ourselves. Physicians tend to neglect their own need for psychological, emotional, or medical help and are more critical than most people of both others and themselves. Stress and burnout may be added risk factors for all suicide rates among male physicians and female physicians in relation to the rates in the general population of the same sex.

11.14 Conclusion

1. It is found to be unacceptable to try to justify direct or indirect killing without having a sound knowledge of ethics. Consensus or the democratic majority rule principle fails. Matters of killing or letting die especially ought never be left to majority rule or popular vote. They have already been left to the majority with disastrous results.
2. Because people in general are anti-inquiry and will not consider the necessary requirements and arguments before killing or letting die, they are insufficiently qualified to make such judgments. We should try to save all desperate people in the world and be pacifistic until we have attained the ethical knowledge to arrive at any other less humanistic conclusion.
3. Life is not the same in terms of the consequences or value to the person or to others. As we are often required by our vote or medical resource limitations to determine (in)directly who dies, ethical models were presented by means of which we can do so.
4. The reasons typically given for killing are fallacious although killing is now regarded as a normal everyday foreign policy. If one reason is given to promote it, the door is open for any reason whatsoever. Pacifism is recommended to eliminate fabricating any reason or rationalization for killing. The medical profession has a pacifistic view against killing, but actual practice and letting die even undermines their idealistic claim to never kill. A sound ethics is needed to resolve the cases in which the risk to health and life is present.
5. Killing is a self-defeating position because if we justify killing others, others can justify killing us. It is similarly self-defeating to block medical research and then expect to benefit from such research or to thereby allow others to die.
6. One cannot decide the issue of justifying killing x number to save y number on the basis of numbers alone. The utilitarian theory is not a genuine ethical theory, but a mechanical formula which cannot be applied unless a more adequate ethical theory is employed to do so. A holistic naturalistic humanistic ethics was suggested for this. One cannot decide the issue of justifying killing, because one first needs knowledge about ethical justification. For most people the term "justification" may be used, but it is a cultural and an empty term for them without an ethical basis.

7. If killing even one person is the ultimate tragedy, the killing few to save more, becomes irrelevant. The problem is how to achieve this ideal of causing no one's death. The problem is more how to save lives without killing.
8. Often even educated people lack understanding and sensitivity regarding killing and death. One can see the need for adding philosophy as critical thinking (speaking), and education about ethics and emotion to the curriculum at every level of general and medical education. Critical philosophy is beginning to be regarded by some as necessary for an adequate medical education [126]. Henk ten Have, Editor in Chief of *Medicine, Health Care and Philosophy*, wrote, "One of the astounding innovations of medical education has been the introduction of philosophy and bioethics into the preclinical, clinical and post graduate curricula of physicians, nurses, dentists, and allied health professionals. In the Netherlands, all medical schools provide ethics and philosophy teaching, often supported by medical ethics chairs and departments. In Germany, professorships in medical ethics are now being recognized and established as necessary infrastructure for contemporary medical education" [127].

It is for reasons such as the above that it is stated, "Medical records should also briefly sketch the patient's life, plans, hopes, fears and ultimate wishes" [128]. Philosophical Counseling would also require such adequate information as a foundation for sound ethical decision-making regarding patients. In the determination regarding who receives medical care and how it is received, Philosophical Counselors may be used in place of Ethical Committees. The function of the latter is said to be educational, to evaluate hospital policy, clarify ethical issues regarding patients, and determine ethical options, especially in an advisory capacity. "Ethics committees and consultants can, however, also be quite counterproductive, actually producing more harm than good. This can occur, specifically, whenever such entities (a) see themselves constituted to enforce a particular religious point of view, (b) allow themselves to be co-opted by the institution in which they work, seeing themselves as but an arm of that institution" [129] (See Chapter 17). Ethics and bioethics committees can also be merely political committees. Preferable would be not to have scarce resources and have treatment available for all. Also, physicians have the responsibility to advocate for the expansion of insufficient resources. They must be allowed to have and exercise the leadership appropriate to physicians [130].

References

1. Dickinson, E. 1970. *Complete poems of Emily Dickinson*, 3 vols. ed. Johnson, T. London: Faber.
2. Glover, J. 1984. *What sort of people should there be?*, 181. New York: Penguin.
3. Glover, J. 1984. *What sort of people should there be?*, 183. New York: Penguin.
4. Shibles, W. 1987. Reformatory blame. *Journal of Rational-Emotive Therapy* 5:266–281.
5. Ziegler, J. orig. 2005/2007. Das Imperium der Schande. Der Kampf gegen Armut und Unterdrückung. Pantheon (orig. L'Empire de la honte, Fayard, Paris).

6. Glannon, W. 2005. *Biomedical ethics*, 143. New York: Oxford University Press.
7. Shenkin, H. 1991. *Medical ethics: Evolution, rights and the physician*, 428. Dordrecht: Kluwer.
8. Mercier, P. 2006. *Nachtzug nach Lissabon* (Night-train to Lisbon) 12th edn. Random House/BTB: 230.
9. Shibles, W. 1993. Majority rule fallacy. *Indian Philosophical Quarterly* 20:329–347.
10. McMahan, J. 2002. *The ethics of killing: Problems at the margins of life*, 243. Oxford: Oxford University Press.
11. Glannon, W. 2005. *Biomedical ethics*, 145. New York: Oxford University Press.
12. Winslow, G. 1982. *Triage and justice*, 106. Berkeley, CA: University of California Press.
13. Daniels in Beauchamp, T., and Walters, L. eds. 1999. *Contemporary issues in bioethics*, 5th edn, 410. Belmont, CA: Wadsworth.
14. Glannon, W. 2005. *Biomedical ethics*, 152. New York: Oxford University Press.
15. Hope, T., Savulescu, J., and Hendrick, J. 2003. *Medical ethics and law: The core curriculum (for medical students)*, 183. London: Churchill Livingstone/Elsevier.
16. Glannon, W. 2005. *Biomedical ethics*, 144. New York: Oxford University Press.
17. American Medical Association (AMA). 2001. *Principles of medical ethics*. Adopted by the AMAś House of Delegates June 17, 2001.
18. Bahro, M, Kaempf, C., and Strnad, J. 2001. Die Verteilungsgerechtigkeit medizinischer Leistungen. Ein Beitrag zur Rationierungsdebatte aus wirtschaftsethischer Sicht. *Ethik in der Medizin* 13:45–61; esp. 45.
19. Glannon, W. 2005. *Biomedical ethics*, 146. New York: Oxford University Press.
20. Glannon, W. 2005. *Biomedical ethics*, 159. New York: Oxford University Press.
21. Glannon, W. 2005. *Biomedical ethics*, 145, 162–163. New York: Oxford University Press.
22. Myers, K. 2001. Medical errors: Causes, cures, and capitalism. *Journal of Law and Health* 16:255–288.
23. Glannon, W. 2005. *Biomedical ethics*, 161. New York: Oxford University Press.
24. Bircher, J., and Wehkamp, K.H. 2006. *Das ungenutzte Potential der Medizin. Analyse von Gesundheit und Krankheit zu Beginn des 21.* Zürich: Jahrhunderts. rüffer & rub Sachbuchverlag.
25. cf. National Health Care for the Homeless. (nhchc.org).
26. Schwartz, L., Preece, P., and Hendry, R. 2002. *Medical ethics: A case-based approach*, 171. Edinburgh: Saunders.
27. Beauchamp, T., and Walters, L. eds. 1999. *Contemporary issues in bioethics*, 5th edn, 73. Belmont, CA: Wadsworth.
28. Ropeik, D., and Gray, G. 2002. *Risk houghton mifflin*, 28. Boston: Houghton Mifflin *Co*.
29. Ropeik, D., and Gray, G. 2002. *Risk houghton mifflin*, 44. Boston: Houghton Mifflin *Co*.
30. Austria has an EU funded screening for women aged 50–70 years of age these being those at increased risk and not so likely to be screened on their own initiative due to social disadvantages: Mammography-Screening.
31. Glannon, W. 2005. *Biomedical ethics*, 143. New York: Oxford University Press.
32. Herzinger, R., and Parsa-Parsi, R. 2004. Consumer-driven health care. Lessons from Switzerland. *JAMA* 292:1213–1220; esp.1220.
33. Stroble 1966. PhD diss. stroble@hawaii.edu.
34. World Medical Association Statement on *Weapons of Warfare* and their *Relation to Life and Health*. October 2006.
35. See "Requirements before Going to War" in Shibles, 2002 eBook on humor at http://facstaff.uww.edu/shiblesw/ or search: "Shibles humor."
36. Singer, P. 2004. *The president of god and evil*, 44. New York: Penguin.
37. Cf. Miles, S. 2004. *The Hippocratic Oath and the ethics of medicine*. New York: Oxford University Press; Declaration of Geneva and Helsinki 1964. update the oath.
38. *Medical Ethics Today* 2004:771.

39. Council on Ethics and Judicial Affairs of the American Medical Association. Code of medical ethics: Current opinions with annotations. *AMA* 2002–2003:292. www.ama-assn.org/ama/pub/category/2498.html.
40. *Medical Ethics Today* 2004:659.
41. Declaration of Geneva. 1948. Adopted by the General Assembly of World Medical Association at Geneva Switzerland, September 1948. www.cirp.org.
42. Miles, S. 2004. *The Hippocratic Oath and the ethics of medicine*, 61. New York: Oxford University Press.
43. Miles, S. 2004. *The Hippocratic Oath and the ethics of medicine*, 60. New York: Oxford University Press.
44. Miles, S. 2004. *The Hippocratic Oath and the ethics of medicine*, 62. New York: Oxford University Press.
45. Elliott, C. 1999. *Bioethics, culture and identity: A philosophical disease*, 22. New York: Routledge.
46. Cf. Miles, S. 2004. *The Hippocratic Oath and the ethics of medicine*, 67–74. New York: Oxford University Press.
47. Gross, M. 2004. Bioethics and armed conflict. *Hastings Center Report* 34:22–30 esp. 23.
48. Cf. Gross, M. 2004. Bioethics and armed conflict. *Hastings Center Report* 34:22–30 esp. 25.
49. Gross, M. 2004. Bioethics and armed conflict. *Hastings Center Report* 34:22–30 esp. 27.
50. Gross, M. 2004. Bioethics and armed conflict. *Hastings Center Report* 34:22–30 esp. 24.
51. Sidel, V. 2004. Medicine and war. In *Encyclopedia of bioethics*. ed. Post, S., 3rd edn.,5 vols, 2564–2565. New York: MacMillan.
52. Koch, T. 2006. Weaponising medicine: Tutti frattelli (all brothers) no more. *Journal of Medical Ethics* 32:249–252 esp. 249.
53. Koch, T. 2006. Weaponising medicine: Tutti frattelli (all brothers) no more. *Journal of Medical Ethics* 32:249–252 esp. 251.
54. Koch, T. 2006. Weaponising medicine: Tutti frattelli (all brothers) no more. *Journal of Medical Ethics* 32:249–252 esp. 250.
55. Kottow, M. 2006. Should medical ethics justify violence? *Journal of Medical Ethics* 32:464–467; esp. 464.
56. Kottow, M. 2006. Should medical ethics justify violence? *Journal of Medical Ethics* 32:464–467; esp. 466.
57. The American Association for the Advancement of Science "NIH Budget Slips." 2006:617.
58. Schweitzer, A. 1923/1996. *Kultur und ethik*, 348–349. München: Becks̍che Reihe; *Ethik geht nur so weit, als die Humanität, das heißt die Rücksicht auf die Existenz und auf das Glück des einzelnen Menschenwesens geht. Wo die Humanität aufhört, beginnt die Pseudoethik.*
59. Dirie, W. 1998. *Desert flower*, 18. London: Virago Press.
60. Thomson, J. 1990. *The realm of rights*, 176–202. Cambridge, MA: Harvard University Press.
61. Unger, P. 1996. *Living high and letting die: Our illusions of innocence*, Ch. 4. New York: Oxford University Press.
62. Ziegler, J. orig. 2005/2007. Das Imperium der Schande. *Der Kampf gegen Armut und Unterdrückung*. Pantheon (orig. L'Empire de la honte, 2005, Fayard, Paris)
63. Taurek J 1977. Should numbers count? *Philosophy and Public Affairs* 6:293–316.
64. McMahan, J. 2002. *The ethics of killing: Problems at the margins of life*, 189. Oxford: Oxford University Press.
65. Glover, J. 1977. *Causing death and saving lives*, 59. Pelican: Harmondsworth.
66. Glannon, W. 2005. *Biomedical ethics*, 145. New York: Oxford University Press.
67. Winslow, G. 1982. *Triage and justice*, 105. Berkeley, CA: University of California Press.
68. Schweitzer, A. 1923/1996. *Kultur und Ethik*, 313. München: Beck'sche Reihe; *Humanität besteht darin, dass nie ein Mensch einem Zweck geopfert wird. Die Ethik der ethischen Persönlichkeit will die Humanität wahren. Die von der Gesellschaft aufgestellte ist dazu unvermögend.*

69. For an analysis of desire see Shibles, W. 1994. Cognitive-Emotive Theory of desire. *Journal of Indian Council of Philosophical Research* 3:25–40.
70. cf. Singer, P. 1972. Is Act Utilitarianism self-defeating? *Philosophical Review* 81:94–104; Singer, P. 1993/orig. 1979. *Practical ethics*, 2nd edn. Cambridge: Cambridge University Press.
71. Beauchamp, T., and Childress, J. 1994. *Principles of biomedical ethics*, 4th edn., orig. 1979–2001, 225. Oxford: Oxford University Press.
72. Singer, P. 1994. *Rethinking life and death: The collapse of traditional values*, 196. New York: St. Martin`s.
73. Dylan, T. 1957. *The collected poems of Dylan Thomas*. Orig.1939. New York: New Directions.
74. Beauchamp, T., and Childress, J. 1994. *Principles of biomedical ethics*, 4th edn. orig.1979–2001, 219 ff. Oxford: Oxford University Press.
75. Dworkin's view, which is governed by a system of rights and interests. Dworkin, G. 1988. *The theory and practice of autonomy*. Cambridge: Cambridge University Press.
76. Harris, J. 1985. *The value of life: An introduction to medical ethics*, 47. London: Routledge/Kegan Paul.
77. Kuhse, H. 2006. A modern myth: that letting die is not the intentional causation of death. In *Bioethics: An anthology*. eds. Kuhse, H., and Singer, P., 2nd edn. (1st 1999), 315–328 Oxford: Blackwell.
78. Kuhse, H. 2006. A modern myth: that letting die is not the intentional causation of death. In *Bioethics: An anthology*. eds. Kuhse, H., and Singer, P., 2nd edn. (1st 1999), 325. Oxford: Blackwell.
79. Rachels, J. 2001. Killing and letting die. In *Encyclopedia of ethics*. ed. Lawrence, B., 2nd edn., vol. 2: 947–950 esp. 948. London: Routledge; Rachels, J. 1975. Active and passive euthanasia. *NEJM* 292:78–80.
80. Feinberg, J. 1984. *Harm to others. Failure to prevent harm*, 126–186; esp. 171. New York: Oxford University Press.
81. Gratton, C. 1990. Letting and making death happen, withholding and withdrawing life-support: morally irrelevant distinctions. *Journal of Medical Humanities* 11:75–79; esp.75.
82. Harris, J. 1985. *The value of life: An introduction to medical ethics*, 54. London: Routledge/Kegan Paul.
83. Feinberg, J. 1984. *Harm to others. Failure to prevent harm*, 126–186; esp. 127. New York: Oxford University Press.
84. *Jackson v. City of Joliet* 715 F. 2d 1200, 1202. 7th Cir. 1983 supremejustice.com/us/465/1049/case.html.open jurist.org/715/f2d/1200/jackson-v-city-of-joliet-ross-d.
85. Corpus Juris Securidum. 2000. "*Negligence*". Vol LXII, 387. St. Paul, MN: Thompson/West group
86. Ratcliffe, J. ed. 1981. *The good Samaritan and the law*. Gloucester, MA: Peter Smith, 276–277.
87. Alexander, L. 1996. : Affirmative duties and the limits of self sacrifice. *Philosophy and Law* 15:65–74; esp. 67–72.
88. Hope, T., Savulescu, J., and Hendrick, J. 2003. *Medical ethics and law: the core curriculum*, 43. London: Churchill Livingstone/Elsevier; cf. *Medical Ethics Today* 2004: 545 ff.
89. Feinberg, J. 1984. *Harm to others. Failure to prevent harm*, 126–186; esp. 127 ff. New York: Oxford University Press.
90. Feinberg, J. 1984. *Harm to others. Failure to prevent harm*, 126–186; esp. 128. New York: Oxford University Press.
91. Ratcliffe, J. ed. 1981. *The good Samaritan and the law*, 276–277. Gloucester, MA: Peter Smith.
92. Callahan, J. 1995. Ensuring a stillborn: the ethics of fetal lethal injection in late abortion. *Journal of Clinical Ethics* 6:254–263; esp. 261.

93. Singer, P. 1993/orig. 1979. *Practical ethics*, 2nd edn, 169. Cambridge: Cambridge University Press.
94. Singer, P. 1993/orig. 1979. *Practical ethics*, 2nd edn, 168. Cambridge: Cambridge University Press.
95. Singer, P. 1993/orig. 1979. *Practical ethics*, 2nd edn, 181. Cambridge: Cambridge University Press.
96. Feinberg, J. 1984. *Harm to others. Failure to prevent harm*, 126–186. New York: Oxford University Press.
97. Schweitzer, A. 1923/1996. *Kultur und Ethik*, 339, 342. München: Beck'sche Reihe; *Nach der Verantwortung, die ich in mir erlebe, muss ich entscheiden, was ich von meinem Leben, meinem Besitze, meiner Zeit, meiner Ruhe hingeben muss und was ich davon behalten darf.*
98. Schweitzer, A. 1923/1996. *Kultur und Ethik*, 339. München: Beck'sche Reihe.
99. Schweitzer, A. 1923/1996. *Kultur und Ethik*, 59. München: Beck'sche Reihe: *Das Ethische (aber) kommt nur im Einzelnen zustande.*
100. Schweitzer, A. 1966/2003. *Die Ehrfurcht vor dem Leben. Grundtexte aus fünf Jahrzehnten*, 8th edn, 34.. München: Bähr HW, Beck CH; *Wir leben in der Welt und die Welt lebt in uns.*
101. See Shibles, W. "Requirements Before Going to War" in humor eBook on www.drbarbaramaier.at LINK: Warren Shibles.
102. Maier, B. 2005. Unsere Empfindsamkeit für und unsere Sorge um das Leben von Menschen auf dem Hintergrund moderner Biotechnologien. In *Leben nach Maß" – zwischen Machbarkeit und Unantastbarkeit. Biotechnologie im Licht des Denkens von Albert Schweitzer*. ed. Schüz, G.. Beiträge zur Albert – Schweitzer – Forschung vol 10, , 115–142. Frankfurt am Main: Peter Lang Verlag. www.bundeskanzleramt.at/bioethik/Detail: http://www.bka.gv.at/Docs/2006/4/3/BMaier_Schüz.pdf.
103. cf. Schweitzer, A. 1966/2003. *Die Ehrfurcht vor dem Leben. Grundtexte aus fünf Jahrzehnten*.ed. Bähr, H.W., 8th edn, 111. München: Beck CH: *Ich bin Leben inmitten von Leben, das leben will.*
104. Schweitzer, A. 1923/1996. *Kultur und Ethik*, 325. München: Beck'sche Reihe.
105. Schweitzer, A. 1923/1996. *Kultur und Ethik*, 332. München: Beck'sche Reihe.
106. Harris, J. 1985. *The value of life: An introduction to medical ethics*. London: Routledge/Kegan Paul; Nida-Rümelin, J. 1996. Wert des Lebens. In *Nida-Rümelin ed Angewandte ethik. Die Bereichsethiken und ihre theoretische Fundierung. Ein Handbuch*, 382 ff. Stuttgart: Kröner Verlag
107. Schweitzer, A. 1923/1996. *Kultur und Ethik*, 328. München: Beck'sche Reihe: Das *„Grundprinzip des Ethischen. . .ist die Hingebung an Leben aus Ehrfurcht vor dem Leben.*"; 315: Schweitzer's ethics is *„tätige Selbstvervollkommnung von Mensch zu Mensch"*.
108. Schweitzer, A. 1923/1996. *Kultur und Ethik*, 307. München: Beck'sche Reihe.
109. Kuhse, H. 1994. The sanctity of life doctrine. In *Individuals, Humans, and Persons: questions of life and death*. eds. Kuhse, H., and Singer P., 59. Salzburg: Academia Verlag Sankt Augustin.
110. Warren, M. 1997. *Moral status: Obligations to persons and other living things*, 34 ff. New York: Oxford University Press.
111. Schweitzer, A. 1933. *Out of my life and time*, 157, 233. New York: Holt, Rinehart & Winston.
112. Warren, M. 1997. *Moral status: Obligations to persons and other living things*, 37. New York: Oxford University Press.
113. Schweitzer, A. 1933. *Out of my life and time*, 255. New York: Holt, Rinehart & Winston.
114. Post, S. 2004. *Encyclopedia of bioethics*, 3rd edn, 5 vols, 2475. New York: Macmillan.
115. Ferguson, P. 1997. Causing death or allowing to die? Developments in the law. *Journal of Medical Ethics* 23:368–372; esp.369.
116. Ferguson, P. 1997. Causing death or allowing to die? Developments in the law. *Journal of Medical Ethics* 23:368–372; esp.371.
117. Schneidman, E. 1979. *Preventing suicide. Psychology of suicide*. New York: Science House.

118. Durkheim in: Post, S. 2004. *Encyclopedia of bioethics*, 3rd edn, 5 vols, 2476. New York: Macmillan.
119. Benatar, D. ed. 2004. *Life, death, and meaning: Key philosophical readings on the big questions*, 298. Lanham, MD: Rowman and Littlefield.
120. Humphry, D. 2002. *Final exit*, 3rd edn. New York: Delta Trade; Humphry, D. 2005. *The good euthanasia book*. Junction City, OR: Norris Lane Press; and suicide.org. Accessed 2007.
121. Hume, D. 1965. On suicide. In *Of the standard of taste and other essays*.ed. Lenz, J., 151–160. Indianapolis, IN: Bobbs-Merrill.
122. Harding, J. 2000. *Is there a duty to die? And other essays in bioethics*, 129. New York: Routledge.
123. Harris, J. 2005. The age-indifference principle and equality. *Cambridge Quaterly of Healthcare Ethics* 14:93–99.
124. *AARP Journal* Nov/Dec. 2005:12.
125. Harding, J. 2000. *Is there a duty to die? And other essays in bioethics*, 133. New York: Routledge.
126. Cf. Pieringer, W., and Ebner, F. 2000. *Zur Philosophie der Medizin*. Wien: Springer.
127. Ten Have, H. 1999. Medical philosophy and the cultivation of humanity. *Medicine, Health Care and Philosophy* 2:1–2
128. Nussbaum, M., Sen, A. eds.1993. *Quality of life*, 64. Oxford: Clarendon Press.
129. Nussbaum, M., and Sen, A. eds.1993. *Quality of life*, 51–52. Oxford: Clarendon Press.
130. Medical Ethics Manual. 2005. *World medical association*, 73. France: Ferney-Voltaire Cedex.

Chapter 12
A Critique of Autonomy and Patient Responsibility

If a desire, or choice, is not based on a rational evaluation then it is not autonomous. [2]

Abstract Autonomy for patients is respect for the rational decision-making, without coercion, of a patient regarding his or her medical care. Autonomy involves the view that the patient has the capacity: 1. to create ideas and goals for life, 2. for moral insight, self-legislation, privacy, 3. for political involvement and personal responsibility. Self-determination, care and responsibility go together, their separation weakens each of them. Unquestioned autonomy is challenged by criticisms of the principle of autonomy: Autonomy can be a way to avoid decision-making, it can be a way for physicians to transfer responsibility such as leaving patients to their autonomy as a form of defensive medicine. From autonomy follows responsibility. Challenging a severely limited autonomy concept we need rational, informed, critical decision making and a consequentialistic, naturalistic theory of ethics.

Keywords Autonomy · criticisms of the principle of autonomy · paternalism · consent · rational choice · self-determination · responsibility · transfer of responsibility · patient code of ethics · patient duties

12.1 Introduction

Autonomy is one of the four principles often given in the literature on bioethics [1–3]. Arras wrote, "Bioethics as a field has tended to enshrine the value of autonomy, it places individual rights above communal well-being" [4]. Daniel Callahan, on retiring as Director of Hastings Center confessed, "Nothing has exasperated me so much as the deference given in bioethics to the principle of autonomy" [5]. Autonomy is often seen as autocracy, which is the centrality of the patient's interests to the exclusion of all else [6]. "The capacity for and exercise of self-determination can be. . .the fundamental ideal of the person [patient] within medical ethics" [7]. According to the principle of autonomy the patient has the ability and the right to determine his or her own treatment. Some say the ability to be autonomous is the characteristic of

B. Maier, W.A. Shibles, *The Philosophy and Practice of Medicine and Bioethics*, International Library of Ethics, Law, and the New Medicine 47,
DOI 10.1007/978-90-481-8867-3_12, © Springer Science+Business Media B.V. 2011

being a person, i.e., of having the rational and psychological abilities, behavior and knowledge of one's own wishes, values and social abilities [8].

What we may call the "word-field" of autonomy can include many things such as the following:

1. The freedom to choose,
2. A total or partial authority,
3. A selfish decision-making,
4. An uninformed to totally wrong decision,
5. An absolutistic, non-holistic, non-consequentialistic decision,
6. A preference rather than an ethical decision because one has not studied ethics,
7. A decision, which disregards all interests of others.
8. Autonomy could be moral, unmoral, immoral, ethical, or unethical.
9. Autonomy could be a form of appeal to authority, even one's own questionable authority. One can suggest an "appeal to autonomy fallacy" just as we have an "appeal to authority fallacy." If one appeals to a physician as an expert, but evaluates the advice given, it is not appeal to authority. To appeal unquestioningly to someone having absolute knowledge, is an appeal to authority. One could appeal to patient autonomy.
10. Autonomy can include personal liberty,
11. Can allow the individual not to be governed by societal or religious dogmas or dictates.
12. Autonomy can mean to exercise individual rights,
13. To consent to suggestions (e.g., therapy, examinations),
14. To reject paternalistic/maternalistic decision-making.
15. Autonomy might also aim at independent decision-making.

Autonomy for patients is respect for the rational decision-making, without coercion, of a patient regarding his or her medical care [This would be the opposite of absolutist positions]. Autonomy involves the view that the patient has the capacity:

1. To create ideas and goals for life,
2. For moral insight, self-legislation, privacy,
3. For political involvement and personal responsibility.

12.2 Criticisms of the Principle of Autonomy

1. The problem of anti-inquiry. People are usually against critical thinking (speaking) as a challenge of what they are enculturated into and with what they feel comfortable and are deliberately anti-inquiry. As argued earlier, most have little interest or education in inquiry, ethical theory and practice, emotions, philosophy, etc. One may be said to have autonomy to the extent that one is qualified to have it. The informed, not the uninformed, patient is necessary for successful medical treatment. "The comprehension by patients of medical information is not outstanding" [9].

We must consider a person's disability to judge (*Unvermögen zu urteilen*) just as we consider the patient's medical disability. Anti-inquiry is anti-medicine. The so-called "autonomy of the patient" means respect for the patient as a rational agent acting freely and not under constraint. Without a background in critical thinking (speaking) and ethics, full autonomy of patients may be put in question.

2. Autonomy can be a way to avoid decision-making, or to give up on decision-making. It is a way for physicians to transfer responsibility [10]. Leaving patients to their autonomy can also be merely a form of defensive medicine on the part of the physician [11].

3. The execution of irrational autonomy can subject the patient and society to harm. Thus, medical drugs and food are regulated by the government to protect people from their own bad choices. One can consent to one's own demise.

4. The patient is often not interested in information and makes decisions without it and irrationally. "*Independent people may be self-centered, selfish, lacking in fellow-feeling or solidarity with others*" [12]. O'Neill says that autonomy can undermine trust, mutual obligation and recommends "principled autonomy," along Kantian lines, in place of "individual autonomy." It may be merely an adversarial claim. It may lead to failure to treat the mentally ill as they have autonomy to refuse treatment. O'Neill argues on a Kantian position that autonomy should be based on principles, which could apply to us all. This is almost what the word "principle" means. But how or whether this principle is one of reason is not clarified. This helps little if one would will that others also act as selfishly as most do. But what it does is to point out some major problems with mere autonomy and suggests that additional rational and ethical considerations must be involved in decision-making. Kilner, similar to O'Neill, holds the view that autonomy is often problematic because of patient's lack of information, lack of understanding, coercion, mental capacity, selfishness, and irresponsibility [13].

In a democratic society all may have an equal vote, but many vote irrationally. Allowing the uncritical and uninformed to have authority and autonomy, where they cannot have it in terms of their qualifications, is actually not to respect them, not to care about them because of disrespect for their welfare. Autonomy undermines the utilitarian view, and can undermine the benefit of other autonomous people as well as society. 14% of adults in the U.S have a below basic reading level (National Assessment of Adult Literacy). In some nations most of the people are illiterate. What value would autonomy have in such a context? Voting also requires an educated populace. A democracy based on an indoctrinated, uncritical and ethically and emotionally uneducated populace cannot yield a humanistic or enlightened government.

According to the Royal College of Obstetricians and Gynecologists the following guidelines should be followed by physicians: "Obstetricians must respect the woman's legal liability to ignore or reject professional advice, even to her own detriment or that of her fetus." "A competent woman, who has the capacity to decide, [may do so] for religious reasons, . . .irrational reasons, or for no reason at all. . .even though the consequence may be death or the serious handicap of the child she bears, or her own death" [14]. In the U.K., and the U.S., a woman may reject treatment for any, rational or irrational, reason regardless of the consequences, e.g., the death of herself or her child [14]. Patients often make irrational decisions. One needs to deal here as well with emotional and dialogic patient-physician interaction [15]. "If a desire, or choice, is not based on a rational evaluation then it is not autonomous" [16].

Case Example – Autonomy: A pregnant woman in the 35th week of gestation was admitted to the obstetrical ward. She was in much pain especially in the right lower abdomen and so sought help. She was to stay as an in-patient. This was bad news for her and her husband as they had planned the delivery at home when the time was due. Her husband was very nervous and domineering. The routine examinations were performed: vaginal, blood, ultrasound, and cardiotocogram (monitoring of fetal heart beat before birth). An additional ultrasound exam of the kidneys of the pregnant woman was given as well as a urine analysis check for infection. The cause for the pain was not yet determined, but the cardiogram for the fetus indicated that a cesarean section was urgently and immediately necessary to save the life of the child. A conflict arose when the couple would not give consent. In her vulnerable and fearful situation a pregnant woman can be suggestible and find decision-making difficult. This was the case here, as her husband who was fixed on having a natural childbirth at home and also mistrusted the medical system convinced her to refuse the life-saving cesarean. Time to save the fetus had almost run out. The cardiogram results became even more critical. I, her obstetrician, could no longer wait for the fetus to die and at the very last moment convinced her to have the cesarean. As we prepared for the operation the husband aggressively attempted to stop us before the operating theatre by scolding me, and the nurses present. The operation was nevertheless successfully performed. Later the neonatologist on duty informed us that we had saved the child in the very last moment. He also specifically and emphatically stated that to the father. The child was admitted to the neonatal intensive care unit for further treatment, stayed there for about 10 days, and was released in a completely healthy state. I was extremely pleased to have saved the child from their parents' judgments and their negative emotions. I prepared myself for a discussion about autonomy with the parents for the following day. I also recommended that they seek psychological support because of their negative attitudes, which endangered the child, and because of their partnership difficulties, which became apparent. It had taken much effort on my part to convince the mother-to-be about the urgently needed cesarean section against the incompetent counteractions of her husband. Later the couple returned to thank me for not having given up on them and for having the courage to oppose their autonomous choices thereby saving the child. They had both realized now that they were about to make wrong and fatal decisions.

They deliberately chose me as their obstetrician for their next pregnancy. The second birth also required a cesarean, and a healthy girl was born, but not, here again, without protest. The husband again fell back on prejudices against physicians and the medical system. The choice of me as the obstetrician by the woman seemed due to her idea, that she needed protection from herself and fatal decision-making. She even said so. She in effect held, "You are the expert and really care for us, so it is proper that you overrode our fatal demands." With every further appointment with me they also brought the boy who had been narrowly saved and who was now quite intelligent and adventurous.

"Autonomy" is an open-context term, which is situation dependent. There are times when the physician must have the authority and courage to aggressively enter in. It may be noted that it is sometimes necessary for the physician to educate the patients or refer them to therapy or for education. "Competent patients have the right to refuse treatment, even when the refusal will result in disability or death" [17]. Thus, physicians encourage patients to make bad decisions. On the other hand, they also give the qualification that autonomy must be reasoned autonomy [18]. It is important therefore for self-protection for the physician to have patients sign consent forms and attest that they understood the information given them. As a physician one has, also as a part of preventative medicine, to help the patient develop his or her capacity for autonomy [19]. Exercising one's autonomy may mean a bad decision for all involved [20].

5. Suppose a patient refuses treatment or tells the physician that he or she does not wish to know what the ailment or appropriate treatment is.

6. Autonomy has no a priori standing, it is an atomistic myth that one is self-contained, self-sufficient, self-determined instead of living cooperatively [21].

7. The court conspires in supporting bad medical treatment (See also causation in the law, and "legal causality" in the Chapter 3). The court protects a so-called intrinsic value of autonomy even if the consequences and the actual interests of the individual are worsened or neglected. Court interests are in that sense against individual interests. The court stresses a fictional ideal autonomy.

Case Example – Autonomy and the Law: As a physician, one always has to wonder if one could have done something differently or better. Only such an attitude makes one into a physician critical about oneself and his/her actions and offers the chance to life-long learning. If a complication develops the physician tends to blame him/herself even if the complication could not have been anticipated. This is an unfair evaluation after the fact. In lawsuits it seems to be presupposed that there is a perfect principle of autonomy according to which there is a rational, fully informed patient who is able to select perfectly the best and least risky among the different options of therapy. The claim and realization of autonomy of a woman in the pain of labor is problematic. She is potentially misinformed, anxious, vulnerable, and suggestible. It would be difficult for her to make a genuinely autonomous decision. Even the law in specific cases accepts birth giving as an extraordinary and extreme situation. The patient's information in his/her particular situation is always weak, and incomplete and informed consent often only fictionally given. Every anxiety driven decision cannot be considered autonomous or complete. In our experience, the patient at best accepts the physician's advice, or rejects it, without having a critical assessment of the options. The alternative is the inhumane challenge of explaining all the possible risks in detail. At Salzburg's Women's hospital a free information course about birth giving is available once a week. The information about that course is made widely available for all pregnant women planning to deliver their babies at the hospital. They receive the written invitation at their appointments to check before delivery. Almost no one attends the session. This indicates that they do not bother to obtain adequate information upon which to make autonomous decisions. They also seem not to wish to have such important information before giving birth. One would think that this would disqualify from making such decisions. But after difficulties arise people concerned can claim that they would have made a different decision if they only would have known about the difficulties now encountered and tend to blame the physician if anything goes wrong. Also, it may be noted that there is no risk-free option in obstetrics, in medicine, or elsewhere in our lives.

A physician has sometimes the dilemma of choosing between respecting a patient's unwise or even fatal autonomous choices and giving sound, effective and professional medical treatment. This is especially problematic when the patient lacks genuine autonomy of judgment. By definition, genuinely critical choices cannot be autonomous because they must involve inquiry, discussion, expertise, etc. It would not be fair then, if one had not fulfilled these requirements to complain about the decisions made by the physician afterwards or to claim that they did not receive such information, yet some do so. Physicians in such cases should not be blamed for complications during labor or birth. Should uninformed autonomy result in unfair complaint, the physician may be forced to offer fewer choices, for example, offer only cesarean sections and no more vaginal births, which is often the practice. The physician cannot give all possible consequences. If my car dealer does not give me all possible information about car accidents, I could even legally accuse the dealer of my driving on thin ice and falling through [22].

Insurance does not protect the physician if it is imagined that insufficient or incomplete information is given the patient. Lawyers even talk about such approaches as: We cannot get the physician for malpractice, but we can try to get him or her on grounds of giving incomplete information to the patient. Unfortunately anything may be construed as being incomplete. At what point is information complete? The patients even after signing a paper saying they have read and understood the information can in the court claim that they did not. The patients can claim they did not get the information they clearly got and disobey the instructions of the physician and still win a law suit. The legal system is conclusive, irrefutable, beyond question, yet not based on the actual facts of medical practice, and often in disregard of the medical facts. In Austria, the children of the physician may also inherit the lawsuit, consequences and costs. The result is that the physician is required to be extremely defensive also to the detriment of patients and the distancing of the physician from the patient. It blocks caring treatment.

8. Falsely conceived autonomy forces un-professionalism. It would often be ethically and medically unprofessional to accept without question the "autonomy" of the patient or spokesperson regarding the type and availability of medical care. The physician is usually bound by the patient's decision to operate or not. Must the physician have the possibility to give needed treatment even when a patient does not wish or allow it, just as we educate students even if they do not want it, for the benefit of the student and society? Psychiatric patients also often do not want much-needed treatment. "If a patient be under orders, he will not stray; left to himself, he will give up the struggle and depart this life. . .so [the physician must] take

the patient in hand" [23]. The patient is not on the same level in knowledge as the physician or nurse.

Autonomy of the expert can be contrasted with paternalism. "There are situations in which paternalistic behavior is ethically justified" [24]. Also, Loewy states, that unthinkingly and unfeelingly abandoning persons to their autonomy is the flip-side of paternalism. [25] Experts have knowledge others do not have and it is not paternalism to actively use such knowledge and guide the patient or those seeking such advice. It makes sense to: a. present the patient with all reasonable options, b. ensure that s/he understands the outcome, c. deal with negative emotions, e.g., fears. One cannot adequately communicate to the patient all of the reasonable options unless the patient can understand. There are serious problems with the principle of autonomy. Patient Centered Therapy is a form of autonomy. The principle that the therapist cannot help those who do not want to be helped is a rationalization. Not wanting to be helped is part of the disorder. The therapist must find ways around the resistance. But how?

9. Autonomy leading to extreme harmful outcomes of the patient must sometimes be severely limited although this is to be argued carefully in each case. Beauchamp and Walters wrote that rational autonomy must be based on informed choice and should sometimes be restricted on grounds of irrationality and harmful outcomes [26].

10. Unreliability of consent. Patients as well as their families are often not competent to decide about such matters. Such family members in Germany can only make treatment decisions if they have been previously authorized to do so by the family member or court [27].

Similarly, family members, who lack the needed ethical and critical ability background, especially if they are in addition to inherit, cannot be depended upon to make objective ethical and informed decisions about a patient. In these respects, it could be ethically and medically unprofessional to accept the "autonomy" of the patient's spokesperson, but not the one based on a physician's expertise. Family members may be selfish or have antagonistic relations with other family members. It cannot be assumed that because they are family that they support each other's best interest. Therefore, only the patients themselves should be the ones to authorize decision-making for them on the part of a family member or other person. The patient should explicitly in advance authorize decision-making authority of a family member or friend, otherwise friendship or surrogate has to be carefully established.

Some physicians exclude anti-social preferences, unlikely preferences, and restricted actual desires [28]. To give an example, often women desire and even insist on undergoing even the most extreme, invasive reproductive measures to have a child even if they have virtually no chance to have one [29]. They feel they cannot live unless they have a child, yet they do not consider adoption. Some feel they are not really women unless they have a child. They think that somehow the genes must be their own ones among the millions of possible fertilizing sperm. A license is needed to drive a car, but not to have a child. Griffin wrote of an "informed desire" approach [30]. But even informed consent and preferences may not suffice. If each has a right to determine what is to be done with their own body by the principle of autonomy, it would let the addict continue with the addiction, and let people die of easily curable diseases.

11. Autonomy involves unearned respect. It is an ethical fallacy to have respect-as-such for whatever is desired, asked for, demanded, and unquestioningly fulfilled not considering consequences. It is also a fallacy of the argument from authority, in which the patient is falsely regarded as the authority for his or her life or the quality of their lives unless they qualify as authorities. "Respect" is an open-context value term and must be given a meaning to be intelligible. When a meaning is given it will be seen that respect is not innate, but rather has to be earned. One has respect for someone typically because of what they have studied and learned to do, e.g., respect for a surgeon. Autonomy must be earned.

12. Autonomy may be only one self-serving factor and involve the disregard of the interests of others. This is like deciding for oneself without regard for other's interests. Such kind of "autonomous" decision can neither be called ethical nor autonomous in terms of the requirements for autonomy. Autonomy may involve selfishness, concern mainly with one's own desires, but altruism should not exclude one's own interests. The good of the patient, the quality of life of those involved, including the physician who is often overworked and bears extreme liability, in addition to the other factors should not be excluded.

The uninformed cannot be said to give "consent." It is often disallowed for minors, yet allowed for those even less rational and less informed adults. "Consent" presupposes that one can make informed choices. One form of autonomy is consent.

13. Autonomy is relativistic. There is relativism in the concept of each person having free judgment, or autonomy. Self-determination requires expertise and the help of others just as one may need medical advice from others. The assumption is also made in Rogerian therapy and Socratic oriented views of Philosophical Counseling, that the patient can solve his or her own problems and that the therapist need not actively guide and challenge the patient. In opposition to this view, the patients should not be regarded as experts in decision-making in medicine, however they should be encouraged to participate as much as possible in learning about all of the relevant factors.

Jopling opposes the anti-therapeutic, anti-realistic, relativistic approaches of Lahav, Achenbach and others in the area of philosophical counseling. He wrote, "The price of respect for the client's autonomy may be the flourishing of the self-deception and self-illusion" [31].

Case Example – Alternative Medicine: A 44-year-old woman came to the walk-in of the Women's Hospital accompanied by her family physician, a general practitioner of "alternative medicine." The patient told me as a physician that she had abdominal pain on and off, and she mentioned anemia, because of what she imagined was a supposed "self regeneration" of her body. The general practitioner wanted only the alleged anemia to be treated. I asked for the permission to do a full gynecological examination and to order reports from previous hospital stays. From the report as well as from bimanual palpation and ultrasound examination, I learned that she had ovarian cancer (Stage IIIc), was operated on, but refused the much-needed chemotherapy. She was convinced that "self regeneration" had cured her, and the "alternative medicine" physician had reassured her about that, whereas in fact the abdominal tumor she had was more than 12 centimeters in diameter. Thus, sound treatment or at least realistic chances for life were defeated by the patient's fatal, unwise decision not to have chemotherapy a year earlier. I asked also about the role that the general practitioner had played. She said that she as a physician herself would not undergo chemotherapy no matter what chances would be there for life, and that it is too toxic for one's body. Perhaps in this case irrational decision-making on the part of both the patient and the general practitioner were themselves toxic.

A criticism of this account is that autonomy can be a form of relativism according to which one can make an uninformed decision. Active guidance is needed in place of passive minimalism. On a higher level, Jopling states, "Philosophical counselors shoulder a significant burden of responsibility in helping their clients achieve an accurate, defensible, action-guiding and truth-oriented self-understanding" [31].

The more the patients know about their condition and participate in the treatment, the better. Some patients have even found minor and major medical information and cures not available to medical workers. Also, those patients who are experts in a relevant profession, for example, biologists, chemists, physicists, philosophical practitioners, etc., critical thinkers knowledgeable about ethics, etc. can in return offer guidance to the physician who may not be as well-informed in these areas. One surgeon learned about anti-scar salves from a patient who researched it on the internet and the physician revised his treatment procedure. The physician also depends on the patient to report the effectiveness of the treatment given, e.g., reactions to medications and drugs.

It is maintained that the physician or counselor and patient or client must be on an equal basis. If this were true, how could they claim to have expertise or anything to offer to the patient or client? An expert is not on an equal basis with the uninformed, not for information, not for the adequate treatment. Furthermore, the claim that they are equal is undermined by the claim that each person is unique. Our development is guided by competent others, we cannot be on top of everything and have to acknowledge that, but we can aim at finding out as much as we are capable of by the help of physicians, nurses, counselors.

14. Autonomy is neither an ethical theory, nor a theory of ethics. Unquestioned autonomy is not an ethically maintainable position.

15. Autonomy of the person or self requires a definition of these terms. A person for Harris is "a creature capable of valuing its own existence" [32]. On this view, also the patient would need to know about ethics in order to value. What definition of the self is to be given here?

16. Priority of scarce resources should not be based primarily on autonomy. If patients oppose the goals of medicine by their actions, requests (autonomy of patient), and belief systems, the physician may or must withdraw or withhold treatment [33].

17. Autonomy might be based on the supernatural. There is a problem with "ascertaining that the patient's reasons for the choice [is one] that can – in terms of the patient's and not the health professional's value system – be logically defended. Thus, patients are entitled to make decisions, which a majority of us might consider

foolish – refuse transfusions, for example, because their particular faith prohibits it" [34]. Failure of the patient to understand because of "staunchly though unreflectively held opinions may stand in the way of functional understanding... (and be) a form of superstition that may be cultural, religious or idiosyncratic" [35]. There is no genuine autonomy if one's views are dictated by religion. The "Church of Truth" rejects medical treatment of all illnesses in favor of spiritual healing. Their members refused an infant needing medical treatment. The courts tend to favor medical treatment over religious restraints for children, e.g., Jehovah's Witness prohibition of blood transfusion [36].

18. The principle of autonomy is not consistently used in areas other than medicine. Should it apply to all areas of one's life as well as medicine? (See Chapter 8)

19. Patient autonomy and consent blocks some lawsuits in that the decision is made by the patient rather than the healthcare worker, but promotes other lawsuits based on subjective authority. A patient's consent is needed even when taking the pulse. [37] Courts recognize implied consent so doctor need not ask for consent. But the patient may always sue. Coming to a hospital or doctors office does not imply consent to being touched. Even touching without consent may constitute battery.

20. Autonomy is important in the area of abortion and even late abortion or feticide as it allows the individual to make choices for their future quality of life. It is interesting that in such cases the autonomy of women is often especially put in question.

21. From autonomy follows the patient's future quality of life with or without the required responsibility for his/her health. If patients are to have autonomy it is required that they be capable of and fully responsible for their decisions. Responsibility for oneself and others must be based on a sound ethics. It is almost identical with ethics, for example, on a naturalistic ethics whereby to be responsible is based on doing so because it is rational to bring about one's holistic and informed wants and likes. An adequate naturalistic ethics leads to pragmatism and humanism. Also, Lenk and Maring have defined moral responsibility as a relationship between people [38]. Responsibility is the result of self-obligation, which corresponds to all people [39]. Self-obligation and social ascription are then overlapping [40].

"Self-determination, care and responsibility go together, their separation weakens each of them" [41]. The person who has the freedom for self-determination, has to take over the responsibility for such decisions. Autonomy is self-determination (*Selbstbestimmung*). Autonomy presupposes competence in decision making and acting. Autonomy without responsibility is half hearted, nothing, empty. It is valueless value.

22. Even with regard to a patient's autonomy, physicians must have a possible professional discretion and action (*Handlungsspielraum*) due to their high degree of responsibility and the possible courses of complications of medical treatment, and their foresight because of experience. Good and open communication between patient and health care worker must be maintained before, during and after the treatment. This must be a central ingredient to patient care and especially so for

patient autonomy to be meaningful. Another word relevant to this whole structure is trust [40].

23. Autonomy may also be seen as autocracy: absolute power, dictatorship, and unlimited power over one's own decisions including the impact on others. This may be as true of the patient as well as of the healthcare worker, but is selfish and not ethical.

24. Autonomy is often clouded by bias, negative emotions such as, anger, fear, envy, revenge, thought-feelings of inferiority or superiority, burn-out, pain, etc. Frequently female patients ask for a male physician to do the operation, without realizing they could thereby be choosing the least experienced surgeon assistant instead of the experienced senior female physician.

25. Psychiatric patient autonomy. Mitterauer, Maier, and Griebnitz argue that informed consent in the case of psychiatric patients is not to be determined merely because of the psychiatric diagnosis or classification alone, but preferably on a case by case basis for understanding of consequences of the medical intervention for the body, quality and happiness of life [42]. Note that the assessment is holistic involving the positive emotions and happiness of one's life. Such patients may have diminished capacity for: (a) autonomous evaluation (e.g., delusions) (b) understanding facts and causal procedures, (c) resolving a conflict because of their personal evaluation, e.g., due to a psychotic state [43]. Criteria for determining the ability to understand (become informed) involve the following factors: cognitive, emotional, communicative, and behavioral. The actual capability of consent must also be assessed [44]. The above criteria and diminished capacities regarding autonomy apply to the non-psychiatric patient as well, but in lesser degree. They may have prejudices and dogmatic views, which prevent them from understanding information and arguments or drawing sound conclusions. Another problem is that a patient may deceive a physician or therapist by giving (e.g., DSM IV or learning) symptoms of a certain disease and pretending to have them.

In sum, challenging a severely limited autonomy concept we need rational, informed, critical decision making and a consequentialistic, naturalistic theory of ethics and education about emotions.

12.3 Patient Responsibility and a Patient Code of Ethics

> Medical ethics is one-sided. It dwells on the ethical obligations of doctors to the exclusion of those of patients [45].
>
> There is no patient autonomy without patient responsibility.

The physician's Hippocratic oath may be applied to everyone, rather than remain the exclusive domain of the healthcare worker alone. Everyone is or might (and will) become a patient in his/her life.

The American Medical Association states, *As a member of this profession, a physician must recognize responsibility to patients first and foremost* [46].

All of the burden is put on the physician. One could also state, *As a benefiting recipient of medical care, a patient must recognize a responsibility and gratefulness to healthcarers first and foremost.*

According to David Resnik patients have an obligation to be responsible enough to follow physician's orders and physician-patient contracts should be considered [47]. Although patient's rights have been extensively defended, "Surprisingly, the bioethics literature also [as with AMA and AHA publications] has very little to say about the ethical basis of the patient's duties" [48].

Draper and Sorell argued for the development of patient's responsibilities in medical ethics. "Duties fall mainly on doctors and only exceptionally on patients. . .[and] exempt patients from obligations" [49]. Patients also have responsibilities to doctors. It is as if patients can do no wrong. "Autonomy must go hand and hand with taking responsibility for what is chosen" [50]. "Autonomy without responsibility is not autonomy." [51] If things go badly the physician may be blamed for the autonomous patient's decision. Patients are often negligent, contributing to or causing their bad health, often opposing medical research for theological or other reasons, not supporting funding for medical services, failing to obey physician's directions, failing to show up for appointments, not telling the truth to physicians, refusing to contribute organs after death, being not grateful for their care, being difficult or non-cooperative. "Competent patients have a responsibility to look after their own health" [52]. In England as well as America patients have the right to refuse treatment for irrational reasons or no reason at all [53]. Patients actual as well as future also have a civil obligation to promote health care for all as well as for themselves. If and to what extent they do so may well be questioned. Patients have an obligation not to misuse or overuse medical resources or expect others to pay for their poor lifestyle [54].

The National Health Service (NHS) in Great Britain is dropping patients from healthcare lists if they frequently do not show up for appointments [55]. Proposals are also made to charge patients who do not show up. "Delivering the NHS Plan," stresses patient responsibility so as to use health resources fairly and appropriately.

The National Patient Safety Foundation [56] suggests ways patients may help prevent medical error. NPSF suggests these steps: become a more informed health care consumer, seek information about illnesses or conditions that affect you, research options and possible treatment plans; choose a doctor, clinic, pharmacy, and hospital experienced in the type of care you require; ask questions of your doctor, nurse, pharmacist, or benefits plan coordinator; seek more than one opinion, write down your medical history including any medical conditions you have, illnesses, immunizations, allergies, hospitalizations, all medications you're taking, and any reactions or sensitivities you've experienced; write down the names and phone numbers of your doctors, clinics, and pharmacies for quick and easy reference. Work with your doctor and other health care professionals as a team, make sure you understand the care and treatment you'll be receiving. Take medications exactly as prescribed. Buetow and Elwyn hold that, "Patients appear to be morally responsible for the avoidable errors they make, contribute to or can influence" [57]. Patients are also responsible to conserve resources.

The American Medical Association's Code of Medical Ethics [58] regarding patient responsibilities states that it is the responsibility of patients to:

1. Be truthful and express their concerns clearly to their physicians.
2. Provide as complete a medical history as possible.
3. Request information or clarification when they do not fully understand their health status or treatment.
4. Cooperate with agreed-upon treatment plans and appointments.
5. Take personal responsibility, when they are able, to prevent the development of disease.
6. Consider participating in medical education by accepting care from medical students, residents, and others.

Thus, the patient is expected to maintain "health-enhancing behavior." They should help prevent disease and consider adverse effects on others [59].

12.4 Patients Duties

Each of the following is a responsibility and duty of the patient [60]:

Check Diagnosis. Critically evaluate the diagnosis and treatment of the healthcare worker.

Communication. Communicate fully and well with the healthcare workers.

Communication. Express your medical goals and desires clearly to the healthcare worker.

Dependability. Show up promptly for appointments.

Designate Representative. Designate which, if, any family member you wish to medically represent you if you are not able to represent yourself.

Ethics Committee. In problematic cases seek the advice of an ethics committee.

Harm. Do no harm to oneself or the healthcare worker.

Healthy Lifestyle. Live a healthy lifestyle. Patients should tell the healthcare worker their idea of what a healthful lifestyle is and whether or not they have been living one. Such a lifestyle may help, but cannot be expected to protect one from all illness and disease.

Honesty and Comprehensiveness. Honestly and completely inform the healthcare worker about one's condition to be treated.

Humanism and Altruism. Be humanistic and altruistic. Physicians are supposed to act for the society and humanity as a whole as well as for the individual [61]. Presumably, patients should do so as well. If the patient acts against humanity or society the physician is also defeated in fairly treating the patient.

Misuse of the Law. Do not use the legal system merely for financial gain in malpractice suits. Do not engage in unfair and unnecessary litigation.

Negative Emotions. Express no negative emotions against the healthcare worker, do not hold your illness as the physician's fault.

Nutritive Responsibility. Patients should be informed about nutrition and follow a healthful nutritional plan.

Obey Instructions. Follow the medical and treatment instructions given by the healthcare worker precisely and in a timely fashion.

Obtain Information. Obtain as much information as reasonably possible about one's own condition and treatment. Autonomy is not autonomy to refuse to obtain information.

Organ Donation. Sign to offer to contribute organs upon death.

Prevention. Preventative overall health measures should be taken at all times.

Question Decisions. Patient should question decisions at every step. If needed they may obtain an advocate [62].

Rationality. Make medical decisions on an ethical and rational basis. Autonomy is not autonomy to make bad decisions.

Reproductive Responsibility. Fertile women who have any chance of becoming pregnant should refrain from smoking, take folic acid, and live a healthy lifestyle. After conception they should immediately follow the obstetrician's plan of care. In certain countries if the doctor is negligent thereby producing a damaged child the doctor can be sued. The mother, however, may be completely negligent as to the damage of the fetus [63].

Required Information. Whatever the healthcare worker is required to offer, the patient must make an informed and rational decision about. For example, as doctors are by law required to offer prenatal diagnosis the patients should also avail themselves of prenatal diagnosis as well.

Respectfulness. Show respect and gratefulness to the caregivers. Try to especially reward them for their care.

Responsibility to Advance Medicine. Financially and politically support medical research and treatment institutions. Leave some of your will to such causes. It is also a duty not to undermine or block medical research and treatment, e.g., stem cell research, national and international healthcare systems, physicians without borders who volunteer to help people in the most dangerous areas of the world, etc.

Responsibility to be Informed. Attend relevant health information sessions offered.

Responsible Lifestyle. Avoid having others to pay or provide treatment for one's own risky or poor lifestyle.

Risks. Avoid risk taking which might endanger yourself or others. This includes risky sports and extra driving (See Chapter 16).

Second Opinion. Seek a second opinion if there is doubt about the treatment to be given.

Self-Care. Patients should actively participate in their own care (participatory treatment).

Sexual Responsibility. Use preventative measures to prevention of AIDS and other sexually transmitted diseases.

Unnecessary Treatment. Do not misuse or overuse medical resources by risk-taking, poor lifestyle, demanding unneeded care.
Will. Make a "living will" so that if in a coma, etc. the healthcare worker will know how to proceed.
World Healthcare Responsibility. Support medicine financially and by acting and voting to promote national and world healthcare systems.

12.5 Case Report: Patient and Legal Irresponsibility

A Salzburg gynecologist examined a pregnant woman and determined that further examination of her fetus was required. He referred the woman to the center of prenatal medicine in the hospital. The woman did not follow the instructions. After additional urging from the physician the woman went for the hospital examination only 2 months later. It was found then that the baby would be a child with Down-Syndrome, severely disabled, and too late now for an abortion. The woman, though she failed to follow the physician's instructions, sued the physician for the support of the child. The physician is thereby effectively made bankrupt for his lifetime though he has worked all of his life successfully. The lower courts ruled in favor of the physician, but the high court said the physician's referral was not sufficient in terms of parent information. The high court also powered to make judgment. The experts in Prenatal Care determined that the physician had made no error in judgment or treatment, but in fact had been especially perceptive to ask for the referral, which others might have missed. The legal case shows that patients need not be responsible and that they do not accept the consequences of their own actions. If things go wrong they have the power to sue physicians and ruin their careers and lives [64]. The *Ärztekammer* for Salzburg had to distribute special notices and display posters in July 2006 telling patients that they must follow physician's orders and be responsible enough to obtain and understand the information given them, otherwise ask for further explanation. This notice came as a result of a suit by a patient who did not follow the physician's instructions, but the high court on appeal in 2006 injudiciously supported the patient.

References

1. Parts of this chapter were published in Maier, B. 2006. *Autonomy: On decision-making in prenatal diagnosis*, In Parental Testing Individual Decision or Distributed Action. 179–189. München, Wien: Profil Verlag.
2. Hope, T., Savulescu, J., and Hendrick, J. 2003. *Medical ethics and the law: The core curriculum*, 34. London: Churchill Livingstone/Elsevier.
3. Maier, B. 2000. *Ethik in Gynäkologie und Geburtshilfe. Entscheidungen anhand klinischer Fallbeispiele*, 14–16. New York, Tokio: Springer, Berlin.
4. Arras, J. 2002. Pragmatism and bioethics: Been there, done that. *Social Philosophy and Public Policy* 19:29–58; esp.29.

5. O'Neill, O. 2002. *Autonomy and trust in bioethics*, 36. Cambridge: Cambridge University Press.
6. Loewy, R. 2000. *Integrity and personhood: Looking at patients from a bio/psych/social perspective*, 48. Dordrecht: Kluwer.
7. Nussbaum, M., and Sen, A. 1993. *Quality of life*, 127. Oxford: Clarendon Press.
8. Quante, M. 2002. *Personales Leben und menschlicher Tod: Personale Identität als Prinzip biomedizinischer Ethik*, 174. Frankfurt am Main: Suhrkamp.
9. Jonsen, A., Siegler, M., and Winsdale, W. 1982/1998. *Clinical ethics*, 4th edn, 56. New York: McGraw-Hill.
10. Maier, B. 1999. Ärztliche Aufklärung in ethischer Perspektive. In *Ärztliche Verantwortung und Aufklärung*. eds. Harrer, F., and Graf, A., 165–178; esp.176. Wien: Orac Verlag.
11. For a discussion of responsibility see Maier, B. 2000. Ethik in Gynäkologie und Geburtshilfe. *Entscheidungen anhand klinischer Fallbeispiele*, 14–19. Berlin, New York, Tokio: Springer.
12. O'Neill, O. 2002. *Autonomy and trust in bioethics*, 24. Cambridge: Cambridge University Press.
13. Kilner, J. 1994. Allocation of health-care resources: Microallocation. In *Encyclopedia of bioethics*. ed. Post, S., vol 2, 1107–1116; esp. 1081. New York: Macmillan.
14. Hope, T., Savulescu, J., and Hendrick, J. 2003. *Medical ethics and the law: The core curriculum*, 123. London: Churchill Livingstone/Elsevier.
15. Berghmans, R., and Widdershoven, G. 2003. Ethical perspectives on decision-making capacity and consent for treatment and research. *Medicine and Law* 22:391–400.
16. Hope, T., Savulescu, J., and Hendrick, J. 2003. *Medical ethics and the law: The core curriculum*, 34. London: Churchill Livingstone/Elsevier.
17. Medical Ethics Manual. 2005. *World medical association*, 44. France: Ferney-Voltaire Cedex.
18. Medical Ethics Manual. 2005. *World medical association*, 64. France: Ferney-Voltaire Cedex.
19. Maier, B. 1999. Ärztliche Aufklärung in ethischer Perspektive. In *Ärztliche Verantwortung und Aufklärung*. eds. Harrer, F., and Graf, A., 165–178; esp.172. Wien: Orac Verlag.
20. Glover, J. 1977. *Causing death and saving lives*, 77. Harmondsworth: Penguin.
21. Loewy, R. 2000. *Integrity and personhood: Looking at patients from a bio/psycho/social perspective*, 53. Dordrecht: Kluwer.
22. Hasenöhrl, G., Maier, B., Steiner, H., Stempfl, A., Georgoulopoulos, A., Zuchna, C., and Staudach, A. 2004. Sectio oder nicht Sectio – oder sollen wir uns das noch antun? *Speculum* 22:1–3.
23. Hippocrates. 1957. Jones, W. tr. *Precepts IX*, vols 2. Cambridge, MA: Harvard University Press.
24. Jonsen, A., Siegler, M., and Winsdale, W. orig. 1982/1998. *Clinical ethics*, 4th edn, 51. New York: McGraw-Hill.
25. Loewy, E. 1996. *Textbook of healthcare ethics*, 61. New York: Plenum.
26. Beauchamp, T., and Childress, J. 1999. *Contemporary issues in bioethics*, 5th edn, 19; 20. Belmont, CA: Wadsworth.
27. Reiter-Theil, S. 2001. The Freiburg approach to ethics consultation: Process, outcome and competencies. *Journal of Medical Ethics* 27, suppl I: i21–23; esp. i22.
28. Nussbaum, M., and Sen, A. 1993. *Quality of life*, 186 f. Oxford: Clarendon Press.
29. Maier, B. 2000. *Ethik in Gynäkologie und Geburtshilfe. Entscheidungen anhand klinischer Fallbeispiele*, 63. Berlin, New York, Tokio: Springer.
30. Griffin, J. 1986. *Well-being: Its meaning, measurement and moral importance*, 33. Oxford: Oxford University Press.
31. Jopling, D. 1996. Philosophical counselling, truth and self-interpretation. *Journal of Applied Philosophy* 13:297–309; esp. 304.
32. Harris, J. 1999. The concept of the person and the value of life. *Kennedy Institute of Ethics* 9:293–308.
33. Jonsen, A. 1998. *The birth of bioethics*, 104. New York: Oxford University Press.

34. Loewy, R. 2000. *Integrity and personhood: Looking at patients from a bio/psycho/social perspective*, 33. Dordrecht: Kluwer.
35. Loewy, R. 2000. *Integrity and personhood: Looking at patients from a bio/psycho/social perspective*, 36. Dordrecht: Kluwer.
36. Post, S. 2004. *Encyclopedia of bioethics*, 3rd edn, 5 vols, 1260. New York: Macmillan.
37. Hope, T., Savulescu, J., and Hendrick, J. 2003. *Medical ethics and the law: The core curriculum*, 67. London: Churchill Livingstone/Elsevier.
38. Lenk, H., and Maring, M. 1995. Wer soll Verantwortung tragen? Probleme der Verantwortung in komplexen Systemen. In *Verantwortung: Prinzip oder Problem*? ed. Bayertz, K., 286–366; esp. 347. Darmstadt: Wissenschaftliche Buchgesellschaft.
39. Lenk, H. 1998. *Konkrete Humanität. Vorlesungen über Verantwortung und Menschlichkeit*. Frankfurt am Main: Suhrkamp Taschenbuch Wissenschaft.
40. Maier, B. 2000. *Ethik in Gynäkologie und Geburtshilfe. Entscheidungen anhand klinischer Fallbeispiele*, 14–15. Berlin, New York, Tokio: Springer.
41. Maier, B. 2000. *Ethik in Gynäkologie und Geburtshilfe. Entscheidungen anhand klinischer Fallbeispiele*, 14. Berlin, New York, Tokio: Springer.
42. Mitterauer, B., Maier, B., and Griebnitz, E. 1999. Probleme der Aufklärbarkeit und Einwilligungsfähigkeit psychisch kranker Patienten bei klinischen Studien zur Prüfung von Psychopharmaka. In *Ärztliche Aufklärung und Verantwortung*. eds. Harrer, F., and Graf, A., 155–184; esp. 155, 159. Wien: Orac Verlag.
43. Mitterauer, B., Maier, B., and Griebnitz, E. 1999. Probleme der Aufklärbarkeit und Einwilligungsfähigkeit psychisch kranker Patienten bei klinischen Studien zur Prüfung von Psychopharmaka. In *Ärztliche Aufklärung und Verantwortung*. eds. Harrer, F., and Graf, A., 155–184; esp. 160. Wien: Orac Verlag.
44. Mitterauer, B., Maier, B., and Griebnitz, E. 1999. Probleme der Aufklärbarkeit und Einwilligungsfähigkeit psychisch kranker Patienten bei klinischen Studien zur Prüfung von Psychopharmaka. In *Ärztliche Aufklärung und Verantwortung*. eds. Harrer, F., and Graf, A., 155–184, esp.161. Wien: Orac Verlag.
45. Draper, H., and Sorell, T. 2003. Patient's responsibilities in medical ethics. *Bioethics* 16: 335–352.
46. American Medical Association. Principles of Medical Ethics. Adopted by the AMAś House of Delegates June 17, 2001 in its Preamble to the Principles of Medical Ethics.
47. Resnik, D. 2005. The patient's duty to adhere to prescribed treatment: An ethical analysis. *Journal of Medicine and Philosophy* 30:167–188; esp.179 ff.
48. Resnik, D. 2005. The patient's duty to adhere to prescribed treatment: An ethical analysis. *Journal of Medicine and Philosophy* 30:167–188; esp.172.
49. Draper, H., and Sorell, T. 2003. Patient's responsibilities in medical ethics. *Bioethics* 16: 335–352; esp.335.
50. Draper, H., and Sorell, T. 2003. Patient's responsibilities in medical ethics. *Bioethics* 16: 335–352; esp.338.
51. Draper, H., and Sorell, T. 2003. Patient's responsibilities in medical ethics. *Bioethics* 16:335–352; esp.340.
52. Draper, H., and Sorell, T. 2003. Patient's responsibilities in medical ethics. *Bioethics* 16: 335–352; esp.341.
53. Draper, H., and Sorell, T. 2003. Patient's responsibilities in medical ethics. *Bioethics* 16:335–352; esp.343.
54. Draper, H., and Sorell, T. 2003. Patient's responsibilities in medical ethics. *Bioethics* 16: 335–352; esp.344.
55. National Health Service. 2002. Engendering patient responsibility. *NHS* 2002:26.
56. www.npsf.org.
57. Buetow, S., and Elwyn, G. 2006. Are patients morally responsible for their errors? *Journal of Medical Ethics* 32:260–262; esp.260.

58. American Medical Association's Code of Medical Ethics 2009 Chicago: American Medical Association Opinion 10.02. Patient Responsibilities http://www.ama.assn.org/ama/pub/physician-resources/medical-ethics/code-medical-ethics/opinion1002.html
59. Medical Ethics Today 2004. 63.
60. For an account of how patients can be more responsible about their healthcare, see www.familydoctor.org/736.xml.
61. e.g., *German Code of Ethics.* www.bundesaerztekammer.de.
62. Schneider, S. 2005. *The patient from hell: How I worked with my doctors to get the best of modern medicine and how You can get too.* New York: Da Capo Press.
63. Hope, T., Savulescu, J., and Hendrick, J. 2003. *Medical ethics and the law: The core curriculum,* 123. London: Churchill Livingstone/Elsevier.
64. *Salzburger Nachrichten.* July 13th, 2006:6.

Chapter 13
Philosophy and Ethics of the Body

Abstract The medical system as part of culture is based on culture and morals, and so are the concepts of the body. Man has a body and is his/her body at the same time. Plessner called it the "excentricity of man," the double aspect of one's existence as body and as "*Leib*." The latter is being one's body in all dimensions conceivable, living to the full, stressing social relations, communication and exchange. Body language is expressing the body, implies language, is "*leibhaftig*", language embodied. It is narrative medicine of bodies. One can speak of the personally understood body in relationships, as one's own body, as the body of a beloved person, a person you care for (as a mother for her child, a physician for his patient, etc.), the body of every human person.

Keywords Body · person · paradigms of the body · self · selves · body language · personality · the un-philosophical body · the philosophical body · *Leib* · ethics of the body

13.1 Introduction

How should medicine, healthcare workers, and patients, deal with the philosophy of the body in medicine if philosophy itself hardly does? There seems to be little interest in finding out about body concepts used in medicine except the narrow medical one according to which body is what we treat. We then do not know what it is what is thereby treated. We know about outcome of treatment in terms of repair of a dysfunctional or defective machine. Philosophers concentrated on the cognitive forgetting about the body. They at best, like physicians, stressed the material body, but then did not analyze the body philosophically (See the analysis of self in this chapter). They thought of embodiment, but embodiment of what? We clearly in medicine do not deal with embodiments behind or within of which would be people or could be somehow found. Whenever we treat bodies, we treat people. So simple is the fact and so hidden by culture. Why? And what are the consequences of that? Our bodies are not separate of anything, which is us. Everything we experience, we experience bodily. "The mind is inherently embodied" [1]. Everything we experience, we experience in language. "Thought is mostly unconscious" [1]. Language is

B. Maier, W.A. Shibles, *The Philosophy and Practice of Medicine and Bioethics*, International Library of Ethics, Law, and the New Medicine 47,
DOI 10.1007/978-90-481-8867-3_13, © Springer Science+Business Media B.V. 2011

conscious. Thought is mentalism. What does it mean in and for our lives? "Abstract concepts are largely metaphorical" [1]. We never leave body or language. We never go beyond body or language. We only *understand* body in flesh and in words.

13.2 Definition of Philosophy and Body

This chapter focuses on the *clarification of the concept of the body*. We may define the term in accordance with the analysis of definition in the Chapter 2. One aspect of defining is to make definite, to de-fine, which is to reduce the abstract to the concrete. Arguments in medicine as elsewhere have little credibility if we do not question the very words they contain. In the philosophy of medicine, as with philosophy in general, it is useful to initially take the position that words are meaningless and statements false until defined and defended. Put in more basic language we could say that with an uncritical language we do not know what we are talking about.

The most basic terms in medicine as well as in ethics are usually undefined or defined in unacceptable ways, such as: body, disease, health, medicine, embryo, fetus, person, life, death, value of life, quality of life, meaning of life, (bio)ethics, etc. In general, philosophy may be defined as honest, open, critical inquiry. The American pragmatist philosopher, John Dewey, argues, that *philosophy is inherently criticism* [2]. It involves criticism of culture and humans in all ways. It may also be defined as criticism of the concepts and methods in the various disciplines, such as law, medicine, science, psychology, etc. Thus, we can only give a critically useful, contextual definition, a definition we can work with, never an absolute one. In conformity with contextualist and ordinary-language views e.g. of pragmatists and Wittgenstein [3], essentialistic definitions are not to be had. They are dangerous in terms of their rigid and inadequate application in the various specific situations and lead to inhumane consequences. We can virtually never find a definition of any term which will be true in the various senses of "true" which will apply to all contexts, times and places whatsoever. There is no absolute or fixed definition of even such concrete things as the body. "The imaging as well as the imagining of the body is never only descriptive, it is always a matter of perception and perspective, a seeing-as" [4]. The analysis of the nature of the body indicates that a person can be described in an infinite number of ways: as a body, an individual, a subject, a client, a mother, a patient, etc. The person is a physical, social, psychological, economic, philosophical, etc. "entity." Definitions may be rather regarded as perspectival seeing-as. As argued in the Chapter 2, to define non-circularly is to relate different things. To define is to take a model or metaphor. We will not, therefore, be able to conclude that the real definition of a word, e.g. body, is such and such. There is no literal definition. Each history and novel could have been written differently, and so each body as a text.

Upon examining definitions of the body, we find a large number of metaphors. Bodies are machines, or "Body is genes." We will not, therefore, be able to conclude that "body" has a fixed metaphorical definition. These metaphors may also be termed models. The terms may be seen as uses in a language game. The meanings are determined by the uses, according to ordinary language philosophy. Such metaphors and models do not describe reality. They construct reality. There is no reality as such. We therefore have to carefully examine the meaning of body in a certain language game, in a certain setting. To understand one's body therefore centers around our definitions.

13.3 The Scientific Method: Medicine as a Science

Medical science as other sciences is thought being based mainly on perception, observation and statistics (See Chapter 19). But this is naive empiricism. Science, including mathematics, is based on and presupposes language. If there is no language there is no science and no mathematics. Perception and observation are derivative and dependent notions. Without language there is no perception. Body and disease are linguistic concepts. Am I ill or am I told I am ill, or do I tell myself that I am ill? I am characterized and even stereotyped by the name of "my" illness. Does the name make me feel bad or the pain? Science and medical science especially, is based on language concepts and theories. Without language there is no theory. There is no objective scientific method as such, but rather there are ways of reaching a goal. As we choose concepts to solve certain problems as well as ways to their solution – it is more like ethics. It is also why ethics is a science. However, ethical science is contextual and critical, cultural science is normative and uncritical.

Our perception already is selective. We solve problems for certain purposes, look for methods to gain desired knowledge. Sometimes we gain what we have not been looking for. The language games between a physician and a patient about the patient's symptoms is a paradigm of a selective, contextual, purpose-oriented inquiry, often reduced to question-answer sheets. Body and disease are not final objective data, but subjective data, and contextual language constructs. Medicine becomes a narrative, not mere physical reality, but expressed physical reality, not just imaging technique, but conversation about the body. But in modern medicine imaging, visualization prevails over the narrative.

There are also non-scientific approaches to the body: cultural, religious, transcendental, superstitious, which come into play in medicine as well. In addition to well confirmed hypotheses and statistical correlations, there is clinical experience. This combines science with the experiential approach. It is in this sense that medicine is a science. Medicine aims at problem solving. The problems of a patient's body, in fact are a patient's problems. What are the goals of treating bodies in medicine? It is to restore or supplement decreased functions and destroyed body parts, by the use of intensive care to save the lives of people extremely traumatized, to prolong lives endangered by cancer through treatments, chemotherapy, radiotherapy, to substitute substances their bodies do no longer produce, such as insulin in diabetic people, serotonin in depressive people, to help infertile couples in having their own children, etc. The question is, does it add to the capacities of a person to fully live his or her bodily life under the conditions given? Does it enhance the dimensions of his or her existence?

Paradigms are models of explanation and thus guide actions, constructions, abstractions, reductions. There are various paradigms of the body in medicine, differing from each other often also contradicting each other. There is constant change of such paradigms. The paradigmatic change in gene technology perspective holds that our body is our genes. There is no body as such. It is neither a thing nor a fact. It is language. It is a model, a fiction. The same is true for substance, material, genes. It is in this sense that the body experiences "real" pain and suffering, which appear so devastatingly real (See also analysis of pain in the Chapter 7).

13.4 A Naturalistic Ethics of the Body

> Ethics must confront with three enemies:
> thoughtlessness, egotistical selfishness, and society [5].

As argued in the Chapters 5 and 6, philosophy is a critique of culture. Virtually every culture is based on absolutistic, traditional, supernatural and irrational thinking. A sound philosophy is an attempt to bring humanistic, practical, critical, rational reasoning into play in order both individually and holistically to create a society and institutions including medical practice towards humanistic goals. Medicine does not exist in a vacuum. It is part of society. If the society is immoral, medicine is immoral, unless it takes on the task of influencing and guiding society in a more ethical direction. We can speak of normative uncritical morals as moral contamination as pointed out by the distinction made between morals and ethics (See Chapter 5) Medicine must set policies, which genuinely promote health and prevent disease and change the culture to conform to this professional medical goal. Our view of ourselves, our bodies is basically indoctrinated. It will be seen how a philosophy of medicine, a criticism of cultural practices can improve our views of medicine and of the body. Fletcher says, "Whatever we are compelled to do [e.g. by culture, dogma, military, or religion] is amoral [unethical in terms we have distinguished between morals and ethics]" [6]. Albert Schweitzer thought, "The progress of ethics consists of our decision to think pessimistically of the morals of society" [7]. In some societies, people still consider it to be morally necessary to remove all or part of the genital organs. To take one example of how culture is harmful: 80 million to 114 million women were brutally circumcised including 80% of the girls in Alexandria (Discussed fully in the Chapter 6). So culture is a problem. People blindly accept tradition and culture and regard it as the standard of morality they have to live up to, not questioning what it means in terms of consequences and so reject anything which deviates from their particular belief systems. In terms of definition they take their cultural beliefs literally. They are held captive by a single picture of the world, not realizing they by being enculturated have created it. They accept such beliefs regardless of the harm to their bodies and those of other people. Even scientists are often captivated by their models in spite of the lack of evidence for them. They seldom have a background in the philosophy of science or ethics of science. Medical practitioners typically know little about the philosophy of medicine or ethics of medicine. Dewey and Tufts wrote, "Ignorance is the root of all evil" [8].

If people wish to keep their bodies they will have to become acquainted with them, learn how to properly nourish them and care for them as much as they can.

Schweitzer's thought is radically individualistic and critical of society. "The crash of culture was caused by leaving ethics with society [morals]. The recovery, the renewal of culture only would be possible if culture would become again the concern of the thinking individual person" [9]. Nietzsche states, "What is needed above all is an absolute skepticism toward all inherited concepts" [10]. Desires, habits and cultural tendencies must be recreated to include ethical consequences [11]. This is a task of the healthcare worker and of medicine, which cannot continue to be ignored. Medicine is not to be a mere servant of whatever culture happens to prevail. Medicine is practiced in a certain social environment. Should medicine continue to blindly comply with and encourage each religious or culture-based medical practice regardless of the naturalistic consequences? In an age of scarce resources, should it follow those who wish to oppose medicine itself in favor of dangerous cultural beliefs regardless of the negative consequences these beliefs might carry? How should we deal with non-ethical considerations? "Non-ethical" is used here in the sense that one simply does not know about the importance of questioning morals in their societies (See Chapter 5 for further clarification).

We ignore our bodies. We worship them. There is at the same time ignorance and cult of the body in one and the same society, in one and the same person. And no one even notices the inconsistencies.

Kossek and Block state, "Morality comes about as a result of the codification of traditional behaviors, conventional wisdom, particular familial or social orientations, and current public opinions. Morals are not subject to intense scrutiny, they do not require a sound philosophical foundation. ... Ethics on the other hand, demand a supportable philosophical foundation" [12].

The medical system as part of culture is based on culture and morals, and so are the concepts of the body. The treatment of the body follows unthinkingly and unfeelingly in that direction. So the body is enculturated as much as our perception of the body is. The popular, normative and common enculturated morality is played off as ethics, which it is not. People have familiar beliefs, which, regardless of how absurd, set their standard for what is right or wrong. They tenaciously hold on to their beliefs simply because they are used to them. This involves the fallacies of argument from familiarity, argument from tradition and argument from authority. Breakthroughs in research have to go beyond such traditions, trespass the borders of moralization towards ethics, trespass the tradition towards new finding (See the Chapter 15).

13.5 The Value of Life in Terms of the Body

By "life" in value of life we may mean mere bodily biological life or our experiential life, being alive, having a life, which is to have desires, goals and ethics. The term is used here to refer to the self, which includes both bodily feelings and our emotional and intellectual experiences. Life is not just body. One can have a happy life with or without a healthy body or a healthy body without or with a happy life. There is a significant difference between body and experience. Experience cannot be reduced to bodily functions. Life alone has no value. Cancer cells are alive. What is the meaning of life of a cancer cell? To *ex-ist* means for a person, to be also able to go beyond his or her body – though not without one's body. On a narrow or patchwork medical model, the person is regarded only as body, so medical treatment is only to repair the body. You are only your organs, bones and muscles. But the medical body is a meaningless body in any other regard except the medical. This also means that there is no value or meaning of life beyond the medical. The only value or meaning of life is that which we give to it. The question is not, "What is the value of life?" but "How can I create it?" For example, the more fully we ethically develop and use our abilities, the more meaning and value our life will have. If we do not give life meaning, it will not have any. It cannot come from the outside. Only the patient, not the healthcare worker, can give the patient a meaning of life. However, the healthcare worker can greatly contribute to the basis of the meaning of life and give hope. If we do not know about ethics, our lives cannot be ethical. If we do not know how to treat our bodies well, we will not have healthy bodies. It does not depend upon a disability, a gene problem, a disease – one can do one's best only on the basis one has. The value of life is determined by one's value system. This is true by definition.

We are subject to the lottery of nature (genes, body characteristics, shape, etc.) as well as to the social lottery, the family one is born into, the culture, time, location, etc. But we also can take a stand against whatever we were given. The body involves lifetime learning. Life is finding out about the meaning/value of life and how to create it. This involves the exploration of the language (games) regarding body. The body is also a poetic and literary creation.

The Catholic view in spite an ongoing tradition of the neglect of the body and harmful medical practices tries on the other hand to overcome that in telling believers that the human body is a gift of god, and therefore one should not treat a gift from the highest authority poorly. Logotherapy deals with finding meaning in one's life. Whatever bodies we have got, they are a value to ourselves, to others, to society, only if we value them.

Business has also invaded medicine. Body is reduced to monetary worth (See economic decision-making in Chapter 8). Those who can afford treatment in the U.S. and even with nations having national healthcare systems, receive it, the uninsured for the most part do not. In a sense, bodies are bought and sold in the Western world. Hospital administration and political economic considerations often determine the availability of healthcare. In war, bodies are viewed as costs per kill [13]. This is of course not treating others as human beings. We rather treat them as economic factors, war targets, sex objects, consumers, supernatural souls, but virtually never humanistically as humans.

Statistical as well as economic models refer to unidentified lives whereas a humanistic ethics always deals with identified lives, the lives and bodies of people. There never should be a gap, whether we are talking about the individual person, we know or any other person of the world we do not know.

13.6 The Mind

Mind is fully critiqued in the Chapters 7 and 18. The problem of the self is already fully discussed in the Chapter 10. Our view of the self influences what the body is thought to be and how it is treated. Unfortunately the commonly held view is quite problematic. Much of our ordinary thinking is conducted in internal monologue or silent soliloquy…in talking to oneself [14]. The therapist and healthcare worker can now deal with the self-talk rather than with a fictive mind and metaphysical thoughts. The physician, the psychiatrist need not treat a supernatural mind. However, how would they do that – by material drugs, injections, etc.?

The churches with various supernatural belief systems separated body from mind (spirit, *Geist*) and thus stressed the inferiority of the body to a constructed opponent, the mind. Neglecting the body meant a virtue, but was in fact harmful and dehumanising. Sexuality was thought to be sinful except for procreation, so even a glance at the sexual organs might be not accepted to god. The body is supposedly a gift of God and so to be taken care of, but the sexual organs are to be denied rather than honoured. One is not to be seen naked. This in former times as well as nowadays in conservative settings misled women not see a doctor, a gynecologist when they needed to, and so caused them delay in discovering diseases and eventually led to death. With such a belief one is not in control of one's body, one cannot be. The concept of sin labels all bodily needs negatively, needs which naturally require to be met and if they are not, lead into a distortion of "body and spirit."

13.7 The *Self* as a Language Construct

The meanings of the term "body" are determined by their various usages in language (See Chapter 18 and discussion of the self in the Chapter 10). Another way of viewing this is in phenomenological or experiential description. We approach and experience the body in everyday language, not in medicine's codes, especially not in medical Latin. In the current medical system the body has epistemological supremacy. The body in a medical body machine concept is the body without a person, without communication, self-talk or interpretation. The critique of psychosomatic medicine holds that a patient's body cannot be treated fully and well without the patient him- or herself. There is not one body there are many bodies as there are many language games, contexts and language users. Am I aware that and how my body is talking to me? My body is self-talk as well. This latter approach we have in psychosomatic medicine aiming at overcoming the body and mind separation concepts. Body is a word existing only as language allows. Body is information. In modern medicine the body is regarded as "our genes." People seem to think our genes determine who we are. Thus, our bodies are estranged from us. We are exposed to a genetic fate. We are led to think we cannot do much about it. So we fail to try to take care of our bodies. But in fact, we can live a healthy lifestyle and even thereby to a certain extent influence our genetic structure. Genes interact with other genes, which interact with surrounding influences and can be altered. How we speak about our body tells us how we think and feel. The physician must listen to the way in which a patient describes his or her body. Often, it seems, the language is so different from that of the physician that they would not appear to be describing the same body. We see our body as..., we feel our body as..., we deal with our body as..., we talk about our body as...[4]. As noted in the above types of definition, language is not only descriptive and concrete, but more often evaluative. Elliott, a physician and philosopher of medicine, presents a holistic, naturalistic, pragmatic view combined with ordinary language philosophy [15]. The self is a verbal picture we create for ourselves, a biography, a story of our life. It is not something we are born with. It is not something, which remains about the same throughout our life. It constantly changes. We are fictive constructs and realize ourselves through language constructions. As stated above, the perception of our bodies is seeing as, smelling as, feeling as... From body cult to body denial there is a wide range of the way in which people deal and feel about their bodies.

Knowledge of self is required for the understanding of what a person or human is, and so also of what harm, death and killing would mean to such self. The concepts of killing, harming, and letting-die can be no clearer than with what is meant by "self." And what is lost by losing oneself? This is a definition of death. The value of life and death depend upon which self we are talking about. Are we or should we be good to ourselves? As indicated above, people often regard their body as a separate, often foreign, self. But rather our bodies are ourselves. We are there, bodily selves. Medicine often treats all the person's selves as only the physical bodily self in order to create the narrow "medical model." For Leon Kass, a chair of President Bush's bioethics commission, the goal of medicine is biological health without concern for the quality of life or happiness defining goals such as happiness and gratifying...desires as false goals for medicine [16]. Medicine on this model does not treat patients, but bodies. A person is accordingly regarded as matter and function. Even this is not quite the case as there is no non-verbal body-as-such, only our verbal conceptions and constructions of body. Mere body, like mere sensation, is naive empiricism. Body is one self and, as indicated above, there are many possible body selves. Sontag put it metaphorically: we are citizens of two worlds, citizens of the world of health, citizens of the world of illness [17]. Insofar as the health-care worker uses language, the various linguistic selves of the patient are treated depending on the extent of the language used. We speak of "talking medicine," talking already is often medicine. Do we in so-called school medicine discover the patients' body talk? Or do we rather have fixed ideas of symptoms, concepts and therapies, often not regarding the individual, but the human body as something general? Generalisation, on the one hand, needed to learn about medicine and its implications can be harmful for the individual. On the other hand, recent research in proteogenomics aims at providing individualised drugs and medicine for individuals.

On the narrative theory, each self is seen as a narrative (See Chapters 9 and 17 for a full discussion of narrative). We are the bodies, which lived and live the stories of our lives [18]. Medicine is natural science, but storytelling as well. One may analogize that data are to scientific theory as a patient's experience is to a narrative.

Stories render emotions, values, goals in and for life. They can be harmful of helpful. We are said to need a unity of life, or rational plan. "Stories are like a metaphorical method involving compelling examples and possibly insightful juxtapositions" [19]. It is estimated that 80% of general diagnostic knowledge could be obtained through thorough anamnesis of the patient.

My "body" is the narrative of my life, my expression of my life's story. My body is my text, not just some texture. We may read the text, understand, misunderstand. It is a changing, asking, answering, opening, closing, expressing, hiding text. If I do not understand my "body," I do not understand myself. How do we find out about our "bodies," and how do we fail to do so?

Our bodily cells as well as our thoughts change rapidly. We tend to think of people as more similar than they are and fallaciously regard them as being equal. We cannot avoid asking the question as to which self has what worth, because our decisions depend on it. The special treatment of a specific patient is successful because of the specific approach. In medical decision-making this has been standard medical practice, only the criteria have always been a problem. This is especially necessary because the medical systems in the various countries are typically in crisis and do not come close to providing the needed services. This means that such prioritizing decisions are already being made, but without an adequate basis.

The problem of self in philosophy puts altruism and egoism into question (See Chapter 10). The self may become the other (altruism) in several ways. It may refer to any part or all of one's belief system. We may, for example, be altruists intellectually, but egoists behaviorally. One may concern oneself with one identity (e.g., one's medical profession) more than another identity (one's physical self). Thus, healthcare workers often neglect their own health and cause their own burnout.

The value of human life is one of the most important concepts to become clear about. In medicine, as elsewhere, decisions are primarily made without either a philosophy of medicine or a philosophy of life. A naturalistic, humanistic theory of ethics was proposed as a basis for clarifying the value of life. When we are clear about the value of life, physicians will be able to treat not merely the narrow physical body dimension of patients, but address themselves as well in various ways to the value of their lives. They could aim at *healing people, not just bodies.*

13.8 The Un-philosophical Body

Especially lacking in most of the cultures is a philosophical concept of the body, a holistic not reductionist, concrete not abstractionist, enhancing not harming concept. That is, body is usually subject to culture and not to rational comprehensive care. People are concerned with the experiences and pleasures a body produces so they hardly think of having a body – except when the body begins to fail them. We treat our bodies as if we would be independent of them, as objects just at our disposal and even as enemies, if they do not provide what we demand of them (e.g. babies, certain shapes, etc). And then if our body does not function and we become ill, we blame our bodies or genes or even the physician for our condition. Insensitivity to body is insensitivity to letting die, to killing... others as well as oneself. In any case, people know very little about what their bodies consist of and how they function. They do not genuinely care for or attend to the body's needs, but mainly use up their bodies, exploit them until they can no longer go on.

We are unconscious of our bodies for most of our lives. We hardly realize the body is there. It is like a stranger. In our society body awareness in a healthful sense is kept hidden. We think bodies can be seen. They can not. They presuppose our language. To see our bodies without language would be a kind of magic we do not possess. We cannot non-linguistically see ourselves. We speak bodies into existence. At some time we were perhaps told, "You have a body." You may have replied in surprise. "Oh." But having a body is only one part of the story about my body. I am my body... in pleasure and in crisis.

The human body in the medical context is also a medical narrative of symptoms and treatment.

What are basic bodily, physical needs? We do not only have bodily needs, we also have wishes. Bodies cannot have desires or needs, only people, who use language, can. To live, we need water. But some deliberately refuse to drink. To live, we need sleep. But strangely enough healthcare workers are forced into long lasting night shifts and work even sometimes over 80 h a week, to the detriment of their bodies. Basic bodily needs seem to integrate into psychic and social dimensions, in many ways and holistically. Conceptually as well as practically the body is never separate from language use. The one cannot be reduced to the other. The so-called medical causes for illness are only positioned after factors such as education, social status, environment, etc.

The concept of anybody else's pain is reconstructed from the way we similarly talk about pain though I cannot have another's pain. Rather we can merely speak in a similar way about the pain [20]. Therefore, we have to carefully listen to what the other, the patient, is saying in a specific situation. It is incompetent to apply theories or abstract terms before contextual inquiry. The physician analyses the dysfunction of the body, the patient expresses pain, his/her pain. The individual's pain has to be traced and treated. Thus, the method of giving case examples and clinical experience are especially useful for explanatory analysis. I have my body, I am my body. I have an illness, I am sick. The concept of the I is discussed in detail in the Chapter 10. The I has many selves, the body is many embodiments. Anatomically and functionally, but also psycho-socially we define and get defined. We differentiate between sex and gender in language games. Sex is our sexual organs, hormones, our sexual make up and appearance of the whole body. But man and woman cannot be reduced to their sexual organs and hormones. Sexual identity, gender is also produced by education, enculturation, socialization, moralization. We not only are of a certain sex, we became engendered as well. And then we perform gender. "I am the result of my bodily performances" [21].

13.9 Outward Physical Appearances: Beauty

Fulfilling the norms of society might involve harm to the body. We do this according to the internalized norms of the society we live in or by opposing them [22].

We shape our bodies we shape ourselves. There are eating disorders because of wanting to be slim at any cost or overeating as a protest against slim body demand, bulimia as a corrupt answer to the challenge of having too much to eat, and nevertheless remain slim etc. Even body protest is normative: as conforming tattoos, following piercing fashions.

A woman does not consist merely of her body parts and shape. There is a bias or prejudice regarding the body. External beauty is of little consequence and relative to taste and culture. It is often to advertise a lack of personality. The ethics and aesthetics of the body go together, as a matter of how much we feel at home in our body, of how much we are identified with our body. Beauty is not only a matter of shape, but of personality, integrity and balance, of "homeostasis" as ancient Greek philosophers called it in a holistic way. Beautiful, ugly, are open context terms and as such they have to materialise in certain defined situations. These terms are dependent upon the (mis)use of language in a certain cultural background. Our beliefs, our cultures are often at war with our bodies. And so, we are again puppets of our culture. Mastering the body by starving is some kind of severe neglect of its needs and a disaster for one's entire life. Girls are anorexic or bulimic having rejected their bodies and female appearances. The anorexic actually sees a stick body as being bloated, like a trick mirror, and thinks one cannot ever be thin enough against all factual evidence. Instead of cultural prejudices, the criterion of beauty can instead have a rational and humanistic basis. One self may be regarded as constituted by personality. In this sense, the healthcare worker would treat not just the body, but the person, even her personality. On this assessment a beautiful body with an irrational and negative personality cannot create a beautiful person, for example a selfish person with a beautiful face. A positive personality may be defined here as having positive emotions, critical thinking, and knowledge of ethics and as acting accordingly. It is in this sense that we can speak of the personality of the body.

13.10 The Face

The face is not a part of the body like other parts, as the brain is not an organ like other organs. The face is central in showing emotions, in communicating, in representing a person. The face is "persona," mask, role, representation, means for communication. You can read in the face about happiness, sorrow, pain, about finest details of the person (Cf. artistic portraits). "The other Face" by Hannah Rheinz describes dramatically what having a face is about and what it means to lose it: "*A hole, where her mouth had been. . . .No nose any more. Only eyes, eyes in distortion left. . . .They had cut the tumour out of her face, yet not leaving a face to be called face any more. She had lost her face. . . .She could not mirror any emotions in that face. The mutilated cannot express it any more. No dialogue any more. . . .Loss of one's face, loss of oneself." (Translation by the author) [23]*

The error is to totally identify ourselves with the body we have. We are more than that – but without it we are not. It is true: our bodies that are we, ourselves. But this is not all to the story. Our body is not just ourselves. We can go beyond our bodies to create ourselves. We might be familiar with our bodies in some context and feel strange in other context, even alienated. What does it mean, what does it do, if my body feels strange to me, is different to other's around me, if I feel stigmatized or physically excluded?

The Face

Two faces
one invisible
in the crowd,
face among faces
no face at all;
one which
speaks
listens
looks
caresses
kisses –
the one which is
you and I.

Barbara Maier

What do people not knowing me see looking into my face, what does my partner? My/your face – open context? What does he experience when I have lost my right breast by cancer, what do I (re)present to him? Will he be seeing me as the woman I have been to him before that, or did I change for him into somebody not any more familiar and desired?

Body threats are threats to oneself. It happens in many forms. One learns to love your diabetes causing body, only because your HbA1C for blood sugar measurement is within a good range and your management of blood sugar allows you to perform the way in which you should. And if the body does not obey and betrays you, it gets to be hated, separated. That additional chronic dependence upon one's dysfunctional body is a constant psychological challenge for a person with diabetes. It is a constant challenge for psychological balance.

13.11 The Body as a Whole and Body Parts: Organs and Transplantation Medicine

The brain is held constitutive of a person therefore no transplantation of the brain is performed. Organs, which are essential for bodily functions, but in function replaceable with success are, for example, kidney, liver, pancreas, skin, etc (For a full analysis see Chapter 14). Sexual organs and their products like sperm, egg, embryo are overloaded with un/moral concepts. A woman might think and some actually do, "I am not a real woman without my womb" (although this is not at all medically correct as the ovaries produce the female hormones). As indicated in the Chapter 14, people wish organs when they are in need of one, but especially in the U.S. do not wish to be organ donors. They think their organs are somehow spiritual or they will have an afterlife with only one kidney. They think that by contributing an organ the physical body or organ will be mangled. In fact, the organ will survive longer and not deteriorate so soon. With scarce resources a fair division of organs is problematic, but it should not be based merely on bodily criteria as is argued in the Chapter 14.

13.12 Reproduction of Bodies?

People do not know about anatomy and physiology, and so seldom take care of their bodies, and are ignorant of the birth-giving process. They do not have real babies, but dolls, fantasies and miracles. They deliver wishes and supernatural entities, not babies. This happens in spite of the proclaimed biotechnological age. A specialty ability of women's bodies is possible pregnancy and the intense change they then undergo. In modern debates about pregnancy and about abortion women often seem to get disenfranchised concerning their bodies. Pregnant women have pregnant bodies, but still it is their bodies, no-body else's, not an embryo's or a fetus' or a child's body. The woman's body should not just be regarded as a fetal container, [24] the pregnant woman is not just a mother machine, [25] her womb is not made out of glass for everyone's introspection [26], the woman's body is not a public place [27]. From men (human beings) stem men (human beings), from bodies stem bodies. We are not pregnant with persons. We do not give birth to persons, personhood is developed to whatever low or high degree later. Pregnant or not, the woman is her body. Women like men want to have control and choices regarding their bodies. What happens to their bodies happens to them. This includes all issues about reproduction and choice. To look upon a pregnant woman as a surrounding for a fetus, a matrix, is to focus on the fetus and put the woman into the background. This is to depersonalize the woman, is to reduce her as a woman, is to hide her body behind a concept of an embryo, a fetus, a "life" [27]. It is to neglect her bodily wants and needs. The question is not pro life or pro choice, the question goes deeper and is about whether the woman "owns" her body or not. If not, she has become a slave, somebody else owns her body, maybe her husband, her family, a religious group or the society, she lives in. "That life, which is predominating with moral hybris over contemporary discourse, belongs to the history of fallacy and insanity – or perhaps of religion – but not to the history of the body" [28].

13.13 *Leib*: Living to the Full

Erotic and well-developed sexuality bring joie de vivre into our lives and allow us to experience our bodies to the full. The Greek word *Erotic* does not refer to just bodily sex as it does in English, but includes a full relationship, a holistic concept, something more like the aesthetics of sensuality. I feel my body and my body is there. Sexuality is however usually portrayed in culture in the most unaesthetic and unerotic ways. Religion would often eliminate sexuality and information about it and the erotic altogether. Belief systems can destroy the body. We are ashamed to reveal and often even to experience our bodies, ashamed to appear naked. Some even are ashamed to find out about their bodies. Culture has especially held women back from expressing their sexuality. The influence of negative and harmful experiences in that area has deep consequences for those concerned. Prevailing experiences of abuse and denial have caused mental and bodily dysfunctions. There is widespread sexual inadequacy among both men and women. Women often give up on sex, and the number of men who are impotent sometime in their lives is perhaps one in seven in the early and middle years to most in the late years. Sexuality has been and still is cut out – even to the horrific extent of female genital mutilation.

Plessner called it the "excentricity of man," the double aspect of one's existence as body and as "*Leib*" [29].

The value of the body to create may be expressed by the old-fashioned German word *Leib,* which means living to the full with all you have got [30]. *Leib* is the rich, holistic and humanistic meaning. It means much more than the English word "body." It is being one's body in all dimensions conceivable, living to the full. Knowledge about the body would bring about its best realization for individuals, for mutual exchanges on all levels of living together. The *Leib* concept also stresses social relations. *Leib* is communication and exchange. Only an artificial reductionism of the body can isolate the body from the *Leib*. Body language is expressing the body, implies language, is "*leibhaftig*", language embodied. Only with an adequate critical ability, there is a healthy, beautiful body.

We are "*leibhaftig*", bodily there in all dimensions of our life. The *Leib* perspective goes beyond mind-body separation reasoning and avoids any artificial and to our experience secondary splitting which serves more religious and political issues than the fulfillment of individual people's goals. *Leib* is being there to the fullest extent one can be, feeling at home in our bodies and with those of other people. In analogy to Fridolin Wiplinger's *The Personally Understood Death* [31] one can speak of the personally understood body. *Leib* always has been understood personally, in relationships, as one's own body, as the body of a beloved person, a person you care for (as a mother for her child, a physician for his patient, etc.), the body of every human person. With such a "*Leib*" concept war would be impossible. "Bodylessness is insensitivity. Behind every abstraction hides insecurity, coldness, domination and weariness" [32].

13.14 The Philosophical Body: The Body as an Aesthetic Whole

The body has been defined by GH Mead as an ongoing changing relationship with others and not as a static entity. In family therapy it is seen, that one's self can be involved to such an extent into a dysfunctional family, that therapy is needed to free that self from the family. To develop a healthy personality and a healthy body it is necessary to go beyond culture and its manipulations, to critically inquire into enculturation through family, friends, religion, school, and job. Critical thinking as well as personality development, give bodies the best chance to unfold their full capacity.

Now, we can make a final move, one toward humanism. If aesthetic emotion involves value, a theory of value is needed. As argued in the Chapter 5 a naturalistic theory of value, positive value reduces to bringing about our reasonable wants and goals on the basis of deliberate and adequate informed inquiry. The theory is consequentialistic and stresses specific contexts and cases. Psychosomatic medicine is a reaction against the separation of mind, language and body, against abstraction from the concrete. It focuses on the individual, on the person. So it aims at materializing a model of narrative medicine, in which the narrative of a patient is to be understood by the physician as the narrative of his or her body to be treated accordingly. The holistic medicine's approach is to be ethically capable to deal with the individual patient and it also provides us with the understanding of ethics in relationships.

Does my body embody my life; and how? Sammy Molcho put it into a metaphor: body language is like clothes of thinking, which lay themselves over the body. Body language is speaking about the body, is language in *Leib,* language embodied. We (re)present our bodies. The naked body is associated with being either unprotected or especially strong. Clothes cover, protect, present, serve as status symbols, they show and disguise. There are uniforms to de-individualize (make all, who are wearing them, the same, like numbers, bodies, parts of a unit). There are body-shaping interventions like tattoos, piercing, styling, beauty surgery. There are body stigmata, like lip-palatine gap, visible and causing reactions in people, which might lead to certain emotions in those who thus get stigmatised. The bodily disability and the disabled body – how to relate to it as people concerned, as people around them? To treat disabled people as people, support them where they need support, integrate them and focus upon them who and how they are and not on the disability.

13.15 Summary of the Ethics of the Body

The highest level of decision-making is holistic, involving the most comprehensive philosophical thinking of which humans are capable (See Chapter 3). The more we know about our bodies at all levels, the more we will have a body, the more we will become our bodies. The more we live according to the needs of our bodies, the more we will feel at home with them. People can cause their own illnesses by not being aware of their body, by not respecting it and giving it adequate care. Our bodies require a constant improvement of our knowledge of nutrition and other aspects. We have seen that the misuse of the language of and about the body can lead to great harm and inadequate treatment. Philosophy and ethics of the body show what can be creatively known about the body.

References

1. Lakoff, G., and Johnson, M. 1999. *Philosophy in the flesh: The embodied mind and its challenge to western thought*, 3. New York: Basic Books.
2. Dewey, J. 1929/1958. *Experience and nature*, 322. La Salle, IL: Open Court; New York: Dover.
3. Wittgenstein, L. 1958. *Philosophical investigations*, 3rd edn. New York: Macmillan.
4. Hanson, N. 1958/1961. *Patterns of discovery*. Cambridge: Cambridge University Press.

5. Schweitzer, A. 1923/1996. Kultur und Ethik. Beck'sche Reihe, München: 338 *Mit drei Gegnern hat sich die Ethik auseinander zu setzen: Mit der Gedankenlosigkeit, mit der egoistischen Selbstbehauptung und mit der Gesellschaft.*
6. Fletcher, J. 1979. *Humanhood: Essays in biomedical ethics*, 91. New York: Prometheus.
7. Schweitzer, A. 1923/1996. *Kultur und Ethik*, 314. München: Beck'sche Reihe.
8. Dewey, J., and Tufts, J. 1932. *Ethics*, 288. New York: Holt.
9. Schweitzer, A. 1923/1996. Kultur und Ethik. Beck'sche Reihe, München: 351 ff. Cf. Maier, B. 2005. Unsere Empfindsamkeit für und unsere Sorge um das Leben von Menschen auf dem Hintergrund moderner Biotechnologien. In *Leben nach Maß – zwischen Machbarkeit und Unantastbarkeit. Biotechnologie im Licht des Denkens von Albert Schweitzer* 2005. ed. Schüz, G. Beiträge zur Albert – Schweitzer – Forschung, vol 10, , 115–142. Frankfurt am Main: Peter Lang Verlag. www.bundeskanzleramt.at/bioethik/Detail: http://www.bka.gv.at/Docs/2006/4/3/BMaier_Schüz.pdf.
10. Nietzsche, F. 1967. *The will to power*, 221. Trs. Kaufmann, W., and Hollingdale, R. New York: Vintage.
11. Dewey, J. 1939. *Theory of valuation*, 29. Chicago: University of Chicago Press.
12. Kossek, E., and Block, R. 2000. *Managing human resources in the 21st century: From core concepts to strategic choice*, 276. Cincinnati, OH: South-Western College Publishers.
13. Chomsky, N. 2001. *War against people. Menschenrechte und Schurkenstaaten*. Wien: Europa Verlag, Hamburg. (Rogue States, The Rule of Force in World Affairs. By Diane Chomsky. Irrevocable Trust.)
14. Ryle, G. 1984/1949. *Concept of mind*, 27. New York: Barnes and Noble.
15. Elliott, C. 1999. *Bioethics, culture and identity: A philosophical disease*. New York: Routledge; Cf. Elliott, C. 1996. *The rules of insanity: Moral responsibility and the mentally ill offender*. Albany: State University of New York of New York Press; Elliott, C. ed. 2001. *Slow cures and bad philosophers: Essays on Wittgenstein, medicine, and bioethics*. Durham: Duke University Press; Elliott, C. 2003. *Better than well: American medicine meets the American dream*. New York: W Norton.
16. Kass, L. 2003. *Beyond Therapy. Biotechnology and the pursuit of happiness. A Report by the President's Council on Bioethics*. New York: Regan Books.
17. Sontag, S. 1981. *Krankheit als Metapher. Fischer TB, Frankfurt am Main; orig. Illness as metaphor*, 5. New York: Farrar, Straus and Giroux.
18. Maier, B., and Wenger, A. 2009. Zur Bedeutung der gynäkologischen Anamnese für die Arzt/In-Patientin-Beziehung. *Balint* 10:48–52.
19. Brody 1987. *Stories of sickness*, 4, 50, 59. New Haven: Yale University Press.
20. Bauer, J. 2006. *Warum ich fühle, was du fühlst. Intuitive Kommunikation und das Geheimnis der Spiegelneurone*. München: Wilhelm Heyne Verlag.
21. Butler, J. 1991. *Das Unbehagen der Geschlechter*, 205. Frankfurt am Main: Suhrkamp.
22. Pieper, A. 1998. *Gibt es eine feministische Ethik*?, 85 München: UTB, W Fink Verlag.
23. Rheinz, H. 1995. Das andere Gesicht. In *Kursbuch 119 – Verteidigung des Körpers*, 117–132 esp.117–120. Berlin: Rowohlt.
24. Purdy, L.M. 1990. Are pregnant women fetal containers? In *Bioethics* 4:273–291. Reprinted In *Bioethics. An anthology 2006*. eds. Kuhse, H., and Singer, P., 2nd edn, 65–75. Malden, Oxford Carlton: Blackwell Publishing Ltd.
25. Katz-Rothman, B. 1995/1989. On surrogacy. (Orig. Recreating motherhood: Ideology and technology in a patriarchal society) In *Ethical issues in modern medicine*. eds. Hunt, R., Arras, J.D., and Steinbock, B. 4th edn, 472–479. Mountain View/California, London, Toronto: Mayfield Publishing Company.
26. Schindele, E. 1992. *Gläserne Gebär-Mütter. Vorgeburtliche Diagnostik – Fluch oder Segen*? Frankfurt am Main: Fischer TB.
27. Duden, B. 1994. *Der Frauenleib als öffentlicher Ort. Vom Missbrauch des Begriffs Leben*. dtv, München.

28. Duden, B. 1994. *Der Frauenleib als öffentlicher Ort. Vom Missbrauch des Begriffs Leben.* dtv, München: 129.
29. Plessner, H. 1970. *Philosophische Anthropologie.* ed. Dux, G., 47ff Frankfurt am Main: Suhrkamp.
30. Cf. *Leiben und leben:* Boss, M., Condrau, G., and Hicklin, A. 1977. *Leiben und leben. Beiträge zur Psychosomatik und Psychotherapie.* Bern: Benteli.
31. Wiplinger, F. 1984. *Der personal verstandene Tod. (The Personally Understood Death.) Todeserfahrung als Selbsterfahrung*, 2nd edn. Freiburg im Breisgau: Alber. Out of print.
32. Moltmann-Wendel, E. 1994. *Mein Körper bin ich*, 133. Gütersloh: Neue Wege zur Leiblichkeit.

Chapter 14
Organ Donation: Mandatory Organ Donation Declaration

Abstract By not donating or saving the organs for the living we are letting someone die. Presumed donation or presumed refuse? Many more organs are needed than available. How to establish a system of fair distribution? It may be recommended that everyone in a country be required to declare if they wish to donate and only those who are on the list to donate can receive an organ regardless of their wealth, position in society, or political influence. This is not presumed donation, but *mandatory declaration of qualification for an organ*. The AMA *Code of Medical Ethics* emphasizes 5 ethically appropriate criteria for the allocation of any limited medical resource. They include likelihood of benefit, urgency of need, change in quality of life, duration of benefit, and the amount of resources required for successful treatment.

Keywords Organ replacement waiting list · allocation · transplantation medicine · The Uniform Determination of Death Act · death requirement · presumed refusal to organ donation · presumed consent to organ donation · family approval for organ donation · equality · mandatory organ donation declaration

14.1 How Many People Need Organs?

By not donating or saving the organs for the living we are letting someone die which is the same as killing them. The consequences are the same. How many people die as a result of lack of organs?

In 2001 roughly 24,000 transplants were performed. 100,000 were on the waiting list for bone marrow transplants. As of July 2003, around six thousand die in the U.S. each year because of lack of organs and over 82,000 (2005) were on the organ replacement waiting list. Roughly half of these on the organ list each year can be expected to needlessly die due to poor policies and supernatural belief systems. About half of all the children on U.S. organ transplant lists die from not receiving a transplant. In 2004, the presumed consent foundation stated that 85,000 are on U.S. national organ donor lists [1]. In 2003 6,500 died because no organ was available. In 2002 fewer than 6,200 donated organs. In the U.S. roughly 17 die each day due to lack of organ donations. In 2002, 39% of all U.S. and U.K. deaths are caused

B. Maier, W.A. Shibles, *The Philosophy and Practice of Medicine and Bioethics*, International Library of Ethics, Law, and the New Medicine 47,
DOI 10.1007/978-90-481-8867-3_14, © Springer Science+Business Media B.V. 2011

by cardiovascular disease. In Britain, in 2003, 7300 were on organ waiting lists, but many are not put on the list because physicians see no chance to obtain an organ. Half of the relatives of potential patient donors refused to allow the donation.

From 1996 to 2005 kidney waiting list increased from 20,000 to 30,000; pancreas from 1000 to 2000. Over 400,000 Americans suffer from kidney failure (end stage renal disease, or ESRD) and require either kidney dialysis or transplantation to live, and 8 to 20 million Americans have reduced kidney function, due primarily to diabetes and hypertension that can lead to kidney failure. In 2003, more than 340,000 people received dialysis treatments. In 2000 over 195,000 people in the U.S. received dialysis paid by Medicare that is by the public. This is because the public refuses mandated donation and so must pay the extra costs. By not having presumed donation the public spends billions on dialysis. Their selfishness is self-defeating. In 2000, living donors increased from 4,000 to 7,500. Living donors sometimes comprise one-fourth of all organs used. England does not allow the living to contribute organs to a stranger. 21,000 die while waiting for a transplant each year. In September 2006 there were over 100,000 needed organs, 17,600 kidneys needed and 2,883 hearts. Only 16,000–29,000 organs are available [2]. According to *JAMA* 62,000 candidates are on National kidney waiting list [3]. The average waiting time is over 5 years.

The crucial question is how many are dying on the waiting list, how many are worsening in their health so that they get medically disqualified for organ reception? And: What is the outcome in transplantation medicine when there is the chance to perform it?

Transplant survival rates in U.S. 2003 were: heart 86%, heart and lung 71% [4].

In 2004, 2016 heart transplants were performed in the United States. As of July 15, 2005, there were 3142 heart patients on the transplant waiting list.

Waiting list candidates were 90,800 as of noon Feb. 9, 2006 [5].

Transplants between January and November 2005 were 25,951 as of 02/03/2006.

Donors between January and November 2005: 13,331 as of 02/03/2006.

UNOS has developed an online database system, called UNet sm, for the collection, storage, analysis and publication of all OPTN data pertaining to the patient waiting list, organ matching, and transplants.

All were 90,800; Kidney 65,131; Pancreas 1,725; Kidney/Pancreas 2,492; Liver 17,178; Intestine 201; Heart 3,003; Lung 3,127; Heart/Lung 137.

All candidates will be less than the sum due to candidates waiting for multiple organs.

Transplants performed January–November 2005: Total 25,951; Deceased Donor: 19,622; Living Donor 6,329 (Based on OPTN data as of 02/03/2006).

Donors recovered January–November 2005: Total 13,331, Deceased Donor 6,997; Living Donor 6,334.

In 2003, 82,000 were on the waiting list for an organ in the U.S. In 2002 there were only 2002 donors. 94,300 were on waiting list Dec. 2006 [5].

Spain leads in providing organs, has presumed consent and requires three doctors to certify the brain death required. Family permission is also sought. Each transplant saves 200,000 Euros in dialysis. Austria had presumed consent since 1982 and by

1990 nearly everyone needing a kidney received one. Only 30% of organs come from road accidents [6]. Denmark switched back to express consent in 1986 and available organs fell.

One thousand German patients die on organ waiting lists, for kidneys especially. Germany has to import organs from other Eurotransplant nations. The donation rate in Germany is only 50% of the rate of Austria. 70% of Germans say they wish to donate organs after death, but in 2000 only 4% had the support of written donor consent. Hospitals report only 50% of the donors and permissions from the nearest relative are difficult to obtain in a timely way [7].

30% of children under age two needing transplants die because they are not available [8].

14.2 Death Requirement

The Uniform Determination of Death Act of 1980, which is endorsed by the American Medical Association and provides that a patient who has "sustained either (1) irreversible cessation of circulatory and respiratory functions, or (2) irreversible cessation of all functions of the entire brain, including the brain stem, is dead" [9]. Clinically and pragmatically, it is not that complex, the respirator is stopped and the patient is dead within 5 minutes The challenge is the decision-making process. Many states have provisions that amend this general definition. Once the physician declares the patient's impending death, the hospital informs a local organ procurement organization (OPO) of the possible organ donation. Upon death of the patient, usually because of irreversible functions of the brain, an OPO representative secures permission from the patient's family and performs a medical evaluation of the potential organ. The OPO then accesses the UNOS computer to match the donor's characteristics to those of a patient awaiting an organ [9]. For each organ recovered from the donor, the computer generates a separate list that ranks potential recipients using factors such as tissue match, blood type, length of time on the waiting list, immune status, and the distance between the potential recipient and the donor [9]. Donation procedures for all solid organs except for kidneys take the potential recipient's degree of medical urgency into consideration. Once a match becomes apparent, the OPO representative contacts the transplant team of the first patient on each list.

One problem is that the organ must be transferred before there is oxygen loss, but legally the subject must first be dead. It is acceptable to let an extremely premature baby die [10]. Donors are usually dead or organs are taken from brain dead patients. The living, however, can choose to donate an organ such as a kidney. Some reject organ donation because people fear they will be judged dead, when they are not, in order to obtain their organs. The case is, however, the reverse. As there are not enough organs one is more likely now to be judged dead so one's organs can be used, whereas if there were presumed donation there would be enough organs available. This argues for universal presumed donation. Because the body has first to be dead, but if it is dead the organs may not be of use, the American Medical Association Council's Guidelines for the Transplantation of Organs favors organ removal of anencephalics while they are still alive [11]. Comatose and anencephalic babies may be used [12].

According to the Council on Ethical and Judicial Affairs, AMA, consciousless anencephalics (very rare and especially rare in countries with prenatal diagnosis because they are selected out long before birth-giving) usually die within a few hours or a week but cannot be used in some states (e.g., Florida) because the body has first to be dead but if it is dead the organs may not be of use [13]. If death is defined as brain cessation or cardiac cessation the organs could be removed. At the writing of this above mentioned article (1995) many more than the listed 1,500 children needed either heart, kidney or liver transplants and 40–70% of them die due to lack of organs. Benjamin Freedman says organ transplantation is a "moral imperative" [14]. "Ethics requires that persons be dead before..." Ethics does not require anything and so cannot be appealed to. The Council's Guidelines for the Transplantation of Organs favors organ removal of anencephalics while they are still alive especially because they have no consciousness [11].

14.3 Opposition to Organ Donation

U.S. has *presumed refusal* to donate organs. People thus bring upon themselves thousands of deaths a year. Each one seems to think it is someone else who will need an organ. If they each thought they would need one the support of organ donation might be greater. People in the U.S. want to receive organs, but do not want to give them. Even over 20 years ago there was lack of interest in organ donation. Of 2,056 telephoned only 19% carried donor cards. Only 7% supported the concept of presumed consent. Those more educated were almost twice as willing to carry a donor card and to be willing to donate a relative's organ upon their death [15]. In 1993, 70% of the American people said they would agree to contribute organs [16]. Surveys have varied from 7 to 38% accepting presumed donation [17]. McConnell wrote, "One suspects that the policy of presumed consent is not politically feasible in the U.S." [18]. He dismisses the possibility of presumed consent too readily and maintains that Americans are not altruistic enough to donate organs although it is done in many other countries. He may be right if there is not more public education.

In U.K. 70% assent to donation, but only 20% carry a donation card. 30% of relatives refuse donation allowance [19]. Some studies show that up to 70% of Americans have verbally favored donating their organs. In practice, however, presumed donation in the U.S. will not be voted in [20].

People mourn, or pretend to mourn, and have expensive funerals at the same time they oppose donating organs and medical research, which could genuinely have helped people live longer, healthier lives (See also Chapter 21). American Lutherans oppose donation of body parts. Christian Scientists oppose transplantations [21].

Again we see inconsistency, a lack of knowledge of ethics and an absence of ethics as shown above. For the Church to oppose stem cell research is in a way to oppose organ replacement. For example, stem cells could possibly repair the organs of those in need of neurons and brain cells, a liver, kidney, eyes, heart, etc. making organ replacement unnecessary. These are additional arguments for presumed consent and another argument why medicine cannot be bound by irrational cultural practices.

Because of the lack of availability of organs in the U.S. prevention of organ damage becomes essential. One may also think of freezing samples of one's own DNA [22] and get tissue by therapeutic cloning. See Chapter 15 Embryonic Stem Cell Research. A Question of Beliefs? Instead of attempting to inform people that it is to their advantage to contribute organs, extreme measures are sought. One is xenotransplantation. However, it does not really work. Money is better spent elsewhere. Animal pathogens can be transmitted to humans (HIV, Ebola, Hepatitis B, Bird Flu). Pigs have known and unknown bacteria and viruses and some are embedded in the DNA. Stress should rather be on prevention and a better system of organ donation.

14.4 Support of Organ Donation

If one can remove organs in surgery, one should be able to do so for transplantation after death [20].

In any case the harm caused by using the organ is so minor as not to be comparable with not using the organ to save a life. Thus if one is faced with the person who is to die for lack of an organ, it should be very hard on the one who wishes to refuse them life. It is the same as killing them. If we need to operate on a patient in emergency without advance directives we do so. Similarly we should do so in the case of someone who has just died and organs to be of use must be removed immediately [23].

One may obtain convincing information from the various organ donation organizations: Presumed Consent Foundation (OPTN) [24], National Kidney Foundation [25] etc. The lack of each organ is great, for example, one in nine adults is said to have chronic kidney disease. Every organ has its own organization to promote a rational policy of procurement and use. Virtually every such non-religious organizations opposes cultural and religious practices, which block such procurement.

One method to encourage presumed consent is to establish a registry of those who opt out of presumed consent, but also those who are against organ donation. These people should also not receive organ donations. The reader is asked to now check his or her driver's license or other documents to offer organ donation. If it is not signed, one may propose that they should have last priority for receiving a needed organ.

Arthur Caplan, Bioethicist at the Department of Medical Ethics, University of Pennsylvania School of Medicine, suggests that the U.S. should presume that the deceased is a donor unless otherwise specified [26]. This also means that relatives ought not to have to sacrifice an organ for a family member or close friend. In terms of priority in general, one may choose to keep all people alive and to offer organs to all possible, but where we must choose to let die or indirectly kill, the decision may be based on a holistic, naturalistic ethical system. An ethics based on lottery is a failed ethical system. But Caplan and Zink note that presumed consent is very unlikely to be adopted in the U.S., "Americans presume that people do not want to be donors. Many European nations presume that they do" [27] (See also Chapter 10). They suggest aggressive measures of medical personnel to show the life benefits of organ giving, e.g., assume the patient and relatives wish to approve donation, and point to the European models of presumed consent.

In some countries where donation is legally automatically presumed, one can opt out. But only about 2% do. Some opt out for religious reasons, though many or most of the major religions permit organ transplantation. Some opt out because it is thought to be "mutilation of the dead," but death by its nature involves bodily mutilation. Such thinking is cosmetic confusion. It ought to be obvious that organ donation allows the organ to live longer. Open-casket viewing funeral practices use cosmetics to conceal and mask the body deterioration to make it seem alive. Although the medical profession tries to treat patients equally, organ donors often specify restrictions. Some have donated lungs only for non-smokers; some for Catholics only, or for Muslims, or members of a certain race or gender. Feminist donors might wish to specify that the recipient be a woman. Singapore excludes Muslims from automatically participating in their donor program, though they can opt in.

China permitted the organs of executed prisoners to be used. However, because they give the death penalty for small offenses such as petty theft, the question arises as to whether or not this is being done for much needed organs. This is at least another argument against the death penalty and against using only prisoner's organs rather than presumed donation of everyone else. If there were automatic, universal donation the temptation to kill one or let one die to save another or save many would tend to be eliminated. We may also consider research into the human transplant use of animal organs and blank embryonic animal stem cells, assuming there are not the present medical hazards. We may use stem cell research to create organs, and we may also promote the development and use of artificial organs.

14.5 Presumed Organ Donation

Austria and at least 21 other countries presuppose each person is an organ donor by birth (e.g., Belgium, Denmark, Finland, Italy, Norway, Spain) [28]. The United States, one of the wealthiest nations in the world, with no universal health care system and the largest military and prison system in the world, does not have presumed organ donation. The U.S. has a strictly volunteer organ donation system. The problem is that with present egoistic, and inefficient policies of the U.S., the supply will never reach the demand. It is a system in crisis, which costs thousands of lives each year with little hope for improvement.

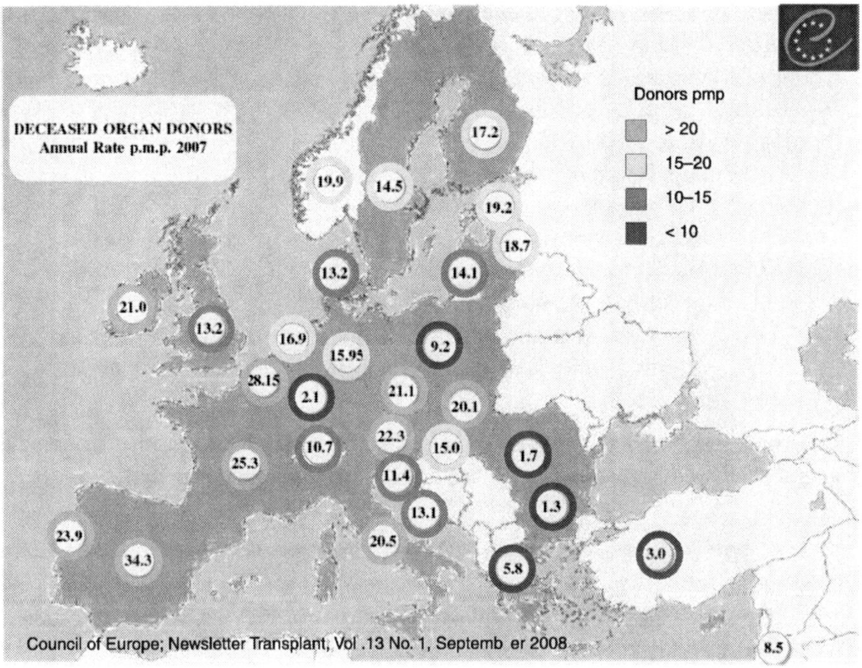

Legislation for organ donation in European Union and European Transfer Area (2008)

Country /Legislation in regard to donor

Bulgaria: presumed consent
Belgium: presumed consent
Estonia: presumed consent
Finland: presumed consent
France: presumed consent
Italy: presumed consent
Latvia: presumed consent
Croatia: presumed consent
Lithuania: presumed consent
Luxembourg: presumed consent
Austria: presumed consent
Poland: presumed consent

Portugal: presumed consent
Sweden: presumed consent
Slovakia: presumed consent
Slovenia: presumed consent
Spain: presumed consent
Czech Republic: presumed consent
Hungary: presumed consent
Denmark: informed consent
Germany: informed consent
Greece: informed consent
Great Britain: informed consent
Ireland: informed consent
Malta: informed consent
Netherlands: informed consent
Romania: informed consent

Source: Austrian Federal Institute for Health. Transplant 2008 (translated by author of the book)

As mentioned above, some countries with the most effective donation policies automatically have presumed consent of each individual for organ donation.

14.6 Family Approval as an Adverse Policy

In some countries, consent of relatives is sought even if not required. But the relatives may have their own biases against donations or not be informed enough to make such decisions. Relatives, in general, may be the worst ones from whom to obtain permission. The deceased donor may have told the family of the desired wishes, but the family may still not allow it because of their own beliefs. Significant others or personal physicians may be better to gain permission from than family members especially when the family members have an unfriendly or adversarial relationship to the deceased or dying. As indicated above, the informed consent policy of, for example, the U.S. and Germany causes well over 40,000 of those needing organs to die each year. Family consent is regarded as a poor authority [29]. It is not clear why the family should have the autonomy and power to give such consent, especially just because they are family members. They may in fact have a hostile relationship. Gundle recommends presumed consent for the U.S [6]. In Spain 200,000 Euros is saved by each kidney transplant, $207,000,000 yearly.

The family is given too much authority even more than the donor card contributor. Relatives cannot change the legal will of the deceased, so why of organ donation? Individual choice should not be made subject to filial approval. Presumed consent protects the individual from questionable contradictory decisions of relatives [30]. The U.S. donor card is sufficient and legal to permit organ use, but family can block donation autonomy. Only a patient should be able to authorize decision-making of a family member or other person.

Gill suggests, however, that in the case donation wishes are unknown the family be consulted [20]. This may be opposed because family is not necessarily a good or even friendly source for the individual in question. In addition, the policy stalls obtaining organs because it leads to always checking with the family about intent even if a donation choice is already made. This is one of the serious problems with presumed donation in Europe, they hold up donation until the family is consulted and this means that far fewer organs are donated. We note that the family unit is neither an ethical unit nor is it necessarily a rational or caring unit. The family may be hostile to the potential organ donor or wish to override the stated wishes of the donor for their own religiously obligatory or private reasons.

Family values are not necessarily ethical values.

Suppose that the organs of the deceased could be sold for large sums of money, as would be the case if it were legal. If a family member were to then have to decide about donating and would thereby inherit hundreds of thousands of dollars, there would presumably be more organ donations.

In many countries a clear legal definition of family is lacking. Which family member? How does s/he qualify? Presumed consent should not require family member approval.

14.7 Recommendations for Obtaining Organs for Transplantation

14.7.1 Lottery

To avoid having to decide whom to choose to receive organs, Harris constructed a system for organ donation whereby people are chosen by lottery to be killed so their organs may be given to a number of people in need of them in order to stay alive [31]. On this view, one may, for example, kill one to save perhaps 20 in this way. Also, Harris sees equality as equal opportunism rather than rational consequentialism [31]. E Cahn following an extreme equality view says we should let all die rather than choose. Daniels wrote, "We can decide whom to save by means of a lottery" [32].

United States versus Holmes (1842) ruled that a lottery system is the best way to decide who must die when due to emergency (e.g., crew ordered to throw 16 from lifeboat) or because of scarcity, someone must die. The theologian Childress, author with Beauchamp of a major bioethics text, also defended random selection [33]. Jury selection also uses random selection. Military draft, on the other hand, is restricted to certain occupation, gender, health and age groups. It is not just random.

One may rather argue that lottery is just another way to avoid decision-making. Kilner states, "It is unacceptable for random selection to be adopted just because other approaches are considered faulty," and "Random selection can be seen as the ultimate display of irresponsibility" [34]. It is to decide not to decide. There is no choice. Singer suggests that those chosen could have volunteered for the organ lottery, but that would not encourage healthy lifestyles because they would know they might always get another organ [35]. And this would not serve to regulate unhealthy lifestyles. He therefore rejects the lottery idea.

The case is that we could, to a certain extent, obtain the needed organs without killing anyone. It is not even necessary to obtain them from the living. The proposal of mandated declaration is later presented to that. Instead of lottery, the nations may rather introduce such a policy of obtaining organs.

14.7.2 The Economic Incentive Approach

To try to deal with the lack of organs crisis a monetary approach has been taken to organ donation in the U.S. The policy is that one cannot sell organs. But if they say you cannot put a price on human life, why do they? Medical treatment is put on a financial basis. Congressional bills proposed tax credits, commemorative medals, etc, for those who donate organs. Pennsylvania offered $5,000 for organ donation after the death of a relative. McConnell proposes a policy of moderate monetary contributions for organs of deceased relatives, but admits that the policy will not produce enough organs to meet the needs [36]. The American Medical Association also only supported a payment scheme to obtain more organs, but opposed presumed consent [37]. Just recently, however, the *AMA* favored presumed consent [38]. One wonders why they were so late in coming to this endorsement. In Great Britain the Presumed Consent Bill 47 was introduced into the House of Commons, only on Feb. 3, 2004.

In India, until a few years ago, people while alive were legally allowed to have their kidneys removed and to sell their organs. The practice, though now illegal, still flourishes. Only Iran allows organs to be sold. People elsewhere might reject this practice, yet they would gladly buy such organs.

It was reported that if all of the body parts of a person were sold it could come to 45 million dollars. Bone marrow $23 million at $23,000 per gram, DNA in every cell equals $9.7 million at $1.3 million per gram, lung $116,400, kidney $91,400, 32 eggs of a fertile woman $224.000, etc (*Wired* August 2003).

For military nations such as the U.S. if they think they are justified to kill soldiers in war, or outside of war, and to kill collateral innocents, they are in effect claiming to own such bodies. In battle no person's autonomy is respected. The organ donation issue shows that people make irrational, anti-humanistic and fatal choices with the power of their autonomy. If our body is inviolable regarding organ donation, then why should soldiers be subject to the draft?

The millions of dollars for dialysis alone for the up to 350,000–400,000 U.S. citizens for kidney sufferers are paid by the public through Medicare/Medicaid because they refuse to allow presumed donation. Americans pay a high financial penalty for not having presumed donation.

14.7.3 Irresponsible Lifestyles and Organ Preference

In terms of the principle of risking x lives to save y lives, it may be that the risk to your life in giving a kidney is 1 in 4,000, whereas the risk for one needing a kidney is certain death without it. This was the thinking of Zell Kravinsky, age 49, who gave a kidney to the needy. He felt that it is not enough to give merely to close friends or the family. It may, however, be argued that the recipient whether family member, friend or stranger should be deserving of. It should produce the most fair, rational or humanistic outcome. Not to aim at that is in a strong sense unethical to do, just as it is to give to a corrupt government to distribute medicine and food to its needy citizens. The government keeps the money or buys destructive weapons with it instead. The reasoning was that one need not just give locally or to a family member, but as a world citizen to help all those in need as much as possible. He also earned money so that he could give $45 million to medical research [39].

According to Walter Glannon, the basic premise is that those responsibly taking care of their health should be given priority over those who do not. Blind equality is bad policy and undermines prevention and distributive justice. This policy encourages healthy lifestyles. Drinking and smoking not only cause problematic conditions for those involved, but also endanger other members of society. It is also maintained that those engaging in risky activities can be taxed, [40] however, they too may by their risk-taking need an organ transplant.

Alcoholics, smokers, drug addicts, excessive risk takers in sports, and those others who have clearly caused their organs to be damaged by their poor lifestyle, may be unacceptable donors and individual decisions will have to be made as to how to treat such donors and those who are in need of organs. Clearly if someone destroys three livers by drinking alcohol, some appropriate measures must be taken. An intense program of education, and therapy, including addiction therapy, may be a required condition before the first transplant in such cases if they had agreed to be donors. Munson argues that alcoholics should be given organ transplants but he does not give any of the conditions mentioned here [41]. That above may be a way in which this can be done.

It is unfair that those who are anti-reason, anti-humanistic, anti-organ donation, anti-medicine, etc. are given the same priority for organ transplants and medical treatment as anyone else. Fair distributive justice would require that if people are not donors, then, one should place them last on the list to receive organs, or presume that they do not want to receive organs at all. It would not be fair to give organs equally to one who refused to donate them or to even prefer non-donors to donors just because they are, for example, rich or famous, friends or family members. But this is practice now. McConnell accordingly questioned, "Why should individuals receive the benefits of a system if they are unwilling to bear their share of the burdens?" [42]. At present, there is an organization of donors who agree to give to other donors: *Lifeshares.com*. Organs and tissues are donated only to *one another* as the members die. *UNOS*, United Network for Organ Sharing, on the other hand, gives organs to all regardless of merit or contribution, however they have medical criteria (compatibility, genetic factors, blood type, etc.) and such factors as time first applied of an organ, length of time on the waiting list, etc. To only use narrow and mechanical medical criteria is to fail to make an ethical decision. It is non-ethical. It is to opt out from decision-making where it is needed most. The result is unfairness, which lets the undeserving live and the deserving die. The argument here is that in those areas where there are criteria and where one must decide between several individuals and where one does have to make fair policy an ethics such as the naturalistic theory of ethics can and ought to be used.

All factors should be included in the consideration. Failure rate of second transplants is high. In a worst-case scenario, one person keeps receiving numerous different organs, which are then rejected while other candidates die. How many organs should be given to a single patient who needs multiple organs? How shall we distribute organs considering the report that organs transplanted from women to men appear to have a higher rate of rejection than from male donors, and autoimmunity is six times higher in women than in men? This means that we do have to consider gender, race, etc. in medical practice because they have direct bearing on diagnosis and treatment. Along these lines, some medical authorities totally exclude alcoholics from receiving liver transplants. It is necessary for the medical profession and legislators to place conditions and restrictions on organ transfer, but these conditions should be based on a humanistic and rational, naturalistic system of ethics. In Europe there are e.g. clear cut counter-indications against HTX (Heart-Transplantation). At present, in effect, in US there is no ethics and no ethical policy in place at all.

Citizens are obligated to render at least minimal assistance if without danger to oneself and without interference to duties to others [43] (Vermont, Minnesota, Massachusetts, Hawaii, Rhode Island). It is argued that posthumous organ donation is a moral duty. As we have duty to help others in need, in some states, so we should have a duty to donate organs after death. It is irrational to worry about the destruction of the organs after death as embalmers destroy the body as well. Would you not allow the use of a dead family member's organs to save another family member? [44] It is argued that the one who refuses to donate organs be confronted by those who are dying due to lack of an organ.

The consideration of numerous factors supports the view that those causing their own illness should be given lower priority for scarce medical resources and should pay for their own care if they can and so as not to increase medical insurance premiums, and decrease insurance availability and burden the medical system, which already cannot adequately care for patients.

Who is the person who would receive the organ? Should it be a rich person with political influence, someone selected just because s/he signed up first on a blind first come, first served basis?

We can rather use a clear and sound humanistic value system as a criterion. Those who are anti-inquiry and cannot or will not acquire such a value system can be given the lowest priority for the transplant or other medical care. Actually possessing a sound ethical system is preferable to merely possessing the potential to develop one. The odds are against the latter doing so. We should give priority, and we often must give priority. There are factors such as: age, gender, conditions of the health of the patient and projected patient longevity, etc. which one must consider. It may be stressed that the physically disabled not only qualify as persons as anyone else, but they may well in individual cases by their own critical education prove to have more social worth and quality of life than the nondisabled. The nondisabled are only temporarily nondisabled especially as they age. On a humanistic ethics all of the factors and consequences one can reasonably obtain would have to be considered.

There is no shortcut or magic formula in ethics. We must use our critical, humanistic informed judgment in each case.

In 2000, 70% of Germans were willing to donate organs, but in practice only 4% of the donations were actually based on the written approval. All other cases are administratively required to be confirmed by relatives. Only about 5% of accident victims have proof of organ donation with them [45]. Germany has informed consent rather than presumed consent. The result was that in 2000, 14,000 were on the waiting list for a kidney and only 2,219 transplantations were performed in 2000 [46]. In 1997 10,000 Germans were on the waiting list for kidney transplants and of these about 2100 received them [45].

Jewish religious law and the Catholic Church opposed organ transplantation because it was regarded as destruction and mutilation of the body. Eventually organ transplantation won at the expense of religion [47]. If one does not take care of one's body, believes in soul, or is pro-war, it is hypocrisy and inconsistent to worry if the body is mangled.

14.7.4 Mandatory Organ Donation Declaration

Not having presumed donation is murder.

Fewer mistakes will be made if it is presumed consent esp. if most people wish to contribute. Now many wish to contribute, but the evidence is not available. Furthermore, if it is more humanistic and rational to contribute then if mistakes are made they will be made on the side of helping people stay alive by organ donation. It would be strange to complain, "Oh, I made a mistake by helping someone stay alive. I did not mean to help them live." One way to make organs more available is to remove all who did not choose to contribute organs off of the list of those to receive organs. As a policy it is necessary that those against presumed donation should not receive organ transplants. At present this may be the majority of people and may be enough so that those needing organs and were willing to donate them will receive them. The autonomy of all of the patients is still preserved.

As we treat emergency patients with presumed consent, we could assume presumed organ donation to supply critically ill patients with organs. A mandated choice list would be immediately available. By mandated choice is meant that one has to choose whether to donate or not. One may change one's option to leave the plan at any time, but not after receiving an organ and not to join when it may be necessary to obtain a transplant. This allows a sufficient supply so that the surgeon may choose the best organ match for a particular patient and one in a nearby location.

If people are irresponsible in not making provisions for donating their organs after death, and have not sufficiently supported presumed donation, they thereby cause their own shortage of organs. People are said to fear that physicians will hasten their death to get their organs. However, the reverse is the case. If there were presumed donation enough organs would be available thereby eliminating that possible unlikely threat. One problem mentioned by Breyer is that the donor has no priority at all for receiving an organ over one who selfishly refused to donate one [48]. The author proposes the solution of the adoption of presumed donation as was found successful especially in Austria and Spain whereby one can opt out if one wishes. This, however, does not solve the latter problem. A system of distributive justice and fairness to responsible donors is needed as opposed to unfair blind equality.

It may rather be recommended that everyone in a country be required to declare (mandatory declaration) if they wish to donate and only those who are on the list to donate can receive an organ. Only those who are on the list as donors can receive organs regardless of their wealth, position in society, or political influence. Those who volunteered to donate, but were turned down because of the unacceptability of the organs may still qualify for an organ on a case-by-case basis. Age is not a major criterion for denying a transplant. One must join the list of donors before one knows an organ is needed preferably already by one's parents in the first year of birth. Infants need infant organs. One cannot later join at an age and under conditions where one may be likely to need a transplant. Insurance companies should not have to pay for transplants for anyone who does not agree to donate organs. Excess organs or organs which cannot be stored could be shared with other countries, which have similar mandated choice donation.

This is not presumed donation, but *mandatory declaration of qualification for an organ*. If they illegally or through error do not declare they must be contacted to declare. They will not be on the list to receive an organ as it is their responsibility to make a declaration. Therefore, a list of non-donors must also be kept so that if one is not on either list they can be contacted to make a declaration. This mandatory system will especially encourage the self-interested ones who have previously refused to donate to donate as well. This will also eliminate the need for the burden and delay caused by permissions of relatives which to a large extent prevents the donation of organs. This will also eliminate the unfairness mentioned by Breyer that one in the Eurotransplant system is not allowed to donate one's organs after death to a relative or friend, but only to the next one having priority on the list. A central registry accessible internationally with no further permissions, including family permissions, is required. Usable organs should be allowed being stored or frozen as appropriate.

In general, the AMA *Code of Medical Ethics* accords with the Final Rule [49]. The *Code* states that organs should be considered a national, rather than a local or regional, resource. That is, geographical priorities in the allocation of organs should be prohibited except when the transportation of organs would threaten their suitability for transplantation. Moreover, the *Code* emphasizes 5 ethically appropriate criteria for the allocation of any limited medical resource. These criteria include likelihood of benefit, urgency of need, change in quality of life, duration of benefit, and the amount of resources required for successful treatment.

One incentive for being registered to donate is that one need not specify the person to have priority to receive the organ, e.g., a relative, because there will be enough organs available for everyone on the list. In fact, for the same reason, there would no longer need to be a priority at all. An additional important humanistic policy would be the attempt to make the number of nations in a transplant exchange system as large as possible including the less developed countries, which have mandatory donor declaration. At present the U.S. would be excluded from such a system because they do not have a humanistic donor system. Presumed or mandatory declaration of organ donation is a good sign of contribution to a humanistic and altruistic society, one, which is not now had in many countries.

Gill argues for mandated, not just presumed, consent in the U.S. and required registration of intent or refusal to donate to be stated on driver's licenses and tax forms. With mandated choice nothing is presumed and autonomy is not violated [50]. The author worries about whether the public will be informed enough to be able to rationally decide whether to allow or refuse donation. This is an added reason why it should be stated on the forms that if one does not agree to donate, one will not be put on the list to receive an organ if needed. If for whatever reason no donation consent is recorded it can be assumed presumed that one wishes to donate as this will produce the most beneficial medical outcome. Presumed consent would meet the stated and unstated wishes of people better than does presumed refusal.

Warren, the co-author of that book did not receive the organ needed to survive, presumably because of ageism. He would have needed a heart-transplantation. But he gave all his organs, which were usable, with his last breath.

References

1. Presumed Consent Foundation Presumedconsent.org, associated with OPTN. (www.unos.org)
2. United Network for Organ Sharing. OPTN/UNOS. optn.org/l.
3. Chalas, A.J., Sutherland, E.R. 2005. The importance of innovative efforts to increase organ donation. *JAMA* 294:1691–1693.
4. Wright, J. ed. 2006. *New York times almanac*. New York: Penguin.
5. Source: United Network for Organ Sharing unos.org/.
6. Gundle, K. 2005. Presumed consent: An international comparison and possibilities for change in the United States. *Cambridge Quaterly of Healthcare Ethics* 14:113–118; esp.115.
7. Breyer, F. 2006. What to do about the organ scarcity. *Europäische Akademie* 62:1–2.
8. Munson, R. 2002. *Raising the dead: Organ transplants, ethics and society*, 84. Oxford: Oxford University Press.
9. The Uniform Determination of Death Act of 1980, which is endorsed by the American Medical Association: http://www.law.upenn.edu/bll/archives/ulc/fnact99/1980s/udda80.htm. Ad hoc committee of the Harvard Medical School. 1968. Report of the Ad Hoc Committee of the Harvard Medical School to examine the definition of brain death. *Journal of AMA* 205:85–88.
10. Munson, R. 2002. *Raising the dead: Organ transplants, ethics and society*, 92. Oxford: Oxford University Press.
11. American Medical Association Council's Guidelines for the Transplantation of Organs. 1995. Opinion 2.16; AMA 1995. Anencephalic neonates as organ donors. *JAMA* 273:1614–1618.
12. Munson, R. 2002. *Raising the dead: Organ transplants, ethics and society*, 144. Oxford: Oxford University Press.
13. AMA Council on Ethical and Judicial Affairs, CEJA Report C–I-88 An encephalic Infants as Organ Donors.
14. Freedman, B. 1988. The anencephalic organ donor: affect, analysis, and ethics. *Transplantation Proceedings* 20(Suppl 5):57–63.

15. Manninen, D., and Evans, R. 1985. Public attitudes and behavior regarding organ donation. *JAMA* 253:3111–3115; esp. 3114.
16. Gallup Poll, Partnership for Organ Donation. Organ Donor Gov. 2005 National Survey of Organ and Tissue Donation Attitudes and Behaviors http://organdonor.gov/survey2005/introduction.shtm
17. Veatch, R., and Pitt, J. 1995. The myth of presumed consent: Ethical problems in new organ procurement strategies. *Transplantation Proceedings* 27:1888–1892.
18. McConnell, T. 2000. Inalienable rights: The limits of consent in medicine and law, 124 ff. Oxford: Oxford University Press.
19. *Medical Ethics Today* 2004: 435.
20. Gill, M. 2004. Presumed consent, autonomy, and organ donation. *Journal of Medicine and Philosophy* 29:37–59.
21. Ross, J. ed. 1986. *Handbook for hospital ethics committees*, 151. Chicago: American Hospital Publications.
22. Kurzweil, R., and Grossman, T. 2004. *Fantastic Voyage: Live long enough to live forever*, 256. New York: Plume (orig. Rodale).
23. Gill, M. 2004. Presumed consent, autonomy, and organ donation. *Journal of Medicine and Philosophy* 29:37–59; esp.49.
24. www.presumedconsent.org (Associated with OPTN).
25. www.kidneywi.org.
26. Caplan, A. 1992. *If I were a rich man could I buy a pancreas? And other essays on the ethics of health care*. Bloomington, IN: Indiana University Press.
27. Caplan, A., and Zink, S. 2004. Is PC the answer to closing the organ availability gap? Internet article available July, 22, 2004.
28. http://www.eurotransplant.nl/files/annual_report/ar_2008.pdf and Council of Europe, Newsletter Transplant 2008. 13 containing figures on organ donation and transplantation 2007:24–41 http://www.coe.int/t/dg3/health/Source/2007transplantNWSLTTR_en.pdf; compare Austrian Federal Institute for Health: see below.
29. Veatch, R. 1995. The myth of presumed consent: ethical problems in new organ procurement strategies. *Transplantation Proceedings* 27:1888–1892.
30. Gundle, K. 2005. Presumed consent: An international comparison and possibilities for change in the United States. *Cambridge Quarterly of Healthcare Ethics* 14:113–118; esp.116.
31. Harris, J. 1975. The survival lottery. *Philosophy* 50:81–87.
32. Daniels, N. 1999. Meeting the challenges of justice and rationing. In *Meaning in medicare: Reader in the philosophy of health care*, ed. Nelson, J., 265–288; esp. 283 New York: Routledge.
33. Beauchamp, T., and Childress, J. 1994. *Principles of biomedical ethics*, 4th edn. Oxford: Oxford University Press.
34. Kilner, J. 1990. *Who lives? Who dies?*, 198 New Haven, CT: Yale University Press.
35. Singer, P. 1977. Utility and the survival lottery. *Philosophy* 52:218–222.
36. McConnell, T. 2000. *Inalienable rights: The limits of consent in medicine and law*, 124 ff. Oxford: Oxford University Press.
37. McConnell, T. 2000. *Inalienable rights: the limits of consent in medicine and law*, 127. Oxford: Oxford University Press.
38. AMA 4/26/04 resolution.
39. Kravinsky, Z. 2005. Niemand sollte zwei Häuser haben. *Geowissen* 35:44
40. Glannon, W. 1998. Responsibility, alcoholism and liver transplantation. *Journal of Medicine and Philosophy* 23:31–49; esp. 44.
41. Munson, R. 2002. *Raising the dead: organ transplants, ethics and society*, 65. Oxford: Oxford University Press.
42. McConnell, T. 2000. *Inalienable rights: The limits of consent in medicine and law*, 125. Oxford: Oxford University Press.

43. Peters, D. 1986. Rationales for organ donation: Charity or duty? *Journal of Medical Humanities and Bioethics* 7:106–121.
44. Peters, D. 1986. Rationales for organ donation: Charity or duty? *Journal of Medical Humanities and Bioethics* 7:106–121; esp.118.
45. Der Spiegel 1997, March 3rd, 10:237.
46. Breyer, F. 2003. Financial incentives for organ donors? *Europäische Akademie Newsletter* 40:1–2.
47. Jonsen, A. 1998. *The birth of bioethics*, 202, 203. New York: Oxford University Press.
48. Breyer, F. 2006. What to do about the organ scarcity? *Europäische Akademie Newsletter* 62:1–2.
49. AMA Opinion 2.03. 2005. *Allocation of Limited Resources*, 9. Chicago, IL: Code of Medical Ethics.
50. Gill, M. 2004. Presumed consent, autonomy, and organ donation. *Journal of Medicine and Philosophy* 29:37–59; esp. 52–53.

Chapter 15
Embryonic Stem Cell Research: A Question of Beliefs?

Abstract As far as we know now, embryonic stem cell research is not very promising in terms of therapeutic gains because of the side-effects but nevertheless not at all neglectable in terms of gaining basic knowledge about the development of diseases and their understanding, trials for toxicity of pharmaceuticals, development of embryos, understanding of fertility, etc. The debate about research on human embryonic stem cells is especially characterized by using value terms to promote certain positions. Humaine medicine is medicine for suffering people. The consequentialistic ethical perspective focuses on affected sick people. Do present and future patients have a right to demand that this research be conducted, and moreover, that it be conducted now and without hesitation?

Keywords Embryonic stem cell research · adult stem cells (AS) · embryonic stem cells (ES) · IVF left over embryos · SCNT (somatic cell nuclear transfer) · cloning · therapeutic cloning · definition of embryo · moralization of embryos · abortion argument · humaine medicine

15.1 Introduction

The debate about research on human embryonic stem cell is especially characterized by using value terms to promote certain positions. The following is an example how rhetoric is shaping approaches.

President Bush announced funding for limited embryonic stem cell research, in a speech, which needs to be commented on: "*As the discoveries of modern science create tremendous hope, they lay vast ethical mine fields*" [1].

Remarkable here is especially the kind of military metaphor used for the work of research.

"*As the genius of science extends the horizon of what we can do, we increasingly confront complex questions about what we should do.*" "*My position on these issues is shaped by deeply-held beliefs*" [1].

Whether deep or not, beliefs might be true or false, founded or unfounded and therefore have to be critically examined. "*Researchers are telling us the next step*

B. Maier, W.A. Shibles, *The Philosophy and Practice of Medicine and Bioethics*, International Library of Ethics, Law, and the New Medicine 47,
DOI 10.1007/978-90-481-8867-3_15, © Springer Science+Business Media B.V. 2011

could be to clone human beings to create individual designer stem cells, essentially to grow another you to be available in case you need another heart or lung or liver" [1]. This, one never could do. It is a false belief, a fear not based upon facts or scientific knowledge. Another "You" is self-contradictory in terms. Another person is never You.

"*We recoil at the idea of growing human beings for spare body parts or creating life for our convenience*" [1]. This is contrasted with the willingness to go to war, destroying lives often just for economic convenience, for revenge, nationalism, oil, pride, life style, etc. President Bush serves as an example of widely held unquestioned beliefs and prejudices about research findings, found in actual debates not only about stem cell research. In ethical analysis we have to critically examine facts, consequences of allowing or prohibiting research, allowing or prohibiting the application of these findings and we have to question irrational beliefs and their consequences [2]. "Those engaged in the ethical and political debate about the generation and use of hES (human Embryonic Stem cells) must understand the science, whether or not they are scientists" [3]. And this also involves an ethical use of language concerning the issues and avoiding misuse to serve certain unscientific and unethical purposes.

15.2 Definitions and Clarifications of Morals and Ethics

Stem cell research and eventual application are subject to ethical questioning because of their influence on human life and their consequences for people. Such questioning first requires an analysis of "moral" prejudices about such research and its application. We adopt the model of the distinction between morals and ethics as laid out in the Chapter 5. In this sense, we may speak of the inner institutional enculturated "morality" of medicine, what "moral thinking" amounts to in the medical system, in science as well as in practice. Morals are all too familiar to us. They are so usual, customary and indoctrinated that we fail to be able to question them. We have become blind to them. They are closely related to what President Bush called deeply held beliefs. Ethics, on the other hand, is not merely morals, but rather the critique of morals, culture, scientific concepts as well as practice, and their theoretical presuppositions. It puts them all into question. We badly want to stick with our beliefs held since generations, with our morals so familiar and comfortable. They save us the trouble of thinking and getting into questions. Ethics is inquiry into the moral prejudice, which cannot be rationally supported by up-to-date sound biological knowledge. Ethical analysis concentrates on rational arguments, and foreseeable consequences. It is working within a contextual framework and examining whether familiar moral principles are at all applicable to specific situations and specific people; and what is going to happen if doing so. The ethical analysis is in development, dynamically changing with new findings of research and science. Ethics means going beyond our morals, means going beyond our polarizations into good and bad, natural and artificial. Philo of Alexandria, e.g., uses the word "nature" for the beliefs in his culture, which he does not dare to question. Morals is often like just dogmatically knowing what is right or wrong, e.g. wrong-in-itself, without inquiring into the facts, details, and contexts of specific situations [4].

Thus, the ethical approach presented here is consequentialistic, evidence based, rational, and anti-dogmatic. It follows the humanism and naturalistic philosophy of John Dewey in stressing the consideration of people rather than abstract or supernaturalistic beliefs. It is contextual, with stress on the specific situation [5].

The term "cloning" is often and erroneously used in heated debates to suggest the duplication of completely formed individuals from which tissues or even spare organs are taken, as did President Bush in his announcement quoted above. These images seem to be taken from science fiction literature, but nevertheless are rather effective in producing fixed ideas about this research. With so-called "therapeutic cloning," which rather is cloning for research, cells are derived from the embryo that, when grown in the laboratory, can be induced to differentiate into cells and ultimately into tissues of therapeutic interest [6].

What can science contribute to the clarification of ethical questions? Science can give information about degrees, stages of development and perhaps suggest recommendations. It can venture prognosis about possible future therapies, which may make the rational evaluations of arguments e.g. regarding the protection of embryonic life versus the healing of deathly sick people easier [7]. It can help us to decide between utopia and biology and to indicate "natural" limitations to the notion of visionary "designer babies."

15.3 Facts and Beliefs About Stem Cells

15.3.1 What We Already Know About Stem Cells

What are stem cells? Cells that have the ability to divide indefinitely and which give rise to specialized cells as well as to new stem cells with their original potential.

Stem cells offer an ideal biological system for gaining basic scientific knowledge about the development of the various human cell types. They give insight in the pathogenesis of diseases through observation of deregulated differentiation processes. They are also useful for the testing and the establishment of drugs for new therapeutic uses.

Several kinds of stem cells are known.

Adult stem cells (AS) are those found in the adult organism (e.g. in bone marrow, skin, and intestine) that replenish tissues in which cells often have limited life spans. They are more differentiated than embryonic stem cells or embryonic germ cells. Embryonic stem cells (ES) are cells that are derived from the inner cell mass of a blastocyte embryo. They retain the special ability to develop into nearly any cell type. Embryonic germ cells (EG) are those that are derived from precursors of germ cells from a fetus and have properties similar to ES cells. Stem cells from the umbilical cord are harvested at delivery from a more or less unproblematic source. Their properties are similar to AS cells.

The potential of stem cells according to their origin is the following:

1. Cells of an embryo until the eight cell stage are held to be totipotent, having an unlimited capacity, even to differentiate into another embryo.
2. Pluripotent are the cells from the inner cell mass of the blastocyte as well as stem cells from the gonad of a fetus. They can generate all of the cell types in the fetus and in the adult and are capable of self-renewal.
3. Adult stem cells are more committed to special types of tissues and organs.

Early in the embryonic development (until about 16 cells) each cell of the early cleavage-stage embryo has the developmental potential to contribute to any embryonic or extra-embryonic cell type. ES cells if placed in the appropriate culture conditions are highly proliferative, maintaining the potential to contribute to all adult cell types. Thomson's achievement (University of Wisconsin-Madison) was getting these stem cells to grow indefinitely in a lab while maintaining normal DNA and without becoming different types of tissue as they would have as part of normal embryonic development [8]. With it came great research possibilities and the chance of elucidating the mechanisms that control differentiation.

Scientifically it makes sense, to repeatedly derive new ES cell lines, because the properties of ES cells differ depending on the methods used to derive them and because prolonged passage in culture reduces the potential of the ES cell population as a whole and because a tremendous amount remains to be learned during the process of derivation itself [9]. Human ES and EG cells are an essential means for understanding the earliest stages of human development, infertility, miscarriage, and birth defects. They are important as a source for understanding how human ES and EG cells differentiate into specific types of cells with the goal of identifying the genetic and environmental signals that direct their specialization into specific cell types and with the goal to discover the disturbing influences. Some speculate that the origins of many human diseases are due to events that occur early in embryonic development, for example, juvenile-onset diabetes [10]. The technique of "homologous recombination" was recently also successfully applied in human ES. This makes it possible to manipulate any part of the human genome to study gene function and mimic human diseases in the Petri laboratory dish [11]. It means coming closer to ES cell-based transplantation and gene therapies although it might as well never be of any clinical relevance.

At the present time, human stem cells can be derived from the following sources: human fetal tissue following abortion (EG cells), human embryos that are created by IVF (In-Vitro-Fertilization) and that are no longer needed for reproductive purposes (ES cells), human embryos that are created by IVF with gametes donated for the sole purpose for providing research material and, potentially, human embryos generated asexually by somatic cell nuclear transfer (SCNT) in which the nucleus of an adult somatic cell is introduced into an enucleated ovum (ES cells). As promising this procedure, also known as therapeutic cloning, might be, the fact that mitochondrial DNA of this enucleated ovum cell might influence the genetic make up of the cells to be established, is to be taken into consideration. There remains the potential for the transmission of mtDNA (mutant) from the host ooplasm, which could lead to mitochondrial dysfunction and possible disease [12].

15.3.2 The Promise of Stem Cell Research in General

Potential medical applications of human ES cell and EG cell research were hoped for many diseases, such as Parkinson's disease and juvenile-onset diabetes, which result from death or dysfunction of just one or a few cell types. Most of these diseases are neither curable nor satisfactorily treatable. The replacement of those cells could offer effective treatment and even cure. HES are tumorigenous and likely to import infectious agents, so their immediate therapeutic potential as far as we can see now is neglectable [13].

But their contribution to research, examination of toxicities of substances and understanding of early developments in embryos and disease origination is prominent [14]. ES cells derived through SCNT even offer the possibility that therapeutic material could be developed from a patient's own cells. This essentially would make an autologous (same donor and recipient) transfer possible and perhaps avoid graft versus host reaction. But at the same time tumors might develop.

Potential medical applications of human ES cell and EG cell research are expected for cancer therapy. Human ES and EG cells may be used to reduce the tissue toxicity brought on by cancer therapy. Already, adult stem cells like bone marrow stem cells, representing more committed stem cells, are used to treat patients after high-dose chemotherapy. Diseases of the nervous system are especially problematic and often fatal. Many of them result from the loss of nerve cells, and the fact that mature nerve cells cannot divide to replace those that are lost: in Parkinson's disease nerve cells that make the chemical dopamine die; in Alzheimer's disease, the cells that make acetylcholine die; in multiple sclerosis, the cells that make myelin die; and in amyotrophic lateral sclerosis, the motor nerve cells that activate muscles die. In stroke, brain trauma, spinal cord injury, and cerebral palsy, numerous types of cells are lost with no built-in mechanism for replacing them [15]. It may be possible to use stem cells to colonize damaged parts of the body, including the brain, and to promote the repair and re-growth of damaged tissue [16]. But we do not know the consequences yet.

Diseases of the bone and cartilage, and of the skin could be treated. It would involve transplantation of stemcells to a recipient or genetic modification of a person's own stem cells and returning them to the marrow. Bridging large gaps in fractured bones by those cells would be an option for morphological and functional restitution. Skin transplants could provide help after burning large skin areas, but are also thought to be useful in the treatment of hereditary and harsh diseases like Epidermiolysis Bullosa.

Human ES cell and EG cell research could help to achieve far better results than those usually gained in the area of toxicity and drug testing. They would allow fewer, less costly, and better designed, human clinical trials yielding more specific diagnostic procedures and more effective systemic therapies [17].

It has to be admitted, that there are open questions, questions of large concern, to which answers will be available only if research is allowed. There is e.g. the issue of how normal is the resulting tissue in terms of the rate of aging, risk of tumor formation (from undifferentiated transplanted cells), effects of harmful mutations, contamination of different tissues, cell type for transplantation, genetic compatibility, immunological tolerance [3]. Also there is the question of how great will the risks of transmission of infectious agents be. Will it be possible to generate the number of cells required for treatment purposes? To what extent will repair be possible? Which dosages will be effective? Whether the possibilities can be achieved, only research can reveal. All scientific research is initially basic research which is open-ended in terms of results [18]. If it is prohibited, one will never know. Reasonable regulations are needed, medically appropriate control as well as publication of negative results [19]. Materialization of clinical application will prove extremely challenging [20]. It has proved so already in the few cases applied [21].

But what should be avoided, is that massive amounts of unfounded beliefs, political coercions and prejudices gain power to prohibit the search of knowledge so much needed to help suffering people and prevent them from dying. We are not at the point, where it can be assured that the results of stem cell research will help for therapy, but it seems rewarding enough to pursue the research for gaining relevant knowledge. It cannot be guaranteed, that the help will be greater than the harm, but we will only find out if we are doing research. Harris stresses an interesting point, "Usually we have no rational basis for determining where the balance of risk and benefit lies. The precautionary principle urges us, irrationally, to give more weight to risks than to benefits. But delay in producing benefit is a real risk to those who might benefit from scientific advance" [22]. The issues concerning safety and danger of new techniques are often (mis)used merely to stop research for other than medical reasons. Of course, there is to be had consideration of harmful side-effects, but seriously or lethally ill patients would rather decide to undergo a critical treatment, which might help to overcome the illness than with no treatment die. Gruen stated, "What constitutes an acceptable risk will vary depending on the condition to be treated and the anticipated benefit of the therapeutic interventions" [23] (See Chapter 12).

15.4 The Controversy About What an Embryo Is

15.4.1 Definitions

To answer the question, what an embryo is, is not so easy as might seem at first glance. Let us look at the definitions available in natural sciences and medicine. There an embryo is qualified as the beginning of an organism in the early stages of development or a stage (between the ovum and the fetus) in the prenatal development of a mammal, or in humans, the stage of development between fertilization and eighth week following fertilization. This is what I am used to see as a physician working in reproductive medicine.

We may then have a closer look at the way in which the term embryo is used. According to a British report, "Before implantation the fertilized egg is termed a *zygote* rather than an embryo. Embryo refers to the developing entity after implantation in the uterus until about ninth week after fertilization. Afterwards it is referred to as a fetus until the time of birth. The terms embryo donation, embryo transfer and embryo research are therefore inaccurate, since these all occur with zygotes, not with embryos" [24]. How we deal with those inaccurate terms still without defining them can be exemplified in the legal text of the *Convention on Human Rights and Biomedicine* of the European Union (1997) [25]. Embryo research is allowed or prohibited according to the already existing laws of the member states [26]. The creation of human embryos for research purposes is prohibited [26]. But, no definition is given of what an embryo is.

15.4.2 Embryonic Development in Its Context

To go one step further in the evaluation of what an embryo is requires critical consideration of the biological basis as well as enculturation. Which opinions cannot be maintained biologically? For example, in every generation the variety and three dimensional shapes have to develop anew. They are not already there, hidden somewhere [27]. The development of the primates is not a continuum since fertilization, but works in two phases: 1. The development in the egg: blastocyte. 2. The development in the uterus: actual embryogenesis. Embryonic stem cells from the blastocyte are in the petri dish toti/pluripotent, but in the petri dish no mouse from ES exists in the experiment, because for this absolutely necessary is the orderly influence of the outer cells of the blastocyte and later, after the implantation in the womb, the influence of the female organism [28]. The embryonic development is not to be looked upon without considering the matrix, whether it is reproduction in a woman's womb or stem cell production for research in a dish.

15.4.3 The Moral Status Ascribed to an Embryo

The moral status depends upon what one holds or critically thinks an embryo is. One person may describe an embryo as a cluster of cells. Others, especially religious people believe that an embryo is a person or deserves the same respect as a person. The debate on potentiality and the problem of not taking seriously the gap between actual and potential has been described in detail [29]. The moral status of an embryo is not the same as later when it has become a person [30]. Despite of their developmental continuity, acorns and oak trees are different kinds of things. So are human embryos and human beings [31]. For others, the moral status varies according to the stage of development of an embryo or fetus. According to religious beliefs, moral status is not ascribed, but is already there in the embryo because of a supposed divine authority [32].

It comes to a "moralization" of embryos in situations where:

(a) Embryos in vitro should or should not be transferred into in a woman's womb. Superfluous embryos according to IVF (In-vitro-Fertilization)/ICSI (Intracytoplasmatic sperm injection) exist, embryos with problematic genetic dispositions are also found by PID (Preimplantation diagnosis) and not to be transferred then.

(b) Embryo(s) in vivo from the woman's womb should (should not) be removed (e.g., by abortion).

(c) Embryo(s) which do not have the purpose of being transferred into the woman's womb, because they are not needed any more for reproductive purposes and could be used for research purposes, if such would be allowed.

What is an embryo? And what should be allowed being done with it? An answer, descriptively given is: the result of the combination of egg and sperm cell after fertilization, a 2-, 4-cell unit, a blastocyte. With that, nothing more need to be said than that which exists, whether in vivo, in vitro, etc. Also a SCNT-embryo is descriptively an embryo, but is not emergent from the fusion of egg and sperm cell. It was created asexually through the transfer of a somatic nucleus into an enucleated egg cell. A small amount of DNA (mitochondrial DNA) will be present there from the egg cell donor. And the objection to creating human beings with the intention of destroying them for research purposes does not apply to SCNT for therapeutic cloning issues [33]. There won't be any human being.

15.4.4 Life Is Not Just Life: When Is a Human a Human?

Usually in law we find the graduated protection of life. The concept "life" involves the problem of abstraction. The abstraction, "life," is drawn from the many-faceted forms of living creatures and their development. In reality the notion of life is concrete. Why proceed in talking about living people, animals, plants etc. indirectly and abstractly and with secondary reconstruction? Phenomenologically, we can proceed along a direct path. In biology we describe human organisms, animals, plants, and also embryos as they show themselves unfold before our very eyes (also through the microscope), without having prior abstractions ready that displace such reality. We grasp and experience life not as a noun or substantive abstraction, but as we live, as a verb. Only in this dynamic will the abstract concept become concrete in the context of our experience when we ask: Who lives, how, under which conditions, with which goals? [34].

Can one evaluate life "itself"? If yes, it may be mentioned that the "life" of an embryo is actually and qualitatively different than that of, for example, a chronically sick person. If one basically accepts that all life is the same, just living in the abstract (which is a paradox already), independent of whether it is the life of a sick person or a blastocyte, one has no concern for the evaluation of life and will remain blind to consequences of abstractionist fallacies. We may focus on the main question at stake. At what point one can speak of a human as a human? In the fertilized egg cell the genetic program is completely available. For its development, however, the embryo requires intensive exchanges, symbiosis with another organism – that of the mother. She is the irreplaceable and unconditional. Only with such implantation in the womb is the developmental program complete, and with birth it is completed. With birth the developing entity is a separate, independent organism, which breathes and now has its own oxygen. Clearly the infant is still needy, but it is now nourished from outside and can, if necessary, also live without the specific mother. Then the organism is *ein Mensch, a human being* [35].

We certainly may take the view that we owe respect to life in its various expressions. But is it all the same respect for the various expressions? "We are used to the concept of respect for persons, but is it meaningful to speak of respecting embryonic tissue?" [36]. Definitions of personhood are usually based upon our everyday experience. They are not necessarily at all descriptive. A person or human may be defined as one who can think rationally, possesses a sense of self, has future goals, and only in this way can develop a fear of death, or be vulnerable [37]. Basic legal rights stress this vulnerability only for persons.

Ordinary personhood rests on certain conditions, which correlate with properties of the human brain, in particular cognitive and emotional prerequisites as self-consciousness, ability to reason, sentience, having a history [38], which are totally lacking in the fertilized egg. The criteria are problematic of course, because they are under- or over-inclusive. Cognitive criteria do not include newborns, people with irreversible brain damage, or in coma. Sentience is possessed by animals though through that they cannot be personified.

Intervening in reproduction and creating in vitro embryos establish a situation of decision-making whether or not transferring those embryos into the womb of a woman or better not. In the IVF-case, without transfer into a woman's uterus there is no potential of becoming a human being. The same lack of potential applies to embryos obtained by SCNT, which cannot be implanted under the present prohibition of human reproductive cloning. Reproductive cloning should be prohibited anyway because of the problematic consequences for the child to be [39].

15.5 Ethical Issues in Stem Cell Research

15.5.1 How to Perform an Ethical Examination?

First of all it is important to examine the nature of moral reasoning and the background beliefs that underlie such reasoning. Primarily one has to consider the consequences if stem cell research is allowed as well as the consequences if it is prohibited, outweighing benefits and risks. In other words, to be fair, one also has to ask the following question: How can we legitimize not to perform stem cell research? Is it clearly legitimate to do so because of unfounded or questionable beliefs? Those who argue against embryo stem cell research often rest their case on the *supernatural, devine* "nature" of an embryo as an ensouled person or as a potential of becoming a person. In distinct contrast and opposition, the scientific nature of an embryo is *descriptively* a cluster of cells. For the super-naturalist, nature is not nature by definition. Supernatural goes somehow beyond nature. As a basis of ethical analysis, one should strictly use medico-scientific distinctions. Medical ethics, like medicine, is a *science* [40]. It is based upon reason, scientific findings, inquiry, and the contemplation of the consequences of applying or withholding results of research. Ethics is not based on dogma nor would it be ethical to do so.

In ethics we ask critically and responsibly about consequences. Who does benefit, from what, at which cost? Diseased people, millions all over the world, could possibly benefit from the new findings of stem cell research, including embryonic research. Responsibility for and solidarity with suffering people compels stem cell research. "Those who inveigh against the derivation and use of pluripotent stem cells make the assumption that an embryo has not only the moral status of a human person, but also a sort of super status that outweighs the needs of others in the human community" [41]. They in terms of consequences show no compassion with those who suffer.

And in addition, what if one is suffering from Parkinson's disease, or is crippled by an accident, and does not hold the belief that an embryo has the moral rights a person has? Will he or she be able to understand that certain unfounded beliefs are more protected than his or her own life or quality of life? There is injustice and unfairness in a society, which gives priority to such beliefs rather than to people.

According to Matthiesen "A duty of solidarity with individuals suffering from disorders prohibits any attempt to hinder research and penalize sufferers. The risk of doing so is all the greater because the predictable future shortfall of spare IVF embryos will make it more difficult to base research on their availability" [42]. What is at stake here? Causing the suffering and dying of real people for the purpose of the supernatural protection of fictional spirits called "embryonic people." What is destroyed in human embryo research? A 100-cell human blastocyte. The personifying feature of a 100-cell blastocyst is its DNA. Cryopreserved embryos, if thawed and transferred into a woman's womb, have no better than a 20–25% chance of successful implantation. To find out what the ethical impact might be, it is necessary to consider the sources, the future fate of embryos, and the context [43]. Embryonic cells from supernumerary IVF embryos are available only if the couple does not wish to have any further children (truly "surplus") and so embryos would be discarded if decided by the couple not to further store them. Embryonic cells from supernumerary IVF embryos are also available, if embryos are left over because of insufficient quality to be used for infertility treatment and would be discarded.

The options are: the embryos are either discarded, which means destroyed through the thawing process or donated for research, which means destroyed through research procedures. The specific context of frozen embryos has to be taken into consideration. Whatever embryos are "this in no ways entails the right of a frozen embryo to transfer, gestation, or to a risk-free pathway into maturation. Adults' and childrens' 'right to life' is, considered constitutionally and as a moral problem, at best a negative right against unwarranted violence by the state or individuals" [44]. One can be drafted, exploited by extreme working conditions, etc. thereby losing one's so-called "right to life" and also to health as defined by WHO as a state of holistic physical, psychological and social well-being.

15.5.2 Inquiry into Language

Moralizing statements are brought into biology, which are phrased in a way only to reinforce prejudices or unquestioned judgments. Moralizing terms are transposed, which are not based on biological inspection. As a result, there are numerous misleading metaphors created as phrases or book titles: "The Dominance of the Gene," "From family tree to stem cell," "The new human," "The Cloned Paradise," "The End of Humanity." How rational and critical are such book titles? Which metaphors thus will be superimposed on biology? What fears will be thereby aroused? People discover, they could surely also have "something genetic", which then leads to a further emotionalization of the debate. An analysis of our everyday speech is given by B. Duden, "The everyday gene"... and its "Context in life," as is used in our language games [45].

Therapeutic "cloning" is a misleading term regarding the adjective: it is research cloning what is in fact done. Also the term cloning itself is loaded with science fiction associations, a second You, a second class human as organ donor [46].

The title "Please, Don't Call it Cloning" (*Science* 2002) shows how difficult it is to render facts when already prejudices have been created and suffused into public (mis)apprehension [47].

15.5.3 The Abortion Argument All Over Again?

Often it is argued that one's position on stem cell research largely boils down to one's position on abortion. There is truth in this, but it is too simple a framework for our issues here. Also, the specific context of abortion when obtaining fetal aborted material for research has to be carefully examined. The moral acceptability of deriving EG cells from the tissue of aborted fetuses is, for some, closely connected with the moral acceptability of abortion. "The abortion question is not a *moral* question, it is a question of examining the facts, actions and consequences for the needs and beliefs of oneself and others in society" [48]. Certain associations of stem cell research with abortions are claimed, such as "causal responsibility." The mere possibility of donating fetal tissue for research according to such claims would lead to abortions, which pregnant women otherwise would not have. This would in the opinion of those against this kind of research result in a more permissive attitude of society towards abortion and induce some kind of "moral complicity of researchers" [49]. Mere "symbolic associations" of stem cells with abortions are even claimed. The option of donation should only be discussed after the abortion decision. Pro-life hardliners question the aborting woman's moral qualification to make the donation decision. The Report of the Human Fetal Tissue Transplantation Research Panel stated that a woman's choice of a legal abortion does not disqualify her legally and should not disqualify her morally from serving "as the primary decision-maker about the disposition of fetal remains, including the donation of fetal tissue for research" [50].

The pregnant woman might rightly choose the abortion for her own sake, which is part of her right to self-determination. Why should not the pregnant woman (and respectively her partner) choose to donate fetal tissue following abortion for stem cell research, not choose to donate left over embryos after IVF procedures, which would otherwise be discarded? Should she/they not feel good because of donation? Is it not also possible to show respect for what we destroy for sound reasons? [51].

15.5.4 *Adult Stem Cell Research-an Alternative to Embryonic Stem Cell Research? Other Alternatives?*

Adult stem cell research is promising and is of specific interest in certain areas [52]. But is pursuing research on only adult stem cells and prohibiting embryonic research at the cost of giving up possible clinically valuable research findings legitimate to appease those who believe embryos to be persons? [53]. The harvesting of AS cells (bone marrow, brain biopsy procedure) is technically difficult, painful, even risky for human beings. AS research is an important branch of research on stem cells, but should not be regarded as *the* alternative to embryonic stem cell research [54].

Stem cell research involves tumor formation, contaminating animal products, questions of genetic (in)compatibility, selecting and generating the right cell type for transplantation, and new approaches to generating ES cell lines. One of the most promising avenues for circumventing the ethical and scientific roadblocks of stem cell research is to continue and expand hES cell research under the thorough and thoughtful oversight of diverse committees [55].

To continue their research after the Bush ban, some researchers tried alternative approaches to satisfy the conscience of those protecting the embryo over research advancements. They thought of politically palatable alternatives [56]. The cell lines that had been created by August 9, 2001 were no longer useful, but rather became really abnormal. So they materialized a procedure, which was called altered nuclear transfer. It consisted of taking away the Cdx2gene responsible for the creation of the trophoectoderm. Without that gene the embryo cannot create a placenta and so not implant into a womb and not develop further. These so-called "knockdowns" were still capable of producing embryonic stem cells [57].

These alternative ways of harvesting embryonic stem cells did not placate conservative critics. It seemed to be a waste of time, effort, and resources to try to please what dogmatically never can be pleased: a mythical view of the embryo. Absolutist views here as elsewhere are characterized by not allowing any rational arguments and so not any considerations of consequences.

Also, the PGD (Preimplantation Genetic Diagnosis) approach by taking one cell from an eight-cell embryo to foster embryonic stem cell lines from that very cell whereas the embryo with its seven cells left can continue developing, did not soften the conservative restrictionism. What really could? So finally, as it was summarized by David Solter in the *New England Journal of Medicine* "Playing politics for the sake of science is probably necessary and sometimes noble; manipulating science for the sake of politics is usually a waste of time" [58]. This is by far understated. Five years in delay for eventual treatment of chronically ill people to circumvent the destruction of embryos cost many lives (See Chapter 11).

15.5.5 *IVF "Left Over" Embryos Versus "Created for Research" Embryos*

Those who are against embryonic stem cell research on grounds of the protection of the embryo cannot be for any performance of reproductive medicine like IVF or ICSI either. Embryos are involved in these procedures, not all of them transferred into a woman's womb – either they are surplus or of insufficient quality. Why then take on a debate whether left over embryos would be morally less problematic than created ones for research? This ethically is contradictory. What characterizes absolutism is that it has little concern for being contradictory.

Professional standards need to be established to avoid the situation that infertility clinics will increase the numbers of embryos remaining after infertility treatment for research purposes. This would mean avoiding an increase of the physically dangerous Ovarian Hyperstimulation Syndrome in order to harvest more eggs for creating embryos, which might in some cases even be life threatening for the women concerned. A morally relevant difference might be discovered between generating embryos for the sole purpose of creating a child, which might be left over when the desire is fulfilled and secondarily be instrumentally used in research and producing an embryo with the prospect for research, a "research embryo" with primary instrumental use. Research with ES cells derived from embryos solely created for research could make sense in spite of the instrumentalization issue. It might come to a shortfall of embryos left over in IVF procedures if the techniques become more and more efficient and also if one needs well-defined embryos, not only the ones, which are just available by chance after IVF-treatment. The use of SCNT (somatic cell nuclear transfer) to produce ES cells would provide a special advantage, the avoidance of graft rejection if the donor nucleus were taken from the transplant recipient. The procedure is likely also in humans to produce an embryo (as in sheep or cows). Perhaps in the future one could obtain an autologous transplant without the creation of an entire embryo. Issues of consistency with attitudes we have towards assisted reproduction are the following: Spare embryos are produced, never used for procreation, stored, thawed and thus destroyed. One might speak of reproductive medicine as an embryo consuming technique. "Only those who think that it is more important to create new humans than to save existing ones will be attracted to the idea that sexual reproduction is permissible whereas the creation of embryos for therapy is not" [59].

There also is the problem of whether or not to compensate egg donors for expenses or for their eggs as such. The hidden agenda of blowing up this issue of compensation seems aiming at not making eggs available for creating research embryos at all. The argument used is the supposed protection of especially desperate poor women to take risks in order to obtain money. Regulation by law is required.

15.5.6 Public Funding?

The ban on embryo research under the Bush administration in America reflects a mere moral, but not ethical, point of view either that embryos deserve the full protection of society because of the moral status as persons or that there is sufficient public controversy to preclude the use of federal funds for this type of research. Under Obama administration the approach towards research including embryonic stem cell research has fundamentally changed [60]. The above mentioned ban conflicts with several goals of medicine such as healing, preventing and doing research characterized by the ethical principles of beneficence and non-maleficence, the *"do no harm principle"* [61]. In comparison to that, we may consider a European perspective: Embryo research is permitted: (a) to promote advances in the treatment of infertility; (b) to increase knowledge about the causes of disease, including congenital disease; (c) to increase knowledge about the causes of miscarriage; (d) to develop more effective contraceptive techniques; (e) to develop methods for detecting gene or chromosome abnormalities in pre-implantation embryos [62].

In January 2001, regulations were made extending the purposes for which embryo research should be licensed: to increase knowledge about the development of embryos, to increase knowledge about serious diseases, to enable such knowledge to be applied in developing treatments for serious disease.

Removal and cultivation of cells from a donated embryo could be regarded as being analogous to tissue donation. Because there are sufficient donated embryos from IVF at present, there is no compelling reason to allow additional embryos to be created solely for research purposes. The embryo donors should be asked explicitly whether or not they consent to such research and the subsequent therapeutic use of the cell line and informed consent must be obtained.

What is the moral and ethical meaning in regard to providing or withholding federal funding? Those, opposing stem cell research, would withhold federal funding because they hold the opinion that federal funding would make taxpayers complicit in causing the destruction of the embryo. Providing federal funding would not only give the opportunity for developing cures for many diseases, but also give the opportunity to regulate and supervise research by the state. It is also an obligation of a socially balanced society to take care of just distribution of potential benefits from stem cell research. This involves appropriate prioritization of this research relative to other social needs in health care, but also the concerns of all other fields of society, e.g. weighing against military and many other problematic expenses. In ethical perspective, one has to examine the inconsistencies in our societies.

There is also independent research promoting endeavours, for example, the California Institute for Regenerative Medicine providing three billion dollars for stem cell research [63]. It was immediately confronted with law suits, one of them in federal court asking for recognition of embryos outside the body as full citizens of the United States with all the rights of children [64].

Due to these debates the focus remained on embryos, no interest was shown for dealing with questions how to perform embryonic stem cell research, how to transfer research findings into clinical practice and how to get consent from both gamete donors or how to regulate the procedures of donation.

Research donors are a special group of donors to be dealt with specifically. There is the issue of compensation for eggs, of the procedures to undergo, the time spent, the hormones taken, the side-effects encountered. If a woman can make the decision to go through IVF/ICSI procedures to become pregnant, why can she not make it for research purposes? Is it, because the result is not immediately hers, not for so-called therapeutic goals of hers? This is certainly also not the case if a fertile woman undergoes fertility treatment like ICSI for conceiving from an otherwise infertile partner. To have a child is a gift to the world. Not to have child is a gift to the world. To give your eggs, your sperm to research purposes might be an incommensurable gift to suffering people.

The embryonic stem cell issue has become a question of beliefs, politics, and power forces. The essential question still would be the medical, which is almost forgotten. It is the task of ethical inquiry to put it into the right perspective again. The question at stake is not to whether or not stem cell research as such should be funded publicly, but rather how funds are to be prioritized. It means to address the issue whether or not embryonic stem cell research carries the promise of helping potential counteracting illness and disease and for how many people it will be available if so.

15.5.7 The Ethical Challenge of Research

Can we be sure what this research will produce? Of course, not. The results of research are open-ended. If we knew the results we would not have to do the research. Even when embryonic stem cell research does not produce every hoped for result, the attempt is still necessary. We obtain results only if we do the research. Medical research is not isolated, not something that can be developed apart from medical-therapeutic goals. Moreover, it is interwoven with practice. Regarding embryo research based on the conviction of the problematic equivalence: "lives of embryos" = "lives of sick people," can burden us with guilt [65]. Merkel goes a step further and makes clear that hesitation to do research in this area presents an ethical problem. For him, the bill for the lack of medical research is not paid with just challenging the patience of researchers, but possibly by the sacrifice of human lives [66]. It is argued, that according to the precautionary principle more weight should be given to the dangers forthcoming with the introduction of a new, one might even say experimental technology than to eventual benefits. Harris says, "This is irrational" [67]. "We are talking about being so cautious as to deprive people of the possibility of therapies for crippling and lethal conditions and standing by while victims mount up year on year" [68]. The weighing is between the harm of side effects and the harm further provoked, if the new technology is not introduced at all. Also we do not have rational measures to generally decide and it is a case-by-case decision-making, which is appropriate here. One of the fatal misunderstandings of how research, especially medical research, works is the idea of separation of research from clinical application. So also it is not clear what is called therapeutic and for what reason. The goal of clinical practice is patient care and the goal of research is to find knowledge and the procedures of how to apply knowledge gained by research for the sake of patients' cure. The separation idea seems to work for moralization only to stop or delay research. There is a legitimate concern about clinical application of recent findings, e.g. about the tumorigenic potential of transplanted stem cell-derived tissues. But such a concern is to be decided by patients who often do not have much choice because of severe or fatal illness and so would prefer a risky procedure to no help at all. Taking on a risky treatment or being likely to die from a fatal disease is the patient's decision and not one of politicians or religious people through withholding such choices by stopping such vital research. Three recommendations have been proposed and would make sense: [69] (1) a national body should oversee stem cell transplantation studies. (2) a mandatory international registry of studies should be established. (3) a national panel should be created to set guidelines for studies. This is a step regarding how research should be performed and critically evaluated. It is a step away from the fatal debate whether at all embryonic stem cell research should or should not be performed.

15.6 Conclusions: Humaine Medicine – Medicine for Suffering People

The consequentialistic ethical perspective focuses on those suffering from chronic and not genuinely curable diseases. Researchers have an enormous responsibility. Most of them agree upon the need of research using embryonic stem cells as well as adult stem cells. As far as we know now, embryonic stem cell research is not very promising in terms of therapeutic gains because of the side-effects of these stemcells (tumorigenity, translation of infectious agents, etc.) but nevertheless not at all neglectable in terms of gaining basic knowledge about the development of diseases and their understanding, trials for toxicity of pharmaceuticals, development of embryos, understanding of fertility, etc.

Embryonic stem cells seem promising perhaps for curing numerous different diseases. By favoring the embryo over people in need of cures, suffering and dying people are prevented from a realistic chance to get cures, improvements of their quality of life. Who is taking responsibility for that? Do affected sick people present and future patients have a right to demand that this research be conducted, and moreover, that it be conducted now and without hesitation? [66] Similarly Harris calls embryonic stem cell research as well as therapeutic cloning mandatory [70].

"Where humanity stops, pseudo-ethics begins" [71]. This is true especially where rigid societal moral concepts lead to inhumane results for actual people. Humanism and ethics are largely interchangeable concepts. They come together for one another. We must judge our principles and our ideals on the basis of humanity, if they are to prove to be ethical. "Bring the principles, opinions and ideals into agreement with humanity. In this way we will establish ethics according to reason, for the true ethics is truly rational".[72] It is humaine to have solidarity with ill and suffering people and it is inhumaine to retreat to unquestioned morals and beliefs. The ethical courage is needed to think and act after critical analysis of familiar moral assessments. It is necessary then to dare the adventure into the incompletely known for the benefit of humanity.

References

1. PBS. August 9, 2001 "News Hour." www.pbs.org/search
2. Weissmann, J.L. 2002. Stemcells – scientific, medical, and political issues. *NEJM* 346:1576–1579 esp.1578.

3. Gruen, L., and Grabel, L. 2006. Concise review: Scientific and ethical roadblocks to human embryonic stem cell therapy. *Stem Cells* 24:2162–2169; esp. 2164.
4. Shibles, W. 1972. *Philosophical pictures*. Dubuque, IA: William, C. Brown.
5. Dewey, J., and Tufts, J. 1932. *Ethics*. New York: Holt; Dewey, J. 1932/1996. *Theory of the moral life*. New York: Irvington Publishers Inc. edn.; Dewey, J. 1939. *Theory of valuation*. Chicago: University of Chicago Press.
6. Evers, K. 2002. European perspectives on therapeutic cloning. *NEJM* 346:1579–1582; PhRMA Genomics: Cloning and stem cell research. http://www.genomics.phrma.org/cloning.html.
7. Nüsslein-Volhard, C. 2004. *Von Genen und Embryonen*, 42. Stuttgart: Reclam.
8. Thomson, J.A., Itskovitz-Eldor, J., Shapiro, S.S., Waknitz, M.A., Swiergiel, J.J., Marshall, V.S., and Jones, J.M. 1998. Embryonic stem cell lines derived from human blastocysts' *Science* 282:1145–1147; Stem Cell Research News: http://www.stemcellresearchnews.com/ University of Wisconsin: News and background information detailing embryonic stem cell research at the University of Wisconsin-Madison: http:///www.news.wisc.edu/packages/stemcells/.
9. Thomson, J.A. 2001. Human embryonic stem cells. In *The human embryonic stem cell debate: Science, ethics and public policy.* eds. Holland, S., Lebacqz, K., and Zoloth, L., 15–26. Cambridge, MA, London, England: A Bradford Book, MIT Press.
10. Okarma, T. 2001. Human embryonic stem cells: A primer on the technology and its medical applications. In *The human embryonic stem cell debate: Science, ethics and public policy.* eds. Holland, S., Lebacqz, K., and Zoloth, L., 3–13 Cambridge, MA; London, England: A Bradford Book, MIT Press.
11. Zwaka, T.P., and Thomson, J.A. 2003. Homologous recombination in human embryonic stem cells. Published online February 10th 2003; doi:10.1038/nbt788.
12. Hall, V.J., Stojkovic, P., and Stojkovic, M. 2006. Using therapeutic cloning to fight human disease: A conundrum or reality? *Stem Cells* 24:1628–1637; esp.1634.
13. Rosenthal, N. 2003. Prometheus` vulture and the stem-cell promise. *NEJM* 349:267–274.
14. Research on Human Embryonic Stem Cells. 2009. Opinion of the Austrian bioethics commission of March 16th, 2009.
15. Strauss, E. 1999. Brain stem cells show their potential. *Science* 283:471.
16. Harris, J. 2004. *On cloning*, 116. London, New York: Routledge.
17. NIEHS. 1999; *Stem Cells: International Journal of Cell Differentiation and Proliferation.* http://stemcells.alphamedpress.org/; Stem Cell Research Foundation: http://www.stemcellresearchfoundation.org/Aout/about.htm.
18. Research on Human Embryonic Stem Cells. 2009. Opinion of the Austrian Bioethics Commission of March 16th, 2009: 68.
19. Kimmelmann, J., Baylis, F., and Glass, K.C. 2006. Stem cell trials: Lessons from gene transfer research. *Hastings Center Report* 36:23–26; esp. 25 f.
20. Thomson JA 2001. Human embryonic stem cells. In *The human embryonic stem cell debate: Science, ethics and public policy.* eds. Holland, S., Lebacqz, K., and Zoloth, L., 15–26; esp. 22 Cambridge, MA; London, England: A Bradford Book, MIT Press.
21. Understanding stem cells. 2009. An overview of the science and issues from the National Academies. Washington, http://dels.nas.edu/bis/stemcells/basics.shtml visited Sept 25th, 2009; Pecorino, L., Stem Cells for Cell-Based Therapies. An actionbioscience.org original article with links to other stem cells websites and articles: http://www.actionbioscience.org/biotech/pecorino2.html.
22. Harris, J. 2004. *On cloning*, 146. London, New York: Routledge.
23. Gruen, L., and Grabel, L. 2006. Concise review: Scientific and ethical roadblocks to human embryonic stem cell therapy. *Stem Cells* 24:2162–2169; esp. 2163.
24. Proceed with Care: Final Report of the Royal Commission on New Reproductive Technologies 1993: 607. www.fedpubs.com/subject/health/newegno.htm. Federal Publications Inc Toronto, Ontario, Canada.

25. *Convention on Human Rights and Biomedicine* of the European Union 1997. (European Treaty Series No 164) http://conventions.coe.int/Treaty/commit/QueVeuler-Vaus.asp?NT=164&CL=ENG.
26. *Convention on Human Rights and Biomedicine* of the European Union 1997. Art 18. (European Treaty Series No 164) http://conventions.coe.int/Treaty/commit/QueVeuler-Vaus.asp?NT=164&CL=ENG.
27. Nüsslein-Volhard, C. 2004. *Von Genen und Embryonen*, 22. Stuttgart: Reclam.
28. Nüsslein-Volhard, C. 2004. *Von Genen und Embryonen*, 54. Stuttgart: Reclam.
29. see Maier, B. 2005. Unsere Empfindsamkeit für und unsere Sorge um das Leben von Menschen auf dem Hintergrund moderner Biotechnologien. In *Leben nach Maß – zwischen Machbarkeit und Unantastbarkeit. Gentechnologie im Licht von Albert Schweitzers Philosophie und Theologie*, ed. Schüz, G., 115–142; esp.124 Frankfurt am Main. www.bundeskanzleramt.at/bioethik/ Detail: http://www.bka.gv.at/Docs/2006/4/3/BMaier_Schüz.pdf.
30. Brock, D. 2006. Is a consensus possible on stem cell research? Moral and political obstacles. *Journal of Medical Ethics* 32:36–42; esp. 38.
31. Sandel MJ 2004. Embryo ethics: The moral logic of stem cell research. *NEJM* 351:207–209.
32. Farley MA 2001. Roman Catholic views on research involving human embryonic stem cells. In *The human embryonic stem cell debate: Science, ethics and public policy*, eds. Holland, S., Lebacqz, K., and Zoloth, L., 113–118. Cambridge, MA; London, England: A Bradford Book, MIT Press; Mendiola, M.M. 2001. Human embryonic stem cells: Possible approaches from a Catholic perspective. In *The human embryonic stem cell debate: Science, ethics and public policy.* eds. Holland, S., Lebacqz, K., and Zoloth, L., 119–125. Cambridge, MA; London, England: A Bradford Book, MIT Press.
33. Brock, D. 2006. Is a consensus possible on stem cell research? Moral and political obstacles. *Journal of Medical Ethics* 32: 36–42; esp.36
34. Maier, B. 2005. Unsere Empfindsamkeit für und unsere Sorge um das Leben von Menschen auf dem Hintergrund moderner Biotechnologien. In *Leben nach Maß – zwischen Machbarkeit und Unantastbarkeit. Gentechnologie im Licht von Albert Schweitzers Philosophie und Theologie*, ed. Schüz, G., 115–142; esp.118 f. Frankfurt am Main. www.bundeskanzleramt.at/bioethik/Detail: http://www.bka.gv.at/Docs/2006/4/3/BMaier_Schüz.pdf.
35. Nüsslein-Volhard, C. 2004. *Von Genen und Embryonen*, 67 f. Stuttgart: Reclam.
36. Lebacqz, K. 2001. On the elusive nature of respect. In *The human embryonic stem cell debate: Science, ethics and public policy.* eds. Holland, S., Lebacqz, K., and Zoloth, L., 149–162 Cambridge, MA; London, England: A Bradford Book, MIT Press.
37. Merkel, R. 2002. Forschungsobjekt Embryo. Verfassungsrechtliche und ethische Grundlagen der Forschung an menschlichen embryonalen Stammzellen. 135–139 München: dtv.
38. Meilaender, G. 2001. Some protestant reflections. In *The human embryonic stem cell debate: Science, ethics and public policy*, eds. Holland, S., Lebacqz, K., and Zoloth, L., 141–147; 143 Cambridge, MA; London, England: A Bradford Book, MIT Press.
39. Convention on Human Rights and Biomedicine of the European Union 1997. (European Treaty Series No 164) http://conventions.coe.int/Treaty/commit/QueVeuler-Vaus.asp?NT=164&CL=ENG.
40. Shibles, W. 1980. Ethics as a science: Going from is to ought. *Iowa Science Teachers Journal* 3:26–32.
41. McGee, G., and Caplan, A.L. 1999. The ethics and politics of small sacrifices in stem cell research. *Kennedy Insitute of Ethics Journal* 9:151–158; esp. 151.
42. Matthiesen, L. ed. 2002. Survey of Opinions from the National Ethics Committees or Similar Bodies, Public Debate and National Legislation in Relation to Human Embryonic Stem Cell Research and Use. Vol, I. in EU Member States, European Commission Research Directorate-General. Directorate E-Life Sciences: Biotechnology, agricultural and food research.

43. McGee, G., and Caplan, A.L. 1999. The ethics and politics of small sacrifices in stem cell research. Kennedy *Insitute of Ethics Journal* 9:151–158.
44. McGee, G., and Caplan, A.L. 1999. The ethics and politics of small sacrifices in stem cell research. *Kennedy Institute of Ethics Journal* 9:151–158; esp. 152 f.
45. Wittgenstein, L. 1958. *Philosophical investigations*, 3rd edn. New York: Macmillan.
46. Evers, K. 2002. European perspectives on therapeutic cloning. *NEJM* 346:1579–1582.
47. Harris, J. 2004. *On cloning*, 6. London, New York: Routledge.
48. Shibles, W. 1974. Ethics as open-context terms. In *Wittgenstein, language and philosophy*, rev. 3rd edn. ed. Shibles, W., 299 ff. Dubuque, Iowa: Kendall-Hunt.
49. Mendiola, M.M. 2001. Human embryonic stem cells: Possible approaches from a Catholic perspective. In *The human embryonic stem cell debate: Science, ethics and public policy.* eds. Holland, S., Lebacqz, K., and Zoloth, L., 119–125; esp.119 ff. Cambridge, MA; London, England: A Bradford Book, MIT Press.
50. Bjornson, C.R. 1999. Turning brain into blood: A hematopoetic fate adopted by adult neural stem cells in vivo. *Science* 283:534–537.
51. Report of the Human Fetal Tissue Transplantation Research Panel NIH 1988.
52. Meyer, J.M., and Nelson, J.L. 2001. Respecting what we destory: Reflections on human embryo research. Hastings *Center Report* 31:16–23.
53. Drukker, M., Katchman, H., Katz, G., Friedman SET, Hornstein, E., Mandelboin, O., Reisner, Y., and Benvenisty, N. 2006. Human embryonic stem cells and their differentiated derivatives are less susceptible to immune rejection than adult cells. *Stem Cells* 24:221–229; Lamelsky, N., Blondel, O., and Laeng, P. 2001. Differentiation of embryonic stem cells to insulin-secreting structures similar to pancreatic islets. *Science* 292:1389–1394; Vaca, P., Martin, F., Vegara-Meseguer, M., Rovira, J.M., Berna, G., and Soria, B. 2006. Induction of differentiation of embryonic stem cells into insulin-secreting cells by fetal soluble factors. *Stem Cells* 24:258–265.
54. Weissmann, J.L. 2002. Stemcells – scientific, medical, and political issues. *NEJM* 346: 1576–1579
55. Gruen, L., and Grabel, L. 2006. Concise review: Scientific and ethical roadblocks to human embryonic stem cell therapy. *Stem Cells* 24:2162–2169; esp. 2168.
56. Hall, S.S. 2006. Stem cells: A status report. *Hastings Center Report* 36:16–22; esp.20.
57. Meissner, A., and Janisch, R. 2006. Generation of nuclear transfer-derived pluripotent ES cells cloned Cdx2-deficient blastocysts. *Nature* 439:212–215.
58. Solter, D. 2005. Politically correct human embryonic stem cells? *NEJM* 353:2321–2323.
59. Harris, J. 2004. *On cloning*, 142. London, New York: Routledge.
60. Embryonale Stammzellenforschung: Obama erlaubt Staatliche Förderung (embryonic stemcell research: Obama allows for public funding): Spiegel online: http://www.spiegel.de/wissenschaft/mensch/0,1518,612278,00.html (accessed october,1st 2009).
61. Meilaender, G. 2001. The point of a ban: Or how to think about stem cell research. *Hastings Center Report* 31:9–16.
62. Survey of opinions from National Ethics Committees, public debate and national legislation in the UK (*Nuffield Council on Bioethics 2000)* and from the law in the form of the *Human Fertilization and Embryology Authority (HFEA)* (Matthiesen, 2002).
63. Magnus, D. 2006. Stem cell research: The California experience. *Hastings Center Report* 36:26–28.
64. Magnus, D. 2006. Stem cell research: The California experience. *Hastings Center Report* 36:27.
65. McGee, G., and Caplan, A.L. 1999. The ethics and politics of small sacrifices in stem cell research. Kennedy *Institute of Ethics Journal* 9:151–158.
66. Merkel, R. 2002. *Forschungsobjekt Embryo. Verfassungsrechtliche und ethische Grundlagen der Forschung an menschlichen embryonalen Stammzellen*, 259. München: Dtv.
67. Harris, J. 2004. *On cloning*, 16. London, New York: Routledge.
68. Harris, J. 2004. *On cloning*, 15 f. London, New York: Routledge.

69. Kimmelmann, J., Baylis, F., and Glass, K.C. 2006. Stem cell trials: Lessons from gene transfer research. *Hastings Center Report* 36:23–26; esp. 25 f.
70. Harris, J. 2004. *On cloning*, 143. London, New York: Routledge.
71. Schweitzer, A. 1923/1996. *Kultur und Ethik*, 349. München: Beck'sche Reihe.
72. Schweitzer, A. 1923/1996. *Kultur und ethik*, 325. München: Beck'sche Reihe.

Chapter 16
The Philosophy of Prevention

Every day we do not help people who are dying.
This is true – not only for physicians.

Abstract We use the word "prevention" as if its meaning is fully before us, self contained and as if we know what it means. Prevention of what? In the first place, prevention presupposes something to be prevented. If prevention has the goal to eliminate or lessen disease, then what prevention means depends on the definition of disease. If disease is thought of only as a physical dysfunction, then prevention only involves avoidance of that physical dysfunction. If disease is defined as whatever blocks one's total physical, emotional and cognitive well-being and improvement of health, then prevention involves helping people change their lifestyle to move toward a balance of physical, emotional, social, spiritual, and intellectual health. Prevention has the insight to anticipate rather than wait for disaster to happen. Prevention is an active process, prevention is a kind of practical as well as philosophical intervention.

Keywords Hunger · starvation · prevention · poverty · disease · health · aging · age · death · personal lifestyle problems · structural lifestyle problems

16.1 Introduction

Before you read this chapter, ask yourself: What could I have done indirectly to save lives and prevent deaths? Then you will also know how many lives you could have saved. We kill just by being in a position to prevent killing. All not helped starving people are thereby helped to die. Over a billion people in the world are starving. Over 25,000 of these die daily. Millions of people are also dying of various diseases and other causes [1]. Six million children under the age of 5 die every year as a result of hunger while every country in the world has the potential of growing enough food to feed all of its people. Most of these countries are in Sub-Saharan Africa [2]. Hunger and starvation is not fate but caused by people especially the very rich ones of the so called "first world" and their so called "free-market" policies [3]. Every day 140,000 die from the various diseases. The causes and number of these

B. Maier, W.A. Shibles, *The Philosophy and Practice of Medicine and Bioethics*, International Library of Ethics, Law, and the New Medicine 47,
DOI 10.1007/978-90-481-8867-3_16, © Springer Science+Business Media B.V. 2011

daily deaths are: infections 50,000, cardiovascular 35,000, cancer 15,000, violent death 10,000, diarrhea 10,000, maternal death 1,600 (Cf. WHO statistics). The U.S. Center for Disease Control (CDC) reported that the leading causes of death in 2004 were: 1. heart failure, 2. cancer, 3. stroke, 4. lower respiratory disease, 5. injuries, 6. diabetes. Guns killed 28,874 people in the U.S. in 1999 [4]. *What is significant is that most all of these deaths are preventable.* CDC accordingly gives the preventable causes of diseases as follows: 1. tobacco use, 2. inactivity and poor diet, 3. alcohol use, 4. microbial e.g. flu 5. toxic agents (asbestos, pollutants), 6. vehicle crashes, 7. firearms, 8. illicit drug use. Another source gives a projected eighty million people will die of AIDS by 2010: (Russia has 5–8 million, China 10–15 million, Indonesia 20–25 million). WHO has planned to try to treat 3 million with aids by 2005. Compared with those infected, this is a totally negligible number. In 2003, India had roughly 1,027 billion people, 35–40% below poverty level, 65% illiteracy, 7% (ca. 4 million) with AIDS, one million with TB, and 16 million births each year.

16.2 An Analysis of Prevention

Life expectancy for the world is about 65 for men, 70 for women. For industrial Western countries: men 75, women 82 years. The death rate for some countries in Africa is: 45 for men, 48 for women [5].

In 2003 the U.S. death rate was 9.1 per 1,000 people; in 2005, 8.78 per 1,000 (around 2 1/2 million) [6]. Life expectancy in 2005 was highest in Andorra, San Marino, Singapore, Japan, Australia, Sweden, Switzerland, Iceland, Canada, and Italy.

According to the WHO REPORT 2006 there is a life expectancy to age 40 s: in Afghanistan, Angola, Botswana, Central African Republic, Ivory Coast; to the age 80 s in Ireland and U.K.

The probability of dying is in Western Countries 3–5 per 1000 before age 5, in Africa 200+ (e.g., Angola, Niger, etc.)

3 billion people of 6.5 billion in world live in poverty (WHO); 47% of people in Bangladesh, 138 million, are below poverty [7].

In Sudan ca. 6.7 million people need food, mostly in Darfur.

Sierra Leone has 6 million people living in poverty.

Even in the Western world poverty is a big problem. In 2004 poverty in Germany affects one in eight residents or 10.6 million people. The Federal Statistics Bureau in its new study says that the figure includes 1.7 million children.

Panama has 90% people living in poverty in the indigenous areas.

Hunger kills 24,000 each day including 18,000 children, one each 3.6 s, 1,000 per hour. 8.700000 each year. One in six people are hungry [8].

According to the National Coalition on Health Care 2004 (the latest year data are available) total national health expenditures rose 7.9% – over three times the rate of inflation [9].

Total spending was $7100 per American in healthcare costs National Institute of Health (NIH). Total health care spending represented over 16% of the gross domestic product (GDP).

U.S. health care spending is expected to increase at similar levels for the next decade reaching $4 trillion in 2015, or 20% of GDP. Health care spending was 4.3 times the amount spent on national defense but: Although nearly 46 million Americans are uninsured, the United States spends more on health care than other industrialized nations, and those countries, which provide health insurance to all their citizens.

Health care spending accounted for 10.9% of the GDP in Switzerland, 10.7% in Germany, 9.7% in Canada and 9.5% in France, according to the Organization for Economic Cooperation and Development.

24,000 people die from hunger and hunger-related causes daily. Of those, 18,000 are children [10].

One person dies of hunger every 3.6 s. That is more than 16 people each minute; 1,000 each hour; which translates into 8,760,000 every single year! One in six people on the planet is hungry.

This is preventable: people are undernourished because of food-distribution problems, natural disasters, government policies, civil unrest, inequitable trade policies, lack of knowledge and greed [3].

The question arises how all people of this world may live longer healthier lives. Medicine is primarily concerned with treating disease and disorder. We may also be concerned to stress how we may live a long healthy life and determine what we may do to prevent disease and disorders.

We use the word "prevention" as if its meaning is fully before us, self contained and as if we all know exactly what it means. We use the word abstractly without definition as if we already know all that it means and so mis-use the term. To understand what "prevention" means we can look at its uses and misuses in ordinary and medical language.

Prevention of what? In the first place, prevention presupposes something to be prevented. We cannot have prevention or safety as such. Prevention is misused if one does not know what it is to be prevented. One may take a daily aspirin or other pills without knowing what they are being taken for. One may take an arthritis painkiller to relieve pain, but which actually increases the inflammation, that is not to prevent the disease from becoming worse. One may take vitamins to try to prevent disease, but which do nothing to improve health. Abstinence prevents pregnancy and disease, but it also prevents human intimacy, human desires and needs. Prevention can apply to all undesirable areas of our lives, e.g., bad management, irrational thinking, etc. Philosophy may be defined as prevention of mistakes and irrational, uncritical thinking (speaking) in the various disciplines including the philosophy of medicine. It becomes the paradigm of prevention.

If prevention has the goal to eliminate or lessen disease, then what prevention means depends on the definition of disease. This is not agreed upon. If disease is thought of narrowly only as a physical dysfunction, then prevention only involves avoidance of that physical dysfunction. If disease is defined as whatever blocks one's total physical, emotional and cognitive well-being and improvement of health then prevention involves a holistic approach to eliminating those things, which do not lead to that goal. "Health promotion is the science and art of helping people change their lifestyle to move toward a state of optimal health. Optimal health is defined as a balance of physical, emotional, social, spiritual, and intellectual health. Lifestyle change can be facilitated through a combination of efforts to enhance awareness, change behavior and create environments that support good health practices. Of the three, supportive environments will probably have the greatest impact in producing lasting change" [11]. The World Health Organization defines health as "a state of complete physical, mental and social well-being and not merely the absence of disease or infirmity" [12]. On this view, prevention would involve improvement of: critical thinking (speaking), education, the environment, ethical understanding and practice, exercise, an aesthetic life, nutrition, physical health, treatment of physical illnesses, etc. In short, humanistic, holistic and philosophical prevention involves active decision-making at all levels to deal with one's whole personality. There is narrow, limited prevention and broad, holistic prevention, medical prevention and philosophical prevention. "Etiology" is defined as the philosophical investigation of causes and origins. In medicine it refers to the investigation of causes and origins of disease. The issue of prevention is one to be dealt with not just by medicine, but also by philosophy and the philosophy of medicine. If by prevention is meant enforcing narrow and enculturated values, medical prevention as it often now exists is unethical. Thus, there can be rational and irrational prevention depending upon what is to be prevented.

The notion of prevention may be clarified using the Metaphorical Method (See Chapter 1). The first method is to clarify the term by an examination of synonyms. Synonyms of prevention are: accountability, active concern about, analyzing, anticipating, avoidance, block, care, carefulness, detection, deter, diagnosis, dissuasion, education, forecast, foresight, forethought, insurance, not let happen, planning, prediction, precaution, problem-solving, prophesy, protection, prudence, responsibility, safety measure. Prevention has the insight to anticipate rather than wait for disaster to happen. Prevention is an active process. In a sense, intervention is a kind of prevention, and prevention is a kind of intervention.

In terms of synonyms, not preventing or failure to prevent involve: abandonment, blame, carelessness, evasion, fault, inattention, irresponsibility, lack of accountability, lack of foresight, liability, negligence. Failure of prevention is by definition to be irrational and irresponsible. Not preventing a death one could have prevented is like causing the death (See Chapter 11). And this is what we are doing and should be held responsible for – worldwide.

Prevention may entail, then, actively being responsible, having concern to rationally plan in advance in order to achieve desired events and avoid undesired events. Then prevention is active involving deliberately acting by not doing something, or acting by doing something. Deliberately or consciously letting one have a disease is the same as causing and not curing a disease. One still ends up with the disease in both cases. To claim there is a difference is to rationalize. Whether one causes starvation or lets it happen, there is still starvation.

Prevention is by definition to be rational and responsible. "Practice" as in medical practice means positive and aggressive action and application, not just theory. It means active treatment to improve and not do harm. When one can prevent harm and does not, it is not medical practice, but negligence. The physician's task is not just to cure disease, but to prevent it in the first place. Prevention is itself treatment.

On the above view, prevention is rational, but it also often is more effective in achieving one's goals. We may base altruism not on egoism, but because of the overall consequences (See Chapter 10). Preventing heart attacks can be easier than finding, financing and surgically replacing a heart. Prevention of an epidemic can prevent deaths, which are too late for medical treatment. Thus, "An ounce of prevention is worth a pound of cure." (or in German, *Vorbeugen ist besser als Heilen*, "Prevention is better than cure. ") The preventative use of contraception, often called "preventatives," is better than contracting AIDS. Senator Tom Harkin (2004) stated that $1.8 trillion is spent in the U.S. on healthcare each year, 75% on chronic diseases, which were preventable. Only 2% is spent on prevention [13]. This is not a picture of healthcare, but of "sick care." 70% of deaths are linked to poor nutrition. Junk food marketers spent $10 billion in 2002 on advertising. At the same time 36 million people die each year from starvation and associated diseases worldwide [3].

In its negative aspect, prevention presupposes something is undesired. It may be noted that prevention also has a positive aspect in that it presupposes something desired. We do not desire disease, but we do desire not only good health, but better health. Medicine also may cure disease, but also be used to promote the best health possible. The philosophy of medicine is not just to maintain a status quo of minimally controlling disease, but of maximally improving the whole person. This is the positive task of prevention. One of the best methods of holistic prevention is by means of education in philosophy and by philosophy of medicine. This is what sound philosophical counseling is based on.

We have spoken of holistic and deep philosophical prevention versus narrow or superficial un-philosophical prevention, as well as ethical prevention versus mere enculturated prevention. On the basis of these distinctions we may also derive the necessity of prevention, which involves everyone in the world versus only a chosen few. Thus, holistic prevention must involve full concern with universal world consequences. Limited prevention becomes universal prevention. Limited concern becomes universal concern. Local ecology becomes "human ecology" as well as "environmental ecology." Normative, common morality is often confined to one's immediate family, one's immediate environment, or one's nationalistic borders. This is, of course not ethics or ethical. One places oneself, or slightly extended self, at the center of the universe regardless of the consequences for others. It is egoistic prevention, sometimes referred to as "national security." Groups not included in one's own group become enemies. Nationalism leads directly to war. In terms of number and scope ethical prevention must consider universal prevention. This relates to issues such as world starvation, world poverty, not just national, but worldwide healthcare systems. The EU, U.N., Peace Corps, Peace organizations, Greens organizations, and "Doctors without Borders" organizations, Red Cross, etc. are just a few examples of organizations, which move against narrow prevention toward world-wide prevention. All such organizations tend to be outside of the scope of individual nationalistic governments. Medical care can thus involve either limited prevention or world-wide prevention. This is similar to limited utilitarianism versus universal utilitarianism. According to the World Peace Forum 2006 "We will declare war as a crime against humanity and demand an end to war" [14].

16.3 Unethical Behavior and Irresponsible Lifestyles

Many diseases are preventable, which means that they are caused by people themselves or people around them either directly or indirectly. They may thus be termed, lifestyle diseases of comfort or in the other direction diseases by lack of what is unconditionally necessary for life. Disease in this regard is mainly caused by ignorance, lack of knowledge, irresponsibility and poor management.

The following statements illustrate:

Largely preventable chronic diseases cause 86% of deaths in Europe [15].

As many chronic diseases are closely linked to lifestyles, an estimated 80% of heart disease, stroke and type 2 diabetes, and 40% of cancer, could be avoided if common lifestyle risk factors were eliminated.

U.S. diseases 2.3 million (2003): HIV, 13,658, alcoholic liver disease 12,360, viral hepatitis 5,431, anemia 4,594, asthma 4,099, malnutrition 3,153, SIDS 2,162, TB 711, pregnancy and birth 545, appendicitis 439, West Nile virus 264, anorexia 79, salmonella 43, measles 1. Flu kills 36,000 Americans each year. Cholesterol kills 700,000 Americans each year [16].

Today, non-communicable diseases cause 86% of deaths and 77% of the disease burden in the WHO European Region. This group of conditions includes cardiovascular diseases, cancer, mental health problems, diabetes mellitus, chronic respiratory disease and musculoskeletal conditions. Cardiovascular diseases are the number-one killer, causing more than half of all deaths across the Region, with heart disease or stroke the leading cause of death in all countries. Seven leading risk factors – high blood pressure, tobacco, alcohol, high cholesterol, overweight, low fruit and vegetable intake and physical inactivity – account for almost 60% of all ill health in the Region. The leading risk factors are high blood pressure for death, and tobacco for ill health. Alcohol is the leading risk factor for both ill health and death among young people in the Region.

"Approximately 70% of all premature deaths before the age of 65 could be accounted for by lifestyle and environmental factors" [17].

WHO claims that 80% of diseases are preventable [18].

According to the *UN Chronicle,* [19] if the major risk factors are eliminated, 80% of heart disease, stroke and type 2 diabetes, as well as 40% of cancer would be prevented. 21 million Americans have diabetes (type 1 and 2) (WHO). Fewer than 12% of people with diabetes meet the recommended goals. They fail by e.g., too little exercise, by not lowering the fat intake, etc [20].

According to Barondess, 80% of people in the US over age 65 have at least one chronic disease. 48% have three or more [21]. The latency developing period was often very long. Prevention is often made difficult because people may presently feel "healthy" regardless of however bad their actual physical health condition may be. Unhealthy plaque is even found in children's arteries. That is, prevention is life-long. It begins before birth.

Half of the people over age 65 have high blood pressure [22]. 30% of adults in most countries have high blood pressure or hypertension. Half of all people over age 65 take some medication [23].

Tobacco causes 5 million deaths a year (WHO, CDC). By 2020 diseases from tobacco are expected to exceed all other causes of disease [24].

25–40% of all U.S. hospital patients are undergoing treatment related to alcohol consumption, and 200,000 deaths yearly result from alcohol related conditions.

Responsibility requires people to consider, as much as reasonably possible, all of the consequences of their actions. Those who abuse their health should themselves be held responsible and they should not shift their responsibility and financial burden on others such as taxpayers or healthcare workers. Proposals were made in the Netherlands to give incentives to people who use the healthcare system only minimally [25]. Punishment and ascription of guilt must be avoided – it does not help the condition of ill people and in addition undermines societal solidarity. This would ruin a system of shared responsibility and solidarity. Also, it is not at all possible to fully understand what has caused detrimental conditions and what people contributed to them or if they did at all.

The more the technology advances, the more people may be negligent regarding lifestyle because they might think that the medical care can repair any damage they do to themselves. As cars become safer, people undo the safety by driving more and faster. This is the nemesis of progress, which needs challenge in terms of ethical progress evaluation.

16.4 Lifestyle as Philosophical and Critical Thinking

People cause themselves to have poor health because of failure to develop critical philosophical thinking, healthy behavior and rational and ethical judgments and policies. Although this is the case, they should not be blamed, but educated.

They may contribute to having a healthful and holistic style of life and, if gifted with reliable genes, to a long and healthy life. Responsible medicine requires knowledge and practice of the philosophy of medicine. On this view, physicians need to do more than merely conduct physical examinations. Hyman and Liponis state, "Lifestyle and other details are essential to an understanding of the patient's condition" [26]. This is true in an even larger philosophical sense than the authors intended. Those who are uncritical, anti-philosophy and anti-inquiry cannot have a healthful lifestyle. The typical society and culture is not a healthful one. People should not be angry because the absence of medical care and cures, because they by their own beliefs and lifestyles caused this situation.

16.5 Areas of Prevention

The following will deal with some of the various problems of prevention in medicine. A full treatment of the positive and negative forms of prevention would, however, include all of medicine.

16.5.1 Education: The Greatest Preventative of Disorder

Disease prevention is a cognitive matter. It is based upon our ability to think critically, analyze causes, and solve problems. The greatest enemy of such prevention is faulty thinking and supernatural beliefs. For these reasons, education in critical thinking (speaking) and ethics are the major ways in which disease, physical, and psychological abuse may be prevented and treated.

Motor vehicle accidents rank as the third most frequent cause of death [27]. The first leading cause of death for ages 1–44 in the U.S. is accidents including auto and sports [28]. In 2003 40% of auto accidents involved drunk driving [29]. Deaths per 100,000 registered vehicles: Austria 21.2, Hungary 42, Poland 41.7, Portugal 29.5, Spain 22.9, U.K. 11.2, U.S. 18.9 [30]. Husak reported that in 2001, an estimated 6,323,000 crashes injured 3,033,000 people. The number of fatalities in the U.S. rose in 2003 to 43,200 [31].

Husak argued that philosophy has been negligent in failing to address the unethical use of automobiles especially as they cause more deaths than perhaps any other cause. He argues especially that reasonable and non-negligent people should not jeopardize the welfare of others by driving crash-incompatible vehicles for frivolous purposes [31]. Pickup trucks and SUVs (sport utility vehicles) are called crash incompatibility vehicles because they are known to cause more damage and death than the usual passenger car. The author states that it is unethical to use them as they cause undue risk and death to others and they are rarely necessary for those driving them. "In America more than a trillion miles per year is purely recreational and cannot be justified under any plausible interpretation of a necessity requirement." This means that personal and impersonal risk analysis would require that one use automobiles as seldom as possible.

To promote automobile safety and prevent accidents involves every level of decision-making, from lessening pollution through clean fuel, environmentally favorable policies, better public transportation systems, and restrictions on unnecessary driving to making road signs clearer. Every aspect of transportation is involved. The goal is to have responsible, safer, humanistic use of autos.

16.5.2 Cancer Prevention

Cancer is second most frequent cause of death [32].

According to the WHO Report dietary factors account for 30% of cancers in industrialized countries [33].

According to the American Lung Association, each year smoking is directly responsible for 87% of all lung cancer deaths in America [34]. 20% of all U.S. adults in the US smoke. In 2006, 20,000 US nonsmokers are expected to develop lung cancer, partly due to secondary smoke, asbestos exposure, air pollution, workplace toxins, family history, gender (women have three times more risk than men among nonsmokers) etc [35].

Cancer Stat Fact Sheets. The following information is based on NCI's SEER Cancer Statistics Review and other statistics from NCI's analysis of SEER incidence data and NCHS mortality data.

From 1998 to 2002, the median age at diagnosis for cancer of all sites was 67 years of age. Approximately 1.1% were diagnosed under age 20; 2.9% between 20 and 34; 6.2% between 35 and 44; 13.5% between 45 and 54; 20.2% between 55 and 64; 26.3% between 65 and 74; 22.5% between 75 and 84; and 7.4% 85+ years of age.

The age-adjusted incidence rate was 469.7 per 100,000 men and women per year. These rates are based on cases diagnosed in 1998–2002 from 13 SEER geographic areas.

From 1998 to 2002, the median age at death for cancer of all sites was 72 years of age. Approximately 0.4% died under age 20; 0.9% between 20 and 34; 3.0% between 35 and 44; 8.7% between 45 and 54; 16.3% between 55 and 64; 27.2% between 65 and 74; 29.7% between 75 and 84; and 13.8% 85+ years of age.

The age-adjusted death rate was 197.8 per 100,000 men and women per year. These rates are based on patients who died in 1998–2002 in the US. Death rates by race and sex were: In 2005 new cancer cases and deaths for men were: Lung cancer 94,000 cases, prostate 30,350. New cancer cases (more men than women) 1,372,910, death by cancer 570,280 [36]. The chance for males to develop cancer is 50%, for women is 33%. Over 77% of cancers are diagnosed in people over 55. Since 1971 cancer rates have remained the same except for the fall in stomach cancer occurrence [37]. This suggests little progress in cancer prevention.

Most all of the information on the prevention of cancer is speculative, but some suggestive information is nevertheless available toward the prevention of cancer. Because people have a limited understanding of their lifestyle factors as they relate to cancer risk an interactive web tool [38] was developed to provide them with personalized cancer risk assessments [39]. The website received close to 10 million viewers in 3 years and links over 1,000 websites link to Your Cancer Risk.

On the other hand, in terms of clinical experience, cancer might be preventable to some extent. Fuhrman states, "Cancer is a disease of maladaptation," and "Cancer is much more preventable than treatable" [40]. Epidemiological studies also have shown that diets low in animal fat, meat, alcohol, and/or calories have a reduced risk of cancer. Physical activity for whatever reason reduces the risk of breast, colon and possibly other types of cancer. That cancer is preventable is suggested by the fact that smoking is directly responsible for 87% of all lung cancer deaths in America each year [34]. Genetically, often it is not.

Biesalski states that one-third of world's cancers, 70% of strokes, 90% of type II diabetes is preventable by lifestyle and diet change [41]. In the analysis examining studies of dietary fat and risk of breast cancer, the authors went on to combine case-control and cohort studies and concluded that "higher intake of dietary fat is associated with an increased risk of breast cancer" [42]. Cancer may be more due to high concentration of pesticides than due to fat itself. Only 5–10% of cancers are said to be hereditary.

The best way to reduce cancers is through changes in lifestyle and environmental improvement. There were 7 million cancers in 2001.

Questions of the cause of cancer remain: Do cell phones and chemicals from the dry cleaners cause cancer? Does eating green vegetables binds with carcinogens? Is Insel correct in asserting that even a few drinks a week in women causes cancer? [43] Do fiber, whole grains, Lycopene cut cancer risk? Ensminger states that most preventatives are unproven [44]. It may be noted that there are many kinds of cancer and many kinds of treatment. Cancer is not just one disease. "There may be at least one type of cancer for every cell in the body" [45]. Cancer is a *group of* diseases characterized by uncontrolled growth and spread of abnormal cells [20]. To say that a substance prevents cancer is therefore unspecific. The terms "uncontrolled" and "abnormal" are also value terms. Thus, to ask, "Does x substance help prevent cancer?" commits the many question fallacy. It is also more accurate to say that a substance may, rather than does, prevent cancers. It is also not very informative to state that, because one could equally state that "X substance may not help prevent cancers." It gives little information to say that a substance may or may not help reduce the risk of cancer. Every advertiser could claim as much.

We are left with uncertainty regarding the various alleged causes of cancer, but one certainty is that there are definite things we can do to increase the risk of cancer.

Cancer and its cause(s)? Some causes yield a high percentage of the cause, e.g. cervix uteri cancer, which is a also sexually transmitted disease, endometrial cancer which is related to overweight, lung cancer, which is due to smoking, and so on.

Cancer is responsible for 10% of death in developing countries and 20% of death in industrial countries [46]. Most prevention suggestions of cancer have insufficient data to back them up. There is only convincing evidence for the following: Increased physical exercise is anti colorectal cancer, breast cancer and avoiding being overweight. Other research is only suggestive or simply insufficient. 25% of people in the U.S. develop a cancer in their lives. 70–80% result from chemical exposure, environment, 40% from cigarette smoking. In the U.S 500,000 die of cancer each year [47].

High level of physical activity may reduce the risk of colon and breast cancer by 50% [48]. Cancer, breast. 1 in 9 women. Only 5% due to genetic predisposition.

Epidemiology. Stomach cancer in Japanese doubles when they move to the U.S. in 1–2 generations [49].

Obesity increases most cancer risks [49]. As much risk of cancer due to poor diet/obesity as smoking [50]. Lower income people smoke more. One-third of people in poverty smoke. 44% of the high school educated smoke, 8% of masters and PhD professionals smoke [50]. One in seven Americans will get skin cancer (1 in 2 who live to 65 will) [51]. Alcohol from first drink increases risk of cancer [51].

16.5.3 AIDS

There are many types of HIV: Type I, II and ten subtypes of type I. There are 40 million new AIDS cases in the world. 1 million in US. One in ten nurses in Africa will die of HIV/AIDS. 20% of the population in Africa will die of AIDS and one in five children will be orphaned [52]. AIDS/HIV in Africa make up 70% of those in world with the disease [53]. Over 3 million died of it in 2004. Up to 2 million in the U.S.; SE Asia 5–11 million [53]. More than 65% of people living with HIV/AIDS (25–28.2 million) live in Sub-Saharan Africa; another 16% (4.6–8.2 million) live in South and Southeast Asia. AIDS is highest in South Africa 5,300,000, followed by India. Zimbabwe had 24.6% AIDS, Zambia 16%. More than 80% of all adult HIV infections have resulted from heterosexual intercourse. More than 90% of HIV infections in infants and children are a result of mother-to child transmission [54].

AIDS is an almost totally preventable disease by means of sex education, availability of contraception and a worldwide health care system. AIDS/HIV involves all aspects of society, not just a disease, which one can cure directly. There are 100 million sex encounters each day producing 910,000 conceptions and 350,000 sexually transmitted diseases [55].

It is a strong argument for a worldwide healthcare and educational system and for the elimination of poverty everywhere. Without such holistic policies trillions will die, and especially the richer nations, which could have prevented it bear the main responsibility for such deaths. There is an almost complete failure of the U.S., and most other rich nations, to have a Department of International Peace and Prevention (cf. Peace Corps), which is as large and powerful as the military and which deals with all aspects of every society to produce humanistic goals and healthcare, livable incomes, rational education, and opposes all forces, which undermine such goals.

16.5.4 Alzheimer's Disease (AD)

Half of those over age 85 have some form of Alzheimer's disease [56].

Scientists think that as many as 4.5 million Americans suffer from Alzheimer's disease (AD). The disease usually begins after age 60, and risk goes up with age. While younger people also may get AD, it is much less common. About 5% of men and women ages 65–74 have AD, and nearly half of those age 85 and older may have the disease [57]. 20 million cases can soon be expected in the EU. Every year Germany has 200,000 new cases of dementia [58].

Statements that Alzheimer's is due to age are not correct. Aging and age are not diseases anymore than time is a disease. We do not cure the past, present and future as such. Time causes nothing. Alzheimer's disease is rather bodily changes, although it may be more prevalent in older people especially if they have not practiced preventative medicine. The National Institute on Aging says that Alzheimer's is not a normal part of aging (NIH). Yet, over 30% of adults over age 85 have dementia.

Research has turned up little in identifying causes of Alzheimer's. In 2005, $60 million was federally funded for the Neuroimaging Intiative to identify by brain images people at risk for Alzeiheimer's [59]. Such images would have little practical value in preventing or curing the disease. The money would have been better spent on cure and prevention.

16.5.5 Lack of Exercise: The Obvious Escapes Us

The public waits for new technology and drugs to cure disease whereas by simply engaging in exercise for 1 or 2 h daily one can preserve health and prevent many diseases. Thus Hyman and Liponis regard poor fitness itself as a disease [60]. The body is so structured that if one does not use it one loses it. It deteriorates. Daily exercise is one of the most important things one can do for one's health. Yet, half of the U.S. population never exercises [61]. This is largely due to psychological factors such as ignorance about health, and negative emotions such as laziness. Some others exercise too much and engage in risky sports. The argument is clear that in these ways people also cause their own ill health.

16.5.6 Sexually Transmitted Disease (STD)

AIDS/HIV will soon claim over a billion lives. In some nations 25% of the people have AIDS. Nearly everyone sometime in their lifetime will have a STD. More U.S. women than men have heterosexual AIDS, but more men have AIDS: men: 749,887; women 170,679 (2003) [62]. There are about 15 million new STD cases each year in U.S. The most common infectious disease in 2004 is gonorrhea, fourth is syphilis [63].

Human papillomavirus is a group of viruses including more than 100 different strains or types. More than 30 of these viruses are sexually transmitted. In May 2004, approximately 20 million people were infected with human papillomavirus HPV. At least 50% of sexually active men and women acquire genital HPV infection at some point in their lives. By age 50, at least 80% of women will have acquired genital HPV infection. About 6.2 million Americans get a new genital HPV infection each year. 74% or 20 million people are currently infected [64]. There is a vaccine nowadays against the four main types of the virus. Human papillomavirus is an inducing factor of cervical cancer and dysplasia. Vaccination of girls and boys before beginning sexual activity would probably eradicate cervical dysplasia and cervical cancer for most cases [65].

Should be vaccination mandatory? Should big money be spent on that kind of prevention when we still have to do Pap smear evaluation for many other reasons and are confronted with scarce resources?

Roughly three-quarters of those aged 15–22 do not use contraception in the U.S. Is prevention of pregnancy not even an issue?

Sexual diseases are prevalent because of poverty, unavailable medical resources, and cultural and religious practices, which suppress sexuality and prevent sex education and the use of contraceptives. Europe has legal houses of prostitution where safer sex is practiced than elsewhere. People cause their own sexual diseases by not providing for the full availability and open expression of sexuality and by opposing education about healthful sex practices.

16.5.7 Longevity

Lifespan may be increased beyond the current 80–100 years by curing and treating diseases, and by preventing disease and accidents through improvement in our lifestyles and changing our anti-medicine, anti-health belief systems.

By 2020, 20% of the population will be aged over 65 [66]. In the U.S. a billion or more may be expected by 2050.

According to Marshall aging theories give the following as causes of aging: poor diet, overeating instead of caloric restriction, limitations to the number of times cells can divide, gene self-destruction, telomere shortening causing cell damage, molecular damage, exposure to toxins, decrease in estrogen and testosterone, harmful by-products of metabolism, damage caused by oxygen radicals, increased rate of oxygen metabolism, a complex integrative combination of factors (systems biology approach), and lack of stimulation of "longevity genes" [67]. "Currently, no treatments, drugs, or pills are known to slow aging or extend life in humans" [68]. This means, that there is no food substance that will slow or extend aging although many claims are made to the contrary. There has been a failure to find a useable definition of aging [69]. "Presently, we have little information about the ageing process and the ability of current or future interventions to alter this process, and appear to be socially unready for anti-aging technology" [70]. There is a societal and culture block to medical advancement in the area of life-extension [71].

The upper limit of life expectancy is open and depends upon advancements in technology including especially nanotechnology, artificial intelligence, gene technology, genomics, bioengineering, bioinformatics, and general medical research [72]. Aubrey de Grey of Cambridge University states, "Our life expectancy will be in the region of 5,000 years. . .by the year 2100" [73]. His suggestions how to reach that, are: 1. DNA/RNA manipulation, 2. eliminating damaged cells and immune system improvement, 3. provide backup mitochondria genes for damaged ones, 4. replace damaged cells including cloning of our own cells. Let`s wait and see. . ..

The general factors thought to increase lifespan are: favorable genetics, caloric restriction, and a healthy diet and lifestyle [74]. There is a myth of genetic determinism. Disease is not just a matter of having favorable genetics. Twin studies suggest that genetics contributes less than 25% of human lifespan variation. Identical twins develop different phenotypes.

Genes. online Mendelian Inheritance database (OMIM) shows 7,500 single genes predisposed to cancer, 400 for heart disease, 300 for diabetes and thyroid, 100–150 for dementia and arthritis [75].

Genetic Epidemiology. Genes cannot tell of personality inheritance. The environment and lifestyle often can modify or prevent the genetic effect. Genetic study is in its infancy and answers few questions while raising many [76].

Genetic testing is available for Huntington's disease and some cancer genes. Breast, ovarian, colon cancer, all are caused by the mutations of the same cancer genes. A child has a 50/50 chance of getting them. A BRCA 1 or 2 carrier has 56–85% chance for breast cancer in a lifetime as opposed to the average 11–12%. Cancer inherited mutations account for only 5–10% of all cancers. Thus, many cancers are preventable [77].

Some of the oldest people are supposedly the Georgians of the Soviet Union, people of Vilcabamba, Ecquador (Andes elevation 4,600), Hunzas of Pakistan (at high elevations and who eat much yoghurt), Masai tribes of Africa (who walk 11 miles a day), etc. but the claims were found to be exaggerations, based on no or faulty documentation. Sweden may actually have the largest number of people over age 100, Japan the longest life expectancy. We often learn little by asking why they lived so long. Epidemiological studies of elderly people in Japan and Okinawa especially those based on evidence-based medicine (EBM) are problematic as will be seen in the Chapter 19. Eva Morris, to date the oldest documented British woman, lived to 114 in 2000. Jeanne Calment of Arles, France, perhaps the longest well-documented life of a human, lived 122 years and 5 months, until 1997. She smoked until 120 and attributed her long life to olive oil, wine and a sense of humor. She often ate two pounds of chocolates a week. Each year before her death she would give reporters a different reason for her longevity. In a similar vein, nutritionists and herbalists often speak of one or a group of substances as promoting health or retarding aging or disease. Such claims are often unfounded and commit the fallacy of simplistic thinking. Many, covertly or openly, oppose extending health and lifespan esp. e.g. in retirement debates [78]. The increase of lifespan also depends upon our willingness to extend it.

16.5.8 Death and Disease

The following are the 10 leading causes of death in the world: 1. Ischemic heart disease 12.6%, 2. Cerebrovascular disease 9.7%, 3. Lower respiratory infections 6.8%, 4. HIV/AIDS 4.9%, 5. Chronic respiratory diseases 4.3%, 6. Diarrheal diseases 3.2%, 7. Tuberculosis 2.7%, 8. Malaria 2.7%, 9. Lung cancer 2.2%, 10. Road traffic accidents 2.1% [79].

In the U.S. 1 in 5 suffer from autoimmune disease, e.g., diabetes type I, or rheumatoid arthritis. Arthritis is the leading cause of disability and affects one in six people, including 300,000 children. The fourth leading cause of disease death in U.S. is diabetes. 20.8 million American children and adults have diabetes. 7% of Americans, 14.6 diagnosed, 6.2 undiagnosed, 54 million with pre-diabetes. $32 billion economic cost of diabetes in 2002.1 out of every healthcare dollars is spent on diabetes and complications. 1 in 500 kids and teens has type 1 diabetes [80].

Cardiovascular disease accounts for 40% of yearly deaths in the U.S. Smokers in U.S. (2003) constitute 22.7%. 440,000 Americans, and 4.9 million people will die from smoking. U.S. alcohol deaths yearly are 100,000. Suicide U.S. 30,000 yearly [81]. 1,100 US college students commit suicide each year [82].

The U.S. alcoholic rate for men is 23.8%, women 4.7% (NIH) One-quarter of the population is alcoholic and most of the rest use alcohol regularly. Smoking and drinking alcohol are implicated in other diseases as well so that if these were eliminated from one's lifestyle perhaps many deaths and illnesses in the world would be prevented. Pneumonia causes 98% of deaths of children in developing countries, and this could have been prevented. Failure to provide enough food for adequate living conditions, a safe environment, help prevent disasters and accidents, give priority to health policies and politics in order to not expose poor people to disease and death is our responsibility. Every starving person or person in need of medical care could have been provided food and care according to the enormous wealth of the rich countries of the world.

16.5.9 Hand-Washing: The Obvious Escapes Us Again

Nosocomial infections (hospital acquired infections, iatrogenic causes) are those, which are a result of treatment in a *hospital,* but secondary to the patient's original condition. They are infections first appearing 48 h or more after hospital admission.

The advance of medicine took a quantum leap forward when the germ theory was discovered. This meant that conditions were required to be as sterile as possible to avoid spread of disease and infection. It meant that healthcare workers as well as patients should wash their hands. This is so well known that it is a form of obviousness humor to even mention it. But it seems too obvious to comprehend. The Center for Disease Control hand-washing guide says hand-washing to be observed only 40% of time. Fry and Burger stated, "Healthcare workers only perform hand hygiene when [not] directed only about 40% of the time" [83]. The result is lower medical success rates and high hospital and iatrogenic disease. Bischoff observed, "Hand-washing compliance of health care workers is unacceptably low" [84]. It was even noted that education, intervention and patient awareness programs failed to increase hand-washing rates. The authors' conclusion was that, the introduction of an easily accessible, alcohol-based waterless antiseptic product significantly improved hand-washing rates. However, the attitudes of patient's and healthcare workers seem resistent to attitude and lifestyle changes [85].

We engage in complex and expensive research to find the cures of diseases, when they could be prevented, by merely washing our hands. The cause of disease in this case is failure to understand the obvious but also one's attitude.

16.5.10 Drugs and Toxins

Drugs. To license a medicine in the U.K can take up to a decade and cost millions of pounds. No drug is absolutely safe. The effects of drugs are largely unknown.

Adverse drug effects are the most common iatrogenic causes of patient injury. The present U.S. drug testing and follow-up system is inadequate [86]. Wolfe and Sidney report that adverse drug reactions cause 100,000 deaths each year and afflict 2.2 million others in the U.S. alone [87]. 61,000 persons have drug-induced Parkinson's disease, 163,000 drug-caused memory loss. In 2003 3.4 billion prescriptions were filled, an average equivalent of 11.7 for each person [88]. This is an indication of the failure of successful disease prevention, poor lifestyles and the failure to cure diseases. [Half of older Americans take five or more medications daily. Only 11% of Americans 65 and older do not take any prescription medications.] Drugs are often prescribed when rather nutrition and lifestyle changes should instead be made [89]. Many diseases are caused, by taking prescribed drugs. Also unhealthful marketing on the part of pharmaceutical companies causes many problems. The American Pharmaceutical Association even attempted to prevent the FDA from requiring correct package information on each prescription [90].

Goozner reported that the U.S. Federal Drug Administration voted to approve a drug for diabetes even though it doubled the risk of heart attack and stroke in clinical trials [91]. There is lax monitoring after drug approved, backlog, conflicts of FDA and industry. The FDA does not do the testing, drug companies do. FDA gets $2 billion of its annual budget from industry. Also, once approved by FDA drugs can be used for anything. High cholesterol drug may be used to treat cancer.

Canada's Adverse Drug Reaction Database (CADRIS) uses data collected from 1965 to Sept. 30, 2003, and contains information from all adverse drug reaction reports (ADR) currently held in Health Canada's CADRIS database. It is a voluntary system, but pharmaceutical companies are legally bound to report any ADR. However, only 10% or less adverse events are reported to the government. The Canadian public has access to all the adverse drug reaction reports [92]. Because the government does not effectively follow up on the effects of drugs taken and reported, each person must individually determine the effect of any drug taken and gain access to the effects of others who have taken the drug by consulting physician's clinical experience and reliable internet sources. There are often supportive and discussion groups concerned with specific drugs. Again patients must share the responsibility for their own healthcare.

Statin drug use within 14 days following acute coronary syndrome does not reduce death, myocardial infarction, or stroke up to 4 months. This critical period was often excluded in trials [93].

Drug lowering of cholesterol may lower heart attack, but decrease longevity [94].

Drug dosages are questionable and often arbitrary. They are problematically determined by manufacturer. The gender aspect is almost always underrepresented. The purchaser is just to follow the recommended dose.

Toxins. We only know we are poisoned when we are poisoned. It is then too late. The journals on toxicology are of no everyday use. We need to ourselves know what is toxic in our environment. Not finding out is another way in which we cause our own illnesses. Most people are not aware of the following claims of toxicity. Nearly every household-cleaning product is exceedingly toxic and its use should be avoided in place of less harmful products, e.g., even the vapor of ammonia, paints and bleach are toxic. Alfalfa sprouts are a leading cause of harmful E. coli. Allergies are especially caused by peanuts, eggs, milk products, wheat, soy, fish. About 1 in 4 Americans suffers from food poisoning each year, and 5,000 die from it [95]. There are over 200 food borne illnesses, many not yet identified. Irradiation does not eliminate food borne illnesses, though it reduces them.

Most all hair dyes release poisons into the body. It is said to cause cancer. Thus, prevention requires that pregnant women and women of child-bearing age are advised not to have their hair dyed at all (as well as to take folic acid and not smoke). The list of possible toxins is too great to present, but every aspect of one's environment may be investigated for toxins.

Research indicates that, 400 micrograms of folic acid, if taken before conception and throughout the first trimester, can prevent up to 50–70% of neural tube defects such as spina bifida. Unfortunately two-thirds of women in the United States do not consume enough folic acid (cdcfoundation.org).

There is also a literature on detoxification foods or supplements, which claim to detoxify, e.g. sulfur-containing amino acids or broccoli, brussels sprouts, cabbage, cauliflower, bok choy, onion. The value, if any, of such detoxification substances requires investigation. Toxification prevention is required to avoid such self-intoxication.

16.6 Hidden Prevention Possibilities

Autonomy. Prevention involves giving patients sufficient information about their illnesses, their drug side effects, and their medical choices. It does not involve taking legal action against the healthcare worker when the patient does not bother to try to understand the information or fails to follow the physician's instructions. Prevention often involves dissuading patients against their own harmful medical decisions. There cannot be intelligent patient autonomy without reason and patient responsibility.

Intellectual Illiteracy. 22% of Americans are functionally illiterate [96]. In 2000, over age 15, illiteracy in India was 57.2%, Senegal 37.3%, Egypt 55.3%. World literacy (reading and writing ability) rate is 77% [97]. The problem is that illiteracy blocks people from gaining access to needed information and thinking. What is not understood is that most of what is written also blocks people from gaining access to needed information and thinking. Newspapers, magazines, popular literature, have little critical value, yet are what are almost exclusively read. The greatest illiteracy is intellectual or discussion illiteracy whereby one cannot present rational, philosophical arguments on any particular question. It is only academic journals and texts and some academic internet-sites, which attempt to do so and these are not read by most all of the general populace. They are not educated in critical thinking (speaking). It was mentioned elsewhere that the public is not educated about emotions thus producing emotional illiteracy. Thus there is illiteracy, intellectual and discussion illiteracy and emotional illiteracy.

Heart disease prevention
Mammals get about one billion heart beats.

Heart disease is the most frequent cause of death. Each year 1,250,000 Americans have heart attacks and 600,000 Americans die each year from it [98]. According to the American Heart Association coronary heart disease (CHD) is the number 1 killer of males and females in America [99]. CHD accounts for 19% of disability allowances by the Social Security Administration. In 2003 71,300,000 American adults had CVD (Cardiovascular disease). One in three adults has CVD. Of those with CVD 53% have CHD [100]. The estimated cost of CVD in 2006 is $403.1 billion. 50% of men and 64% of women who died suddenly of CHD had no previous symptoms of this disease. Within 1 year of a recognized heart attack, 25% of men and 38% of women will die. Lifetime risk of CVD for men is 2 in 3, for women at 40 is 1 in 2. Nearly one, in three adults, has high blood pressure. International CVD per 100,000 of population is highest in Russia, Bulgaria, Romania, Hungary, Poland, Argentina, Czech Republic, China. Lowest is in: Japan, France, Switzerland, Australia, Spain, Canada, Italy, Korea, Norway.

It is estimated that up to 90% of the CVD patients had preventable heart diseases due to poor lifestyles, e.g., 65–90% of the US population is overweight. Risk factors include, cigarette smoking, obesity, lack of exercise, poor diet, alcohol use, psychosocial issues, etc [101]. One could add the unnecessary prevalence of negative emotions causing stress and hypertension.

U.S. Health and Human Services released the *Public Health Action Plan to Prevent Heart Disease and Stroke* to improve cardiovascular health through prevention. The American Heart Association Guide recommendations also stress prevention, which include environmental change, minimizing risk factors, and lifestyle changes (inadequate diet, sedentary lifestyle, tobacco and alcohol use, hypertension, etc.).

Physicians have been captivated by the mythical metaphorical model that heart attacks are caused by a simplistic plumbing model of calcified cholesterol plaque build-up, which blocks arteries. It is thought that the arteries become hardened. Heart attack is rather thought to be caused by inflammation of soft plaque. It is no longer just fat build-up.

A national study of physician awareness and adherence to cardiovascular disease (CVD) prevention guidelines, conducted in late 2004, showed that fewer than one in five physicians knew that more women than men die each year from CVD [102].

Half of all heart attacks have symptoms, which are known hours to weeks in advance, e.g., chest pain; shoulder, neck, jaw, or arm pain or tingling, anxiety, nausea, shortness of breath, fatigue, swollen ankles, lightheadedness, feeling faint, paleness, increase of heart rate. Heart muscle has no physical pain sensation, only nerves near heart. There is a squeezing sensation. It can be triggered by emotional stress, extreme heat or cold. Congestive heart failure is the leading cause of hospital admissions if over age 65. 85% of those dying of coronary heart disease (CHD) are over age 65. CHD is 2x–3x higher in postmenopausal women than men or younger women apparently because of decreased estrogen levels [103]. Heart disease is not a "hazard" or "risk," but rather preventable outcomes. Heart disease is result of poor diet, lack of exercise, smoking, and to a lesser extent family inheritance and genetics, depending on the specific ailment [104].

17.5 million die of stroke and heart attack each year (WHO, CDC, National Center for Health Statistics, Center for Disease Control, Washington, DC).

30% of adults in most countries have high blood pressure or hypertension (>140 systolic, >90 diastolic are already too high, normal is 120/80). 2/3 of heart attack victims have mildly high blood pressure (HBP). Optimal are 120/80 or below, but not too low. HBP may be due to cold or flu decongestants or excessive salt. HBP is more common with lower education or lower income levels.

"The best way to lower blood pressure is to alter the lifestyle factors under your control" [105].

Cholesterol. It is not the lowering of cholesterol alone that is important, as that may be just the lessening of the symptom, which alerts us to the causes involved. Cholesterol may not be the cause or the only cause. Of cholesterol we only need small amounts, less than 200 milligrams per day. It is carried through the bloodstream. LDL, low-density lipoproteins, is bad because it releases cholesterol too easily allowing it to build up in artery walls. HDL (high-density lipoproteins, which is good because binds more closely to cholesterol and even collects the cholesterol released by LDL. Lowering LDL is more important than raising HDL [105].

Prevention. 10% reduction in total cholesterol level means 30% reduction in the chance for a heart disease.

Atherosclerosis is the buildup of cholesterol, which blocks the flow of blood.

Disease. Myocardial infarction is the death of some portion of muscle tissue of the heart due to inadequate blood flow.

There is clearly disagreement about what constitutes prevention and treatment of heart disease, but there is a strong agreement that improving one's lifestyle is the main preventative measure one can contribute oneself to one`s health.

Mistakes and errors (unneeded operations)

Errors claim 44,000–98,000 U.S. hospital patients each year. They could have been prevented to a large extent [106]. The rankings of a few medicine related causes of death are: surgery 8th, x-rays 9th, vaccinations 25th, prescription of antibiotics 29th [107]. Drug errors and medical mistakes are high on the list (See full analysis of mistakes in the Chapter 3). In this brief analysis will be shown how faulty beliefs in culture and religion create harm and death in medicine. The case is clearest with the practice of female circumcision. What at the "present" time is "in", and the time is left open as it is believed that it will continue, is rejuvenation, female genital cosmetic surgery as a surgical temptation for better sex [107].

It is a violation of the physician's oath to do no harm.

Anesthesia protocols reduced deaths from 1 in 20,000 to 1 in 200,000 [108].

A hysterectomy is the second most common surgery among women in the United States (The most common is cesarean section delivery). 600,000 hysterectomies are performed annually in the United States One in three women in the United States has had a hysterectomy by age 60. For every 10,000 hysterectomies performed, 11 women die. Approximately 660 women die. Every 10 min, 12 hysterectomies are performed in the United States. According to a report published by Obstetrics and Gynecology, nine of them probably didn't meet the guidelines set out by the *American College of Obstetricians and Gynecologists* for hysterectomy. The non-profit research RAND Corporation found little justification for at least 25% of the hysterectomies and no justification for another 16% [109].

If you do need a hysterectomy, you generally should not have your ovaries removed unless you have ovarian cancer or other diseases or tumors of the ovaries. However, a hysterectomy candidate whose mother or sister had ovarian cancer may want to discuss the option of preventive removal of the ovaries with her doctor and possibly with a genetics specialist. Further, find out whether you're a candidate for a vaginal hysterectomy, which is less invasive than the traditional open abdominal procedure.

Nutritional Abuse. Nutritional abuse is one of the major reasons why people cause their own illnesses and diseases. In addition, anorexia and bulimia are so well known that no further account need be given here.

Almost everyone is undernourished although they overeat and it is now well established that most people in the rich nations are overweight or obese. For example, over 80% of Americans are malnourished [110]. In 2003, an estimated 136,500,000 American adults were overweight, and 64,000,000 were obese. Nutritional abuse in the world is completely avoidable if people wish to avoid it. "For just about all common diseases the messages on prevention...are remarkably similar": have a good diet [111]. Insel states, "Most weight problems are lifestyle problems" [112].

Occupational and Employment Safety. Diseases and accidents caused by unsafe and unhealthful employment conditions are to a large extent preventable. In the healthcare field itself, the requirement to work over 80–100 h a week is one of the most flagrant unhealthful and dangerous practices of any employment situation in the world. Employment safety organizations have documented the thousands of ways in which working conditions are unhealthful and may be corrected.

Organ Donation Failure. There were in the U.S. 27,029 organ transplants in 2004. The number on the organ waiting lists was 88,176 [113]. Although some were double listed, many were not put on the list because there were no organs available. The mathematics is simple: 61,147 patients requiring organs did not receive them and will die and/or suffer disabilities. The U.S. more than any other wealthy nation has failed to provide organs for transplantation to the needy. The people (and government) have taken the selfishly irrational view that they do not want their organs to be donated after death. The problem of severe organ shortage is theoretically one of the easiest to solve. It is in effect, no problem at all. The law can require that each person (or surrogate for children) officially declare whether or not they wish to donate organs. Their names are put on a list and only those on the list for donation will be able to receive organs for transplantation. Organs not used may be exchanged internationally with countries having similar presumed donation policies for people who have not opted out of donation. The policy would appeal to the self-seeking interests of Americans who, if they were rational, would vote for a required declaration organ donation policy.

Problematic Therapy About 450 million people suffer from mental disorders according to estimates given in World Health Report 2001. One person in four will develop one or more mental or behavioral disorders, at least one during each lifetime. One fifth of teenagers under the age of 18 years suffer from developmental problems. Five of the ten leading causes of disability worldwide are psychiatric conditions, including depression, alcohol use, schizophrenia and compulsive disorder problems [114].

Mental disorders can to a large extent be prevented and corrected and to a large extent they are not. Emotional and intellectual disorders are found in every person, including therapists, insomuch as they lack the necessary education to prevent and correct such disorders. Therapists, like physicians, often treat only the symptoms of the mental disorder rather than the causes. The very term "mental disorder" is a naming fallacy, because there is no evidence for mind and so no mental disorders as such. The official classification of the so-called mental disorders, *DSM IV*, is unacceptable in terms of the philosophy of medicine and the philosophy of psychology. This has been discussed throughout this book. Adequate therapy must include the critical analysis of the concepts and methods of therapy. Mental problems may be prevented, but the failures in the correction and treatment of mental problems must also be addressed See especially the Chapter 7.

Risky Behavior. Risk gets needed. Like any addictive substance, adrenaline creates a potent or strong high. The body becomes tolerant to the dosage; in other words, the body needs a higher dosage to produce the high that was obtainable at the lower doses. The risks must increase to send the levels of brain chemicals higher and often the rate of occurrence goes up. Thrill seeking becomes a way of life. People are not aware of the danger to themselves and are oblivious of the threat to others. For some there is also the desire to standout, to stand above and to dominate and have control in some aspect of life.

Normal competitive sports such as boxing, hunting, long distance running, gymnastics, football, soccer, baseball, etc. are actually dangerous extreme sports. Olympic sports are typically life threatening. Actually reported injuries in the U.S. excluding unreported injuries are: basketball 593,713, bicycle 535,100, football 389,261, baseball (softball) 278,315, horseback riding 71,772, volleyball 55,504, In-line skating 45,856, chain saw use 27,601, razor blades 36,612 (See auto risks discussed above). The actual unreported injuries are hundreds of times greater as most injuries are not reported. There are 144,000 snowboard emergency room admissions per year at a cost of billions [115].

Injuries and deaths from sports can be almost entirely eliminated by playing only healthful, recreational sports according to safety regulations. An over-stressed medical system should not have to divert its resources to treat those who deliberately and needlessly put their lives and the lives of others at risk. To suggest the elimination of competitive sports and extreme sports goes against the one of the cherished values of society and culture. This practice shows the way in which culture and society give medicine low priority. Athletes receive salaries in the millions, but educators, surgeons, physicians, and healthcare workers thousands.

Scarce medical resources. Lack of an International Healthcare System. The cause of scarce medical resources is simple. It is the failure of the public to sufficiently fund healthcare and medical research and of physicians to support it. Their priority is elsewhere so they cannot complain about not obtaining adequate treatment. Besides, the rich have all the treatment they need. In Central/East Africa there is one physician for 14,000 people [116]. Americans are getting quality care only 55% of the time, according to the results of the Community Quality Index Study, 2004 RAND health study on the state of health care in the United States [117].

Inadequate solutions to healthcare scarcity are usually given. The achievement of a consensus on principles of distributive justice is seen as a main goal. However, it is a fallacy to base distribution on majority rule on consensus. Nor should healthcare be based on market accountability, but rather on reasonableness and critical thinking.

Another reason for scarce resources is lack of a well-funded national healthcare system. Since 1960 in Canada all people are entitled to free comprehensive healthcare for life. The U.S. has no such national healthcare system. There are over 45 million Americans who have no health insurance at all. The lack of health insurance is estimated to cause 18,000 deaths a year (Institute of Medicine; U.S. Census Bureau 2004). Texas had perhaps the highest number of uninsured: 24.6% of the people as opposed to the usual 15% [118]. One proposal for nearly all citizens to have health coverage is to require them by law to purchase it or they will be penalized on their state income taxes. Government subsidizes some plans to enable the working poor to buy insurance. The plan is expected to cover up to 1% of the now unprotected state population. However, for adequate preventative medicine in addition to national healthcare, we need international healthcare. Healthcare must go beyond each culture and border. Disease does not stop at the border. According to the *American Medical Association Principles of Medical Ethics*, a physician shall support access to medical care for all people [119].

Starvation and poverty and need for medical treatment. We would think it completely unacceptable to let a family member die of starvation, but people show by their inaction that it is completely acceptable to let a billion people starve. This is the inconsistency found if one has normative, enculturated morality, but no critical ethics. As argued in the Chapter 11, all three could be eliminated without sacrifice of anything but poor priorities if people had the intelligence, rationality and humanistic willingness to do so. All three of these are almost totally preventable and correctable [120].

Twenty percent of the world's population lives on less than $1 per day (One person in five according to the World Bank). One-half of the world's population lives on less than $2 per day. The top 20% of wealthiest people in the world have 86% of the GNP, the bottom 20% have 1%. According to the World Bank, 80% live on only 15% of the world's riches, 86 nations are low income, food deficient countries. Africa has over 30 million starving people. In Swaziland 2/3 are below the poverty rate. Sub-Saharan Africa has the highest number of starving, about 50%. In Africa generally, close to half of the people are starving, live in slums; have no access to clean water, sanitary facilities, or education. Nearly 20% of the children die before age five. South Asia has similar statistics and the largest number of poor.

The number of starving in the following countries is: Eritrea nearly 70% of population (yet it spends 16% of its GDP on the military), Ethiopia 14 million (Burundi and Ethiopia also spent the large amount of 8 and 6.2% of their GDP respectively on the military), Zimbabwe 50% (7 million) and has one of the highest AIDS rates. Over 40 million people in the world have AIDS. In Swaziland at least 40% of the adults have AIDS, and the infant mortality rate is five times the national average, yet King Mswati III ordered a $45 million dollar jet. Each year two million children in the world under the age of five needlessly die of pneumonia. One million people contract malaria each year. A billion people are illiterate (WHO). World Bank says that 19 die per minute of preventable diseases. It is estimated that over 1 billion people are starving and many more are in need of unavailable medical care (World Health Organization).

The richest as well as other nations have enough resources to feed the starving, provide medical care worldwide for every person, and eliminate poverty. It is not a question of resources. It is a question of willingness to do so [121].

Millennium Development Goals Report 2006 has the aim to save the lives of 6 million children by 2015 by obtaining 5.1 billion dollars annually for preventative measures in 42 countries responsible for 90% of child deaths, e.g. water sanitation (about 9 times the next largest item), immunizations, etc. About 1% of the U.S. GNP would be required to obtain the capital needed. However, their proposal to prevent 6 million deaths is meager because over a billion children are starving and in need of medical treatment in the world. 1% of the world GNP (450 billion) is now being recommended to counteract global warming. It is not being asked to aid those in desperate need for food and medicine.

There is a failure to realize that one cannot prosper as a nation while billions of people are dying. What kind of prosperity would that be? Preventative medicine cannot neglect to care for all the desperate people of the world. It can only be genuinely preventative on an international, as wall as national, local, and individual level. International health insurance could be taken out by the richest nations to protect the desperate people of the world. "The United States gave only 0.11% [of its gross national income, or just over $10 billion,] in foreign aid – the lowest proportion of all developed nations" [122].

The U.K., France and Germany each gave three times as much as the U.S. Johannes Raw, President of the Food and Agriculture Organization of the U.N. said on Oct. 16, 2001, "Every person who starves is a judgment on us. For starvation is not an inevitable fate." The UN General Assembly agreed that the richer nations are recommended to give 0.7% of their GNP for development aid in the poor countries [123]. The percent of the Gross National Income (GNI) given in 2005 was: U.S. 0.22% GNI, Norway 0.93, Netherlands 0.82, Sweden 0.92, Luxembourg 0.87, Denmark 0.81 [123].

Most countries have not complied with this agreement. Denmark, on the other hand, gave over 1%. Furthermore, the U.S. aid is geared to U.S. self-interest and its aid does not go to the neediest nations, as does that of Denmark. The U.S. knows that it should give a minimum of 0.7% of its GNP, but deliberately and with full knowledge of the consequences refuses to do so.

Percent of GNP spent on healthcare: U.S. 16–20, Ireland 6.8, Germany 10.3, Finland 6.9, Canada 9.6, U.K. 7.7. Japan spends roughly five times more on medical research and development than does the U.S. The U.S. has no national health care and a bankrupt welfare and social security system for the retired and elderly.

Roughly half of U.S. budget is used for the military [124]. 24 billion is all that is needed to reduce the number of starving people by 400,000 by 2015. Less than three billion dollars could immunize three billion children to save their lives. If we can spend trillions to kill, why not spend trillions to prevent death and keep people alive and improve their education and lives? The U.S. deficit-spends so much borrowed money on war and trivial matters that it can no longer help the needy people of its own country or the world and has become a bankrupt nation.

16.7 Summary

According to a humanistic-naturalistic approach prevention is a holistic, all areas of life encompassing, effort. Education and ethics are required to unravel personal as well as structural life style problems leading to disease, lack of access to a well functioning healthcare system and unnecessary deaths.

Prevention is a task in everyday life of the individual as well as a political challenge.

References

1. *CIA World Fact Book* (internet) reports the following approximate death rates per 1,000 people in 2004: Austria 10, Chad 16, China 7 [unbelievably low], India 9 [unbelievably low], Mosambique 23, Somalia 17 [unbelievably low], Zimbabwe 23, United States 8, World average is 9 (of 6.4 billion people in the world) [again this unbelievably low as the U.S. has 8]. In 2002 the life expectancy age (at birth) according to WHO was: Afghanistan age 35, Angola 33, Ethiopia 41, India 53, Lesoto 31, Malawi 34, Sierra Leone 28.6, Somalia 36.8, Zimbabwe 33.6. In Western countries the age is around 70–76 with women living a few years longer. The authors of this book believe that such published figures as the *CIA World Fact Book* presents are far from being accurate. http://www.fao.org.
2. CARE: care.org.
3. Ziegler, J. 2002. *Wie kommt der Hunger in die Welt? Ein Gespräch mit meinem Sohn*, esp. 58 and 62 f. München: Goldmann Verlag Orig.: *La faim dans le monde expliquee a mon fils 1999*. Paris: Editions du Seuil.
4. Center for Disease Control (CDC) http://www.cdc.gov/
5. *Encyclopedia Britannica Almanac 2006*: *275*. See updates at FAO: http://www.fao.org.
6. Wright, J.W. ed. *NY Times Almanac 2006*: *702–703*, New York: Penguin Books.
7. For further details see Ziegler, J. 2007. Das Imperium der Schande. Der Kampf gegen Armut und Unterdrückung. Pantheon. (orig. L'Empire de la honte, 2005, Fayard, Paris)
8. fh.org.
9. nchc.org/facts/cost.shtml.
10. Food for the Hungry. fh.org Accessed Sept. 16, 2006 The Hunger Project. UN 2006.
11. O'Donnell M.P. 1986. Definition of health promotion. *American Journal of Health promotion* 5:10
12. WHO. 2001. *Basic documents*, 43rd edn, 1. Geneva: WHO.
13. Harkin, T. 2004. *American Journal of Health Promotion* 19:1–2.

14. worldpeaceforum.ca/ accessed Dec. 2005.
15. Press Release EURO/05/06 Copenhagen, 11th September 2006.
16. Time Dec. 4, 2006:66, 68.
17. Pualani, L., and Sanchez-Johnson, L. "Health Promotion and Disease Prevention." II, 495–497 cf. *Healthy People 2010*, 2nd edn, 2 vols. Washington, DC: Dept. of Health and Human Services (DHHS), Government Printing office. Nov. 2000.
18. preventdisease.com. National Institute of Health and Nutrition.
19. *UN Chronicle* no. 1, 2006. un.org/chronicle.
20. *Diabetes Dateline* Spring 2004:1 Most people with diabetes do not meet treatment goals. NIH Publication No 4-4562 May 2004
21. Barondess, J. 2005. On the preservation of health. *JAMA* 294:3024–3026.
22. Fuhrman, J. 1995. *Fasting and eating for health: A medical doctor`s program for conquering disease*, 102. New York: St Martin`s Griffin.
23. Fuhrman, J. 1995. *Fasting and eating for health: A medical doctor`s program for conquering disease*, 192. Griffin, New York: St Martin`s.
24. Gordis, L. 2005. *Epidemiology*, 3rd edn, 68. Philadelphia, PA: Elsevier.
25. Nys, T., and Nys, M. 2006. Psychiatry under pressure: Reflections on psychiatry`s drift towards a reductionist biomedical conception of mental illness. *Medicine, Health Care and Philosophy* 9:107–115; esp.108.
26. Hyman, M., and Liponis, M. 2003. *Ultraprevention*, 168. New York: Scribner.
27. Collman, J. 2001. *Naturally dangerous*, 22. Sausalito, CA: Science Books.
28. Wright, J.W. ed. *New York Times Almanac 2006*. 2006:384–385. New York: Penguin Books.
29. Wright, J.W. ed. *New York Times Almanac 2006*. 2006:415. New York: Penguin Books.
30. Wright, J.W. ed. *New York Times Almanac 2006*. 2006:414. New York: Penguin Books.
31. Husak, D. 2004. Vehicles and crashes: Why is this moral issue overlooked? *Social Theory and Practice* 30:351–372.
32. cancer.org.
33. WHO Report 2003. #916:95.
34. American Lung Association 101 Fact Sheet Nov 2004. www.lungusa.org/lung-disease/lung-cancer/resources/facts-figures/lung-cancer
35. *Time* Dec. 4, 2006:66.
36. American Cancer Society. Wright, J.W. ed. *New York Times Almanac 2006*:381. New York: Penguin Books.
37. Kurzweil, R., and Grossman, T. 2004. *Fantastic voyage: Live long enough to live forever*, 233. New York: Plume (orig. Rodale).
38. www.YourCancerRisk.harvard.edu
39. Stein, C., and Colditz, G. 2003. Risk index expanded to reinforce disease prevention messages on the web. *Journal of General Internal Medicine* 18(Supplement 1):138.
40. Fuhrman, J. 2003. *Eat to live*, 53, 58. Brown, New York: Little.
41. Biesalski, H., and Grimm, P. 2005. *Pocket Atlas of nutrition*. Junkermann, S. tr, 2nd edn. Stuttgart: Thieme.
42. Shelford, G. 1998. Meta-analysis spurious precision? Meta-analysis of observational studies. *BMJ* 316:140–144.
43. Insel, P., Turner, R., and Ross, D. 2003. *Discovering nutrition*, 189. Sudbury, MA: Jones and Bartlett.
44. Ensminger, A. 1995. *The concise encyclopedia of food and nutrition*, 157–158. Boca Raton, FL: CRC Press.
45. Brody, T. 1999. *Nutritional chemistry*, 2nd edn, 907. San Diego, CA: Academic .
46. *Cancer Detection and Prevention* 29, 5 2005. iii–iv
47. Martini, F. 2006. *Fundamentals of anatomy and physiology*, 137. San Francisco, CA: Pearson.
48. Colditz, G. 2004. Cancer and diet. *Encyclopedia of Health and Behavior.* Anderson, Norman, edn. 2 vols, 119–123. Thousand Lakes, CA: Sage I.

49. Lang, R., and Hensrud, D. eds 2004. *Clinical preventative medicine*, 186. Chicago, IL: AMA.
50. Ropeik, D., and Gray, G. 2002. *Risk*, 344. Boston, MA: Houghton-Mifflin.
51. Ropeik, D., and Gray, G. 2002. *Risk*, 346. Boston, MA: Houghton-Mifflin.
52. Kunz Howard, P. 2005. Access to care: A human right. *Journal of Emergency Nursing* 31:419–420.
53. WHO. Nov. 2005; UN Chronicle, no. 2, 2005:27.
54. Wright, J.W. ed. *New York Times Almanac 2006*. New York: Penguin Books.
55. Wright, J.W. ed. *New York Times Almanac 2006*:489. New York: Penguin Books.
56. Martini, F. 2006. *Fundamentals of anatomy and physiology*, 543. San Francisco, CA: Pearson.
57. National Institute on Aging. NIH. Alzheimer's Disease Fact Sheet: http://www.nia.nih.gov/Alzheimers/Publications/adfact.htm
58. Huber, J., and Buchacher, R. 2005. *Das Ende des Alterns*, 93, 97. Berlin: Econ/Ullstein.
59. Source Alzheimer's Association. http://www.alz.org/research/science/earlier-abbeimers-diagnosis.asp#Blank
60. Hyman, M., and Liponis, M. 2003. *Ultraprevention*, 197. New York: Scribner.
61. *Encyclopedia Britannica Almanac 2006:303.*
62. Wright, J.W. ed. *New York Times Almanac 2006*:382. New York: Penguin Books.
63. Wright, J.W. ed. *New York Times Almanac 2006*:379. New York: Penguin Books.
64. The FUTURE II Study Group. 2007. Quadrivalent vaccine against human papillomavirus to prevent high-grade cervical lesions. *NEJM* 356:1915–1927.
65. Sawaya, G.F., and Smith-McCune, K. 2007. HPV Vaccination – more answers, more questions. *NEJM* 356:1991–1993; Rothman, S.M., and Rothman, D.J. 2009. Marketing HPV Vaccine. Implications for adolescent health and medical professionalism. *JAMA* 302:781–786.
66. Morris, B. 2005. A forkhead in the road to longevity: The molecular basis of lifespan becomes clearer. *Journal of Hypertension* 23:1285–1309.
67. Marshall, J. 2006. Life extension research: An analysis of contemporary biological theories and ethical issues. *Medicine, Health Care and Philosophy* 9:87–96.
68. Marshall, J. 2006. Life extension research: An analysis of contemporary biological theories and ethical issues. *Medicine, Health Care and Philosophy* 9:87–96; esp. 89.
69. Marshall, J. 2006. Life extension research: An analysis of contemporary biological theories and ethical issues. *Medicine, Health Care and Philosophy* 9:87–96; esp. 91.
70. Marshall, J. 2006. Life extension research: An analysis of contemporary biological theories and ethical issues. *Medicine, Health Care and Philosophy* 9:87–96; esp. 93.
71. Marshall, J. 2006. Life extension research: An analysis of contemporary biological theories and ethical issues. *Medicine, Health Care and Philosophy* 9:87–96; esp. 94.
72. Kurzweil, R., and Grossman, T. 2004. *Fantastic voyage: Live long enough to live forever*, 15. New York: Plume (orig. Rodale).
73. Kurzweil, R., and Grossman, T. 2004. *Fantastic voyage: Live long enough to live forever*, 24. New York: Plume (orig. Rodale).
74. Westendorp, R., and Kirkwood, T. 1998. Human longevity at the cost of reproductive success. *Nature* 396:743–774.
75. Lang, R., and Hensrud, D. Eds. 2004. *Clinical preventative medicine*, 86; 469. Chicago, IL: AMA.
76. Gordis, L. 2005. *Epidemiology*, 3rd edn. Philadelphia, PA: Elsevier.
77. Anderson, N. ed. 2004. *Encyclopedia of health and behavior*, 2 vols, I, 369. Thousand Lakes, CA: Sage.
78. Kurzweil, R., and Grossman, T. 2004. *Fantastic voyage: Live long enough to live forever*, 8. New York: Plume (orig. Rodale).
79. World Health Organization Accessed Sept 20th 2010 http://www.infoplease.com/A0779147axzz1057ax.html

80. *Encyclopedia Britannica Almanac 2006*:275. See updates at FAO: http://www.fao.org.
81. *Encyclopedia Britannica Almanac 2006*:248, 280–281, 390–391.
82. Time May 22, 2006.
83. Fry, W. Jr., and Salameh, W. eds. 2005. *Handbook of humor in psychotherapy*, 12–15. Sarasota, FL: Professional Resource Exchange.
84. Bischoff, W. 2000. Handwashing compliance by health care workers: The impact of introducing an accessible, alcohol-based hand antiseptic. *Archives of Internal Medicine* 160:1017–1021.
85. Centers for Disease Control and Prevention http://www.cdc.gov/handhygiene/ See also CDC's Guideline for Hand Hygiene in Healthcare Settings.
86. Strom, B. 2006. How the U.S. Drug Safety System should be changed. *JAMA* 295:2072–2075.
87. Wolfe, S. 2005. *Worst pills, best pills*. New York: Pocket Books.
88. Wolfe, S. 2005. *Worst pills, best pills*, xxv ff. New York: Pocket Books.
89. Wolfe, S. 2005. *Worst pills, best pills*, xxiii. New York: Pocket Books.
90. 1FDA MedWATCH 1fda.gov/medwatch/ Reporting of adverse effects of drugs. 7.
91. Goozner, M. 2006. *AARP Bulletin*:10, 33.
92. cbc.ca/news/adr/database/
93. Strom, B. 2006. How the U.S. Drug Safety System should be changed. *JAMA* 295:2046–2056.
94. Graedon, J., and Graedon, T. 1998. *The people`s pharmacy*, 252. New York: St Martin`s Press.
95. Ropeik, D., and Gray, G. 2002. *Risk*, 97. Boston: Houghton-Mifflin.
96. National Institute for Illiteracy 2000. nifl.gov/reders/lintro.htm#D.
97. *Encyclopedia Britannica Almanac 2006.*
98. Kurzweil, R., and Grossman, T. 2004. *Fantastic voyage: Live long enough to live forever*, 8, 197. New York: Plume (orig. Rodale).
99. americanheart.org. April 2006.
100. Source CDC/NCHS. www.cdc.gov/mchsl
101. "Heart Disease and Stroke Statistics—2006 Update." Feb. 14, 2006 *Circulation* published online: circ.ahajournals.org.
102. "Heart Disease and Stroke Statistics—2006 Update." 111; 499–510, Feb. 14, 2006 *Circulation* published online: circ.ahajournals.org.
103. Ropeik, D., and Gray, G. 2002. *Risk*, 361. Boston: Houghton-Mifflin.
104. Ropeik, D., and Gray, G. 2002. *Risk*, 346–349. Boston: Houghton-Mifflin.
105. Ropeik, D., and Gray, G. 2002. *Risk*, 354. Boston: Houghton-Mifflin.
106. 1999 National Institute of Medicine 'To Err is Human: Building a Safer Health System. books.nap.edu/books/0309068371/html/index.html. 376. More die of medical errors than car accidents, breast cancer and AIDS. 1999.
107. Maier, B. 2010. Körperbild(er) und Leibbewusstsein. Von der weiblichen Beschneidung bis zur Genitalkorrektur. In: Gender Medicine. Ringvorlesung an der Medizinischen Universität Innsbruck (Hochleitner, M. ed) Vol 3: Sexualität, facultas.wuv: 85–98.
108. Collman, J. 2001. *Naturally dangerous*, 226. Sausalito, CA: Science Books; Ropeik, D., and Gray, G. 2002. *Risk*, 378. Boston: Houghton-Mifflin.
109. cf. "Needless Surgery Hysterectomy." *Consumer Reports on Health*. http://www.consumer reports.org/health/search.htmlquery=hyperlink
110. Hyman, M., and Liponis, M. 2003. *Ultraprevention*, 177. New York: Scribner.
111. Bagnardi, V., Blangiardo, M., La Vecchia, C., and Corrao, G. 2001. A meta-analysis of alcohol drinking and cancer risk. *British Journal Cancer* 85:1700–1705.
112. Insel, P., Turner, R., and Ross, D. 2003. *Discovering nutrition*, 281. Sudbury, MA: Jones and Bartlett.
113. Wright, J.W. ed. *Time Almanac 2006*. New York: Penguin Books.
114. "Preventing Mental Disorders." WHO Report. 2004.

115. *Health Affairs* 25, 1 2006. 293.
116. *Encyclopedia Britannica Almanac 2006*. See updates at FAO: http://www.fao.org.
117. Community Quality Index Study 2004. RAND health study on the state of health care in the United States. Kerr, E.A., McGlym, E.A., Adam, J., Keerey, J., Profiling the quality of core in twelve communities: results from the CQI study. Health Affairs 2004 healthaffairs.org
118. Wright, J.W. ed. *New York Times Almanac 2006*. New York: Penguin Books.
119. American Medical Association Principles of Medical Ethics Adopted by the AMA's house delegated 17, 2.
120. Some Agencies accepting Aid Donations to the World's Needy: American Friends Service Committees afsc.org; Amnesty International; CARE care.org Direct Relief International directrelief.org; Doctors without Borders www.doctorswithoutborders.org; www.msf.org; Food for the Hungry fh.org; For a World Without Hunger fao.com; Habitat for Humanity: Habitat.org; Human Rights Watch; hrw.org; International Rescue Committee the IRC.org; MEDAIR International Humanitarian Aid Organization www.medair.org; OCHA UN Office for Coordination of Humanitarian Affairs http://ochaonline.un.org; OXFAM Oxfam.org.uk; Peace Corps; Red Cross: Redcross.org; Save the Children savethechildren.org; UNHCR UN High Commissioner for Refugees www.unhcr.ch; United States Fund for UNICEF unicefusa.org; WHO World Health Organization who.int/en/; World Emergency Relief worldemergencyrelief.org; World population is 6.5 trillion as of January, 2007 in 2010–6.8 billion, 2020–7.5 billion, 2050–9 billion. (census.gov/main/www/popclock.html).
121. Ziegler, J. 2007. *Das Imperium der Schande*. Pantheon. (orig. L'Empire de la honte, 2005, Fayard, Paris) : Der Kampf gegen Armut und Unterdrückung.
122. Singer, P. 2004. *The president of good and evil*, 122. New York: Penguin.
123. First pledged 35 years ago in a 1970 General Assembly Resolution, the 0.7% target has been affirmed in many international agreements, including the March 2002 International Conference on Financing for Development in Monterrey, Mexico and at the World Summit on Sustainable Development held in Johannesburg later that year. Millennium Development Goals (WHO): May 2005 affirmed the .7% target.
124. warresisters.org.

Chapter 17
Ethics Counseling: Philosophy of Medicine Counseling Instead of Medical Ethics Counseling

Ethics Counseling: Philosophy of Medicine Counseling Instead of Medical Ethics Counseling

The art of morality is the practice of it, not rules laid down. [1]

Abstract The philosophy of medicine is a critique of the concepts and methods of medicine. Ethics Committee often only represent the enculturated views of its members, often only reflect the morals of a society. Who would qualify to be on such a committee and what should be the requirements? This is not at all clear, not even dealt with. Medical ethics counseling is not enough when only dealing with certain bioethics directions like principlism, Kantian deontology, situational ethics, case method, etc. Philosophy of medicine counseling is asked for when dealing in depth with the theory and practice of medicine and bioethics.

Keywords Philosophy · philosophical counselling · Philosophical Practice · ethics committees · consensus · ethics counseling (EC) · Humanism · philosophy of medicine counseling · philosophy practitioner · emotion

17.1 Introduction

There are now thousands of ethics committees in the medical field. In the U.S. institutions receiving federal money for biomedical research were required to establish Institutional Review Boards (IRBs). In the recently developed areas of bioethics, medical ethical counseling and ethics committees, there is great confusion as to what ethics is, and what the duties of the related thousands of committees should be [2].

People generally are not educated in either ethics or philosophy, therefore it is understandable and predictable that there will be such confusion about ethics in medicine and bioethics (See Chapter 5). Philosophy may be described as the clarification and critique of concepts, methods, and their applications in the various disciplines. Thus, there is the philosophy of medicine, political philosophy, philosophy of biology, philosophy of chemistry, philosophy of law, philosophy of history, philosophical psychology, philosophy of religion, etc. Philosophy may be

B. Maier, W.A. Shibles, *The Philosophy and Practice of Medicine and Bioethics*,
International Library of Ethics, Law, and the New Medicine 47,
DOI 10.1007/978-90-481-8867-3_17, © Springer Science+Business Media B.V. 2011

characterized as honest, open inquiry, with special emphasis on the use and critique of argument, reason, evidence (epistemology being the study of evidence and nature of knowledge), and methods of reasoning, which avoid fallacies in thinking. Philosophy is not an abstract, relativist discipline, but rather the most critical discipline requiring it to be better grounded and more careful than even science as is the case in the philosophy of the natural and social sciences.

"In Germany ethics committees give advice about individual cases, provide guidelines for conflicts, and organize training in medical ethics". Physicians are often against ethics committees. "The underlying ethical arguments of physicians' decisions are rarely communicated or justified" [3].

This is because they cannot be as they have virtually no background in ethics. The case with professional ethics committees is the reverse. They supposedly are able and do justify their opinions. The author states that although the ethics committee evaluates ethical problems the individual physician has a responsibility to have a sound knowledge of ethical reasoning as well [4].

Ethics committees also are not genuine *ethics* committees at all, but rather miscellaneous people selected to make consensus statements based on normative practices. They serve only to achieve a consensus of religious, political, business, legal, and other cultural groups regardless of how irrational and opposed to sound medical practice they may be. They often go along with prevailing morals in a society, not endeavoring the task of critical ethical analysis.

In philosophy, consensus and majority rule are fallacies. "Consensus is a notion more used in bioethics than understood. There is, as a result, no informed, well-grounded consensus about the nature, value and limits of consensus" [5]. Consensus, like democracy, presupposes the existence of educated participants, and those interested in the benefits to all, not just to oneself or one narrow group of people, economic or political unit. A broad democracy or consensus would include the holistic welfare of all people, e.g. as represented by international bioethics organizations. It is not merely a survey of uninformed individual or local preferences [6].

Ethics committees should deal with ethics. They therefore should consist only of people who are qualified to deal with ethical questions. Ethics committees are not for the establishment of a mere consensus by collections by normative, societal and special interest groups, religious and political interests and trade-offs. This is a conflation of ethical decisions with normative decisions fallacy. Representatives from law, hospital, parents, healthcare workers, those who will significantly be affected by the decision may be consulted, but are not part of ethical decision-making any more than other people having to decide ethical questions.

An Infant Care Review Committee was proposed (ICRC) to decide difficult neonatal issues [7].

But no philosopher or ethicist was included, although a lay community person was.

"Medical philosophical theory should become an integral part of everyday clinical work and scientific medical research" [8]. Fulford speaks of "biophilosophy." He advocates studying the actual use of ethical terms in context rather than mere abstract or usual uncritical definitions. For example, the definition of delusion as false factual belief directly contradicts the uses of the term in practice [9]. However, he does not use or apply ordinary-language philosophy adequately and commits the mentalistic fallacy by speaking of "mental disorders" [10].

Even the best doctor often knows only the tip of the iceberg in medical cases. This is part of the challenge but one has to be aware of it. There are diverse general contextual and individual differences. Medical science like all science is incomplete and many of the underlying causes are beyond our knowledge. We deal with only proximal causes and general probabilities. We do not know much about how a single cell, atom or nerve works. We do not know what energy is, or what quarks are. We know practically nothing of the smallest and largest possible worlds. In these senses, medicine and the natural sciences rest on the unknown. A sounder basis is the philosophy of science, which is a critique of the concepts and methods of science.

For example, philosophy does not automatically accept concepts such as value, equality and person, but requires these notions to be clearly and well defined, and it examines the various theories of concepts. In the area of ethics, it does not just espouse one theory or a normative or culture-bound religious perspective of bioethics as is often done in society, but must adequately and fairly examine the philosophy (concepts and methods) of the various ethical theories, belief systems, and religions to see what does and does not have a humanistic, rational and practical basis. Qualified philosophy and ethics committees cannot just uphold uncritical cultural and normative values. Singer, for example, stated that philosophy is the criticism of our prevailing moral standards [11]. The move is from religion to the philosophy of religion, and from medicine to the philosophy of medicine, and then on to a more comprehensive philosophy. Caplan stated, "Few physicians or those who work with them have any sophistication about the philosophy of science... [or the]...philosophy of medicine" [12]. He states that few healthcare workers can tell the difference between scientific statements and religious or faith-based statements. "Medical schools and academic research centers do not...make the philosophy of medicine a part of the culture of academic medicine" and this results in the undermining of medicine as a critical science and turns it into religious and political propaganda. Of course philosophy must be critical of its methods and concepts as well.

The appearance of National Ethics Committees, as opposed to mere advisory chancellor or president's advisory committees, can be a significant step toward putting the ethics of medicine and bioethics on a higher and critical philosophical level, rather than on a normative legal, religious and uncritical cultural level [13]. However, at present it is noted that such committees cannot define what ethics is [13].

A presentation of some problems with the present state of ethics counseling in medicine will now be given. It will be seen that it is unclear what it is, what its functions are, and who should be ethical counselors [14]. What ethics counseling is usually said to be is inadequate. Toward correcting this situation it will be argued that ethics counselors should be mainly replaced with philosophical counselors (philosophical practitioners, experts in practical fields as well as in philosophy) who are able to deal with all of the relevant issues in the philosophy of medicine. Singer objected, "It is absurd to leave the moral philosopher out of the debate. His/her training makes him/her more than ordinarily competent in assessing arguments and detecting fallacies" [15].

Philosophers are also already established in a discipline so that a new, narrow discipline of "ethical counselor" or ethics committees need not be established. Philosophy of medicine practitioners will then not be limited to a few methods or a narrow set of rules or principles, but be able to apply all of the knowledge of philosophy to the area of medicine. Ross, for example, gives too narrow a definition: "The bioethicist's area of expertise is philosophical analysis of health care decisions that involve ethical conflicts" [16].

Philosophers cannot be expected to know all of philosophy, just as no physician can be expected to know all of medicine. Those philosophers having expertise in philosophical practice and areas more relevant to medicine, e.g., ethics, death and dying, emotion, decision making, fallacious thinking, philosophy of law, philosophy of medicine, the problem of the self, philosophies of life, philosophy of management, value of life, etc. would be more appropriate as philosophy of medicine consultants. The philosophical counseling approach extends beyond ethics to include clarification of concepts and methods in every area affecting medicine.

17.2 What is Ethics Counseling (EC)?

In attempting to answer this question some of the key texts on bioethics and the philosophy of medicine will be examined and summarized. Hospitals and healthcare institutions in the U.S. are required to have ethics counseling. Other countries are also developing them. Ethics committee consultation is also encouraged by the courts. The goals of ethics consultation are at present said to be too diverse and lack standardization. Tulsky and Fox state that there is "no formal consensus on these goals" [17]. *There is no philosophical or overall basis of such counseling.* "Perhaps the greatest impediment to research [on evaluating EC] has been the absence of a coherent conceptual framework and a deliberate, systematic integrated, farsighted approach." (ibid.)

17.2.1 Task Force on Standards of Bioethics Consultation (USA)

The Task Force on Standards of Bioethics Consultation gives the following description and report [18]. Bioethics Consultation (EC) is said to deal with certain ethical issues: abortion, reproductive techniques, organ donation and transplantation, genetic testing, STD, autonomy, patient rights, truth telling, consent, competence, resource allocation, scarce resources, decision making, grief, interpersonal relations, burn-out, death, political and social issues, conflicts of interest, ethics of medical research, managed care, government regulations, codes of ethics, accreditation, bioethics law, institutional administrative law, legal aid resources, euthanasia, DNR (do not resuscitate) policy, and medical futility policy.

It may be pointed out that the philosophy of medicine deals with many more and broader issues. In addition, ethics involves the use of terms such as good and bad and these terms may be used regarding any problem whatsoever. There are no ethical, moral, or value questions as such. There are no ethical reasons for anything, only reasons. Accordingly, there are no medical or bioethical ethical questions as such. Anything may be turned into an ethical question. Ethics is about the use and misuse of ethical terms in any context. It cannot adequately be limited by topic or issue. Subject matter or special topics do not distinguish an ethical statement from a descriptive one.

The ethics counselor is said to have the task of 1. gathering information, 2. clarifying value and conceptual issues, 3. relating normative values (including institutional and legal values), 4. proposing morally acceptable options, 5. building consensus among all involved, and 6. facilitating discussion. If there is no consensus they are to seek appropriate decision-making authority. If there is none, one may turn to the court [19]. The task of the counselor is so limited here that it precludes a satisfactory outcome. If the philosopher of medicine had such a limitation, no satisfactory result could be achieved. To analyze issues in ethics one needs more than state normative values and give opinions. Furthermore, building consensus is an appeal-to-majority fallacy, not an acceptable method of dealing with ethical issues. Again, this is a call for relativistic, arbitrary decisions. Unfortunately, consensus is used to try to justify a particular culture and its medically harmful practices. The method here of "moral" decision-making is not acceptable.

The Task Force rejects certification or special educational program accreditation, but instead recommends voluntary participation. This may be because they have not decided what ethics counseling is. If it were recognized that it is primarily a philosopher's expertise, then they would have to be trained as philosophers. They take an un-philosophical and anti-critical position in stating that ethics decisions are just a matter of consensus, so the average untrained or uninformed person should be included in the committee to represent normative or cultural values. The task force also places stress on patient autonomy even over consensus. But the principle of autonomy is problematic and is discussed in the Chapter 12.

Questions are raised as to whether or not to have a committee or an individual ethics counselor, whether or not to record the consultation (e.g., as part of patient's record); who should be allowed to request an ethics counselor: physician, member of the ethics committee, the patient, family of patient, nurse, or staff member? [17]

Aulisio states that the duties of the ethics counselor should be the following:

(a) To deal with present ethical issues: autonomy, medical futility, resource allocation, etc.
(b) To deal with guilt and grief. The ethics counselors themselves are supposed to possess tolerance, patience, compassion, courage, "good" character, humility, prudence, integrity, etc. [One could also argue that these qualities should be possessed by the physician and healthcare worker, but also by the patient. No method is given for acquiring such qualities.]

But the problem is that, as they state, "There is neither an agreed-on set of competencies nor any standard education or training for those who provide ethics consultation." Also, they say that individuals have the right to live by their own values no matter what these values are [20]. This is a harmful relativistic morals, not ethics. It is often spoken of as the autonomy principle.

Case example: "The Bioethics Commission at the Austrian Chancellor"

The Austrian Commission for Bioethics at the Chancellor (2001–2010) consisted of 18 members/4 women and consists now of 24 members/8 women representing the fields of medicine, biochemistry, molecular biology, law, information technology, sociology, psychiatry, theology, and philosophy [21]. A considerable part of the members represent their strong religious backgrounds. As a member of the commission, the author of this book has a background as a physician, surgeon and possesses a doctorate in philosophy. However, philosophy as such is not much represented on the commission nor is a professional background in ethics. The task of the commission is restricted to discussion about important results of medicine and biology, information technologies,. . . and their ethical impact on society and to reporting findings to the Chancellor and other politicians. Nothing has force or is binding. Thus, the Commission is basically powerless. There are no concrete goals, and the Commission's advice could be completely ignored. Vague terms are used such as discussions, recommendations, proposals, and opinions. The members basically spend their time reading through many documents. It meets once a month. It is deeply divided especially on religious lines, thereby preventing consensus. The discussions often show that the models for decisions were not models of objective and professional communication but often consisted of mere rigidly held opinions. The commission has basically no expertise in ethics, uses unclear procedures, does not define its goals, etc. After many tactical moves the commission managed to present two opposed written views on the subject of pre-implantation diagnosis for consideration of the politicians. The reports also appear on the inter-net. The religiously held group was the minority opinion. As with the American President's Commission on Bioethics, the Austrian Commission seems to basically serve as a steam valve or a kind of repair-morals group to calm down what people get excited and threatened about. It should avoid challenges and seek consensus, as if ethics were a matter of consensus. Mittelstrass states that we can have consensus about decisions causing the most "wrong" and harmful consequences of actions [22].

The commission is what its members are.

17.2.2 Basic Ethical Principles in European Bioethics and Biolaw

Dealing with principles one has to consider: "No formula can take into account of the infinite range of such exceptional events" [23]. To give an outlook to what is taught in the area of European Bioethics and Biolaw it is possible to check: *Medical Ethics Today: The BMA'S Handbook of Ethics and Law* [24]. The book intends to be the standard guide for physicians to follow.

It uses the Beauchamp-Childress four principles approach of beneficence, non-maleficence, autonomy, and justice, which has been highly criticized in the present book on the philosophy of medicine.

Principles are given for application or performance. Principles may be required or optional, binding or non-binding. Synonyms of a required "principle" are: decree, directive, law, order, protocol, regulation, rule, and set formula. These tend to be required orders, commands, instructions, and directives. The healthcare worker is controlled, managed and overpowered by such principles. Synonyms of a non-binding or non-required principle are: advice, analysis, approach, clarification, design, explanation, guideline, information, method, plan, policy, practice, recommendation, report, strategy, suggestion, and system. Binding principles take away reasoning and judgment on the part of the healthcare worker. Non-binding principles allow them. One can easily equivocate between binding and non-binding principles. For example, the principle of autonomy may be regarded as binding, when it is only a guideline. "Practice" may refer to the descriptive performance or the required method, or both. Because there is a standard practice does not mean it is the best one or that it always should be followed.

As a rule could be put the following: Never give a rule, in any case, at least not an unchallenged one, unchallenged by the individual and the context (self-referential and self-binding)!

The following principles presented in Munson are critically examined as follows [25]:

1. Non-maleficence. Do no harm principle [26]: is violated in medicine by requiring excessive 80–100 h work weeks, bad management, failure of physicians to advise regarding prevention, treating symptoms rather than causes, failure to deal with treatment holistically, etc.
2. Beneficence [27]. This is thought to be a principle but actually is no principle as it means to do good and good is an empty term. One would not know what to do. Beneficence today often means to provide a medical repair or business service to clients. Scarce resources and inhumane working conditions of healthcare workers do not show beneficence to staff or patients. Benevolence and beneficence can refer to anything. They may partly refer to altruism (See Chapter 10).
3. Principle of Utility [28]. Produce the greatest benefit. It is unclear what benefit means as it is an open-context term. Does it mean cost-benefit analysis? If utilitarianism is implied a critique of it should be given.
4. Distributive Justice [29]. This is an important principle as it qualifies blind equality, which is so often prevalent.
5. Equality. The important insight is given that equality presupposes responsibility. "Everyone must bear an equal part of the social load" [30].
6. Principle of Need. Those in greater need obtain a greater share of health resources. This principle is not followed in the U.S. where millions do not have health insurance.

7. Principle of Contribution. What one puts in is rewarded. On this principle one could give medical priority to those who have contributed to society as opposed to those who did not.
8. Principle of Effort. This would reward those who contributed most.
9. Autonomy. "Rational individuals should be permitted to be self-determining" [31]. The stress is on rational. If one is irrational, or is uncritical and/or knows nothing about consequences, confidence in autonomy is baseless. There is then no evidence-based autonomy. This means that such people would not qualify for autonomy.

Representing the collective European context of bioethics and biolaw, Rendtorff and Kemp have given the following account founded on the four principles of *autonomy, dignity, integrity* and *vulnerability* [32]. To this is added a fifth: "Solidarity and Social Responsibility." These principles are supposed to guide us in regard to decisions in medicine, bioethics and biolaw. The principles allegedly are descriptions of the present basis of bioethics and are intended also to serve as guidelines [33].

Beauchamp and Childress grounded bioethics on four principles: *autonomy, non-maleficence, beneficence* and *justice* rather than on a philosophy or ethics [34]. Childress was a religious ethicist so it was a way of combining religion with secular thinking through principles. But autonomy is not good in-itself, and to say that "good is beneficence (good making) or non-maleficence (not bad making)" is circular. As "justice" is a value term, it is also circular to base "right on justice (right)." Justice is not a principle anymore than ethics is a principle. These "principles" are based on the lack of a rational foundation, and exclude the analysis of emotion and ethics [35]. Ethical principles are not ethics. Ethics is a critical process, not just laws or rules to follow. Principlism is like the "bag of values" approach in education according to which certain values are randomly selected, e.g., obedience, honor, spirituality, etc. Engelhardt stated, "Beauchamp and Childress assert that there is a common morality disclosable in some philosophical manner such as an appeal to moral intuitions, ... [these are] thin theories of the good" [36]. Wesley Smith, who supports an absolutistic Christian bioethics, nevertheless says there is nothing wrong with these guidelines [37]. This all only indicates a fairly complete lack of knowledge of ethics. The following is an analysis of the account by Rendtorff and Kemp [32]

17.2.2.1 Autonomy [38] (See also Chapter 12)

Autonomy is respect for the rational decision making, without coercion, of the patient regarding his or her medical care. [This would be the opposite of absolutist positions.] Autonomy involves the view that the patient has the capacity:

1. to create ideas and goals for life,
2. for moral insight, self-legislation, and deserved privacy,
3. for political involvement and personal responsibility.

A criticism of this account is that autonomy can be a form of relativism according to which one can be uneducated, uninformed, selfish and self-righteous with no concern for others or the improvement of society. It could be ethically and medically unprofessional to accept the "autonomy" of the patient or spokesperson regarding the type and availability of medical care. The physician must have the option to attempt to give needed treatment even when the patient does not wish or allow it. Psychiatric patients, and those needing therapy, also often do not want or are not able to agree to much-needed treatment. Patients as well as their families are often not competent to decide about such matters. Also, without a background in critical thinking (speaking) and ethics, full autonomy of the patient may be put in question as indicated by the following statements:

"If a patient be under orders, he will not stray; left to himself, he will give up the struggle and depart this life. . .so [the physician must] take the patient in hand" [39].

Autonomy has no *a priori* standing. It is an atomistic myth that one is self-contained, self-sufficient, self-determined instead of living cooperatively [40].

"There are situations in which paternalistic behavior is ethically justified" [41].

Beauchamp and Walters wrote that rational autonomy must be based on informed choice and should sometimes be restricted [42].

"Discussions of autonomy are riddled with paradoxes" [43]. "Coherence does not seem to be a defining characteristic of autonomy" [44].

David Jopling wrote, "The price of respect for the client's autonomy may be the flourishing of the self-deception and self-illusion" [45]. The physician may take the view that before granting blanket autonomy one first needs to assess whether or not the patient is capable and informed enough for autonomy.

It is a fallacy of the argument from authority, in which the patient is falsely regarded as the authority. "Respect," and respect for "autonomy," is an open-context value term and must be given a meaning to be intelligible. Autonomy can be the centrality of the patient's interests to the exclusion of all else [46]. Autonomy might only be selfishness, concern mainly with one's own desires. We do not hold that altruism should exclude one's own interests. The good of the patient, the quality of life of those involved (including the physician who is often overworked and bears extreme liability) in addition to the other factors should not be excluded. The other people involved also have autonomy so there must be a balancing of autonomies. The autonomy of the patient can destroy the autonomy of all of the others involved.

Autonomy is often falsely contrasted with paternalism. Every expert has knowledge others do not have and it is not paternalism to actively use such knowledge and guide the patient or those seeking such advice. In some cases, the patients are more informed than the healthcare worker.

17.2.2.2 Dignity

"Dignity" is a value term like "good" and, as such, it is an open-context term, which could mean anything, and usually does. Rendtorff and Kemp also partially recognize this in saying, that on Kant's view: "The *dignity* of man consists precisely in his capacity to make universal [*ethical*] law" [47]. [This is true by definition, which is described by Kant as an analytic statement, that is, it is circular]. For the Stoics, dignity (*dignitas*) rested on rationality and inquiry [48]. This view gives dignity a meaning and is closest to the approach of the naturalistic philosopher.

But then the authors write, "Dignity is something to be acquired by moral effort, rather than something absolute that is present in the human from its very beginning" [49]. One must earn dignity. On the other hand, dignity is dependent upon others and being regarded as dignified by unethical, uneducated masses is only political popularity or enculturation conformity, not genuine dignity. Dignity could rather come from oneself as an educated, inquiring and ethical person. We may then speak of "self dignity." Dignity should not be a matter of consensus, but be based on rational thinking and action.

Dignity is said to be:

(a) Dignity is said to be the "intrinsic value of the human being in society" [50]. [Fallacy of intrinsic value. Also it is circular: dignity means value.]
(b) Dignity is said to be the "moral agency of the human subject." (Circular: dignity means moral.)
(c) Dignity forbids any commercial use of the human body. For dignity, humans cannot be "objects of trade or commercial transactions" [51]. It may be noted that the entire medical profession in the US is a commercial system. We cannot sell organs because it is supposedly not dignified [49]. But physicians, miners, police, fire people, etc. are paid to risk their lives.
(d) Dignity is said being based on the indeterminate position of the individual in the universe (It is not clear what they mean by this).
(e) "Self-esteem, proudness, shame, feeling of inferiority and degradation are essentially matters of human dignity." (Circular: It says dignity is not being undignified.)
(f) Dignity means the equality of all humans (Misuse of the abstract word equality. Equality may be unfair in matters of merit or distributive justice).
(g) "Dignity includes the individual's openness to the metaphysical dimensions of life" [51]. "Important values in medical prioritization are equal human dignity, solidarity, confidence and security, respect for freedom, vulnerability, self-determination and metaphysics" [52]. [It is not clear what "metaphysical" means. It is often the same as supernatural. Mystical beliefs are not dignified, but typically irresponsible and harmful.]
(h) "*Dignity* is... acquired by *moral* effort," is also circular [49]. Most of the statements about dignity are either circular or fallacies of thinking. Nothing can be intrinsically good, or good-in-itself:

    1. "Dignity is the intrinsic value and responsibility of every human being" [51]. [Circular and intrinsic (in-itself) fallacy.]
    2. "Intrinsic dignity belongs to every human being" [53].
    3. "Dignity is a virtue" [54]. (circular)
    4. Dignity is the "intrinsic value of a species of animal or plant...in its evolutionary story" [55].
    5. Dignity is the "intrinsic value of the humanity of the person" [56].

6. If dignity implies others (society) it is circular to say that dignity implies society. Could one be dignified outside of society? Or dignified in one's own eyes?
7. Humans are regarded as ends-in-themselves [57]. (Misuse of the term: "ends-in-themselves," as an in-itself fallacy.)
8. "As a member of the human species, every individual has inalienable dignity" [56]. Could this be said of anyone, even cruel and destructive people? Being a member of the species does not guarantee dignity. Species classification as classes is value neutral.
9. "There is no consensus about the status of the human embryo in Europe" [58]. But the authors then go on to say that it is a "potentiality of person" [58]. They also speak of the "dignity of the embryo" [59].

17.2.2.3 Integrity [60] and Narrative Analysis

See also narrative analysis in Chapter 9.

"The aim of medical practice is, if possible, to restore or to make whole the damaged physical and psychological unity of the patient" [61]. In general, integrity seems to stress wholeness, but if so, philosophy would be especially relevant because of its concerns with both depth and wholeness of analysis. The coherence of the individual's life (life story) or unity of culture is supposedly not to be destroyed. Doctors would focus on the "narrative coherence of the patient's life story" [62]. Treatment must preserve and restore "the personal identity as narrated integrity" [55]. One's life story includes one's hopes and values. This is to treat the patient as a whole person.

Narrative analysis was anticipated by the Structuralists, who tried to find deep signs and structures, a semiotic. Several other accounts of the life story narrative may be interjected here. According to Kemp and Rendtorff [63] narrative is a primary and radical sense of truth involving our motivating imagery, models, attitudes, choices, authority of the story itself, images such as the perfect baby, a morality which comes out of the story itself [64]. John Arras regards "narrative ethics" as 1. a close reading of literature whereby the text gives moral knowledge, 2. a means to give the finer particulars of moral situations. It can emphasize concreteness, context, detailed consequences, and connectedness. 3. one's account of one's social group, 4. authenticity of the storyteller's testimony, 5. a closer look at the context [65]. Narrative ethics opposes the search for universals and the application of theories. It is thought to be more persuasive than theory. It is also a view of neo-pragmatism. However, storytellers are not likely to be critical thinkers or philosophers. Life stories may be autonomous and relativistic. Which stories are worth telling? We see also how ambiguous the notion of the life story narrative approach can be. Rosemarie Tong argues that narrative knowledge only gives hints at universal truth [66].

We are always "woven" into stories [67].

Our experience, our language is part of a net of stories. Even before we are born stories are made up about us, who we will become, etc. A moral and interpreted and critical context of action is provided from the view of the storyteller. Integrity of life is created. Goals of those involved can thereby be hypothesized, created, compared, and evaluated. An initial definition given is: Narratives are foundations of time sequences of actions, and in the act of telling the person is brought to language. This point combines with the theme throughout this book on bioethics that language has epistemological primacy and that medicine is grounded in language.

The story, the language, and language style create the world [68].

Michela Galzigna, who teaches at the medical school in Padua, Italy, is influenced by Wittgenstein's notions of language-game concept and "forms of life" and extends this to develop a notion of narrative medicine and narrative ethics, which give epistemological primacy to language in medical treatment: "Medical practice is characterized by different verbal constructions," The physician is spoken of as a "narrative physician," whereby s/he tells a story and the patient tells a story, rather than just having the physician give some information based solely on "evidence-based medicine" [69]. We supposedly are created by our fictive stories and conversations. On this view, knowledge arises out of our conversations. She thinks the placebo effect of language accounts for up to 70% in psychiatry of the effect of the medication prescribed. This would also give support to the cognitive theory of emotion (See Chapter 7). She quotes one of the most important and generally least comprehended passages in Wittgenstein's *Philosophical Investigations*: "I shall also call the whole [of language], consisting of language and the actions into which it is woven, the 'language game'... [this] is part of a frame on whose basis our language operates. . .the term language game is meant to give prominence to the fact that the speaking of a language is part of an activity, of a form of life" [70] (See also the Chapter 18). One problem remains with conversations and stories being uncritical and culture-bound. She therefore suggests that the conversation be critiqued and deconstructed and the meaning be negotiated. She is aware, however, that the definition of narrative is somewhat obscure. To the language game account we may also note that much of the meaning of a sentence or story comes from the intonation, not just the words (See also following section on *Medical Language: The Ordinary Language Approach* for case method).

The philosophical counselor, Günther Bachmann, posted on his web site: "Who changes his/her language, changes also his/her thoughts and feelings."[71] This also coheres with the cognitive theory of emotions.

We may now return to the account of integrity given by Rendtorff and Kemp:

1. The created and narrated coherence of life as a wholeness of a life story is not to be violated [55]. "Manipulation of the human body that substantially changes personal identity should be prohibited" [72].
2. Stress is on personal self-determination as a form of integrity.
3. Integrity is seen as good character, e.g., honesty, reliability, fairness, truth, sincerity, candor. Integrity involves moral coherence of legal and medical systems and values of society (e.g. trust between physician and patient, ensuring privacy, etc.)

4. There should be integrity of national territory, and international law [55]. [Dealt with in the area of political philosophy.]
5. "Personal integrity includes both a psychological and a corporeal dimension" [55]. [This raises the mind-body dualism problem and the problem of self and identity, which would need philosophical analysis and critique. Hume wrote, "With regard to the mind. . .we have no notion of it, distinct from the particular perceptions" [73]. Mentalistic explanations have also been argued against. Yet few people or academicians are aware that ideas, mind, imagination and memory as descriptions of entities are pseudo-psychological concepts.] (See Chapter 18)
6. One has to deal with the integrated wholeness of a person [*Lebenszusammenhang* [74] "Life context"] [62] "The ethical principle of integrity may even be extended to be applied to animals, the vulnerable life of nature and even the whole living world" [75]. [This would seem to argue for world healthcare, world citizenship, and the responsibility to care for all of the starving and diseased of the world.] What is interesting in this account is that the person as a whole is to be dealt with and the whole personality involves oneself as a world citizen. Philosophy traditionally deals with the person in a holistic way in the most comprehensive contexts. Now this notion is being brought into the philosophy of medicine.

17.2.2.4 Vulnerability [76]

Vulnerability to disease means that we are all subject to illness, especially the weak and poor [77]. It is regarded as the basic principle of medical ethics [78]. It is not clear why this is a principle. Certainly, if there is no disease there is no need for treatment of disease. "Advocatory ethics" is where human beings must be advocates for beings not able to communicate [79]. Those who can should help those who cannot help themselves, especially in the last phase of human life. With vulnerability the importance is given to care, responsibility, and empathy [80]. "Human life would lose all sense if it lasted forever" [81]. It is circular to say that if we do not have death we could not "sacrifice our life for a good cause" [81]. Human life would certainly not lose all sense if it lasted forever. It would just not be the nature of human life as we know it and not in the realm of known possibility.

We could say that vulnerability is: 1. due to physical defects or debilitating influence of culture, poor belief system, etc., 2. due to ourselves, e.g., unhealthy lifestyle, taking unnecessary risks, uncritically accepting biased beliefs. We may also be vulnerable because tempted to do what is against our reason.

17.2.2.5 Solidarity and Social Responsibility [76]

This is common working together and taking responsibility for the protection, welfare and prosperity of the individual in the society and state. It is also an attempt to enhance ourselves and, this way, improve societies. It is asked, "What do we want humanity to be like?" [77]. "They [the solidarity principles] are oriented towards the good life in a civilizatory process" [77]. This has some similarity to humanism, which has been discussed earlier.

17.3 Criticism of Bioethical Principlism

> Just in the degree in which. . .s/he subordinates the individual case to some classification of diseases and some generic rule of treatment s/he sinks to the level of the routine mechanic. . .rigid, dogmatic, instead of free and flexible [82].

The moral principles themselves need critical philosophical analysis. Several forms of principlism must first be distinguished. In the usual accounts of principlism it is often not clear which types are being used. It may be noted that any set of principles presupposes a certain philosophical theory. They can be principles which are phenomenological, Kantian, utilitarian, existentialistic, scientific, relativistic, etc. Law often is thought to be a moral minimum of consensus to be aimed at. It may be noted in the following that the philosophy of law, which is a critique of the concepts of and methods in law, may be used to give insight into the theory of principlism.

(a) *Logico-deductive system of principles.* One simply deduces from the written law with a minimum of interpretation and without concern for the consequences, wishes or needs of the parties involved. One follows the letter of the written law. The first law is to obey all laws.

(b) *Positivist theory in the philosophy of law,* e.g., by the neo-Kantian, Hans Kelsen [83]. (See also bioethics in Austria cited below). One merely deduces from the law or other principles without consideration of the particular case, social or other consequences. Theory deduces from a limited number of cases and reduces it to a deductive geometry. The legal system is just norms and commands. Positivist law is law as imperatives and for the British philosopher, John Austin, law is commands of the political system. The positivist says one should obey law as law. Kelsen treats all law as a formal, autonomous, logical, consistent system without concern with consequences, wants or needs. Law is imperatives and to be based on imperatives (cf. Kant's categorical imperative), which Kelsen calls "norms" which are supposedly ideal, "pure" forms. Law ends up being abstract and transcendental logic. There is no overlap between law and morality, which is called the "separability thesis." The law cannot say what is good or bad or why. Thus, a law can be completely legal, but unethical. One eliminates all that is not strictly law, does not mix law with ethics, causality or sociology. This is, then, opposed to a sociological theory of law. The sociological theory of law involves balancing interests, needs, desires, and consequences. The principles or laws may or may not be based on reason, nor need they be beneficial. The legal notion of precedent (*stare decisis*) fits well with such a system, as does any system of absolute, fixed truth, such as religious dogma. An abstract principle is one which cannot be reduced to particular cases, a general principle may. The principles or laws may also be based on mere consensus, which may or may not have a rational basis. Consensus does not substitute for sound arguments, but may be based on the *argumentum ad populum,* the fallacy of appeal to the masses.

(c) *Principles and laws may be based on various theories of scientific method.* We may base principles on a scientific method, which gives epistemological primacy to language rather than to mere sense perception, the latter often being referred to in the philosophy of science as "naive empiricism." One problem with principles is that each principle can be infinitely re-described. We must choose the best perspective or seeing-as. The possibilities are limited by our language-games and rhetoric.
(d) *Principles may also be inductively arrived at on the basis of rational, consequentialistic and humanistic thinking.* In this case, principlism can combine with the clinical case method, which is critiqued later. The principles are derived from experience with specific, practical, concrete, cases. Facts and laws are seen as hypotheses with a certain degree of confirmation. They are not absolute.
(e) *Pragmatically based principles.* Laws are regarded as constructive, stipulative, and outcome oriented in terms of concrete situations (case method). On the pragmatic view, law is just a working hypothesis in need of constant testing in actual practice. It is flexible and experimental. These principles are based on the pragmatic and humanistic philosophy of John Dewey. Dewey states, "Rigid habits sink below the level of any meaning at all" [84]. Law is a kind of re-description or reevaluation. Law must rest on a sound philosophy and humanism, but often does not. Wittgenstein wrote, "Not only rules, but also examples are needed for establishing a practice. Our rules leave loop-holes open, and the practice has to speak for itself" [85]. The analysis of a case would then be determined by a full critical consideration of all of the factors and consequences and principles involved in a situation in terms of a naturalistic theory of ethics in order to obtain humanistic and desired results. Justice Richard Posner advocates "pragmatic adjudication" according to which the judge does not stick with just following rules, authorities and precedent cases rigidly, but rather employs human reason to determine the concrete situational facts and the human consequences involved now and in the future. Adjudication means to use reason, not to just follow rules or principles [86]. Dewey notes that Pragmatism changes judgment in view of new facts, whereas religion and positivistic law do not. "There is such a thing as faith in intelligence" [87].
(f) Principlism has many of the same drawbacks as formal logic: over-abstraction, reductionism, failure to consider contexts and cases, situational irrelevance, rigidity, meaninglessness, etc (See critique of formal logic in the Chapter 18).

In themselves, rules and laws are neither good not bad. Law is not ethics nor is it ethical as such. But ethics applies to law as a meta-statement. Laws can be good or bad. What makes a law moral (though not ethical) is that it may be seen as a duty, command, or imperative. But laws are not imperatives. If one violates a law then, if caught, one may be given a penalty. But one has the right (the ability) to disobey all laws. In regard to punishment, law is based on the lowest level of ethics: do x or you will be punished. In this sense, physicians are unfairly treated when inevitable mistakes are made. Correction, not punishment, is a more intelligent and appropriate method. In a sound ethical system one does things because of the desired consequences and because it is rational to do so. Laws do not even seem to imply an ethical system or philosophy of life. One may think law presupposes a common morality. Some opt for legal paternalism, others for legal autonomy. This has a direct parallel with medical treatment.

Some criticisms of Principlism are given in the following:

1. What is interesting and important is that the authors, Rendtorff and Kemp offer criticisms of their own views (included here), to which others are added in this book. They think, however, that the objections can be overcome. There are presented more counter arguments and also cited others who would disagree.
2. We would need another theory of ethics to apply the principles [88]. Thus, it may be noted that the principles are not themselves a theory of ethics, but presuppose unspecified ones.
3. Principles may be taken dogmatically or as being more than guidelines. They are like a search for the ten commandments of the ethics of medicine. Principlism is often an attempt to impose or find universal or fixed laws, which can unite the various belief systems of society. The authors themselves realize that there are no fixed, true, objective principles or laws and thus hold such principles to be only guidelines.
4. There are too few principles [89]. The principles are just expansions of the meanings of a few value words such as autonomy, dignity, integrity, solidarity, vulnerability. Why were just these words chosen when thousands of others could also easily apply? Why were these terms not more carefully analyzed and critiqued? Certainly one can give philosophical insight by the method of analyzing issues in terms of certain words or basic metaphors. This work must, however, be done carefully and critically and the negative and mistaken meanings must also be presented. This method is developed in the Chapters 18 and 1.
5. There are too many interpretations of the principles.
6. Too few philosophies are represented: Heidegger, Sartre, Habermas, utilitarianism, christianity, Kant, Paul Ricoeur, etc. Generally left out are, for example, pragmatism, contemporary humanism, John Dewey and ordinary language philosophy.
7. Principles are oversimplifications. Equality, autonomy, etc., are oversimplifications and have the effect of avoiding rational judgments. Noddings wrote, "The removal of judgment institutionalizes abuse and leaves the whole system open to a charge of absurdity" [90].
8. The principles are pre-given and so block ethical decision [91]. We like to just follow rules rather than think. Two things are wrong with this: a) it is just following, and b) they are just abstract rules.
9. The principles interfere with judgments in concrete situations [91]. General principles are used instead of a detailed and adequate examination of specific cases. They are not contextual or case oriented enough for the consequentialist

or casuist [92]. They are correct in this criticism. Their view is referred to as the "tyranny of principles" [93].

10. The assumption of absolute rights (e.g., autonomy of persons, rights of an embryo to life) is an absolute way to escape decision-making. In addition, right is empty, open context term.
11. Stress on equality often overrides other principles and factors of a situation in many unfair ways. Some say let all die equally rather than make decisions as to whom to treat. Some hospitals refused dialysis to all equally, rather deciding which few to treat [94].
12. The autonomy principle may override sound medical practice in terms of consequences.
13. To say that the considerations of quality of life are too difficult to decide on is another excuse not to decide. The result is unfairness and mismanagement of resources.
14. The principles approach is relativistic. Consensus, autonomy, rules of society, laws, codes of ethics, autonomy are relativistic. Principles are not a theory of ethics, they do not construct cohesion and rational argumentation.
15. Principlism is often based on normative, uncritical cultural beliefs and practices. Culture (custom, tradition, religion, ritual, normative values, etc.) has too many definitions (often contradictory) of good and so cannot be used as a standard or as a substitute for ethics. It generates relativism.
16. Principlism is not comprehensive or holistic though they refer to holism [89].
17. The principles are not integrated [89]. The relationship between principles is unclear [89]. They often conflict [91].
18. The principles are not related enough to practice [89]. An alternative approach would be pragmatism, which includes holistic humanism.
19. Principles are too vague and abstract [89]. In this sense, they are too unrelated to societal contexts, although they are used in society [89].
20. The principles seem to be "good in themselves" or intrinsically good, rather than being based on reason or rationality [91].
21. Consensus facilitation appears to be the main function of most ethics committees. It is another form of the search for true and universal decisions. John Kilner, for example, looks for consensus [95].
    Freiburg has a Center for Ethics and Law in Medicine (ZERM) instead of establishing ethics committees. However, it is based primarily on arriving at consensus [96]. But, as was previously discussed, consensus itself must be critiqued. "Consensus is still described as the goal of pragmatic deliberation, but only consensus that 'can withstand moral [ethical] scrutiny'" [97]. John Fletcher wrote, "I do not here or anywhere else assume or imply that the voice of the majority makes anything moral…Public opinion is not a legitimate source of ethics" [98]. Consensus, like democracy and Parliamentary Procedure, presupposes the existence of educated participants concerned with the interests of all, not just one narrow group of people or political unit. A broad democracy or consensus would include the representation of all people, e.g., as represented by international bioethics organizations. It is not merely

a survey of uninformed individual or local preferences [99]. But the participants are not informed, have virtually never been taught about emotion, ethics and critical thinking (speaking) in the schools, and are not all concerned for the good of all or of the betterment of society. Consensus is a way to avoid decision-making. What is evident are attempts to control and to gain or keep power (See Chapters 3 and 8). Rather each person should present arguments for objective consideration and discussion.

"Ethics committees and consultants can... be quite counterproductive, actually producing more harm than good. This can occur, specifically, whenever such entities (a) see themselves constituted to enforce a particular religious point of view, (b) allow themselves to be co-opted by the institution in which they work, seeing themselves as but an arm of that institution" [100]. Committee members are haphazardly chosen sometimes requiring an average person, and religious members usually predominate. But one average person cannot represent all average people, and even many religious members cannot represent all religions. The goal, in any case, would be rather to represent the best decision for all concerned and for the betterment of society, not just to support an unethical society. Consensus cannot guarantee to achieve this goal.

22. The philosophical theory, which grounds the principles should be explicitly stated and promoted.
23. Principles and laws are not ethical theories. The ethical theories, which they imply should be clearly stated.
24. The degree to which the principles or laws are to be complied with must also be stated as well as the consequences of not complying with them. The present criminal system, as it requires more abusive punishment, would be one of the worst models of consequentialism. The consequences should rather aim at education and intelligent correction. Members of ethics committees may be found liable. Ethics committees (EC) are advantageous because the court typically gives special weight to their deliberations.
25. What are the types and ranges of decision which the EC may make? Is the EC decision a recommendation, opinion or viewpoint, advisory, or binding decision, guiding principle, etc.? Which type of decision would apply to which involved party, e.g., to the physician, patient, etc.?
26. According to some philosophical counselors, there is to be no theory, only methods. This is called the "Socratic Method" which is much like relativistic autonomy, and is unfortunately widely held by many philosophical counselors. This seems like the case method, but it is just subjective relativism. But why just these methods? Methods can be indoctrinated also. And why should all of the theories in philosophy be swept away as being valueless?
27. We simply do not just think in principles or laws, any more than in syllogisms or symbolic logic (See Chapter 18). Thinking involves all of our reasoning and emotions in language, which contains much more than mere principles.
28. It is one summary statement to note that the search for principles seems to be a way to avoid considerations of ethics, philosophical theory, and scientific

method, as well as to avoid individual decision-making by the use of reason to solve bioethical problems.

29. John Dewey wrote, “A person's duty is never to obey certain rules.” Duty comes only from “concrete relations to people and things.” “A person has not to do Justice and Love and Truth; s/he has to do justly and truly and lovingly” [101].

17.4 Case Method of Clinical Ethics

The good is never twice alike. . . .It is new every morning [102].
Theory is itself theoretical.
You cannot swim outside of water.

Theory often refers to contextless general principles or laws irrespective of the particular concrete situation. It is a vague term. The case method involves actual clinical cases. The terms “Clinical ethics” seem to have appeared in 1976 by Joseph Fletcher [103]. Firstly, the case method is not a theory of ethics. That one deals with cases is no more an ethical theory than is performing an operation. Ethics can be about medical cases or anything else. The clinical case method presupposes a theory of ethics, which it does not provide. Secondly, “clinical” and “case” sound scientific, but they are not precise terms. It is unclear just what methods are being referred to. Little states, “The clinical process is fuzzy and messy and imprecise” [104]. There is not one case or one clinical method. Thus, the case method of clinical ethics is neither method nor is it ethics.

1. Casuistry. One form of the case method is called casuistry and is the view of Jonsen and Toulmin who argue for principles, which apply to, but also come out of specific cases. This idea of casuistry is based on the method in fifteenth and sixteenth century theology. It is a method of classifying cases into similar cases and then relating them to paradigm cases and principles. A map of differences and likenesses between cases is called “moral taxonomy” [105]. This is not a surprising approach as Father Albert Jonsen, Society of Jesus, approaches bioethics from the Catholic tradition. He seems to be trying to integrate the religious perspective into the secular tradition by seeming to use the scientific approach. Thus, he with others wrote the text, *Clinical Ethics*.
   Thus, casuistry is a form of principlism, the very view the case approach allegedly opposes. However, on the view of Jonsen and Toulmin, “Moral wisdom consists, not in a hardline commitment to principles, which we accept without qualification, but in understanding the human needs and relations that are nurtured by a life of reflective moral action” [106]. The case method is not a search for super-principles [107]. These, however, may be empty words considering the hidden religious agenda, which co-opts sound medical decision-making.
2. Casuistry is also defined as: “Decision-making using particular cases, where the judgments reached rely on judgments reached in prior cases” [108]. They call

them "impartial, universal action-guides" [109]. This is yet another way to avoid making decisions. They suggest that we collect cases, which represent a consensus in society [108]. It is a search for paradigmatic ethical cases. Yet, case theory implies an ethical theory, which it does not provide. Like legal precedent, it is based on arbitrary rules. Casuistry involves basing a case on prior cases [110]. This means that it is subject to all of the criticisms that precedent is subject to, for example, whether the prior case is judged wrongly, and whether the cases are similar, and whether it actually lacks concern for consequences in the specific case at hand, etc.?

Because similarity of cases is involved, the case method suffers also from the same criticisms as there are to the principle of equality (e.g., blind equality, disregard of merit, too narrow or unspecified sense of similarity, etc.). A second objection is that "each case must be considered on its merits rather than be automatically approved because of its type" [111]. While this opposes the imposition of abstract principles from above, it suggests that there are "merits" in the case itself. They are not there. The case presupposes merits, that is, an ethical system outside of the cases.

Tomlinson entitles the issue: "Casuistry in medical ethics rehabilitated, or repeat offender?" The answer is "repeat offender" [112]. The casuistic principles are often abstract themselves. Furthermore, such method of precedent suffers from the same problems it does in law. The precedent may have been misguided from the start. Casuistry may also be associated with absolutistic moral systems [113]. We do not know how to balance one case with another [114]. Paradigms remain unchallenged [114].

3. Cases without principles or unprincipled cases. This method often involves subjectivity, what is called "intuitionism," whereby the healthcare worker without knowledge of ethics just comes up with a decision. They often take a medical decision as an ethical one. We may interpret the case method as one in which an ethical theory somehow mysteriously or epiphenomenally arises out of the particulars. It does not. Typically in science there are only data or facts in relation to a certain scientific theory or principle, not free-floating data as such. And there are no theories or principles without data. One cannot, for example, intelligently deal with emotion in a clinical case setting unless one knows about the theory of emotions. Without knowledge of emotions one cannot be caring and one cannot be a good administrator who can manage crisis and deal with the emotions of the staff and patients (See Chapter 8).

   But, furthermore, clinical case analysis involves all of our thinking, not just data and theory. We sometimes use principles and usually do not. Thinking involves all of the techniques of language and emotions. Furthermore, each case can be infinitely re-described. There is no one description of a case. We may often decide to act against principles. The narrative method is a form of case method involving concern with a perspectival, but not full, detailed story of a case. Narratives, novels, and biographies, include much more than mere principles and data. On the other hand, narratives, although rich in context and detail, may be poor and philosophically naive, life stories.

Another form of case method is the "situational ethics" of Joseph Fletcher which involves the view that whatever is done under the influence of the specific situation out of love is ethical [115]. His view coheres with the view that ethical terms only have meaning if something concrete is substituted for them. Unfortunately love and emotions are not analyzed and love, his underlying principle, is left too vague to generate an ethics. He even asserts that love is indefinable, that is, not definable in terms of something else. But this would suggest that love is unintelligible. Also we can define every term, and if we cannot define love in terms of something else, we would just generate the circularity that love is love. If love means good, it is circular to say all that is done out of love is good. He does regard love as *agape,* which often means love of a metaphysical or supernatural sort. He commits the fallacy saying love is "intrinsically good." His view is actually a sort of utilitarianism: to produce the greatest love for the greatest number. We are to love everyone non-reciprocally and unconditionally whether deserving or not. Sometimes, for Fletcher, ethics is based on reason and consequences and sometimes not. According to the similar existentialist view of Sartre, and Simone de Beauvoir, in *The Ethics of Ambiguity*, each situation is radically unique and the world is basically incoherent. There are no valid principles at all [116]. This leads to self-righteous relativism. This view connects with the contemporary view of deconstructionism. In any case, the basic view is that there are, in general, no absolute truths or principles. Judgments must be made in terms of each unique, concrete situation.

4. Principles without cases. The philosophy of medicine is attacked by Ezekiel Emanuel who, having a background in political science, predictably favors a political approach, though he refers to himself as a bioethicist: "Medical ethics must stop being case oriented and become institutionally oriented. We bioethicists must stop approaching problems from a philosophical perspective and adopt a political science perspective" [117]. "The context of medical ethics can no longer be cases, but institutional structures" [117]. "This focus on ethical principles and rules is no longer tenable—if it ever was" [117]. "Managed care has two primary techniques to control costs: financial incentives and guidelines or protocols" [117]. He wants to create incentives rather than just financial reward: professionalism, disclosure, competition, review of financial incentives and guidelines, mediation and appeals procedures [118]. He is, in effect, saying that philosophy and ethics do not apply to medicine. He opposes case-by-case decisions, which come out of ethical committees and favors a more deductive pre-given or political policy. The question arises as to which ethics such political and economic policy is to be based on? It is a blind or floating ethics, that is, no ethics at all. Medicine is often reduced to cost accounting and politics. It often is already just that.
5. Pragmatic Case method. What I term "pragmatic case method" is also the scientific method of examining the specific details and context of the specific case. There is no dogma in advance of information and facts. Dewey wrote, "In quality, the good is never twice alike.... It is new every morning" [119]. Scientific method and ethics both involve problem-solving to bring about our goals, wants

and desires. Thus, this is genuinely *scientific clinical case ethics*. Ethics means for the pragmatist, problem-solving ability to put a plan to work. One must first clarify and sort out the facts. Adequate theory and philosophy must include practice. Theory and principles can be bottom up, as opposed to top down abstract reasoning. For example, we cannot just deduce from patient autonomy alone. We need the whole story. Dewey presented how such reasoning is to take place in his works on logic. He has in these works vigorously argued against formal, abstract systems such as symbolic and Aristotelian logic (See critique of formal logic in the Chapter 18). For Dewey, the valuation is formed in the process, not something taken from outside. It must meet the needs of the particular situation. "We believe in moral laws and rules to be sure, but they are in the air. . .no working contact with the average affairs of everyday life" [120].

Schneider records how he had to break protocols and challenge treatments to regain his health. Physicians may follow protocols rather than the particular case [121]. But stress should rather be on the specific individual, not just on statistics [122].

It may be noted that even the principles of the informal logical fallacies must be treated concretely as they are not always fallacies. For example, the *tu quoque fallacy* (or "you-also fallacy") is relevant if equality is at stake, and circular definitions are helpful if one does not know the meaning of a term, but does know the definition of its synonym. It is no defense to say, "But you violated the rule as well." The search for universals or principles in bioethics can be a form of the all-fallacy. Consensus seeking can be a form of *argumentum ad populum fallacy*. By following established principles a medical judgment may commit the *appeal to authority fallacy*. Informal logical fallacies must virtually always be embedded in and assessed in each particular context. They are not fallacies absolutely or abstractly "in themselves." This view fits well with the following statement: "Moral knowledge is essentially particular" [123]. This is to say that there is no abstract logic and no abstract moral knowledge. Thus, an ethics committee (involving philosophical practitioners) may be called on for clarification in a case to case basis.

6. Ordinary language case method [124]. According to this theory the meaning of language is its use embedded in a specific language-game, such as greeting, describing, excusing, etc (See also the account of ordinary language in the previous section on integrity). To understand the meaning one can only look at the full context of the usage. The meaning does not involve words, which represent pseudo-psychological "ideas" in us. Rather, language and specific context are all that are involved. We can only use language, not get outside of it or beyond it. It has epistemological primacy. There is no further understanding to be had. This means that meaning depends on examining specific cases. The case determines the meaning. It is in this sense that one cannot go beyond cases. Fulford recently attempted to use Ordinary-Language Philosophy in medicine [125].
7. Philosophical Practice is practical philosophy, which involves an analysis of specific cases. "Practical philosophy" is redundant, practical metaphysics a contradiction. Both healthcare worker and philosopher can learn by the experience

of Philosophical Practice. It involves the overall perspective of the philosophy of science and philosophy, both theory and practice. It considers all of the relevant perspectives, factors and consequences [126]. It involves probability and uncertainty, such as clinical biostatistics, decision analysis, epidemiology, etc. "Different clinicians...made very different medical judgments about the same sort of problem" [127]. How the physician will deal with a patient with chronic pelvic pain is very much dependent upon whether s/he has only a background in classical medicine with a surgical orientation, or a more holistic psychosomatic approach [128].

17.5 Holistic Philosophy of Medicine

Canton sees holistic or humanistic medicine as the attempt to attain a better quality of life. Then contradictorily the author states that the physician and patient should accept their anger and deal with it constructively. However, they cannot do so without knowledge about emotion, and none is proposed. "The doctor should...indulge in occasional fantasies of anger against the patient who frustrates him" [129]. Holistic medicine does not mean here irrational, holistic, or spiritualistic or alternative medicine. Holistic medicine supposedly derives its name from biologist-philosopher, Smuts's [130]. This cannot be true as the philosophy of humanism existed long before that. The author also traces it back to Hippocrates. One could add Galen, Paracelsus, Avicenna as well as many philosophers. "Humanistic medicine" is narrowly defined, as if it were the literal definition, as "the relationship between physician and patient." This is not at all what humanism means in philosophy [131]. The author then defines holistic medicine as treating each person as a "unique individual made up of body, mind and spirit" [132]. This is a mentalistic, uncritical and philosophically unacceptable definition. It has no use in practical treatment. In addition, the treatment of physicians and nurses requiring long, exhausting working hours more than in nearly every other profession and with low wages (outside the U.S.) are certainly the opposite of humanistic and holistic medicine. The author thinks holism involves merely appealing to cultural and religious groups, family practices and pastoral therapy [132]. The author points out that the patient must be held responsible for his or her own active part in treatment. This is an important point because the prevailing trend is to give the patient much autonomy without the corresponding responsibility. Patients, for example, often wish to take a pill instead of eating well or exercising, or live a lifestyle, which causes illness. One could say then that holism partly involves including patient responsibility in treatment. The author appeals to Freudian association, without critique, as being useful for holism. Self-regulation is said to have a placebo effect, but placebo effect is undefined. The author mentions self-actualization as extraordinary joy and creativity [133]. This is an important point as it means one can treat even those who are well but wish to develop further. Unfortunately, no method is given as to how this may be done. Philosophy, however, deals with how one may live an aesthetic and ethical life [134]. One of the tasks of Philosophical Counseling is how to live a more aesthetic, critical and rational life with more positive emotions, and sound life goals. One cannot have a holistic life without holistic philosophical thinking. Wellness centers are incomplete without philosophical counselors. It is in philosophy itself that humanism was defined and developed. The author however states that holistic medicine is merely an attitude, with no theory of emotion (attitude) being offered [135].

It is "patient oriented" as opposed to disease oriented, for example, therapy, nutrition, psychology, sociology, etc. It may be noted that the authors do not mention philosophy, philosophical psychology, ethics, philosophy of emotion, philosophical counseling. What is "patient oriented" to mean? We have to treat people (personalities) as well as diseases.

This would certainly have to include philosophical counseling, although it is unknown in the medical profession. Humanism is similarly unknown as far as definition is concerned. Medicine is mainly concerned with causality, but causality is one of the most misconceived notions in science which philosophers, not physicians, have clarified extensively (See the Chapter 4). Thus the medical model oriented physician is not even scientific about causality. The reference to psychotherapy as going outside of the medical model is also misconceived. The DSM IV classification of mental disorders, like those classificatory systems before it, and alternate international classificatory systems have been seen to be largely unscientific and unacceptable. Psychology is also to a large extent based on the medical model, use of statistics, experiments with mice, and a narrow EBM approach. Humanistic psychology is an entirely different, non-experimental oriented model. Only a philosophical approach, philosophy of medicine, of psychology can be called an adequately comprehensive medical or psychological model. It includes ethics, philosophical psychology, philosophy of emotion, philosophy of science, humanism, pragmatism, analysis of fallacious reasoning, critical thinking (speaking), philosophical counseling, etc. In order to treat the whole person medicine needs a philosophy, and goals, which are based on a sound ethical system.

The oath of physicians as given in the World Medical Association Declaration of Geneva: "Solemnly pledge myself to consecrate my life to the service of humanity." points at an approach of holistic medicine [136].

The physician needs to be aware of the social context and the environment of the patient [137].

And also "No doctor can truly know his or her patients without understanding their inner lives" [138]. Finding meaning and purpose in life is important to prevent or deal with illness and death [139]. An HMO Insurance (Wisconsin) states: "Health is more than the absence of disease or infirmity, it is a feeling of overall physical, mental, and social well-being" [140]. Part of health club and fitness memberships is reimbursed. The American Holistic Medical Association states that "optimal health" is the primary goal of holistic medical practice. It is the conscious pursuit of the highest level of functioning and balance of the physical, environmental, mental, emotional, social and spiritual aspects of human experience, resulting in a dynamic state of being fully alive. This creates a condition of wellbeing regardless of the presence or absence of disease [141]. Holistic medicine is a conscious holistic pursuit view of optimal health regardless of the presence or absence of disease. It stresses love as a powerful healer.

The World Health Organization defined health as "a state of complete physical, mental and social well-being and not merely the absence of disease or infirmity." The constitution recognized "the employment of the highest attainable standard of health...as one of the fundamental rights of every human being." "This includes the right to adequate food, water, clothing, housing, health care, education, [and] security in event of: unemployment, sickness, disability, old age" [142]. The 1998 WHO statement included spiritual or supernatural dimensions of health [143]. It mentioned "The need to respond to each individual's spiritual quest for meaning, purpose and belonging" [142].

Tulsky and Fox, for example, say that ethics consultation involves increasing job satisfaction (good) of physicians, nurses, etc [144]. This is especially appropriate because of the excessive exhausting hours physicians, nurses and other healthcare workers must work. Night attendance policies are in crisis requiring some physicians to work over 100 h a week, as opposed to the normal 40 h week. Surgeons may be required to operate when they are totally exhausted. This is one reason why the fourth or fifth leading cause of death is medical mistakes. This is partly an administratively programmed mistake.

We may define mental health as including rational thinking (healthful, sound, scientific) as opposed to irrational thinking (confused, disturbed, disordered, supernatural or delusional). It may include knowledge of ethics and being ethical, critical thinking (speaking) and a holistic consequential approach to problem solving. It may include knowledge about emotions and the ability to minimize negative emotions and maximize positive ones.

It is for reasons such as the above that it is stated, "Medical records should also briefly sketch the patient's life, plans, hopes, fears and ultimate wishes" [145]. Philosophical Practice (Counseling) would also require such adequate information as a foundation for ethical decision-making regarding patients. Thinking is not unrelated to the body. The field, which most deeply and adequately deals with thinking, is philosophy.

The various methods presently proposed in bioethics are often mechanical ways and avoid making adequate, qualitative decisions. "There is no mechanical substitute for judgment amid a welter of particulars" [146]. The following is a brief presentation and analysis of such methods.

Ethical methods used in bioethics are: case method, casuistry (different than case method), care theory, principlism, religious views, narrative ethics, legal rules, administrative rules and codes, feminist gender theory, phenomenology; Kantian deontology, universalism and Categorical Imperative; deconstructionism, normative or culture-based consensus, subjective intuitionism. To this we may add pragmatism [147]. Preferable to these models one may suggest the model for bioethics being the *philosophy of medicine*. In terms of ethics, one may argue for a scientific basis of ethics based on critical descriptions of the uses and misuses of ethical terms and on a naturalistic theory of ethics on the lines of the humanism-pragmatism model. Moreno even argues that bioethics is naturalism [148]. John Fletcher chooses the methods in bioethics of principlism, casuistry, care and of the philosophy of pragmatism [147]. Engelhardt wrote, "Bioethics has focused on disclosing a content-rich moral vision that can justify an all-encompassing global approach to health-care policy" [149]. The problem often is that the management, as is the case usually, has no knowledge of what humanism is and is ironically locked into a superficial medical, non-humanistic and highly political, unenlightened model. The attempt to adopt a holistic, humanistic approach then fails due to lack of understanding of what humanism is and because of resistance to what people think it is.

The close connection between the philosophy of medicine and the scientific medical model is solidified by the adoption of a naturalistic theory of ethics based on science, such as is Dewey's. One may speak of "ethical diagnosis" just as one speaks of "clinical ethics", which brings ethics together with scientific medicine. Arras concludes, "Thus, if bioethicists are seeking a larger philosophical account that will effectively frame and justify their emerging pragmatic and interdisciplinary roles in academia, medical clinics, and policy councils, Dewey's work would be a natural starting point" [150]. Whether or not and to what extent a naturalistic theory of ethics and/or the pragmatic model is adopted, at least one may argue for bioethics being based on the philosophy of medicine. The natural extension of this is that Ethics Committees, Institutional Review Boards (IRBs) and Ethics Counseling in medicine consist of or include philosophical counselors and practitioners knowledgeable about the philosophy of medicine and ethics, ideally philosophical practitioners. This allows various models of philosophical analysis and ethics to be considered and applied.

17.6 Brief Description of Dewey's Pragmatism and Naturalistic Ethics

All knowledge is practical. Dewey rejects intuition, formal logic, abstraction, metaphysics, supernaturalism, and all, what is fixed: ideas, principles, rules, knowledge; and also rejects commandments, universals (e.g., Kantian or religious), duties, indoctrination, dogma, absolute truth, absolute certainty, mind or spirit-body dualism, mentalism (mind and ideas as such; mind is only acting and speaking, inner and outer conversation, not a spiritual entity), nonparticipatory education, appeal to authority, absolute ethics or right-in-itself or wrong-in-itself, intrinsic values, mysticism, unscientific or impractical medicine, atomistic rather than dynamic thinking, a priori reasoning rather than a posteriori reasoning, formal logic [151].

Dewey supports the scientific method, a philosophy of change and function (process theory), the full contextual examination of each case (case method), the practicality of all knowledge, practical experience, evidence-based problem-solving, outcomes research, concern for all in social relationships as a basis of humanism, interdisciplinary inquiry, situational and contextual nature of knowledge, constant reevaluation of values and knowledge, evidence-based reason and open inquiry, knowing as doing, and the concern for making life here and now better for all. He is for naturalism, consequentialism in problem-solving and in ethics, communication as a source and form of knowledge, participatory practical education (cf. internship, practicum), an ethics based on scientific problem-solving to satisfy human informed desires and needs (of the betterment of both the individual and society at large) with full consideration of consequences in each particular case, combining theory and practice, with adequacy of analysis to include all of the specific relevant factors. He stresses care for language using gerunds (active thinking rather than static thought). Meaning and knowledge is had only in and for its use. There is a coherence rather than correspondence theory of truth, and all acts are seen as interactions. Creative inductive logic of discovery and inquiry, and comprehensive decision-making are favored. It is not the case that pragmatism is just relativism and that important principles will be broken just to obtain results. Relevant laws, rules, principles, emotions, theories and their criticisms, etc. are included among the things one must consider. The philosopher could also give further understanding about what wants, needs and likes are.

17.7 Humanism Contains Many of the Elements of Contemporary Definitions of Philosophical Practice (PP)

Not the negation of life in itself, rather only that is ethical, which stands in the service of world affirmation and becomes appropriate to it [152].

All of the positive, reasonable, basic bioethics principles mentioned above are already in humanism, which is clear and extensive and is grounded in a naturalistic philosophy, e.g., the pragmatism of John Dewey. There are numerous characteristics of humanism, which are like the prevailing definitions of PP, e.g., avoiding indoctrination. "Secular humanism is the attempt to justify and elaborate a common moral framework grounded in what we share as persons." No claim to truth is made [153].

More careful would be a guiding or setting of a model, rather than a common framework. It would be too much to say that this is a search for universals. Few know what humanism is. It is here basically the philosophy of John Dewey [154].

Dewey stresses bringing about one's informed wants and likes deliberately on the basis of inquiry and adequate consideration of consequences. And, as pragmatic, Dewey's philosophy is a paradigm case for practical philosophy. What Barbara Norman calls "ecological philosophical counseling," [155] may be more adequately rendered as humanism. Whereas humanism was once taught at Columbia University, University of Chicago and elsewhere in the U.S., the Dutch seem to be now the leaders in the field of humanistic PP. It is therefore noteworthy that in Utrecht is the Universiteit voor Humanistiek (University for Humanistic Studies). They train philosophical counselors and have practiced PP for some time, especially in the areas of bereavement and offering alternatives to religious belief (For a brief comparison of humanism and religion see the Chapter 6). Albert Schweitzer even speaks of the lack of humanity of people (*Humanitätslosigkeit*) [156]. The philosopher, Hans Lenk, gives his own constructed 13 features of humanism: to recognize humane limitations/boundaries [of ethical behavior], to recognize the contexts and limits regarding the respect of others and ourselves, to try to get to know the whole human as a person, to argue personally, to allow open free space for decision making and acting, to see justice as fairness, to focus on compassion; to think of ourselves not only as rational, thinking selves, but also as compassionate, communicative and sensitive, human beings; to be concerned with our ecology, to materialize our own responsibility in our own places of possible acting, to be aware that the respect towards other existing life is part of the respect towards my own, to cultivate oneself; to treat ourselves, other people and all living things well [157].

Religious groups have tried and been very effective in internationally establishing their religious beliefs in medical institutes and counseling groups, as was shown in the Chapter 6. Ethics committees and the philosophy of medicine must be divorced from any such attempt to indoctrinate into a particular religion such as Christianity, Judaism, Islam, Buddhism, etc. Philosophical Practice and Ethics Committees should not be for the promotion of religion. "The ultimate bioethics agenda is startingly radical: dismantling traditional Western values and mores and forging a new ethical consensus based on values most people do not presently share" [158]. Of course, the consensus part may be omitted.

17.8 The Present Definitions and Methods of Philosophical Counseling (PC) are Too Restrictive

There is no dogma, and we have it.

Anti-philosophy. Braddock and Tonelli also attempt to limit the scope and usefulness of the philosophy of medicine: "The medical establishment resisted the idea that philosophers, theologians, and lawyers could contribute anything useful to patient care" [159]. "Attempts by non-clinician consultants to participate directly in the medical decision-making for an individual patient will continue to be viewed by physicians, patients, and the public as an intrusion" [160]. "Non-clinician consultants must remain outside the therapeutic relationship, but from that position they may facilitate and educate and thus fulfill some of the goals of ethics consultation" [160]. According to the statement, one may also argue that medical clinicians should not make ethical or philosophical counseling or holistic decisions. This would mean that physicians are not qualified to make ethical decisions without the appropriate education in ethics.

Ordinary Language Approach. The ordinary language approach in philosophy deals with what the healthcare worker and others actually say. Davies and Hudson presented actual statements of physicians who were in important decision making positions as follows: "I'm the judge and jury...and that may be the patient's bad luck" [161]. "At the risk of sounding arrogant, I really don't care what they [ethics committees] think" [161]. "The physicians often defined ethical *dilemmas* as situations with '*no real answer*.'" (underlines by the authors of this book.) [162] This last statement is both circular and false. "'I think it's a matter of finding something that a family can live with and you as a physician feel is justified'" [162]. This stresses a mere consensus rather than informed argument. We need to know what the basis for the justification is and if one has knowledge of ethics upon which to make such judgments. "Medical ethics is not a useful field of study" [162]. This was the common view of the physicians. "'I don't know what an ethicist is, and I think that's a made-up jargon term.'" It was noted that opinions about medical ethics were extremely varied [163]. Galen even believed that ethics is irrelevant, unessential and external to medical practice [164].

Kantian Approach.

Kant's philosophy, the Categorical Imperative, and the view that each person should always be regarded as an end, never only as a means, need interpretation and critique. Post-Kantians and others have themselves given extensive critiques of these notions. These technical theories are not ready to be just taken over as part of ordinary language or interpreted as an absolute commandment. Kant's principles are often principles of non-contradiction. They are not based on a consequentialistic use. What Kant means by Reason and Will is abstract, obscure, essentialistic and mentalistic. He speaks of our having mentalistic faculties and a Will and Reason as such. We supposedly know the moral law by "intuition." This is a subjective and problematic criterion not an acceptable epistemological method of knowing. Schopenhauer, for example, attacks Kant's Categorical Imperative as being like the "Thou shalts" of religion. Kant's terms (and ethics) are technical terms not ready to be taken over into everyday practical life. The point is not to critique Kant here, but rather to point out that ethics counselors are needed to critique and sort through the ethical theories and critiques of the various philosophers, as well as others. We may for example, contrast the above with Dewey's views: "The 'desirable' or the object which should be desired [valued], does not descend out of the a priori blue nor descend as an imperative from a moral Mount Sinai" [165]. "The Golden Rule gives me absolutely no knowledge, of itself, of what I should do" [166] (See also critique of Kant in the Chapter 5).

Existential Theory Approach. Several existentialist writers report six stage theories. This runs into epistemological problems with stage theories in general. Colin Clayton bases his views on Martin Heidegger's *Being and Time*, and so concepts are frequently employed such as: "Being-in-the World," "Daseinsanalyse," and "lived understanding." The goal is, "To conduct an enquiry, a seeking, leading to an understanding of one's Being-in the-World, one's own *Dasein*" [167]. The counselee's revelation is said often to go through the following stages: distressed concern, despair, meaninglessness, emotional deprivation, nothingness, and finally awareness of duty.

Narrative Method. Narrative Method is about patterns of dialogues. For Hoogendijk, Philosophical Practice (PP) is openness, critical inquiry, and pluralism of styles of dialogues [168]. For Eite Veening, PP is critical analysis of one's characteristic patterns of conversations ("metalogues"), which block us. Stress is on only the self-investigation of the actual statements made by the counselee [169]. The Wittgensteinian and pragmatist could support this aspect of the approach, but once again, emotions and their analysis are to be disregarded [170]. Ethical Counseling (EC) undermines the individual as narrative and life story.

For Zoë PP is only to assist the narrator (counselee) to construct a story or narrative. The philosophical counselor must avoid narrative closure [171]. PP on this view becomes fictive storytelling. Although constructivism is an important position especially in respect to the Wittgensteinian view of the epistemological primacy of language and the notion of language games, there is nevertheless a problem of relativism here. Ethical counseling also stresses the patient's narrative and life-story.

Relativism. Philosophical Practice indoctrinates relativism. This also applies to EC. The literature on PP almost univocally states the following points:

No answers or conclusions are to be given [172]. If so, we may be indoctrinating relativism: that one argument is as good as any other, that all is subjective and arbitrary. Ethical counseling also often limits itself to a mere recommendation or prohibits members from giving their conclusions or viewpoints.

Barbara Norman says that Philosophical Practitioners should not be confrontational.[173] But confrontation of ideas is what philosophy is about. In opposition to this, Elliot Cohen states, "It is possible for a PP to also attack the counselee's irrational evaluations" [174]. It is not just to make us comfortable, but to make us uncomfortable with our most cherished, uncritical beliefs, emotions and life. Ethical counseling similarly opposes confrontations or challenges thereby restricting the effectiveness of philosophical practice.

Schefczyk states that the is-ought fallacy should be avoided because values are subjective, not factual [174]. But the critical literature shows that the is-ought fallacy is not a fallacy at all. Also, ethics is not just subjective, anymore than the scientific method is. On the contrary, Dewey holds that if ethical terms are to have any meaning they must be reduced to naturalistic, scientific terms [175]. From medical facts and consequences we can, on a naturalistic philosophy, generate what we should do or what is best to do. In summary, in their attempt to be fair and non-indoctrinative, the PP and often the ethics counseling literature in effect state, *there is no dogma, and we have it.*

17.9 Philosophical Counseling or Philosophical Practice (PP)

The following is a proposal for a more adequate definition of ethical counseling as philosophical practice. It is a return to critical philosophy.

Several of definitions of philosophy will first be applied to Philosophical Practice (PP). One model of philosophy is the clarification of concepts and methods in the various disciplines. On this view, PP would be rendered by the formula: "The philosophy of x" where x is a discipline, for example, philosophy of psychology, science, medicine, etc., but also the philosophy of therapy itself. That is, we would not do therapy, but the philosophy or critique of therapy; not psychology, but the philosophy of psychology. PP would not, then, diagnose according to the classifications given by psychotherapists in DSM IV, but instead give a critique of the methods and concepts of DSM IV [176]. Two thirds of all psychotherapy patients put on drug treatments for depression do as well as or better with a placebo than those treated with an active medication [177]. Whether or not this is true, one would come to PP to learn not about medicine, but about the philosophy of medicine. As the physician gives medical diagnosis and treatment, the PP gives philosophical analysis and clarification.

On a second model of philosophy, philosophy may be practical and problem-oriented, as opposed to being merely historically oriented. In this respect, it clarifies topics such as abortion, argument, autonomy, care, death, duty, emotion, good, grief, lying, medicine, mind, person, reason, scientific method, spirit, thought, value of life, etc. One would then go to PP to clarify such issues. In this respect, philosophy is about the most significant aspects of one's life. Philosophy investigates, creates and preserves the value of life. This is especially true for the philosophy of humanism.

It ought not to be surprising that almost everyone in any society is to a large extent indoctrinated, enculturated and emotionally dysfunctional. PP can address such issues in the way philosophers traditionally have. What will be gained are ideas, which change our lives such that we will not be able to go back to our previous way of thinking or our dysfunctional emotional life. It is in this sense that philosophy is therapy. It cures lack of both reason and knowledge. Wittgenstein similarly regards philosophy as therapy. He wrote, "There is not a philosophical method, though there are indeed methods, like different therapies" [178]. James Peterman has written a book on Wittgenstein's philosophical therapy [179]. But philosophy does not just do therapy anymore than it does science. It must be more critical and careful than therapy or science, although it may contribute to both therapy and science. It is not merely another six-stage therapy. PP and EC should not, then, give up on philosophy, limit or restrict it, but instead we should have more philosophy in PP and allow its full range of contribution.

In the following we will argue that we should use the more comprehensive philosophy of medicine counseling, instead of the present badly defined medical ethics counseling. Philosophy of medicine counseling is done by the Philosophical Practitioner (Philosophical Counselor). Philosophers already strongly influence ethics in medicine [180]. Many journals exist on the philosophy of medicine and bioethics, and articles on the subjects can be found in nearly any philosophical journal. Some specific examples of philosophies stressed in various countries follow in the reference [181].

We have seen the international need for a more comprehensive, holistic and critical approach to ethics, and medicine. PP is deeper and broader than medical ethics committees. This book itself shows the kinds of clarifications PP can give. Alternative views could, of course, have been given. Philosophy is the only study that presents the various holistic philosophies, offers a critique of the concepts and methods of the various relevant disciplines, and presents and analyzes ethical systems. In short, PP can do all that philosophy can do. It is not limited to the narrow area of ethics alone or to a set of narrow vague principles. It includes the philosophy of x where we substitute a discipline for x, e.g., law, medicine, political science; social science, not religion—but the critique of religion. Philosophy of medicine is also part of the philosophy of the natural and social sciences. It meets the requirement of raising questions, which are not usually raised. The philosophy of medicine is guided by argument and rationality, not by politics, cultural practices, dogma, corporate domination, mere consensus, or special interest groups. The latter are rather brought into question. Disapprovingly, Wesley Smith wrote, contemporary bioethicists "accept no moral standard or ethical rule, no matter how deeply valued, as self-evident. Every moral principle must be reassessed and deemed 'rational' if it is to pass muster" [182]. Philosophy is in direct opposition to Smith's position and encourages what Smith disapproves of.

Philosophical counselors are already used in business and medicine as well as for individual concerns and personal development. This addresses the question, "How can we develop more humanistic people?" Although PP may not support mere business and corporate profit in medicine, it could be used to support efficient, ethical and enlightened corporate goals.

Regarding the amount of knowledge of medicine philosophers need to know, one has to acknowledge philosophers as philosophers will not know what doctors know, and doctors will not know what philosophers know. "Doctors are no better qualified to make ethical decisions than most people" [183]. In spite of this lack, the American Medical Association sets ethical standards. On the other hand, some physicians are obtaining advanced degrees in ethics in medicine and in philosophy. Officials do not know who should be on ethics committees. Hosford recommends 1/3 doctors, 1/3 nurses, 1/3 other (law, philosophers, administrators?) [184] Those philosophers having expertise in philosophical practice and areas more relevant to medicine, e.g., ethics, death and dying, emotion, decision making, logical thinking, philosophy of law, etc. would be especially appropriate as philosophy of medicine consultants. Each member can benefit from the others. The philosophers selected for ethics committees should be practical, and problem-centered, rather than metaphysicians, super-naturalists, historians of philosophy, or religiously or theologically oriented. Nor should the members be watchdogs of the dominant indoctrinations of culture or special belief systems. Accordingly, John Fletcher wrote, "A high-order mistake is to choose or wrongly defend an indefensible world view that conditions one's basic perspective in ethics" [185].

17.10 A Proposal to Change the Title of Philosophical Practice to Philosophy Education

Because philosophers do more than therapy and counseling, why would they want to make philosophy something it isn't, restrict it and so reduce it to non-philosophy, to the uncritical and limited field of therapy or merely ethics? What philosophers do is to inquire more deeply and broadly than is done in any other discipline, create new knowledge and educate others about philosophy. It is what they are trained and qualified to do. They can then justifiably claim to be Philosophy Educators and Advisors (PE). They cannot so easily, or justifiably, claim to be Philosophical

Therapists or Counselors or Ethics Counselors (for it is not clear what this is). They are not trained to be. But there is another reason not to use the terms "therapist" and "counselor" or "ethics counselor": The Philosophical Counselor or Philosophy Therapist or the ethics counselor could possibly be sued for not having the specified qualifications of a therapist. "Therapy" and "counselor" are legally protected terms and occupational practices in the various states in the United States as well as in Europe.

Hosford says the ethics counselors should be only advisory and for educational purposes [186]. Hendrick advises that the functions of ECs should be nonbinding, only advisory and ECs should be only educational groups, if they wish to avoid vulnerability to lawsuits. Giving specific physician-binding advice about the treatment of a specific patient may result in a negligence suit if the outcome is negative [187]. The physician can choose to override the EC or ask for a binding decision to share the liability. One function, then, of the EC is to allow the physician to share responsibility and liability in problematic cases.

We need not reinvent the wheel to create a new area of ethics counseling (EC), one, which is not good philosophy and not good counseling – the worst of both worlds. There is instead a simple solution. We can change the name and practice from Ethics Counseling or Philosophical Counselor to Philosophy Educator and Philosophy Advisor and do so with full justification. People who have philosophical problems can go to the Philosophy Educator or Practitioner. Few physicians or patients have time to attend a full course in philosophy, but nevertheless have philosophical problems to resolve. In fact, every problem can be in some fundamental way a philosophical problem. As a Philosophy Educator or Practitioner one can also recommend texts, and assign homework. The Philosophy Educator or Practitioner does not claim to cure, only to educate. Whatever philosophy does the Philosophy Educator can claim to do. It should be made clear that when the PE or PP deals with an issue, for example, suicide, abortion, or killing or letting-die, etc. it is not done so as a therapist, physician or health professional, but as a philosopher. This should help avoid lawsuits. One might, of course, be sued as a philosopher.

17.11 The Philosophy Practitioner and Emotion

The literature on ethics commissions shows no concern with the theory or education of the emotions. Statements made about emotion also show an un-therapeutic lack of knowledge of emotion. The Task Force on Standards of Bioethics Consultation discussed earlier says EC should deal with the cognitive *and* affective part, but gives no theory or way in which this can be done. Aulisio, Arnold and Youngner state that the duties of the ethical counselor are to deal with guilt and grief [188]. On their view, the counselors themselves are supposed to possess tolerance, patience, compassion, courage, good character, humility, prudence, integrity, etc. But these qualities should be possessed by all healthcare workers and also by the patients. Yet, no way is provided by means of which this idealistic goal may be achieved.

This requires training in emotion, and such education is not as yet to be found in our educational institutions. This training in emotion may be provided by the PP. A critical analysis of emotions is needed. In-service training is recommended.

The PP literature states that PP requires *empathy*, avoidance of confrontation to *prevent negative emotions*, dealing with *emotional dysfunction* and the promotion of *enthusiasm* [189]. These claims are unfounded because one does not find in the PP literature only knowledge or theory of emotion according to which one can do such counseling. One brief exception is the article by Elliot Cohen, one of the early advocates of PP, in support of Rational-Emotive therapy [190]. Roger Paden wrote, "I believe the philosophical counselor might be able to learn a great deal from the Rational-Emotive therapist" [191].

Marks states, "The dominant view in contemporary analysis of emotion is a cognitivist one" [192]. In philosophical literature the cognitive-emotive theory is prevailing, and its counterpart in therapy is the Rational-Emotive Theory (RET or REBT). There is a fortunate collaboration here because philosophers can work out the theoretical exploration while the therapists provide the actual clinical experience. Yet, there is some unfounded opposition to the cognitive theory of emotion, which is given by philosophical counselors [193]. Neither philosophers nor PP or EC have shown much of any understanding or knowledge of emotion. "Dignity is based on self-other relations of shame and proudness" [194]. This is a faulty view. On the cognitive theory of emotion, negative emotions are due to faulty thinking. Rendtorff and Kemp state, "Immortal life would increase boredom" [195]. But boredom is caused by people, not by things or events. It is to encourage a dysfunctional emotion. Nevertheless, there is a philosophical literature on emotion and philosophy of emotion which can and should be brought into practice. Physicians and healthcare workers say, "The patient seems depressed," or "If the patient only had a visitor things would go better," or "I better go in and try to cheer her up." But if one does not know about emotion theory, how is one to attend to any of these things effectively?

Because emotions are such a central part of one's life and personality, without knowledge of them the Philosophy Educator (PP) may be seen to be negligent. Emotions can no longer be left out of philosophical education. Dealing with emotions also goes a long way toward making philosophy more practical as well as more adequate and humanistic. Positive emotions create the aesthetic. Aesthetics is ethics and so is bringing about one's informed wants and likes deliberately on the basis of inquiry and consequences. The more we know the more ethical and aesthetic we become. It is in this sense that the practice of medicine is an art.

My argument is now that if philosophy and ethics counseling (EC), PP, or PE are to have credibility and practicality they must include the education of emotions. And with such knowledge the philosopher, EC and PE can educate the client, the health care worker and themselves about emotions in general as well as about particular emotions (See also the Chapter 7).

17.12 Summary

As stated at the beginning of this chapter, the recent appearance of National Ethics committees, as opposed to mere advisory chancellor's or president's advisory committees, can be a significant and empowering step toward putting the ethics of medicine and bioethics on a higher and critical philosophical level, rather than on a normative legal, religious and uncritical cultural and political level.

In the five levels of decision-making in medicine itself which are 1. Medical, 2, Ethical, 3. Interdisciplinary, 4. Philosophy of Medicine, 5. Philosophical... the philosophy of medicine is an essential component for the creation of a humaine medicine and society.

References

1. Dewey, J. 1971. *The study of ethics: A syllabus. The early works of John Dewey, 1889–1892*, vol 4, 220–362, esp. IV:226. Carbondale, IL: Southern Illinois University Press.
2. Sources: AMA "Guidelines for Ethics Committees in Health Care Institutions." On the history of ethics commissions see Jonsen, A. 1998. *The birth of bioethics*, 362–367. New York: Oxford University Press; Electronic communications regarding Ethics Committees is being especially enhanced by the internet. Eiser, A. 2001. Electronic communication in ethics committees: Experience and challenges. *Journal of Medical Ethics* 27(suppl I):i30–i32. See also the Joint Commission on Accreditation of Healthcare Organizations, Society for Health Care and Human Values, American Medical Association, American Society for Bioethics and Humanities (ASBH See especially their "Core Concepts for Health Care Ethics Consultation," etc.) See also the bioethics and philosophy of medicine journals and references in the bibliography. Eckstein, S. ed. 2003. *Manual for research ethics committees*, 6th edn. London: Centre for Medical Law and Ethics, Kings College; England: Cambridge University Press.
3. Dorries, A. 2003. Mixed feelings: Physicians' concerns about clinical ethics committees in Germany. *HEC Forum* 15:245–257; esp. 251–252.
4. Dorries, A. 2003. Mixed feelings: Physicians' concerns about clinical ethics committees in Germany. *HEC Forum* 15:245–257; esp. 255.
5. Benjamin, M. 1996. No consensus about consensus. *Hastings Center Report* 26:39–40; esp. 40.
6. Cf. Arras, J. 2002. Pragmatism and bioethics: Been there, done that. *Social Philosophy and Public Policy* 19:29–58; esp. 52–53.
7. Sklansky, M. 2001. Neonatal euthanasia: Moral considerations and criminal liability. *Journal of Medical Ethics* 27:5–11; esp.8.
8. Fulford, K. 2001. Philosophy into practice: The case for ordinary-language philosophy. In *Health, science and ordinary language*, ed. Nordenfeldt, L., 171–208; esp.171 GA: Rodopi, Amsterdam/Atlanta.
9. Fulford, K. 2001. Philosophy into practice: The case for ordinary-language philosophy. In *Health, science and ordinary language*. ed. Nordenfeldt, L., 171–208; esp.177, 201 GA: Rodopi, Amsterdam/Atlanta.
10. Fulford, K. 2001. Philosophy into practice: The case for ordinary-language philosophy. In *Health, science and ordinary language*. ed. Nordenfeldt, L., 171–208; esp. 191ff GA:

Rodopi, Amsterdam/Atlanta; Hyman, M., Liponis, M. 2003. Ultraprevention. Scribner, New York: 326.
11. Singer, P. 2002. *One world: The ethics of globalization*, 62. New Haven, CT: Yale University Press; Singer, P. 2002. *Unsanctifying human life: Essays on ethics*, ed. Kuhse, H., 62. Oxford: Blackwell; See also Maier, B. 2003. Das Vertraute und das Fremde in ethischer Perspektive. Das Vertraute und das Fremde als moralische Begriffe. In *Psychosomatische Gynäkologie und Geburtshilfe*, 291–296. Gießen: Beiträge der Jahrestagung 2002 der DGPFG, Psychosozial-Verlag.
12. Caplan, A. 2006. No method. No madness. *Hastings Center Report* 36:12–13; esp.13.
13. Such national ethics committees are found in Belgium, Denmark, Finland, France, Greece, Italy, Norway, Portugal, but not in others, e.g. Austria. Germany, Ireland, Netherlands (though the People's Health Council is similar). Rendtorff, J., and Kemp, P. 2000. *Basic ethical principles in European Bioethics and Biolaw*, 163, 174. Copenhagen, Barcelona: Centre for Bioethics and Law.
14. Maier, B. 2010. Moral und Ethik als kontextoffene Begriffe. Ethik transdisziplinär. In *Rechtsphilosophie: Vom Grundlagenfach zur Transdisziplinarität in den Rechts-, Wirtschafts- und Sozialwissenschaften*. eds. Wöhle, C.B., Augeneder, S., and Urnik, S., 205–216 Frankfurt am Main: Peter Lang.
15. Singer, P. 2002. *Unsanctifying human life: Essays on ethics*. ed. Kuhse, H., 64ff Oxford: Blackwell.
16. Ross, J.W. 1986. *Handbook for hospital ethics committees*, 40. Chicago, IL: American Hospital Publications.
17. Tulsky, J., and Fox, E. 1996. Evaluating ethics consultation: Framing the questions. *Journal of Clinical Ethics* 7:109–115; esp.111.
18. Source: "American Society for Bioethics and Humanities (ASBH) Task Force on Standards for Bioethics Consultation 1995": Report: Core Competencies for Health Care Ethics Consultation (ASBH 1966–1998). Adopted by the American Society for Bioethics and Humanities May, 8, 1998, which is also available on the internet) See also Aulisio, M., Arnold, R., and Youngner, S. eds. 2003. Health Care Ethics Consultation: Nature, goals, and competencies. American College of Physicians – American Society of Internal Medicine: 166–202. The task force includes 21 members and is sponsored by, for example, the American Medical Association, Society for Bioethics Consultation, the American Society for Bioethics and Humanities ASBH, etc.
19. Aulisio, M., Arnold, R., and Youngner, S. eds. 2003. Health Care Ethics Consultation: Nature, Goals, and Competencies. American College of Physicians – American Society of Internal Medicine: 166–202 esp.:166 ff.
20. Aulisio, M., Arnold, R., and Youngner, S. eds. 2003. Health Care Ethics Consultation: Nature, Goals, and Competencies. American College of Physicians – American Society of Internal Medicine:166–202; esp. 171–173.
21. bka.gv.at/bioethik/ or www.bundeskanzleramt.at/bioethik/.
22. Mittelstrass, J. 1991. Auf dem Wege zu einer Reparaturethik? In *Ethik ohne Chance? Erkundungen im technologischen Zeitalter. Ethik in den Wissenschaften 2*. eds. Wils, J.P., and Mieth, D. Tübingen: Attempto.
23. Gawande, A. 2002. *Complications: A surgeon`s notes on imperfect science*, 42. New York: Picador/Holt.
24. *Medical Ethics Today: The BMA'S handbook of ethics and law*. 2004, 2nd edn. London: BMJ (Also Windows CD ROM) .
25. Munson, R. 2000. *Intervention and reflection: Basic issues in medical ethics*, 6th edn. Belmont, CA: Wadsworth/Thompson Learning.
26. Munson, R. 2000. *Intervention and reflection: Basic issues in medical ethics*, 6th edn, 32. Belmont, CA: Wadsworth/Thompson Learning.
27. Munson, R. 2000. *Intervention and reflection: Basic issues in medical ethics*, 6th edn, 34. Belmont, CA: Wadsworth/Thompson Learning.

28. Munson, R. 2000. *Intervention and reflection: Basic issues in medical ethics*, 6th edn, 36. Belmont, CA: Wadsworth/Thompson Learning.
29. Munson, R. 2000. *Intervention and reflection: Basic issues in medical ethics*, 6th edn, 37. Belmont, CA: Wadsworth/Thompson Learning.
30. Munson, R. 2000. *Intervention and reflection: Basic issues in medical ethics*, 6th edn. 38. Belmont, CA: Wadsworth/Thompson Learning.
31. Munson, R. 2000. *Intervention and reflection: Basic issues in medical ethics*, 6th edn, 40. Belmont, CA: Wadsworth/Thompson Learning.
32. Rendtorff, J., and Kemp, P. 2000. *Basic ethical principles in European bioethics and biolaw. Vol I Autonomy, dignity, integrity and vulnerability*. Copenhagen, Barcelona: Centre for Bioethics and Law.
33. Rendtorff, J., and Kemp, P. 2000. *Basic ethical principles in European bioethics and biolaw. Vol I Autonomy, dignity, integrity and vulnerability*, 357, 56. Barcelona: Centre for Bioethics and Law, Copenhagen.
34. Beauchamp, T., and Childress, J. orig. 1979–2001/1994. *Principles of biomedical ethics*, 4th edn. Oxford: Oxford University Press.
35. Rendtorff, J., and Kemp, P. 2000. *Basic ethical principles in European bioethics and biolaw. Vol. II Partners` Research*, 50. Copenhagen, Barcelona: Centre for Bioethics and Law.
36. In Rendtorff, J., and Kemp, P. 2000. *Basic ethical principles in European bioethics and biolaw. Vol. II Partners` Research*, 38. Copenhagen, Barcelona: Centre for Bioethics and Law.
37. Smith, W. 2000. Culture of death: The assault on medical ethics in America. Encounter Books, San Francisco, CA: 30.
38. Rendtorff, J., and Kemp, P. 2000. *Basic ethical principles in European bioethics and biolaw. Vol I Autonomy, dignity, integrity and vulnerability*, 25. Copenhagen, Barcelona: Centre for Bioethics and Law.
39. Hippocrates Precepts IX 1952. Jones, W. vol 2 Harvard University Press.
40. Loewy, R. 2000. *Integrity and personhood: Looking at patients from a bio/psycho/social perspective*, 53. Dordrecht: Kluwer.
41. Jonsen, A., Siegler, M., and Winsdale, W. orig. 1982/1998. *Clinical ethics*, 4th edn, 51. New York: McGraw-Hill.
42. Beauchamp, T., and Walters, L. eds. 1999. *Contemporary issues in bioethics*, 5th edn, 19, 20. Belmont, CA: Wadsworth.
43. Noddings, N. 2002. *Starting at home: Caring and social policy*, 109. Berkeley, CA: University of California Press.
44. Noddings, N. 2002. *Starting at home: Caring and social policy*, 114. Berkeley, CA: University of California Press.
45. Jopling, D. 1996. Philosophical counseling, truth and self-interpretation. *Journal of Applied Philosophy* 13:297–309; esp. 304.
46. Loewy, R. 2000. *Integrity and personhood: Looking at patients from a bio/psycho/social perspective*, 48. Dordrecht: Kluwer.
47. Kant, I. 1785/1998. *Groundwork of the metaphysics of morals*. ed. Gregor, M. New York: Cambridge University Press; Kant, I. 1788/1997. *Critique of practical reason*. Gregor, M. tr, New York: Cambridge University Press; Rendtorff, J., and Kemp, P. 2000. *Basic ethical principles in European bioethics and biolaw. Vol I Autonomy, dignity, integrity and vulnerability*, 18. Copenhagen, Barcelona: Centre for Bioethics and Law.
48. Rendtorff, J., and Kemp, P. 2000. *Basic ethical principles in European bioethics and biolaw. Vol I Autonomy, dignity, integrity and vulnerability*, 32. Copenhagen, Barcelona: Centre for Bioethics and Law.
49. Rendtorff, J., and Kemp, P. 2000. *Basic ethical principles in European bioethics and biolaw. Vol I Autonomy, dignity, integrity and vulnerability*, 34. Copenhagen, Barcelona: Centre for Bioethics and Law.

50. Rendtorff, J., and Kemp, P. 2000. *Basic ethical principles in European bioethics and biolaw. Vol I Autonomy, dignity, integrity and vulnerability*, 359. Copenhagen, Barcelona: Centre for Bioethics and Law.
51. Rendtorff, J., and Kemp, P. 2000. *Basic ethical principles in European bioethics and biolaw. Vol I Autonomy, dignity, integrity and vulnerability*, 35. Copenhagen, Barcelona: Centre for Bioethics and Law.
52. Rendtorff, J., and Kemp, P. 2000. *Basic ethical principles in European bioethics and biolaw. Vol I Autonomy, dignity, integrity and vulnerability*, 365. Copenhagen, Barcelona: Centre for Bioethics and Law.
53. Rendtorff, J., and Kemp, P. 2000. *Basic ethical principles in European bioethics and biolaw. Vol I Autonomy, dignity, integrity and vulnerability*, 36. Copenhagen, Barcelona: Centre for Bioethics and Law.
54. Rendtorff, J., and Kemp, P. 2000. *Basic ethical principles in European bioethics and biolaw. Vol I Autonomy, dignity, integrity and vulnerability*, 31. Copenhagen, Barcelona: Centre for Bioethics and Law.
55. Rendtorff, J., and Kemp, P. 2000. *Basic ethical principles in European bioethics and biolaw. Vol I Autonomy, dignity, integrity and vulnerability*, 40. Copenhagen, Barcelona: Centre for Bioethics and Law.
56. Rendtorff, J., and Kemp, P. 2000. *Basic ethical principles in European bioethics and biolaw. Vol I Autonomy, dignity, integrity and vulnerability*, 32. Copenhagen, Barcelona: Centre for Bioethics and Law.
57. Rendtorff, J., and Kemp, P. 2000. *Basic ethical principles in European bioethics and biolaw. Vol I Autonomy, dignity, integrity and vulnerability*, 325. Copenhagen, Barcelona: Centre for Bioethics and Law.
58. Rendtorff, J., and Kemp, P. 2000. *Basic ethical principles in European bioethics and biolaw. Vol I Autonomy, dignity, integrity and vulnerability*, 363. Copenhagen, Barcelona: Centre for Bioethics and Law.
59. Rendtorff, J., and Kemp, P. 2000. *Basic ethical principles in European bioethics and biolaw. Vol I Autonomy, dignity, integrity and vulnerability*, 37. Copenhagen, Barcelona: Centre for Bioethics and Law.
60. Rendtorff, J., and Kemp, P. 2000. *Basic ethical principles in European bioethics and biolaw. Vol I Autonomy, dignity, integrity and vulnerability*, 359. Copenhagen, Barcelona: Centre for Bioethics and Law.
61. Rendtorff, J., and Kemp, P. 2000. *Basic ethical principles in European bioethics and biolaw. Vol I Autonomy, dignity, integrity and vulnerability*, 61. Copenhagen, Barcelona: Centre for Bioethics and Law.
62. Rendtorff, J., and Kemp, P. 2000. *Basic ethical principles in European bioethics and biolaw. Vol I Autonomy, dignity, integrity and vulnerability*, 39. Copenhagen, Barcelona: Centre for Bioethics and Law.
63. Rendtorff, J., and Kemp, P. 2000. *Basic ethical principles in European bioethics and biolaw. Vol I Autonomy, dignity, integrity and vulnerability*, 58–61. Copenhagen, Barcelona: Centre for Bioethics and Law.
64. cf. Nussbaum, M. 1990. *Love's knowledge: Essays on philosophy and literature*. New York: Oxford University Press; Brody, H. 1987. *Stories of sickness*. New Haven: Yale University Press; Frank, A. 1995. *The wounded storyteller: Body, illness and ethics*. Chicago: University of Chicago Press; Nelson, H. 1997. *Stories and their limits: Narrative approaches to bioethics*. New York: Routledge. Compare also the notion of "communicative rationality" by Habermas, J. 1981. Theorie des kommunikativen Handelns. Suhrkamp, Frankfurt am Main: 334 He proposes a domination-free situation of dialogues.
65. Arras, J. 2002. Pragmatism and bioethics: Been there, done that. *Social Philosophy and Public Policy* 19:29–58.
66. Tong, R. 1998. The ethics of care: A feminist virtue ethics of care for healthcare practitioners. *Journal of Medicine and Philosophy* 23:131–152; esp.133.

67. Schapp, W. 1953. *In Geschichten verstrickt*. Wiesbaden: Zum Sein von Mensch und Ding.
68. Elliott, C. 1999. *Bioethics, culture and identity: A philosophical disease*, 123. New York: Routledge.
69. Galzigna, M. 2004. Aspects of medical practice: Insight from narrative theory, and Changing narratives, changing through narratives. Internet papers. (Presentation at the International Conference on Philosophical Practice, Copenhagen, August 3rd–6th, 2004).
70. Wittgenstein, L. 1953/1968. *Philosophical investigations*, 3rd edn. New York: Macmillan. Tr. Rev. Anscombe, E., 7, 240, 23Cambridge, Mass: Blackwell Publishers.
71. Bachmann, G., On his website: Wer seine Sprache ändert, der ändert auch sein Denken und Empfinden. Stuttgart, bachmann@textuniversum.de.
72. Rendtorff, J., and Kemp, P. 2000. *Basic ethical principles in European bioethics and biolaw. Vol I Autonomy, dignity, integrity and vulnerability*, 45. Copenhagen, Barcelona: Centre for Bioethics and Law.
73. Ryle, G. 1949. *Concept of mind*. New York: Barnes and Noble.
74. Rendtorff, J., and Kemp, P. 2000. *Basic ethical principles in European bioethics and biolaw. Vol I Autonomy, dignity, integrity and vulnerability*, 38. Copenhagen, Barcelona: Centre for Bioethics and Law.
75. Rendtorff, J., and Kemp, P. 2000. *Basic ethical principles in European bioethics and biolaw. Vol I Autonomy, dignity, integrity and vulnerability*, 45. Copenhagen, Barcelona: Centre for Bioethics and Law.
76. Rendtorff, J., and Kemp, P. 2000. *Basic ethical principles in European bioethics and biolaw. Vol I Autonomy, dignity, integrity and vulnerability*, 361. Copenhagen, Barcelona: Centre for Bioethics and Law.
77. Rendtorff, J., and Kemp, P. 2000. *Basic ethical principles in European bioethics and biolaw. Vol I Autonomy, dignity, integrity and vulnerability*, 56. Copenhagen, Barcelona: Centre for Bioethics and Law.
78. Rendtorff, J., and Kemp, P. 2000. *Basic ethical principles in European bioethics and biolaw. Vol I Autonomy, dignity, integrity and vulnerability*, 450. Copenhagen, Barcelona: Centre for Bioethics and Law.
79. Apel, H.O. 1988. *Diskurs und verantwortung*. Frankfurt: Suhrkamp.
80. Rendtorff, J., and Kemp, P. 2000. *Basic ethical principles in European bioethics and biolaw. Vol I Autonomy, dignity, integrity and vulnerability*, 47. Copenhagen, Barcelona: Centre for Bioethics and Law.
81. Rendtorff, J., and Kemp, P. 2000. *Basic ethical principles in European bioethics and biolaw. Vol I Autonomy, dignity, integrity and vulnerability*, 50. Copenhagen, Barcelona: Centre for Bioethics and Law.
82. Dewey, J. 1920. Reconstruction in philosophy. In *The middle works 1899–1924*. ed. Boydston, J.A., vol 12, 176. Carbondale, IL: Illinois University Press.
83. Kelsen, H. 1945. General theory of law and state. Tr. Wedberg, A., Cambridge, M.A., and Kelsen, H. 1967. *Pure theory of law*. Knight, M. tr., 2nd rev. edn. Berkeley: University of California Press; Kelsen, H. 1957/1971. What is justice? Tr. Knight, M., 2nd rev. edn. Berkeley: University of California Press.
84. Dewey, J. 1920. Reconstruction in philosophy. In *The middle works 1899–1924*, ed. Boydston, J.A., vol 12, 211 Carbondale, IL: Illinois University Press.
85. Wittgenstein, L. 1969. *On certainty*. eds. Anscombe, G., and Wright, G. , #139. Anscombe, G., Paul, D. trs, New York: Harper.
86. Justice Posner, R. in Dickstein, M. ed. 1999. *The revival of pragmatism*, 236–253. Durham, NC: Duke University Press.
87. Dewey, J. 1934. *A common faith*, 26. New Haven, CT: Yale University Press.
88. Rendtorff, J., and Kemp, P. 2000. *Basic ethical principles in European bioethics and biolaw. Vol I Autonomy, dignity, integrity and vulnerability*, 313. Barcelona: Centre for Bioethics and Law, Copenhagen.

89. Rendtorff, J., and Kemp, P. 2000. *Basic ethical principles in European bioethics and biolaw. Vol I Autonomy, dignity, integrity and vulnerability*, 312. Copenhagen, Barcelona: Centre for Bioethics and Law.
90. Noddings, N. 2002. *Starting at home: Caring and social policy*, 232. Berkeley, CA: University of California Press.
91. Rendtorff, J., and Kemp, P. 2000. *Basic ethical principles in European bioethics and biolaw. Vol I Autonomy, dignity, integrity and vulnerability*, 313. Copenhagen, Barcelona: Centre for Bioethics and Law.
92. Rendtorff, J., and Kemp, P. 2000. *Basic ethical principles in European bioethics and biolaw. Vol I Autonomy, dignity, integrity and vulnerability*, 319. Copenhagen, Barcelona: Centre for Bioethics and Law.
93. Jonsen, A., and Toulmin, S. 1988. *Abuse of casuistry: A short history of moral reasoning*. Berkeley: University of California Press.
94. Kilner, J. 1990. *Who lives? Who dies?* New Haven, CT: Yale University Press.
95. Kilner, J. 1990. *Who lives? Who dies?*, 109 New Haven, CT: Yale University Press.
96. Reiter-Theil, S. 2001. The Freiburg approach to ethics consultation: Process, outcome and competencies. *Journal of Medical Ethics* 27(suppl I): i21–i23; esp. i21—i22.
97. Arras, J. 2002. Pragmatism and bioethics: Been there, done that. *Social Philosophy and Public Policy* 19:29–58; esp. 49.
98. Fletcher, J. 1995. *Introduction to clinical ethics*. ed. Frederick, M.D., 266. London: University Publishing Group.
99. cf. Arras, J. 2002. Pragmatism and bioethics: Been there, done that. *Social Philosophy and Public Policy* 19:29–58; esp. 52–53.
100. Nussbaum, M., and Sen, A. eds 1993. *Quality of life*, 51–52. Oxford: Clarendon Press.
101. Dewey, J. 1971. *Moral theory and practice. The early works of John Dewey. 1889–1892*, vol III, 93–109; 106. Carbondale, IL: Southern Illinois University Press.
102. Dewey, J. 1922. *Human nature and conduct*, 211. New York: Henry Holt.
103. Fletcher, J., and Brody, H. 1994. Clinical ethics. In *Encyclopedia of bioethics*, ed. Reich, W., Vol I, , 399–404; esp. 399 New York: Macmillan.
104. Little, M. 1995. *Humane medicine*, 26. Oxford: Cambridge University Press.
105. Jonsen, A., and Toulmin, S. 1988. *Abuse of casuistry: A short history of moral reasoning*, 13. Berkeley: University of California Press.
106. Jonsen, A., and Toulmin, S. 1988. *Abuse of casuistry: A short history of moral reasoning*, 342–343. Berkeley: University of California Press.
107. Jonsen, A., and Toulmin, S. 1988. *Abuse of casuistry: A short history of moral reasoning*, 294. Berkeley: University of California Press.
108. Beauchamp, T., and Walters, L. eds. 1999. *Contemporary issues in bioethics*, 5th edn, 17. Belmont, CA: Wadsworth.
109. Beauchamp, T., and Walters, L. eds. 1999. *Contemporary issues in bioethics*, 5th edn, 18. Belmont, CA: Wadsworth.
110. Schwartz, L., Preece, P., and Hendry, R. 2002. *Medical ethics: A case-based approach*, 5. Edinburgh: Saunders.
111. Kilner, J. 1990. *Who lives? Who dies?*, 70 New Haven, CT: Yale University Press.
112. Tomlinson T 1994. Casuistry in medical ethics rehabilitated, or repeat offender? *Theoretical Medicine* 15:5–20.
113. Tomlinson T 1994. Casuistry in medical ethics rehabilitated, or repeat offender? *Theoretical Medicine* 15:5–20; esp. 13.
114. Tomlinson T 1994. Casuistry in medical ethics rehabilitated, or repeat offender? *Theoretical Medicine* 15:5–20; esp. 14.
115. Fletcher, J. 1966/1997. *Situational ethics: The new morality*, esp. 57. Louisville, KY: Westminster Knox Press, Fletcher, J., and Brody, H. 1994. Clinical ethics. In *Encyclopedia of bioethics*. ed. Reich, W., vol I, 399–404. New York: Macmillan; Shibles, W. 1978. *Rational love*. Whitewater, WI: Language Press.

116. Sartre, J. 1992. *Notebooks for an ethics*. Pellauer, D. tr. Chicago: University of Chicago Press.
117. Emanuel, E. 1995. Medical ethics in the era of managed care: The need for institutional structures instead of principles for individual cases. *Journal of Clinical Ethics* 6:335–338; esp. 335.
118. Emanuel, E. 1995. Medical ethics in the era of managed care: The need for institutional structures instead of principles for individual cases. *Journal of Clinical Ethics* 6:335–338; esp.337.
119. Dewey, J. 1922. *Human nature and conduct*, 211. New York: Henry Holt.
120. Dewey, J. 1909. *Moral principles in education*, 57. New York: Houghton Miflin.
121. Schneider, S. 2005. *The patient from hell: How I worked with my doctor to get the best of modern medicine and how You can too*, 260. New York: Da Capo Press.
122. Schneider, S. 2005. *The patient from hell: How I worked with my doctor to get the best of modern medicine and how You can too*, 258. New York: Da Capo Press.
123. Jonsen, A., and Toulmin, S. 1988. *Abuse of casuistry: A short history of moral reasoning*, 330. Berkeley: University of California Press.
124. Wittgenstein, L. 1953/1958. *Philosophical investigations*, 3rd edn. New York: Macmillan; Elliott, C. 1999. *Bioethics, culture and identity: A philosophical disease*. New York (Wittgensteinian approach): Routledge.
125. Nordenfelt, L. 2001. *Health, science and ordinary language*, 171–208. Amsterdam/Atlanta, GA: Rodopi.
126. cf. Jonsen, A., Siegler, M., and Winsdale, W. orig.1982/1998. *Clinical ethics*, 4th edn, 10. New York: McGraw-Hill.
127. cf. Jonsen, A., Siegler, M., and Winsdale, W. orig.1982/1998. *Clinical ethics*, 4th edn, 19. New York: McGraw-Hill.
128. Maier, B., Akmanlar-Hirscher, G., Krainz, R., Wenger, A., and Staudach, A. 1999. Der chronische Unterbauchschmerz – ein immer noch zu wenig verstandenes Krankheitsbild. In *Wiener Medizinische Wochenschrift* 149/13:377–382.
129. Canton, R. 1980. Wholistic medicine: Reactivation of an old aspiration. *Western Journal of Medicine* 133:171–172.
130. Smuts`s J 1926. *Holism and evolution*. See internet. Ethical counsel ch.
131. See Shibles, W. 1995. Analysis of the definitions of Humanism. *Scottish Journal of Religious Studies* 16:51–61.
132. Smuts's, J. 1926. *Holism and evolution*. See internet. Ethical counsel ch: 547.
133. Smuts's, J. 1926. *Holism and evolution*. See internet. Ethical counsel ch: 549.
134. For concrete hands-on methods see Shibles "Humor" internet site 2002.
135. Gordon, J. 1982. Holistic medicine: Advances and shortcomings. *Western Journal of Medicine* 136:546–551.
136. Medical Ethics Manual. 2005. World Medical Association. Ferney-Voltaire Cedex, France 2005:18.
137. Medical Ethics Manual. 2005. World Medical Association. Ferney-Voltaire Cedex, France 2005:75.
138. Hyman, M., and Liponis, M. 2003. *Ultraprevention*, 325. New York: Scribner.
139. Hyman, M., and Liponis, M. 2003. *Ultraprevention*, 324–325. New York: Scribner.
140. Sept. 2006 letter to subscribers.
141. holisticmedicine.org/.
142. WHO 1948, 191 state memberships; Yach, D. 1998. Health and illness: The definition of the World Health Organization. *Ethik in der Medizin*: 7–13; esp. 8.
143. Yach, D. 1998. Health and illness: The definition of the World Health Organization. *Ethik in der Medizin*: 7–13; esp. 11.
144. Tulsky, J., and Fox, E. 1996. Evaluating ethics consultation: Framing the questions. *Journal of Clinical Ethics* 7:109–115; esp. 110.
145. Nussbaum, M., and Sen, A. eds. 1993. *Quality of life*, 64. Oxford: Clarendon Press.

146. Arras, J. 2002. Pragmatism and bioethics: Been there, done that. *Social Philosophy and Public Policy* 19:29–58; esp. 46.
147. Fletcher, J. ed. 1995. *Introduction to clinical ethics*, 264. Frederick, MD: University Publishing Group; For example: Arras, J. 2002. Pragmatism and bioethics: Been there, done that. *Social Philosophy and Public Policy* 19:29–58; esp. 29; Miller, F., Fins, J., and Bacchetta, M. 1996. Clinical pragmatism: John Dewey and Clinical Ethics. *Journal of Contemporary Health, Law and Policy* 13:27–51; Keulartz, J., Korthals, M., Schermer, M., and Swierstra, T. eds. 2003. Pragmatist ethics for a technological culture. Dordrecht, NL: Kluwer; McGee, G. ed. 1994/2003. Pragmatic bioethics, TN. 2nd edn.: Cambridge, MA: Vanderbilt University Press.
148. Moreno, J. 1995. *Deciding together: Bioethics and moral consensus*. New York: Oxford University Press.
149. Engelhardt, H.T. 2002. The ordination of bioethicists as secular moral experts. *Social Philosophy and Policy* 19:59–82; esp. 82 .
150. Arras, J. 2002. Pragmatism and bioethics: Been there, done that. *Social Philosophy and Public Policy* 19:29–58; esp. 41 .
151. Dewey, J. 1922. *Human nature and conduct*. New York: Henry Holt; Dewey, J., and Tufts, J. 1932. *Ethics*. New York: Holt.
152. Schweitzer, A. 1990/1923. *Kultur und Ethik*, 311. München: Beck'sche Reihe; *Nicht Lebensverneinung an sich, sondern nur das, was im Dienste von Weltbejahung steht und in ihr zweckmässig wird, ist ethisch.*
153. Engelhardt, H.T. 1991. *Bioethics and secular humanism: The search for a common morality*, 138. London: SCM Press.
154. For a recent article defining humanism, see Shibles, W. 1995. Analysis of the Definitions of Humanism. *Scottish Journal of Religious Studies* 16:51–61.
155. Lahav, R., and Tillmanns, M. eds. 1995. *Essays on philosophical counseling*, Ch. 4. Washington, DC: University Press of America.
156. Schweitzer, A. 1966. Die Ehrfurcht vor dem Leben. Reverence for life. *Grundtexte aus fünf Jahrzehnten. Basic texts from five dedcades*, ed. Bähr, H.W., 8th edn, 65. München: CH Beck.
157. Lenk, H. 1998. *Konkrete Humanität. Vorlesungen über Verantwortung und Menschlichkeit*, 90 ff. Frankfurt am Main: Suhrkamp TB Wissenschaft.
158. Smith, W. 2000. *Culture of death: The assault on medical ethics in America*, 7. San Francisco, CA: Encounter Books.
159. Braddock, C., and Tonelli, M. 2001. Too much ethics, not enough medicine: Clarifying the role of clinical expertise for the clinical ethics consultant. *Journal of Clinical Ethics* 12:24–30; esp. 25.
160. Braddock, C., and Tonelli, M. 2001. Too much ethics, not enough medicine: Clarifying the role of clinical expertise for the clinical ethics consultant. *Journal of Clinical Ethics* 12:24–30; esp. 29.
161. Davies, L., and Hudson, L. 1999. Why don`t physicians use ethics consultation? *Journal of Clinical Ethics* 10:116–125; esp.119.
162. Davies, L., and Hudson, L. 1999. Why don`t physicians use ethics consultation? *Journal of Clinical Ethics* 10:116–125; esp.121.
163. Davies, L., and Hudson, L. 1999. Why don`t physicians use ethics consultation? *Journal of Clinical Ethics* 10:116–125; esp.122.
164. Fletcher, J., and Brody, H. 1994. Clinical ethics. In *Encyclopedia of bioethics*. ed. Reich, W., Vol I: 399–404; esp.400 New York: Macmillan.
165. Dewey, J. 1939. *Theory of valuation*, 32. Chicago: University of Chicago Press.
166. Dewey, J. 1971. *Moral theory and practice. The early works of John Dewey*. 1889–1892, vol. III, 93–109; esp.100. Carbondale, IL: Southern Illinois University Press.
167. Clayton, C. 1996. The Practice of philosophical enquiry as a therapeutic experience (PETE), Face to Face Dasein. (logos@facetoface.org.uk) Owl of Minerva Publications, available on the Internet.

168. Hoogendijk, A. 1988. Spreekuur Bij Een Filosoof, Ceen, Utrecht.
169. Veening, E. 1987. Monolog, Dialoog en Metaloog. *Filosofische Praktijk* 2:19–25.
170. Wittgenstein, L. 1953 orig./1968 3rd edn. *Philosophical investigations*. New York: Macmillan.
171. Kristeva, Z. 1995. Philosophical counseling: Bridging the narrative rift. *Society for Philosophy in the Contemporary World* 2:23–28.
172. Prins-Bakker in: Lahav, R., Tillmanns, M. eds 1995. *Essays on philosophical counseling*, 135. Washington, DC: University Press of America.
173. Lahav, R., and Tillmanns, M. eds. 1995. *Essays on philosophical counseling*, Ch. 4. Washington, DC: University Press of America.
174. Schefczyk in: Lahav, R., and Tillmanns, M. eds. 1995. *Essays on philosophical counseling*, 126. Washington, DC: University Press of America.
175. Dewey, J. orig. 1938/1964. *The theory of inquiry*. Rinehart and Winston, New York: Holt. See also Shibles, W. 1978. Teaching ethics and emotions in the schools. *Confluent Education Journal* 8: 30–34; Shibles, W. 1980. Ethics as a science: Going from is to ought. *Iowa Science Teachers Journal* 3:26–32 (Includes critique of L Kohlberg); Shibles, W. 1974. Ethics as open-context terms. In *Wittgenstein, language and philosophy*, ed. Shibles W.. Rev. 3rd edn. Dubuque, Iowa: Kendall-Hunt.
176. Shibles, W. 1974. *Emotion: The method of philosophical therapy*, 492. Whitewater, WI: Language Press; Shibles, W. 1994. Anxiety: A pseudo-concept. *International Journal of Applied Philosophy* 9:43–52.
177. Raabe, P. 2001. *Philosophical counseling: Theory and practice*, 226. Westport, CT: Praeger.
178. Wittgenstein, L. 1968. *Philosophical investigations*, 3rd edn, 133. New York: Macmillan,. The subtitle of Shibles′ work: Shibles, W. 1974. *Emotion: The Method of Philosophical Therapy*. Whitewater, WI: Language Press, also shows the stress on philosophy as therapy..
179. Peterman, J. 1992. *Philosophical Therapy: An introduction and defense of Wittgenstein's later philosophical project*. Albany, NY: SUNY Press.
180. Rendtorff, J., and Kemp, P. 2000. *Basic ethical principles in European bioethics and biolaw. Vol I Autonomy, dignity, integrity and vulnerability*, 147–279. Copenhagen, Barcelona: Centre for Bioethics and Law.
181. Rendtorff, J., and Kemp, P. 2000. *Basic ethical principles in European bioethics and biolaw. Vol. I Autonomy, dignity, integrity and vulnerability*. Copenhagen, Barcelona: Centre for Bioethics and Law, Austria: Hans Kelsen (philosophy of law), "Biolaw and bioethics in Austria are marked by a pragmatic positivistic legal tradition." "Bioethics has been dominated by a Catholic position…In opposition to this approach medical science advocates medical progress and scientific medicine as the basis of bioethics." Kant′s ethics, humanistic philosophy of the Greens. Belgium: Catholic domination of ethics committees, Anti-Catholic pluralism, Emmanual Mounier (personalist), critics of American principlist approach. Denmark: Alf Ross, philosopher of law, humanistic philosophy of the Greens Party. Marxism, utilitarianism, phenomenology, Protestant influence. Finland: Singer′s consequentialism and the American principlist philosophy (Beauchamp and Childress) France: Philosophers: Maine de Brian Bergson, Hans Jonas, Kant, Merleau-Ponty, Paul Ricoeur, Sartre, Levinas. Phenomenology. Strong Catholic influence. Germany: Utilitarian, Hans Jonas, Heidegger, Habermas, Kant, American Principlism, humanistic philosophy of the Greens Party. Italy: Catholic vs. secular bioethics, American principlism, personalism combined with Catholicism, Emmanual Mounier (personalist), Merleau-Ponty, Abompiani, D′Agostino, Catholic opposition to consequentialism and utilitarianism. "There is general disorientation caused by huge contradictions between the Catholic Church and secular modern Italy." Netherlands: Humanistic tradition. Norway: Strong Lutherianism, Kant, Hans Jonas, utilitarianism, French philosophy, Arne Naess, K-E Tranøy (principlism) Spain: Utilitarianism, consequentialism, rights based as by Rawls, Dworkin, Victoria Camps.

Sweden: Knut-Erik Tranøy introduced the four principles approach, ordinary language approach of Wittgenstein, pragmatism.
182. Smith, W. 2000. *Culture of death: The assault on medical ethics in America*, 25. San Francisco, CA: Encounter Books.
183. Singer, P. 2002. *One world: The ethics of globalization*, 63. New Haven, CT: Yale University Press.
184. Hosford, B. 1986. *Bioethics committees: The healthcare provider`s guide*, 85. Rockwell, MD: Aspen.
185. Fletcher, J. 1995. *Introduction to medical ethics*, 264. Frederick, MD: University Publishing Group.
186. Hosford, B. 1986. *Bioethics Committees: The healthcare provider`s guide*, 86. Rockwell, MD: Aspen.
187. Hendrick, J. 1986. Legal aspects of clinical ethics committees. *Journal of Medical Ethics* 27: i50–i53; esp. i52–i53.
188. Aulisio, M., Arnold, R., and Youngner, S. 2000 Internet. Health care ethics consultation: Nature, goals, and competencies. American College of Physicians American Society of Internal Medicine; See also Aulisio, M., Arnold, R., and Youngner, S. eds. 2003. *Ethics consultation: From theory to practice*. Baltimore, MD: John Hopkins Press; Benatar, D. ed. 2004. Life, death, and meaning: Key philosophical readings on the big questions, 841–847. Lanham, MD II: Rowman and Littlefield.
189. Prins-Bakker in Lahav, R., and Tillmanns, M. eds. 1995. *Essays on philosophical counseling*. Washington, DC: University Press of America, 38, ch 10; Hoogendijk, A. 1988. Spreekuur Bij Een Filosoof, Ceen, Utrecht: Ch 12.
190. Lahav, R., and Tillmanns, M. eds. 1995. *Essays on philosophical counseling*, 38, Ch 9. Washington, DC: University Press of America.
191. Paden, R. 1998. Defining philosophical counseling. *International Journal of Philosophy* 12:1–17, see also Shibles, W. 1974. *Emotion. The method of philosophical therapy*. Whitewater, WI: Language Press.
192. Marks, J. ed. 1986. *The ways of desire*, 3. Chicago: Precedent.
193. Jopling, D. 1996. Philosophical Counseling, truth and self-interpretation. *Journal of Applied Philosophy* 13:297.
194. Rendtorff, J., and Kemp, P. 2000. *Basic ethical principles in European bioethics and biolaw. Vol I Autonomy, dignity, integrity and vulnerability*, 35. Copenhagen, Barcelona: Centre for Bioethics and Law.
195. Rendtorff, J., and Kemp, P. 2000. *Basic ethical principles in European bioethics and biolaw. Vol I Autonomy, dignity, integrity and vulnerability*, 50. Copenhagen, Barcelona: Centre for Bioethics and Law.

Chapter 18
Medical Language: The Ordinary Language Approach

Language should be studied by studying language, not by algebraic symbolic logic.

Abstract Medical practice is characterized by verbal constructions. The ordinary language approach deals with what the healthcare workers and patients actually say. The paradigm of language games is positioned against formal logic, which as an abstraction has no relevance neither for our lives nor medical practice/thinking. Anamnesis is dealt with as presentation of the self. The self is seen to be a verbal picture we create for ourselves, a biography, a story of our lives, not something we are born with. We may also say that insofar as the physician uses language, the various linguistic selves of the patient are treated depending on the extent of the language used. From the ordinary language philosophy point of view, the language in and of medicine is examined, fallacies are presented and the metaphorical method introduced for analysis of concepts and settings.

Keywords Language · language game · ordinary language approach · linguocentric predicament · formal logic · informal logic · mathematics · mentalism · self-talk · metaphor

18.1 Introduction

Definition in language was already analyzed in the Chapter 2. "The word 'logic' is never used in the same way by two different philosophers" [1]. In the Chapter 17 Wittgenstein's notions of language-game and "forms of life" were presented and extended to the notion of narrative medicine and narrative ethics which give epistemological primacy to language in medical diagnosis and treatment. Medical practice was characterized by different verbal constructions. We are thereby created by our fictive stories and conversations. Do not assume our forms of life are "correct." Wittgenstein wrote "I shall also call the whole, consisting of language and the actions into which it is woven, the 'language game'... [this] is part of a frame on whose basis our language operates...the term 'language game' is meant to give prominence to the fact that the speaking of a language is part of an activity, of a form

B. Maier, W.A. Shibles, *The Philosophy and Practice of Medicine and Bioethics*, International Library of Ethics, Law, and the New Medicine 47,
DOI 10.1007/978-90-481-8867-3_18, © Springer Science+Business Media B.V. 2011

of life" [2]. Communication is itself a language game. Rationality and humanness are bound up with language. Language game – "meaning the whole of language and the activities in which it is embedded" [3].

Wittgenstein stated, "What has to be accepted, the given, is – so one could say – forms of life" [4]. Reflexivity and self-applicability are forms of metaphor or rhetorical devices. "Use" is itself a use of language. The meaning of the word "game" is itself a use in language game. Wittgenstein wrote, "I could not express what I want to say in any other way than by means of [these words]" [5]. Accordingly, metaphor has no literal translation, although it has interpretations. Judgment is a form of life. Form of life is part of an activity. One cannot understand language in isolation from the activity in which it is embedded. There is not a correspondence or "picture." Hope can only be experienced by those who have mastered the use of language, i.e. as forms of life [6]. Hanfling states, that forms of life are given and created. They include life as well as action [7]. Fingarette explains, "The forms of human existence have their source so profoundly in language that a practical understanding of language – even though it be tacit – is of the essence wherever we are engaged as human beings" [8]. "For all practical purposes, human beings are constituted by language; they exist in it, and also by means of it" [9].

The ordinary language approach in philosophy deals with what the healthcare workers and patients actually say. If thinking is reduced to one's language-use, as some philosophers have suggested, then "self" refers to us as language-users or anything one might say to oneself or aloud. Self is most essentially thought, and thought may be largely reduced to language-use. Thought is really critical speaking. Intonation often tells us what is meant rather than the usual meaning of words. The written word lacks all the meaning of the intonations and so is impoverished meaning. That people are a complexity of linguistic constructs and language games is not surprising because that is also the case with everything else. Scientific theories and the various disciplines are linguistic constructs as well. *Language has epistemological primacy.* The scientific method is not fundamentally based on naive perception or observation, but rather on the language, which they presuppose. "Understanding the world in sense-perception itself involves the use of [ordinary language] symbols or images... for only in this way can the elements in perception be identified" [10]. Language, on the other hand, does not presuppose perception. Language only presupposes language so we are caught in a reflexive, linguo-centric predicament. We are trapped inside our language. This very statement is a linguistic one. Explanations only explain within a language game of explaining. Wittgenstein held that it is a mistake to look for the justification of a language game. This is because there is nothing but another language game of justification in which to do it. It would make no sense to speak of justification by something outside of language. "I cannot use language to get outside of language" [11]. Language is not a rule-governed practice [12]. It is not just a human creation with rules and criteria subject to human choice.

Wittgenstein argues that we cannot observe ourselves thinking [13]. One can observe oneself speaking or writing, but even then not without language. He asks, "Was what I was doing then really thinking; am I not making a mistake?" [14]. We cannot be mistaken about thinking as a non-linguistic process because we have no evidence of thinking as such at all.

The self is basically not what we see or perceive. Thus we are basically language constructions. In the metaphysics of Heidegger there is a transition from a focus on "Dasein" (being) to a focus on language. "We – humans – are conversations. The being of people is found in language" [15]. The self, the I is language. Heidegger wrote, "Language speaks. Man speaks only in so far as s/he carefully 'complies with′ language" [16]. "Language is no longer considered as the exterior reproduction of a thought already formed in the immanence of consciousness. It may be said that thought is accomplished only in and through language." "Only in the word and language do things first become and exist as things" [17]. Jean-Paul Sartre wrote in *Being and Nothingness* [18], "I *am* language" [18]. It will be seen similarly that language cannot be reduced to a system of formal mathematics-like symbolic calculus. There is an attempt to reduce diagnostic language to "fuzzy sets," which is like attempts to reduce it to statistics, mathematics and symbolic logic. Albin applied "fuzzy sets" to medical diagnosis [19]. Steimann and Adlassnig more recently applied it to the clinical setting [20]. Language conceals a richness, which does not enter into the domain of discursive thought. Furthermore, formal logic will be seen not to be discursive. Rather than using a reductionistic logic, the way to study language is to study language. Even logicians argue that standard formal logic sentences must be rejected.

The self was seen to be a verbal picture we create for ourselves, a biography, a story of our lives, not something we are born with. In fact, there is no self, no I, no me as such anymore, no self-in-itself. We are elaborate fictions. The self was earlier analyzed as a construction of our language and as not having a single, literal definition. We have many definitions of "person." We cannot say how many we have. "The lives of even 'normal' beings are largely controlled by words and symbols whose meanings are ambiguous and ill defined. . . . For if they are unclear or confused or inconsistent, then the way of life is so also" [21]. There is no non-verbal body as such, only our verbal conceptions and constructions we call "body." Mere body, like mere sensation, is naive empiricism. We may also say that insofar as the physician uses language, the various linguistic selves of the patient are treated depending on the extent of the language used.

From the ordinary language philosophy point of view, the limits of our language are the limits of our world. Without language we would not "know." Thinking reduces to language-use. Our theories of person and human, and all theories in science, including bioscience, are linguistic constructs. We are what we can say about ourselves, what language permits. In any case, we cannot get outside of or beyond our language. If there were no language, there would be no person. To ask about the value of life is to ask about the possibilities of inquiry, critical and creative "thinking" and so of the possibilities of language. In terms of communication, we cannot assume that others "understand" the same things we do and in the same way we do (See analysis of understanding in the Chapter 8). "Language change produces new people" [22]. We are changed by language and in our communication with others. Knowledge is constituted and altered by the various universes of discourse. Thus, the value of life involves exploring the critique of language. This is one of the central tasks of philosophy and philosophical counseling.

That language has epistemological primacy, may be illustrated by the following authors although of a distinct minority. They suggest various supportive arguments. Logic uses language to claim that language is not usable. It examines language by means of language claiming not to have used language. The "extra-linguistic" itself is not outside language. We cannot get outside of our language to give an explanation of it. What does it mean? What can we say? Our lives are bound and bracketed by sentences with periods at the end. We can only play the language game of explaining within the language itself. We are caught in a linguo-centric predicament whereby we can only define language by means of language.

Our words "reason" and "concept" are created by language. "Language is everywhere primary; the concept arises from the word.... Language created reason; before language man was devoid of reason" [23]. We do not first have concepts, then language. "The content of language cannot be separated from its form in such a fashion that we could imagine that mankind, previously outfitted with all the essential concepts, would need only to invent external signs in order to initiate communication."

"Treat everything – our language, our consciousness, our community – as a product of time and chance" [24]. But we cannot even reduce language to time and chance. It precedes them.

For Heidegger, language is not an interior thing or result of thought or consciousness. The very existence of things is constituted by language. Language uses and defines us rather than we are using and defining language [16]. He wrote, "Language is no longer considered as the exterior reproduction of a thought already formed in the immanence of consciousness. It may be said that thought is accomplished only in and through language" [17]. Barthes opposes fixed essences such as logic, but instead favors the model of the individual in flux. People are too diverse to define. Nor is there a fixed good. He opposes the singular, the coherent, and the dogmatic. He accordingly distrusts the prevailing views of literature. There is no one interpretation of a work. Demystification instead is needed. Literary language should be preoccupied with the actual use of words, not with cold theory. One should sound out language to the full. Meanings are not private. We only use language never expropriate it [25].

In the Chapter 3 it was maintained that the scientific method cannot be based merely on perception, because perception presupposes language. It would be circular to examine perception by means of perception. Geiger thinks, "All thinking has arisen, through the mediation of language, from visual perception" [26]. He realizes that language is needed to mediate, but still gives epistemological primacy to perception without realizing that perception presupposes language. No language, no perception. Perception is also inextricable from language-use. Perception presupposes language also because there are various linguistic theories of perception. We are not clear how perception works. Our knowledge of it is still superficial.

Some philosophers try to use syllogisms, formal logic, or symbolic calculus as a tool of reasoning to deal with problems in philosophy, medicine or bioethics. This is a mistake. Just as language does not presuppose perception, ordinary language does not presuppose formal logic. *Formal logic is rather a misuse of ordinary language.*

The logician thinks in ordinary language about formal logic, but cannot think in formal logic about formal logic, or in formal logic about ordinary language. Formal logic presupposes ordinary language. If there were no ordinary language there would be no formal logic. Mathematics also epistemologically presupposes language. Without the language we speak, mathematics and formal logic would have no meaning and could not exist. "Even those [logicians] who advocate it do not write, or talk, or think, in conformity with its rigid notion of reasoning. They can't" [27].

In short, language not formal logic has epistemological primacy. Formal logic goes the other way. It seeks to limit language to a narrow artificial symbolism. We do not speak symbolic logic. We speak language. We could "speak" logic, but only as we "speak" mathematics. Thus, ordinary language cannot be reduced to formal logic. Systems come from language, not language from systems. *Formal logic will be seen to be another way in which medical theory and practice are dehumanized and made irrational.*

18.2 Formal Logic as a Pseudo-Logical Failure

> Formal patterns. . .lead to a logical vacuum [28].

An examination of the literature shows that for many scholars, formal logic has little scientific foundation or practical use. As will be shown, formal logic commits the ordinary language fallacies (informal fallacies) such as: abstractionism, circularity, reductionism, irrelevance, false cause, simplification, etc. Logicism and its Platonism [abstractionism] are "mathematical alchemy" [29]. Nevertheless, its study is as widespread as is that of astrology. However, it still fails to have a genuine use in problem solving. The concepts that logicians employ are so abstract and obscure that logicians are referred to as metaphysicians. Formal logic "winds up trailing large and frequently uncongenial clouds of metaphysics" [30]. "A genuine logic. . .is not a metaphysic. . .it does not require to take refuge in metaphysics" [31]. Doss regards formal patterns of logic as a form of Platonism (i.e., an abstractionism fallacy) [32]

Frege, Russell, Gödel, Cantor believed in absolute truth. They tried to reduce the world to indubitable, solid, non-contradictory foundations, but failed to do so. It does not work in mathematics and it does not work in medicine. Mathematics should not be independent of the material world. But realist mathematics is independent of mind and language. Mathematics is analytic apriori, absolutely certain. Numbers are supposedly real objects. He is a mentalist in that ideas are psychological and thoughts are contents of ideas. They are things. Sentences (marks/sounds) express thoughts. Mathematics is in a third real realm independent of people. Each sentence supposedly has unique and fixed meaning that is its objective content. Sentences designate propositions with determinate truth value. The senses of sentences are superhuman abstract objects. There is supposedly sense in some platonic realm of meaning. Meanings exist somewhere. Quine had a naturalistic and empiricist, holistic view, and there was no a priori [33]. Mathematics is continuous with science if it is to be true.

"The present work...is an attempt to show that this fundamental abstraction [formal truth irrespective of its truth in point of fact] everywhere leads to failure, failure both to account for the procedures of human thinking and failure to attain even formal consistency" [34].

Schiller states, "It is necessary to pull down the pseudo-science of Formal Logic, and to show what an incoherent, worthless, and literally unmeaning structure it is" [35]. "Formal logic is a meaningless science" [36].

"The hope that philosophical problems can be, by some stereotypical operations, reduced to standard problems in Formal Logic is a baseless dream" [37].

Wittgenstein who himself contributed much to symbolic logic in the early *Tractatus Logico-Philosophicus* [38] and the philosophy of mathematics, later repudiated formal logic. In place of such logic he developed the now famous ordinary language philosophy in his *Philosophical Investigations* [39]. Wittgenstein rejected the *Tractatus* theory of meaning. "The *Tractatus* account of language was a dehumanized one" [40].

What is rather needed is a humanized medicine, so we choose here the ordinary language approach, which is a concrete, personal, full contextual approach dealing with people and not with abstractions.

Ernest, like Wittgenstein, says that to obey a rule is like playing the game as understood by others, a custom. Mathematics is a game with signs according some rule. Rule – following is not just a decision, but built into our communicative behavior. Truths of mathematics are acceptance of linguistic rules. According to Ernest, mathematics is a narrative, conversational rhetoric, dialogue, and is persuasive. Style is a part of mathematics. Mathematics texts comprise imperatives, assertions from writer to reader [41]. "Mathematical proofs are almost never rigorous proofs" [42]. Just as mathematics is not rigid rules, medicine is not either. We may only imagine a world of certain or true mathematics, or medicine. It is in medicine assumed that statistics, evidence-based medicine, and mathematics are the objective hard core of medicine. There are no such worlds.

For Hersh mathematics is seen as a part of the social-cultural-historical side of human knowledge. He has a socio-historical approach against Frege, Russell, etc. "Most mathematicians hold contradictory views on the nature of other's work" [43]. "Mathematicians don't usually discuss philosophical issues...Philosophers have little to say to us" [44]. Physicians often feel the same way about philosophers. "Mathematicians mostly don't want to bother about philosophy" [45]. "An inarticulate, half-conscious Platonism is nearly universal among mathematicians... naïve, uncritical Platonism" [46]. Mathematics is a human, not objective enterprise. It is socio-historical. It is fallible [47]. Forget an immaterial, inhuman "reality" [48].

For Wittgenstein, mathematics is a form of life, a language-game. We must not confuse these games. We can only use mathematics, not understand it [49]. Mathematics is not about mathematics. Mathematics is a form of rhetoric. We can only try to see what we can do with it. It is like: social constructivist, but opposed to: realist, formalist, Platonist, reductionist, fundamentalist in mathematics. Mathematics is an expressive complexity, an exploration of the limits of our language, not just a matter of giving truths. The German, *Bildung*, is mathematics as education, self-development, mapping, giving models, an activity, picturing, without first truths. Mathematics games have no strict definition. Rule is undefined. Rules, and principles, like induction, deduction, logical reasoning, conclusion, and like saying that something follows from something else, are all vague metaphorical activities, not scientific or precise processes. They involve all kinds of assumptions and language uses. They are metaphorical. There is no appeal to primitive rules as logical objects, but there is seeing how mathematics is shaped by evolving language. There are no fixed propositions, sentences, meanings, truths, facts, objects, numbers, logical entailment. He relativized these. 2=2=4 is a recipe for measurement, counting, a rule to "go on." Mathematics is proved only if it has a use. Mathematics often does, logic virtually never. Mathematics is discovery, invention, a way to construct empirical events. To understand mathematics is to use it. Just as language grows, so does mathematics. One can't reduce mathematics to logic. One needs the whole context. Certainty depends on dialogue, not reality. Wittgenstein has a naturalistic and fallibilistic social philosophy of mathematics. His stress is on mathematics practice so to compare medical practice. Mathematics needs no foundation any more than sense impressions need analysis [50]. Mathematics is created and invented by us. We miscalculate [51]. Mathematics is linguistic conventions embedded in our socializing practices. It is normative. Mathematics is persuasiveness, pragmatic, naturalistic, descriptive, not logical necessity. Statistics and mathematics are meant to persuade. Proof is narrative for human consumption [52]. "The kinds of use we feel to be the 'point' are connected with the role that such-and-such a use has in our whole life" [53]. "The mathematician is not a discoverer: he/she is an inventor" [54]. Wittgenstein is not trying, like others, to reform mathematics, but to describe it. "'Mathematical logic' has completely deformed the thinking of mathematicians and philosophers" [55].

However, formal logic as with any model such as Platonism or supernaturalism, no matter how false, is not devoid of all insight. We also learn by the mistakes of the formal logicians. In any case, it will be argued that bioethics and the philosophy of medicine should not be taught or guided by logicians (cf. Chapter 17). Philosophy Professor Barry Smith, University of Buffalo and Leipzig, has recently received one of the largest philosophy research grants ever. It is to clarify medical terminology. Unfortunately, he is a formal logician and metaphysical ontologist, which would preclude success in clarifying medical language [56]. He says for example on his web site that a concept is by definition an abstract concept. This is circular and the term "abstract" suggests the abstractionist fallacy. Concept is also defined as a class of synonym terms, which is also circular. As will be seen below, the logician's logic and faulty theory of meaning cannot deal with concepts in either ordinary or medical language.

18.3 Formal Logic is Irrelevant to Thought, Reason and Emotion

In logic reality is not encountered at all, not even as a problem [57].
It is not possible to discuss in logic about a cup of coffee.

Symbolic and formal logic have so captivated philosophers that it is thought objectionable even to question them, just as it is thought offensive to question the Church, or was once thought radical to question tradition. Gilbert Harman stated, "There is no clearly significant way in which logic is especially relevant to reasoning" [58]. We may thus let the literature speak for itself [59].

Formal logic cannot deal with emotion. It is bloodless. "Formal logic must not abstract from the person side of knowing and purposes. It abstracts from personality" [60]. Formal logic dehumanizes thought [61]. "A logic which repudiates psychology repudiates meaning, and lapses into nonsense" [62].

Whatever symbolic logic "argues" about is never what is important. Arguments depend on the clarification of terms. Reasoning should rather include: "understanding, context, truth, error, relevance, selection, risk, interest and purpose" [63]. Logic does not do so. Formal logic does not improve one's reasoning ability [64]. In this sense, the formal logician is what may be described as "discussion illiterate." It is yet another form of avoiding rational inquiry. Even with ordinary language there are, then, often hollow words in place of genuine understanding. Most people are too enculturated to read or understand arguments. "Philosophy begins with distrusting language" [65] (See Chapter 8 for an analysis of understanding).

Mentalism is the fallacy of thinking that there are such entities as a mind, an imagination, or a memory as entities. This opposes the widely held view that we have a mind and a body. These are pseudo-psychological concepts [66]. "In immediate experience there is no mind" [67]. Thought, deciding, remembering, planning, understanding and all such alleged states are also mentalistic fallacies. There is no mind, so there is no longer a use for the concept "mental." To call inner states "mental processes" may make them sound scientific, but they describe nothing. There is no "mental physiology." There is no "mental illness." Wittgenstein spoke of the myth of mental process [68].

In describing the alleged workings of a person's "mind" we are not describing a set of shadow operations. Thinking is not like digestion. There are no nonlinguistic "ideas" before we are said to think. "Knowledge is not *translated* into words when it is expressed. . .a translation of something else that was there before they were" [69]. There is only language-use [70]. "We have no prelinguistic consciousness to which language needs to be adequate" [71]. One may be astounded to find out that one cannot think before one speaks to oneself. All thought is linguistic. The poet, W. H. Auden once put it this way, "How do I know what I think, until I see what I say?" Susanne Langer wrote, "It is by virtue of language that we can think" [72]. Max Müller regarded thought and language as identical [73].

We do not have nonlinguistic ideas and then express them in words. "The heart of language is not 'expression' of something antecedent, much less expression of antecedent thought" [74]. We do not know what we think until we see what we say [75]. It is a fallacy to think that ideas are "expressed" in language or in anything else. "It would make no sense to talk about thoughts which were not expressed or formulated. If they were not expressed in language, I would not know that I had them" [76]. Dewey and others identify thought with external rather than internal behavior [77].

The paradigm for inner thought is self-talk. What does a silence say? One may speak of an eloquent silence (*beredtes Schweigen*). We carry on a dialogue with ourselves. But if the "internal" dialogue is silent can something be said? "If we had not talked with others and they with us, we should never talk to and with ourselves" [78]. For Ryle, thinking is a kind of silent saying, tentative self-talk. "Much of our ordinary thinking is conducted in internal monologue or silent soliloquy.... talking to oneself" [79]. The therapist and healthcare worker can now deal with the self-talk of a patient, rather than with a fictive mind and ghostly thoughts.

It is for reasons such as these that Wittgenstein wrote that formal logic is "dull and useless" [80]. Logic is not relevant to human thought. Logic will not help physicians or anyone to think or reason and so will not help patients.

18.4 Formal Logic as Irrelevant to Ethics or Bioethics

Formal logicians search for the meaning of ethics in truth tables instead of examining the uses and misuses of ethical language in ordinary discourse and in ethics.

For example, deontic logic is an attempt to reduce terms like obligation and permission to a system of algebraic symbols based on truth tables. It asks whether or not ought-sentences have truth value. If they do not or cannot, formal logicians again fail to account for ordinary language. Instead of logic clarifying ordinary language, it destroys it as a language in order to try to reduce it to a type of mathematics, a calculus of sentences or "sentential calculus." Rules of syllogistic inference are similar to problems in Euclidean geometry rather than elucidation of our language. "Problems in, say, ... moral responsibility are not like this" [81].

18.5 Formal Logic as Formal Fallacy

Formal logic is "a devout determination to replace all concern for subject matter with the concern for form" [82]. Note the term "devout" which suggests that formal logic is like a faith or religion. Like mathematics, all context is eliminated so that one can deal only with a system of number-like symbols. The formal logician thinks there is only one approach to language: an ideal formal logic. But arguments are never formal. They must be concrete and internal to their subject matter. Formal logic is defined as formal, therefore not as practical or relevant. Practical formal logic is a contradiction. Thus, Doss states, "Reasoning ...is not at all a matter of formal patterns, the concern for correctness might now seem to lead to a logical vacuum" [83]. "The formal patterns which the theory presupposes will not fit the way we actually reason" [84]. The logician is un-in-*formed*. Informal logic is about actual language and is to be preferred over formal logic.

18.6 Formal Logic as a Fallacy of Abstractionism

Much ado about nothing (Shakespeare).
Logic is a game of abstraction [85].

Fundamentally, formal logic commits the philosophy of language fallacy of abstractionism. It abstracts from meanings, language, and human contexts to yield artificial symbols having nothing to do with anything. It is not clear what abstraction is, but one meaning is: reduction from the concrete, unfounded vagueness, and this will suit our purposes here. "The more refined formal logic becomes, the less relevant it becomes" [84]. Formal logic is a game in itself played for its own sake. "Formal logic can at best be a game with. . .fictions" [86].

Formal logic seeks empty "abstract universals" without concern for particulars. "The search for universal forms is both misleading and futile" [87]. "No one can ever say what a 'universal' is except by reference to particulars" [88]. Logicians use the metaphysical existential term "exists" in their "universal quantifier": for all x. . .; and the particular existence: there exists an x such that. . . The former also generates an all-fallacy. It is not clear what exists could mean in such an abstraction.

According to the *Unified Field Theory of Science* (of the Vienna Circle) all science must be reduced to the empirical. This is a commendable goal, but as they tried to use the metaphysical theory of formal logic as the ideal language with which to do that they failed. Logicians have a metaphysical use of language [89]. The goal to eliminate metaphysics from science is important, but unfortunately formal logic just adds to it. Nevertheless, against the logicians, the project to reduce science and mathematics to a more solid basis in the philosophy of science or philosophically critical ordinary-language-based empiricism remains a genuine one.

18.7 The Arrogance of Logicians

The more nonsensical logic is the more impervious it becomes to rational objection [34].

"The logician should not scorn actual reasoning, but observe it. . .without arrogance" [90]. The logician "loves technicality, not merely for its own sake, but because it makes him [or her] feel superior and 'scientific'" [91]. A mistaken technology does not provide that.

We cope with the world by means of our language games. For Rorty, like John Dewey, language does not represent the world, but is a tool for coping [92]. "Language, far from serving merely to report facts, is a collective instrument of thought that enters experience itself, shaping and molding" [93]. "If we spoke a different language we would perceive a different world." Logic, on the other hand, is only "domestic" issues among "philosophically-minded logicians" with little or no practical value [81]. Anyone who does not accept their approach is excluded.

18.8 Formal Logic Reduces Language to Mathematics

Informal logic is the practical, concrete analysis of the uses and misuses of ordinary language. Formal logic is the replacement of ordinary language by a mathematical calculus. It is called a "sentential calculus." Formal logic is calculating with the truth value of sentences independent of their content. Formal logic can only determine "logical correctness" as mathematics can only have "mathematical correctness." "Logical reason" is not ordinary reason, but more like mathematical reason. This is not correctness in ordinary reasoning.

Philosophical problems are seen as mathematical problems, mere calculations. It is "for nothing, apparently, but the proliferation of truistic formulae. No philosophical problem of any interest to anyone has yet been solved by reducing it to the shape or size that suits some slot in your slot-machine" [94]. Formal logic cripples reason and language. It is not a debate about such issues, but a debate about whether or not language can be reduced to such an algebra. Again, logic does not clarify terms, it calculates with them. We cannot decide philosophical issues by calculation. Philosophy and mathematics are different things. Philosophers do philosophy of mathematics; they do not reduce philosophy to mathematics any more than they would reduce philosophy to music. When they reduce ordinary language to mathematics, they have just that, mathematics.

18.9 Formal Logic as a Faulty View of Meaning

Formal logic is an impoverishment of language, gives strange stipulations and irrelevant definitions, which enrage the intelligence. Typically, in logic, words just denote objects, sometimes called "simples" which is an obscure notion. The terms must be fixed in meaning like numbers in mathematics otherwise logic would not be able to calculate with them.

Formal logic is like a robot language: rigid and doctrinaire. "The terms [of logic] are conscript terms, in uniform and under military discipline" [95]. Fixed meanings of terms without flexibility are very unlike the richness of ordinary language. On Schiller's view, logic seeks absolute, fixed and literal meanings, but definitions are only for use, not absolute or final [96]. Formal logic gets rid of and abstracts from contextual meaning [97]. "The fixating of meaning is in fact a fiction." "The price of fixity would be unintelligibility" [98]. Waismann also has expressed the same view: "If logicians had their way, language would become as clear and transparent as glass, but also as brittle as glass: and what would be the good of making an axe of glass that breaks the moment you use it?" [99].

Formal logic is like soldiers marching in a line whereby one bullet can wipe out a whole column. "The best conducted drill-evolutions would be the worst possible battle movements" [100]. Language has much wider implications than this. We do many things with language and it has an infinite diversity, rhetoric and pragmatic. *Logos* refers to meaning and formal logic has none.

Logic presents meaning as if it is a primitive culture where words just refer to objects. How does a word name something? Is it a relation? The relationships do not presuppose meanings, but define them. The following writers object to the simplistic view that words just name objects or entities. "The idea that 'the meaning of the word or expression x' referred to any sort of entity became a dead dogma" [101]. On the view of the pragmatists, an object is an artificial aspect after the object experience is already had. Object is future oriented, a set of potentialities, signs of consequences, all we can do with an event, not a fixed thing isolated from experience. Individual objects have no meaning. They are fictions, metaphors.

For Derrida, the sign has epistemological primacy. Writing destroys personal presence. We cannot know things directly, only through language. Thus, an object is nothing one can just point to as such to generate the logician's theory of denotative atomistic meaning. Nothing in itself may be labeled as the meaning. An object is rather a subject of investigation, a relation and function in our total behavior. It is a use of language. "It is logically impossible to give an account of the relationship between language and the world as if this were an account of how symbols relate to the totally non-symbolic" [102]. This is inaccessible to the logician.

On the common theory of meaning, the word stands for an idea. But as ideas do not exist and are mentalistic fallacies, meanings do not exist and are also mentalistic. The traditional theory of meaning is rejected, because "it is heresy to conceive meanings to be private, a property of ghostly psychic existences" [103]. We may rather regard meaning as the complex, non-mentalistic association of the word with the object. 'Meaning' is a mythological use of a noun" [104]. Meaning may also be analyzed as being part of the context of a language structure and part of the use or pragmatics of a language. The notion of language game is to give associations with both the context of the words, rules, grammar, and structure of the language and the game played, e.g., to greet, describe, explain, etc. For Wittgenstein, the meaning of a word is its use in a language game. "The explanation of the meaning of a word is not a causal explanation, but a rule for the use of the word" [105]. Dewey and others try to reduce meaning to some kind of disposition or external behavior. There is "no justification for collating linguistic meanings, unless in terms of dispositions to respond overtly to socially observable stimulations" [106]. Hirst speaks of language as a use, "what words and sentences can be used to achieve," which is a view like that of pragmatism, but is also based on Wittgenstein's notion of understanding as the use of language games [107]. In sum, the logician's view that words just denote fixed objects must be rejected as being a faulty view of meaning.

18.10 "Propositions": A Pseudo-Logical Term

Why do we say that a proposition is something remarkable?
A proposition is a queer thing [108].

One of the most fundamental terms in formal logic is the proposition. What they call "propositions" bear no relation to sentences in language, but are fixed logical formulas. They are supposedly ideal metaphysical representations of reality. Wittgenstein calls them "chimeras" [108]. "When philosophers use a word – 'knowledge,' 'being,' 'object,' 'I,' 'proposition,' 'name,' – and try to grasp the essence of the thing, one must always ask oneself: is the word ever used in this way in the language in which it is at home?" [109]. Logicians do not seem to know what a sentence is and do not explore its possible meanings and real life pragmatics. Sentences are transformed into logical "propositions," and reduced to the symbols: p, q, r, etc. These symbols stand for sentences, which are either true or false. "If p is true..." This is a part of a proposition, but a silly phrase. For the mathematician-logician a sentence expresses a "proposition." What is that? The referent of a sentence is supposedly a truth-value. Language is rather the medium in which we express "truths."

Regarding the ambiguous term "inference" as from propositions, Wittgenstein says, "We readily imagine that inferring is a peculiar activity, a process in the medium of the understanding, as it were a brewing of the vapour out of which the deduction arises" [110]. Logical inference is not normal ordinary language inference, but a kind of mentalistic metaphysical "ultra experience" having little to do with reason or scientific thinking [111].

Bertrand Russell on logical atomism hypothesized a world of so-called "simples" which words refer to [112]. This is a metaphysical notion, which Wittgenstein later exposed as being unintelligible. A world of atomic facts is supposedly expressible with atomic propositions. Thus, Russell's view that a proposition, if true, expresses a fact, is circular. He substitutes logical fictive constructs for entities. This presupposes that there are logical truths. "Propositions...are the fundamental fictions of symbolic logic" [113]. Wittgenstein rejected Russell's atomism. Propositional attitudes = I *doubt* that, I *hope* that... are hypothesized, but no definition of emotion is found in logic. As was seen earlier, logic completely ignores and is unable to deal with emotions.

18.11 Formal Logic as Dogmatism and Misuse of Symbols

> The effort to reduce even a very simple case of ordinary reasoning to a pattern of symbols can lead one to realize...how dogmatic formal logic really is [114].

Even one of the most famous logicians, Bertrand Russell, admitted that formal logic is rules for combining symbols which signify nothing [112]. Logic is just about tautologies, that is, circular propositions. Formal logic says nothing about anything. It commits the informal fallacy of circularity. It is audaciously claimed that all important distinctions in language can be reduced to just a few symbols. Ryle says the logician works by "conscripted concepts," e.g., all, some, none, implies, and, or ... which are then given an arbitrary logical meaning. But such terms are only actually meaningful from the point of view of ordinary language. Virtually all of the terms used in formal logic have different meanings than they do in ordinary language: all, and, argument, conclude, contradiction, deduction, fallacy, false, if-then, incorrect, infer, justification, not, or, proof, reason, sentence, some, true, etc. They are words taken from ordinary language, given an arbitrary logical meaning and are not relevant to ordinary language. The many ideas of language are too complex to be rendered by the simplistic and primitive connectives and symbols of logic. This is what Wittgenstein called a very crude and primitive attempt at language construction [115]. Such notational codifications and theorems are no part of the clarification of our language. Formal logic is irrational and irrelevant.

Formal logic only derives one true (T) or false (F) statement from another, like making change with different counterfeit coins with equivalent value. "But most of the terms of everyday and technical discourse are not like coins" [116]. Formal logic is like "forbidding all trade save money-changing" [117]. On the other hand, we can justifiably use symbols or diagrams, but only if these are connected with and based on ordinary language, and if it is useful to do so. Symbolic logic does not satisfy this requirement.

18.12 Formal Logic Misuses the Term "Truth"

Truth is falsity. T = F
'All truths are fallacies' seems contradictory, but is meaningful (Authors).
Truths are illusions of which one has forgotten that they are illusions [118].

Now we come to truth tables. Formal logic does not concern itself with or establish the truth of either premises or conclusions. The only concern for the logician is to ascribe truth or falsity to sentences regardless of their meaning. Truth and falsity are not defined and their meaning is unintelligible. They are only symbols. Truth and falsity are reduced to abstractionist empty symbols T and F. They could as well be reduced to stimulus S and response R, male M or female F, or 0 and 1 which sometimes, in fact, they are. This sentence is 0, that sentence is 1. We could equally as well say this sentence is orange, that one is green. Symbolic logic is calculation with the nearly meaningless symbols T and F. "The conclusion is round if the premises are round" makes as much sense as saying the conclusion is true. Suppose T = "I like it," F = "I don't like it," then how could we calculate with T and F? A statement is never true as such. Truth and falsity are not defined in terms of a time and a context. Truth is an open-context term. The addition of "true at time t_1" does not solve the problem. T and F are just empty terms treated as if they were numbers.

Truth in ordinary language is not the same as in formal logic [119]. Truth for Dewey and the pragmatists is an action, something in progress, not a static thing. Truth is not a deep matter, but only a property of sentences made by humans [120]. T/F are language games in the sense that they mean, for example, "I agree."

The logician's notion of independent truth and falsity is irrelevant to human thought. "If there can be truth independent of us, there can be unknowable truth" [121]. It is just a game we sometimes play. "Truth can be dehumanized" [122]. "Truth" in ordinary language refers to valuations, approval, right, the good end, what works best, knowing, interest, useful, good consequences, good to believe. "False" refers to disapproval, wrong, not knowing, etc. The meaning of true and false depends upon context, purpose and pragmatics. It would make no sense to simply say that a proposition or sentence is just true or false, and then to calculate with them as if they were numbers. When, for whom, in what sense? There are no necessary truths [123]. "Logical necessity is an illusion" [124]. Logical proof, demonstration, argument, reasoning – these are not ordinary language concepts. Formal logic, for all its alleged basis on truth tables, is not about truth or falsity at all, but only about artificial abstractionistic "consistency" of symbols. Again, what terms mean in logic is not what they mean in ordinary language and in relevance for our lives.

An argument cannot just be a floating true or false. An argument is perhaps never just true or false. Furthermore, with statements we do more than look for their truth or falsity. Sentences and words have many more uses than asserting truth or falsity. We exclaim, inform, choose, decide, joke, persuade, etc. We may never assert any statement that we hold to be absolutely and in all respects "true," especially if it is devoid of content. Rohatyn put it clearly, "In short, the problem of truth has not even been broached. The deep issues have been left aside in favor of an easy and convenient, and ultimately self-deceptive, brand of symbol mongering. Not only is the formulation philosophically trivial, it is of no help in trying to do something, however modest or humble, about the philosophically inescapable problem or problems of life, in small or in capital letters. Tarski and his followers, as well as his detractors, have nothing to say to [people]" [125]. It is ordinary language and our language games and linguistic theories, which constitute the world [126].

A conversation based on the above rules would not make for a stunning exchange. Langacker wrote, "I opt for a cognitively and linguistically realistic conception of language over one that achieves formal neatness at the expense of drastically distorting and impoverishing its subject matter" [127].

In addition, formal logic creates paradoxes, which are only paradoxes for logicians because of their rigid rules which are irrelevant to ordinary language. "This sentence is false," is a paradox for the logician, but not for the ordinary language philosopher for whom it is just equivocation. The liar paradox, for example, is no paradox at all. That is, suppose a Cretan says, "All Cretans are liars." According to Schiller, the Cretan just meant to refer to other Cretans and not necessarily even all of them [128].

18.13 The Useless Syllogism

> You have had experience of many wanderings without having found happiness anywhere, not in syllogisms.. [129].

One structure of formal logic is Aristotelian syllogistic logic. Sextus Empiricus pointed out that: Any syllogism is a vicious circle since the truth of "Socrates is a mortal" must be known if, "All men are mortal" is known (In the syllogism: "All men are mortal, Socrates is a man, therefore, Socrates is mortal," if we know all men are mortal we already know that Socrates is mortal). The syllogism only explores quantity, or logical constants such as: all, some, not. Later Megarians and Stoics explored and, or, if, and connectives. The syllogism is just a quantitative method of class inclusion using terms such as some, none, all, a (one). It is not an examination of meaning or clarification of language. It deals only with quantitative form. The syllogism is worthless [130].

18.14 Formal Logic is Not Philosophy

Bertrand Russell held that logic is the essence of philosophy [112]. However, for Gilbert Ryle, formal logic is at best of "minor use" in philosophy and this may be a negative one such that many philosophers have revolted against it [131]. Formal logic confiscates from language: creativity, clarity, emotion, meaning, practical problem solving, reason, style, and understanding. It violates our humanity. It purges from language that about which we are most concerned. What it creates is a cemetery of symbols where once was a living language. "Nor can the handling of philosophical problems be reduced to either the derivation or the application of theorems about logical constants" [132]. Formal logic is "language" in cobwebs. It is not genuine language, but rather a game to play for its own sake. As such, it is absurd. "It is not our aim to refine or complete the system of rules for the use of our words in unheard-of ways" [133]. It is not philosophy nor is it useful to the philosophy of medicine.

"Typical philosophers [e.g., logicians] are not philo-sophers in the very sense of the word, they are rather philo-logs. They are friends of the logos, but not of sophia, although the word 'philosopher' means friend of sophia" [134]. But they are not friends of language either as they destroy it.

"Questions which can be decided by calculation are different... from the problems that perplex" [135]. Rather, philosophers may do philosophy of formal logic/mathematics, not just logic/mathematics. "It is a logic for everyday use, and not merely for the obscene initiation rites of the philosopher who is 'no philosopher'" [136]. The philosopher should be engaged in doing informal, not formal, logic [137].

Logic is the denial of critical thinking (speaking) and philosophy. It is irrelevant to philosophy. That logicians identify philosophy with formal logic makes philosophy useless, which encourages the public to ignore it and close philosophy departments. The public does so as well because they are basically anti-inquiry. Formal logic gives philosophy a bad name. "The philosopher may and should neglect logic, while it is the logician who ignores philosophy at his [or her] peril" [125].

18.15 The Primacy of Ordinary Language and Pragmatism

Formal logicians are people who are trying to have never learned a language.

It is one thing to describe language and its fallacies as is done in informal logic, and it is another to replace and destroy it. We have seen that formal logic takes the life out of language to produce a dead "language" no one can speak. Formal Logic is both a misuse and abuse of language.

As mentioned earlier, for logic to be intelligible, it must presuppose ordinary language. Formal logic presupposes and is built from ordinary language, as is mathematics. "I am not letting out a secret when I say that the ordinary rules of logic often break down in natural speech – a fact usually hushed up by logic books" [138]. Therefore instead of studying symbolic or formal logic we should study philosophy, rhetoric, and the pragmatics of language. Informal logic uses full-blooded concepts and is not constrained by the straitjacket of a calculus. "A theory of informal logic would be a far more thorough study of what makes some given instance of reasoning fallacious" [139]. A word is not just a word, but an experience. Words are synonyms for experience. Meaning is an experience [140]. Wittgenstein called this "forms of life" [141]. We live inside of words. Medicine is language. Body is language.

Dewey is known for presenting a practical, relevant, problem-solving logic. Schiller calls it a "Humanistic logic" [142]. "It is better to lean on Dewey for support... than to try to justify logic in any number of disappointing ways, which rely or trade upon the methods, which we have already exposed" [143]. Stopczyk also supports a practical pragmatism [144]. This is also like the ordinary language tradition. We do not fully understand our own language. There is more meaning than finds expression [145]. It is like trying to understand thinking by thinking. This is the linguo-centric predicament. We use our language to try to understand our language. It is self-reflexive. The linguistic self tries to understand the linguistic self. Language is meta-language. Walton spoke of the idealization and false assumption that knowledge in analytic philosophy and logic consists of true propositions or justified true belief. He defends a pragmatic view of defeasable knowledge. It is false that knowledge can only be built up or proved from true premises ("knowledge collection") to conclusions. That is not how scientific inquiry actually works. To apply this view to medicine, we may observe that we have only incomplete and pragmatic knowledge, not true and fixed knowledge. Knowledge is based on practical acceptance rather than on absolute proofs or truths [146]. Knowledge is like an on-going search. Arguments often proceed by intelligent guesses. Walton, however, fails to appreciate that the idealized formal logic is itself not just not pragmatic, but actually fallacious.

18.16 Formal Logic Excludes Metaphor and Creative Language

> Many of our practices and much of our discourse in healthcare hinge on metaphor and analogy, whose significance is sometimes overlooked [147].
>
> To know is merely to work with one's favorite metaphors [118].

Formal logic prevents creativity and exploration of language, cannot render ordinary language into an argument, and cannot handle interjections, particles or metaphors. Language is metaphor, metaphor is language. "Intellectual and moral progress [is] a history of increasingly useful metaphors" [148]. Metaphors are un-paraphrasable. We cannot compare anything beyond language only different languages and language games [149]. Lipmann states that our entire possession of language rests on a metaphorical basis [150]. "All thought is metaphorical is itself a metaphor". Thought does not exist as such, so metaphors define thought. Furthermore, the view that metaphor is a transfer of thought (or ideas) is false, because there is no thought. Mental fictions are not transferred. "Induction" and "deduction" are essentialistic and presuppose a literalism. They are also metaphorical. Hirst wrote that the view concepts are acquired by some process of abstraction is "a thoroughly confused notion" [151].

Formal logicians create their own problems and then cannot find their way out of them – like a spider caught in its own web. "Meaning depends upon [language] use" [152]. No use, no meaning. Meaning and use cannot be decided in advance. This is also another argument for the case method in medicine (See Chapter 17). "Any meaning which is worth conveying must be in a manner new" [153]. The "logic of language" is metaphorical for how language works. It is not the application of logic to language. Thinking is basically language-use and as such, to explore the possibility of our language by poetry, critical thinking (speaking), etc. is to creatively explore the limits of our thinking, or more narrowly the limits of our knowledge. In addition to the language game account, we may also note that much of the meaning of a sentence or story comes from the intonation, not just the words. In short, formal logic is senseless because it eradicates the full meaning from ordinary language. It is bad semantics and bad philosophy.

Logic uses unchangeable symbols, whereas ordinary language changes its meanings as it is used. Reading a poem is not a problem in mathematics. Formal logic is too rigid and irrelevant to be able to handle or deal with questions, exclamations, interjections, particles, commands, subjunctive forms, emotion statements, poetry, metaphor, or anything else in ordinary language. Furthermore, formal logic cannot deal with humor [154]. The terms and concepts of logic are themselves metaphors. As was seen logic takes ordinary language terms and uses them in a bizarre sense for own purposes. The terms look like normal words such as implies, follows, proof, satisfy, solution, verify, evidence, true, false, form, logical, etc, but the terms have lost all ordinary language meaning. They are metaphors, which do not mean what they mean in ordinary language. What is proof in science or practical everyday life is not what is proof in logic.

Some metaphors are:

The bioethicist is considered as a "stranger." "The *stranger* has been proposed as a good metaphor for the ethicist in professional education because his or her outside perspective can challenge ordinary assumptions" [155]. The physician may see the philosopher as a foreigner in the area of medicine. What has the philosopher to do with medicine? This issue is dealt with in the Chapter 17.

One of the most outrageous and far-fetched of recent metaphors is to speak of the fetus as a person. On this metaphor one could also speak of the sperm as little swimming people. It reminds us of the antiquated homunculus theory according to which it was thought that there was a miniature fully-formed human in each sperm cell.

Fetus as inhabitant of the uterus.
Fetus as possession of the mother.
Mother as a receptacle for an infant.
Mother as servant of her child.
Pregnancy as accident.

Pregnancy as blessing.
Pregnancy as illness (cf. Morning sickness).
Pregnancy as mistake in contraception.
Illness as a learning experience, teacher or lesson learned.
Medicine as business. Patients as consumers and clients.
Physician as playing God/Allah/Buddha/etc.
Physician as teacher.
"Theory versus practice" is also a metaphor.

As mentioned in the Chapter 3, "diagnosis" from the Greek means to "see through" which is close to the notion of "seeing as" in the philosophy of science. It is inconsistent that many metaphors in medicine derive from the military and one may ask why this is the case. Medicine is pacifism in its emphasis on saving lives as opposed to the militarism, which characterizes most politics in most cultures. "Medical organization, particularly in the hospital, resembles military hierarchy; and medical training, particularly with its long, sleepless shifts in residencies, approximates military training more than any other professional education in our society" [156].

Arsenal of medicine.
Bombardment with toxic radiotherapy.
Fetus as invader.
Heroic medical action.
Kill cancer cells.
Medicine as war against disease.
Patient fighting and refusing to surrender.

Further analysis of metaphor is given in the Chapter 21.

References

1. Russell, B. 1919. *Introduction to mathematical logic*, 25. London: Allen and Unwin.
2. Wittgenstein, L. 1953/1958. *Philosophical investigations*, 3rd edn, 7, 240, 23. New York: Macmillan.
3. Wittgenstein, L. 1953/1958. *Philosophical investigations*, 3rd edn, 1. New York: Macmillan.
4. Wittgenstein, L. 1953/1958. *Philosophical investigations*, 3rd edn, 226. New York: Macmillan.
5. Wittgenstein, L. 1953/1958. *Philosophical investigations*, 3rd edn, 216. New York: Macmillan.
6. Hanfling, O. 2002. *Wittgenstein and the form of life*, 174. London: Routledge.
7. Hanfling, O. 1989. *Wittgenstein's later philosophy*. Albany, NY: SUNY.
8. Fingarette, H. 1969. *Self-deception*, 46. New York: Humanities.
9. Poirier, R. 1992. *Poetry and pragmatism*, 133. Cambridge, MA: Harvard University Press.
10. Hirst, P. 1974. *Language and thought. Knowledge and the curriculum*, 69–83. London: Routledge and Kegan Paul; esp. 76: Outside the effective use of symbols of some sort there is no understanding.

11. Wittgenstein, L. 1975. *Philosophical remarks*. ed. Rhees, R. Hargreaves, R., White, R. trs., Oxford: Blackwell.
12. Hanfling, O. 2002. *Wittgenstein and the form of life*. London: Routledge.
13. Wittgenstein, L. 1953/1958. *Philosophical investigations*, 3rd edn, 327. New York: Macmillan.
14. Wittgenstein, L. 1953/1958. *Philosophical investigations*, 3rd edn, 328. New York: Macmillan.
15. Heidegger, M. 1968. *Existence and being*, 3rd edn, 277. Chicago, IL: Regenery.
16. Heidegger, M. 1957. *Der Satz vom Grund*, 161. Pfullingen: Neske; Heidegger, M. 1979. Unterwegs zur Sprache. Pfullingen (On the way to language. Hertz, P. tr., Harper and Row, New York):12: Die Sprache spricht, nicht der Mensch. Der Mensch spricht nur, indem er geschicklich der Sprache entspricht. *Language speaks, not man. Man speaks insofar as he/she appropriates him/herself in destiny to language.* (Author`s translation).
17. Heidegger, M. 1983. Einführung in die Metaphysik. Klostermann, Frankfurt am Main: Im Wort, in der Sprache werden und sind erst die Dinge. (Author`s translation in the text.).
18. Sartre, J.P. 1956. *Being and nothingness*, Esp. 531–553. Barnes, H. tr. New York: Philosophical Library.
19. Albin, M. 1975. *Fuzzy sets and their application to medical diagnosis*. PhD Dissertation, Berkeley, University of California.
20. Steimann, F., and Adlassnig, K. 1994. Clinical monitoring with fuzzy automata. *Fuzzy Sets and Systems* 61:37–42.
21. Aiken, H. 1955. Moral philosophy and education. *Harvard Educational Review* 25:39–59.
22. Rorty, R. 1989. *Contingency, irony and solidarity*, 9. New York: Cambridge University Press.
23. Geiger, L. 1878. *Der Ursprung der Sprache*, 2nd edn., 1869:141. Stuttgart.
24. Rorty, R. 1987. *Contingency, irony and solidarity*, 22. New York: Cambridge University Press.
25. Barthes, R. 1982. *A Barthes reader*. ed. Sontag, S. New York: Hill and Wang.
26. Geiger, L. 1878. *Der Ursprung der Sprache*, 2nd edn., 1869:184–187. Stuttgart.
27. Doss, S. 1985. Three Steps toward a theory of informal logic. *Informal Logic* 7:127–135.
28. Doss, S. 1985. Three steps toward a theory of informal logic. *Informal Logic* 7:132.
29. Wittgenstein, L. 1978. *Remarks on the foundation of mathematics*. eds. Wright, G., and Anscombe, G., 284. Oxford: Blackwell.
30. Rohatyn, D. 1974. Against the logicians: Some informal polemics. *Dialectica (Swiss)* 28:87–102; esp.87.
31. Schiller, F. 1930. *Logic for use*, 443. New York: Harcourt, Brace.
32. Doss, S. 1985. Three Steps toward a theory of informal logic. *Informal Logic* 7:128.
33. Quine, W. 1969. Linguistics and philosophy. In *Language and philosophy*, ed. Hook, S. New York: New York University Press.
34. Schiller, F.C.S. 1912. *Formal logic*, viii. London: Macmillan.
35. Schiller, F.C.S. 1912. *Formal logic*, viii, ix. London: Macmillan.
36. Schiller, F.C.S. 1930. *Logic for use*, 50 cf. New York: Harcourt, Brace; Schiller, F.C.S. 1912. *Formal logic*, 6. London: Macmillan.
37. Ryle, G. 1953. *Formal and informal logic. Dilemmas*, 111–129; esp.126. Cambridge: Cambridge University Press.
38. Wittgenstein, L. 1961/2000. *Tractatus logico-philosophicus*. Pears, D., McGuiness, B. trs., London: Routledge.
39. Wittgenstein, L. 1953/1958. *Philosophical investigations*, 3rd edn. New York: Macmillan.
40. Hanfling, O. 2002. *Wittgenstein and the form of life*, 89. London: Routledge.
41. Ernest, P. 1998. *Social constructionism as a philosophy of mathematics*, 173. Albany, New York: SUNY.
42. Ernest, P. 1998. *Social constructionism as a philosophy of mathematics*, 184. Albany, New York: SUNY.
43. Hersh, R. 1997. *What is mathematics, really?*, 40 New York: Oxford University Press.

44. Hersh, R. 1997. *What is mathematics, really?*, 41 New York: Oxford University Press.
45. Hersh, R. 1997. *What is mathematics, really?*, 45 New York: Oxford University Press.
46. Hersh, R. 1997. *What is mathematics, really?*, 11 New York: Oxford University Press.
47. Hersh, R. 1997. *What is mathematics, really?*, 22 New York: Oxford University Press.
48. Hersh, R. 1997. *What is mathematics, really?*, 23. New York: Oxford University Press.
49. Wittgenstein, L. 1978. *Remarks on the foundation of mathematics*. eds. Wright, G., and Anscombe, G. Oxford: Blackwell.
50. Wittgenstein, L. 1978. *Remarks on the foundation of mathematics*. eds. Wright, G., and Anscombe, G., 378. Oxford: Blackwell.
51. Wittgenstein, L. 1978. *Remarks on the foundation of mathematics*. eds. Wright, G., and Anscombe, G., 221. Oxford: Blackwell.
52. Wittgenstein, L. 1978. *Remarks on the foundation of mathematics*, eds. Wright, G., and Anscombe, G. 173. Oxford: Blackwell.
53. Wittgenstein, L. 1978. *Remarks on the foundation of mathematics*. eds. Wright, G., and Anscombe, G., 16. Oxford: Blackwell.
54. Wittgenstein, L. 1978. *Remarks on the foundation of mathematics*. eds. Wright, G., and Anscombe, G., 111. Oxford: Blackwell.
55. Wittgenstein, L. 1978. *Remarks on the foundation of mathematics*. eds. Wright, G., and Anscombe, G., 300. Oxford: Blackwell.
56. cf. UMLS – U.S. National Library of Medicine which has one million biomedical concepts and 2.8 million concept names; Medical Fact World, WMLS, and ontologist.com.
57. Nietzsche, F. 1977. *Twilight of the idols. A Nietzsche reader*. Hollingdale, R.J. tr, New York: Penguin.
58. Harman, G. 1986. *Change in view*, 127. Cambridge, MA: MIT Press.
59. Schiller, F.C.S. 1930. *Logic for use*, 31. New York: Harcourt, Brace. "Formal logic is essentially an abstract theory about the formal relations of imaginary objects called absolute truths; it becomes a pseudo-science the moment it tries to be relevant to actual knowledge."30: "Formal logic remains entirely in the air, and its hypothetical deductions have no application to our thinking and reasoning." 31: "It [formal logic] has abstracted from all the essential features of actual knowledge, and from the whole context of reasoning. It ignores that every act of thought is performed by someone, at some time, in some place, from some motive, for some end." vi: 60: "Logic should be a theory of actual reasoning, and... a logical theory...cannot conceivably be applied to life.".
60. Schiller, F.C.S. 1930. *Logic for use*, 167. New York: Harcourt, Brace.
61. Schiller, F.C.S. 1912. *Formal logic*, 11. London: Macmillan.
62. Schiller, F.C.S. 1930. *Logic for use*, 23. New York: Harcourt, Brace.
63. Schiller, F.C.S. 1930. *Logic for use*, 49. New York: Harcourt, Brace; 25: "Formal logic covertly declares war upon actual thinking." 50: Formal logic, which is indifferent to meaning, truth, error, and human problems is "ridiculous.".
64. Goldstein, L. 1988. Logic and reasoning. *Erkenntnis*: 297–320; esp.297.
65. Waismann, F. 1965. *Principles of linguistic philosophy*, 176. New York: St. Martin`s Press.
66. Mead, G.H. 1977/1934. *On social psychology*, 100. Chicago: University of Chicago; The locus of mind is not in the individual. Hirst, P. 1974. *Language and thought. knowledge and the curriculum*, 69–83; esp. 72. London: Routledge and Kegan Paul; "This view regards the term mind as the label of a pseudo-object." Ryle, G. 1949/1984. *Concept of mind*, 40. New York: Barnes and Noble; "The phrase 'in the mind' can and should always be dispensed with." 18: "A ghost mysteriously ensconced in a machine." 15f: "The mind is not a 'man in the machine.'" "The mind is not a 'ghost' in the machine.'" "The dogma of the ghost in the machine.".
67. Mead, G.H. 1977/1934. *On social psychology*, 89. Chicago: University of Chicago.
68. Wittgenstein, L. 1967. *Zettel*. Anscombe, G. tr, ed. Wright, G. , 211 Oxford: Blackwell.
69. Wittgenstein, L. 1967. *Zettel*. Anscombe, G. tr, ed. Wright, G., 191, p 33. Oxford: Blackwell.
70. Ryle, G. 1949/1984. *Concept of mind*, 50. New York: Barnes and Noble.

71. Rorty, R. 1989. *Contingency, irony and solidarity*, 21. New York: Cambridge University Press.
72. Langer, S. 1960. *A modern book of aesthetics*, 3rd edn, 252. Rader, M., Rinehart, Winston: Holt (From her "Problems of art". 1957).
73. Title of article: "The Identity of Thought and Language." 1887. Wittgenstein, L. 1958. The blue and brown books. Harper, New York: 3, 4: There is no mind, only language. Hanfling, O. 2000. Philosophy and ordinary language: The bent and genius of our tongue. London: 111: "'The problem of consciousness' is largely due to the physicalist assumption that consciousness is a property or product of the brain; an assumption that is both unwarranted and incoherent." Wittgenstein, L. (1953/1958) Philosophical investigations. 3rd edn. Macmillan, New York: 329: Meanings do not accompany words.
74. Dewey, J. 1934/1958. *Art as experience*, 179. New York: G Putnam`s Sons.
75. Hartnack, J. 1972. *Language and philosophy*, 59. The Hague: Mouton; "There can be no thoughts and no thinking if there is no language.": 56 "To know something is not a psychological occurrence.": 46 "Language is a condition of consciousness. To conceive of something, it is necessary to conceive of it as something... [in] language." Quine, W. 1966. The ways of paradox and other essays. On mental states. Random House, New York: 213–214 "We repudiate mental states as entities." Wittgenstein, L. 1967. *Zettel*, 211, 39. Anscombe, G. tr, ed. Wright, G. Oxford: Blackwell; "In philosophy one is in constant danger of producing a myth of mental processes. Instead of simply saying what anyone knows and must admit." Hirst, P. 1974. *Language and thought. knowledge and the curriculum*, 69–83 esp.71. London: Routledge and Kegan Paul; "I see no reason to postulate the existence of a peculiar medium in which thought occurs." "We are surely not aware of a non-linguistic or non-symbolic process of thought that is translated into words." All we have is the language. Dewey, J. 1922. *Human nature and conduct*, 561. New York: Henry Holt;"I believe that the identification of knowing and thinking with speech is wholly in the right direction." Quine, W. 1964. *Word and object*, 77, 160. Boston, MA: MIT Press; "The notion of there being a fixed, explicable, and as yet unexplained meaning in the speaker's mind is gratuitous.".
76. Hartnack, J. 1972. *Language and philosophy*, 54. The Hague: Mouton.
77. Quine, W. 1969. Linguistics and philosophy. In *Language and philosophy*, ed. Hook, S., 97 New York: NYU Press; "The old empiricist looked inward upon his ideas; the new empiricist looks outward upon the social institution of language." 98: "Talk of 'ideas' comes to count as unsatisfactory except in so far as it can be paraphrased into terms of dispositions to observable behavior." Skinner, B.F. 1957. *Verbal behavior*, 449. New York: Appleton-Century Crofts; "Thought is simply *behavior*–verbal or nonverbal, covert or overt. It is not some mysterious process responsible for behavior, but the very behavior itself is all the complexity of its controlling relations.".
78. Dewey, J. 1934/1958. *Art as experience*, 170. New York: G Putnam`s Sons.
79. Ryle, G. 1949/1984. *Concept of mind*, 27. New York: Barnes and Noble.
80. Wittgenstein, L. 1979. *Wittgenstein`s lectures*. Cambridge 1932–1935. Lee, D. ed., 139. Oxford: Blackwell.
81. Ryle, G. 1953. *Formal and informal logic. Dilemmas*, 111–129; esp.111. Cambridge: Cambridge University Press.
82. Doss, S. 1985. Three steps toward a theory of informal logic. *Informal Logic* Vol. 7 127–128.
83. Doss, S. 1985. Three steps toward a theory of informal logic. *Informal Logic* Vol. 7 132.
84. Doss, S. 1985. Three steps toward a theory of informal logic. *Informal Logic* Vol. 7 130.
85. Schiller, F.C.S. 1930. *Logic for use*, 24. New York: Harcourt, Brace.
86. Schiller, F.C.S. 1930. *Logic for use*, 27. New York: Harcourt, Brace.
87. Doss, S. 1985. Three Steps toward a theory of informal logic. *Informal Logic* Vol. 7 133.
88. Schiller, F.C.S. 1930. *Logic for use*, 71. New York: Harcourt, Brace.
89. Wittgenstein, L. 1953/1958. *Philosophical investigations*, 3rd edn, 241. New York: Macmillan.
90. Schiller, F.C.S. 1930. *Logic for use*, 442. New York: Harcourt, Brace.

91. Schiller, F.C.S. 1930. *Logic for use*, vii. New York: Harcourt, Brace.
92. Rorty, R. 1989. *Contingency, irony and solidarity*. New York: Cambridge University Press.
93. Waismann, F. 1965. *Principles of linguistic philosophy*, 175. New York: St. Martin`s Press.
94. Ryle, G. 1953. *Formal and informal logic. Dilemmas*, 111–129; esp.113. Cambridge: Cambridge University Press.
95. Ryle, G. 1953. *Formal and informal logic. Dilemmas*, 111–129; esp.117. Cambridge: Cambridge University Press.
96. Schiller, F.C.S. 1930. *Logic for use*, 19. New York: Harcourt, Brace.
97. Schiller, F.C.S. 1930. *Logic for use*, 53. New York: Harcourt, Brace.
98. Schiller, F.C.S. 1930. *Logic for use*, 56 f. New York: Harcourt, Brace.
99. Waismann, F. 1965. *Principles of linguistic philosophy*, 23. New York: St. Martin`s Press.
100. Ryle, G. 1953. *Formal and informal logic. Dilemmas*. Cambridge: Cambridge University Press.
101. Garner, R. 1975. Speech acts. In *Syntax and semantics*. eds. Cole, P., and Morgan, J., vol 3, 316. New York: Academic.
102. Hirst, P. 1974. *Language and thought. Knowledge and the curriculum*, 69–83 esp. 76. London: Routledge and Kegan Paul.
103. Dewey, J. 1958. *Art as experience*, 189. New York: G Putnam`s Sons.
104. Waismann, F. 1965. *Principles of linguistic philosophy*, 154. New York: St. Martin`s Press.
105. Waismann, F. 1965. *Principles of linguistic philosophy*, 119. New York: St. Martin`s Press.
106. Quine, W. 1964. *Word and object*, ix. Boston, MA: MIT Press.
107. Hirst, P. 1974. *Language and thought. Knowledge and the curriculum*, 69–83; esp. 76–77. London: Routledge and Kegan Paul.
108. Wittgenstein, L. 1953/1958. *Philosophical investigations*, 3rd edn, 93, 94. New York: Macmillan.
109. Wittgenstein, L. 1953/1958. *Philosophical investigations*, 3rd edn, 116. New York: Macmillan.
110. Wittgenstein, L. 1978. *Remarks on the foundation of mathematics*. eds. Wright, G., and Anscombe, G., 6 Oxford: Blackwell.
111. Wittgenstein, L. 1978. *Remarks on the foundation of mathematics*. eds. Wright, G., and Anscombe, G., 8. Oxford: Blackwell.
112. Russel, B. 1919. *Introduction to mathematical logic*. London: Allen and Unwin.
113. Schiller, F.C.S. 1932. The value of formal logic. *Mind* 41:53–74; esp. 68.
114. Doss, S. 1985. Three Steps toward a theory of informal logic. *Informal Logic* 7:127.
115. Wittgenstein, L. 1953/1958. *Philosophical investigations*, 3rd edn, 2 ff. New York: Macmillan.
116. Ryle, G. 1953. *Formal and informal logic. Dilemmas*, 111–129; esp.122. Cambridge: Cambridge University Press.
117. Ryle, G. 1953. *Formal and informal logic. Dilemmas*, 111–129; esp.124. Cambridge: Cambridge University Press.
118. Nietzsche, F. 1952. *Werke*. ed. Stenzel, G., 2 vols. Salzburg: Bergland Buch.
119. Schiller, F.C.S. 1932. The value of formal logic. *Mind* 41:53–74; esp.67.
120. Rorty, R. 1989. *Contingency, irony and solidarity*, 21. New York: Cambridge University Press.
121. Schiller, F.C.S. 1930. *Logic for use*, 157. New York: Harcourt, Brace.
122. Schiller, F.C.S. 1930. *Logic for use*, 155. New York: Harcourt, Brace.
123. Schiller, F.C.S. 1930. *Logic for use*, 171. New York: Harcourt, Brace.
124. Schiller, F.C.S. 1932. The value of formal logic. *Mind* 41:53–74; esp.65.
125. Rohatyn, D. 1974. Against the logicians: Some informal polemics. *Dialectica (Swiss)* 28: 88–102; esp.100.
126. Winch, P. 1964. *Understanding a primitive society*. American Philosophical Quarterly I: 300 Reality is not what gives language sense. What is real and what is unreal shows itself in the sense that language has. Further, both the distinction between the real and the unreal and

the concept of agreement with reality themselves belong to our language. Donovan, J. 1985. *Feminist theory*, 153. New York: Ungar; Reality is constructed through language. Schiller, F.C.S. 1930. *Logic for use*, 174. New York: Harcourt, Brace ; The humanist theory of truth stands out among its rivals by its willingness to recognize the actual procedure of human knowing and to make the best of them.

127. Langacker, R. 1988. An overview of cognitive grammar. In *Topics in cognitive linguistics*, ed. Rudzka-Ostyn, B., 13. John Benjamins.
128. Schiller, F.C.S. 1930. *Logic for use*, 63–64. New York: Harcourt, Brace; A full critique of the liar paradox in formal logic is given in Shibles, W. 1985. *Lying: A critical analysis*. Whitewater, WI. German: Language Press; Shibles, W. 2000. Lügen und Lügen lassen: Eine kritische Analyse des Lügens. Tr. Maier, B. Mainz: Lermann Verlag.
129. Marcus Aurelius. 1964. *Meditations*. Tr Long, G. , 85 Doubleday: Penguin.
130. Schiller, F.C.S. 1912. *Formal logic*, 221. London: Macmillan.
131. Ryle, G. 1953. *Formal and informal logic. Dilemmas*, 111–129; esp. 112. Cambridge: Cambridge University Press.
132. Ryle, G. 1953. *Formal and informal logic. Dilemmas*, 111–129; esp. 123–124. Cambridge: Cambridge University Press.
133. Wittgenstein, L. 1953/1958. *Philosophical investigations*, 3rd ed, 51, 133. New York: Macmillan.
134. Stopczyk, A. 1998. *Sophias Leib. Entfesselung der Weisheit. Ein philosophischer Aufbruch*, 11. Heidelberg: Carl-Auer-Systeme Verlag.
135. Ryle, G. 1953. *Formal and informal logic. Dilemmas*, 111–129; esp. 114. Cambridge: Cambridge University Press.
136. Schiller, F.C.S. 1930. *Logic for use*, 454. New York: Harcourt, Brace.
137. Ryle, G. 1953. *Formal and informal logic. Dilemmas*, 111–129; esp. 124. Cambridge: Cambridge University Press.
138. Waismann, F. 1956. How, I. see philosophy. In *Contemporary British philosophy*, ed. Lewis, H. London: George Allen and Unwin.
139. Doss, S. 1985. Three Steps toward a theory of informal logic. *Informal Logic* 7:127–135; esp.132–133.
140. Schiller, F.C.S. 1930. *Logic for use*, 65. New York: Harcourt, Brace.
141. Wittgenstein, L. 1953/1958. *Philosophical investigations*, 3rd edn, 174. New York: Macmillan.
142. Schiller, F.C.S. 1930. *Logic for use*, 94. New York: Harcourt, Brace.
143. Rohatyn, D. 1974. Against the logicians: Some informal polemics. *Dialectica (Swiss)* 28: 88–102; esp.102.
144. Stopczyk, A. 1998. *Sophias Leib. Entfesselung der Weisheit. Ein philosophischer Aufbruch*, 100. Heidelberg: Carl-Auer-Systeme Verlag.
145. Schiller, F.C.S. 1930. *Logic for use*, 66. New York: Harcourt, Brace.
146. Walton, D. 2005. Pragmatic and idealized models of knowledge and ignorance. *American Philosophical Quarterly* 42:59–69 esp.63.
147. Childress, J. 2004. Metaphor and analogy. In *Encyclopedia of bioethics*, ed. Post, S., 3rd edn. 5 vols, 1834–1843; esp.1834 New York: Macmillan.
148. Rorty, R. 1989. *Contingency, irony and solidarity*, 9. New York: Cambridge University Press.
149. Rorty, R. 1989. *Contingency, irony and solidarity*, 20. New York: Cambridge University Press.
150. Lipmann, O., and Plaut, P. 1927. Die Lüge. Zur Sprachästhetik. Barth, J.A., Leipzig: 234: *Unser gesamter Besitz der Sprache beruht auf metaphorischer Grundlage.*
151. Hirst, P. 1974. *Language and thought. Knowledge and the curriculum*, 69–83 esp.73. London: Routledge and Kegan Paul.
152. Schiller, F.C.S. 1930. *Logic for use*, 62. New York: Harcourt, Brace.
153. Schiller, F.C.S. 1930. *Logic for use*, 55. New York: Harcourt, Brace.
154. Schiller, F.C.S. 1930. *Logic for use*, 165. New York: Harcourt, Brace.

155. Churchill, L. 1978. The ethicist in professional education. *Hastings Report* 8:13–15; cf. Maier, B. 2003. Das Vertraute und das Fremde in ethischer Perspektive: Das Vertraute und das Fremde als moralische Begriffe. In *Psychosomatische Gynäkologie und Geburtshilfe. Beiträge der Jahrestagung 2002 der DGPFG*, 291–296. Gießen: Psychosozial-Verlag.
156. Childress, J. 2004. Metaphor and analogy. In *Encyclopedia of bioethics*, ed. Post, S., 3rd edn, 5 vols, 1834–1843; esp.1835. New York: Macmillan.

Chapter 19
A Critique of Evidence-Based Medicine (EBM): Evidence-Based Medicine and Philosophy-Based Medicine

There are experimental methods and conceptual confusion. The existence of the experimental method makes us think we have the means of solving the problems, which trouble us; though both problem and method pass one another by. [1]

Events that cannot be proven in a double-blind study are nonexistent. [2]

Abstract The critique of EBM is not meant to discard EBM in general, rather to challenge its shortcomings and to support its positive intentions by introducing a philosophy of evidence, and providing a constructive critique of the concept, methods, and its findings. The EBM research studies are based largely on complex mathematical and statistical data and data analysis. Statistics do not give clinical results, but only statistical results. When we quantify we typically remove all of the qualities from the individual. Evidence answers the question of how we know something. We need philosophical analysis to determine what evidence is. This is philosophical evidence-based medicine. The problem of placebo in EBM is not resolved yet. Placebo is defined in this chapter as the positive assessments and emotions one has, and these do have a bodily effect.

Keywords Evidence-based medicine (EBM) · philosophy of EBM · evidence – definition · EBM as statistics · clinical experience · confounding factors · rational medicine · placebo as emotion · nocebo as emotion · philosophy of medicine-based medicine

19.1 Does EBM Really Meet the Challenge of Modern Medicine?

The critique of EBM is not meant to discard EBM in general, rather to challenge its shortcomings and to support its positive intentions by introducing a philosophy of evidence, and providing a constructive critique of the concept, methods and evaluation of its findings.

EBM research was found to change a quarter of the diagnoses and treatments of the team members studied. Straus and coauthors give steps of EBM research: (1)

B. Maier, W.A. Shibles, *The Philosophy and Practice of Medicine and Bioethics*, International Library of Ethics, Law, and the New Medicine 47,
DOI 10.1007/978-90-481-8867-3_19, © Springer Science+Business Media B.V. 2011

Convert need for prevention, diagnosis, prognosis, therapy, causation, etc. into an answerable question. (2) Search for the best evidence for an answer. (3) Evaluate the evidence. (4) Apply it to specific case and one's own experience. (5) Evaluate effectiveness of steps 1–4. Relate results to one's own clinical experience. In certain cases, consult with an expert or another physician [3].

Electronic services can be accessed within 10–25 s. Summaries can also be kept on file. Although EBM is an attempt to counteract unfounded or unquestioned traditional decision-making, it is argued here that this five step procedure is not always and fully adequate. Rather the decision-making procedure should be broad enough to include the philosophy of medicine and challenge underlying concepts of medical thinking and acting as well as the procedures of evaluation in EBM.

Medical research today is permeated with the paradigm, evidence-based medicine, a term coined in 1992. What is EBM? What it is not? EBM in the best sense of the concept is about integrating individual clinical expertise and the best external evidence [4].

There is, of course, better or worse EBM.

Physicians are expected to check EBM literature within 1 h of treatment. There is evidence-based mental health, evidence-based nursing, evidence-based practice, evidence-based treatment, and evidence-based evidence (meta-analysis, and reviews of evidence) – evidence-based everything (EB-X) (cf. Evidence Based on Call eboncall.org/content.jsp.htm, Evidence Based on Cardiology evidbasedcardiology.com/, Evidence Based on Pediatrics evidbasedpediatrics.com/).

One might better speak of Language-based Medicine. What we are doing in this chapter is the philosophy of EBM. It will be seen that narrow concepts of EBM also yield a problematic view of medicine. Many of the criticisms of EBM apply also to methods and concepts in epidemiology and statistics (cf. Statistics-Based Medicine, Statistics-Based Management). Note that after one has used EBM, critical thinking (speaking) and rational judgment are still necessary to be used [4]. EBM is a tool, not a criterion. EBM cannot handle fully, let be alone, case example.

The gold standard of evidence-based medical (EBM) research is the double-blind, randomized, placebo-controlled trial: a test in which researchers randomly assign volunteers to one of at least two groups. Placebo group is the control group, "real" medicine the test group.

People in one group get the intervention, substance or treatment being tested, and people in another group get no treatment or a placebo, and during the trial, neither researcher nor volunteer knows which is which. Such procedures supposedly rule out bias and give clear-cut proof that a treatment is effective. Cochrane Collaboration is one of the main sources of EBM approved research.

Cohort studies involve analytic epidemiology. They identify individuals exposed to varying levels of risk in a large population and observe the rate illness due to the condition in question over time. They study many over a long period. Cohort study is concerned with the occurrence of new cases through time compared in exposed and unexposed groups, e.g., lung disease in smokers and non-smokers. They tell what the effects of this exposure are. The Framingham Heart Study, Nurses

Health Study, and Whitehall II Study are several of the most well-known examples of cohort studies and are available on the net [5].

Case-control studies compare past cases of those with (cases) and without (controls) the disease, e.g., with cancer versus without.

The EBM research studies are based largely on complex mathematical and statistical data and data analysis. They are based on the view that science is grounded on sensation, statistics and mathematics. It will be seen that these assertions of EBM sometimes are insufficient. They have an inadequate idea of what evidence is. So these methods are to a certain extent reductionist, simplistic, exclusionary and philosophically naïve. Charlton states that statistics is not scientific, descriptive explanation [6]. Medicine uses math and statistics to present itself scientific. Statistics and mathematics are often just ways of trying to persuade others of a certain view. But what if statistics were not scientific? We will see that this is often the case. "Sometimes subjective knowledge is more reliable than 'objective' experimentation" [7]. And often, of course, it is not.

It is a myth to think the scientific method is based on sensations. This is naïve empiricism. Sensations have no meaning outside of language. It is language, which has epistemological primacy (See Chapter 18). Thus, all statements in science must remain in the form of language, discussion and reason. Critique and discussion is required as a method. Conclusions, evidence, causes, diagnoses, and treatments must be rendered in language, not in sensations, mathematics or statistics. EBM, however, is reductionist when trying to reduce them to statistics or non-language foundations. The name of the EBM resource program STAT! tells clearly by its name the statistical basis of such trials. EBM uses but is not based on our written and spoken language which is much more complex and sophisticated for human reasoning than statistics. EBM is like mathematics, meaningless unless reduced to language and purpose. It is like unqualified numbers: 1, 2, 3, ... without a clear specification or analysis of what is being numbered.

EBM also assumes that there is fixed and real data, facts, which may just be compared and manipulated. There are no objective facts in science or elsewhere in the sense of absolute truths. Facts are only hypotheses with a certain amount of confirmation or disconfirmation. They are only meaningful for certain theories, in Kant's words, conceptualized sensations [8]. Facts are contextual and tentative fictions. Reality is not described by scientific theories, but created and constituted by them. All of the theories are and must be in the form of language, not mathematical or statistical numbers or symbolic logic. The alleged "reality" changes when the theory changes. There is no direct sensual or epistemological access to a so-called reality.

To think there is a single model or method of science, a "gold standard," is to be captivated by a model or metaphor, to take the one model literally. Thomas Kuhn, in *Structure of Scientific Revolutions*, spoke of this as being captivated by a paradigm [9]. It is a metaphor-to myth fallacy. "Active scientists are in general indifferent to the origin and even validity of the conceptual framework that they accept and within which they work" [10]. We cannot say, "It has been scientifically proven that..." Which science? Which proof? If it has been only scientifically proven that is not adequate evidence. It needs to also be shown by sound philosophical reasoning, discussion and argument. "Scientifically proven" are terms of moral persuasion. The particular findings and cases must be given. "Most findings in psychology journals (and other soft sciences as well) have to be looked at with caution" [11]. Thus, EBM claims to have a science basis, but fails to adequately define science and without the philosophy of medicine has no tools by which to do so. In addition, EBM practice employs non-scientific methods and terms. To think that there are scientific facts and knowledge as such is a myth. One challenge to EBM analysis of causes, diagnosis and problem solving in the natural and social sciences generally is by use of the Metaphorical Method (See Chapter 1).

19.2 What Is the View of Evidence in EBM – Is It Left Undefined?

It would seem to be redundant to speak of evidence-based medicine. Is there medicine without evidence which one could still call medicine? Is medicine not a science after all? Medicine based on cultural and societal practices is not based on science. It will be seen that the attempt to reduce medicine to EBM actually makes it into a metaphysics. What is meant by "evidence" in "evidence-based medicine"? "Best evidence available" does not specify what is to be considered as "best." Evidence answers the question of how we know something. It is an epistemological concept and as such is studied in one of the major fields of philosophy, epistemology. When EBM does not refer to philosophy, the philosophy of medicine, and the study of epistemology, it cannot be evidence-based in the very meaning of the word. It cannot be clear what evidence is for EBM research. Synonyms of "evidence" are: amplification, attestation, authentication, basis, causes, certification, clarification, confirmation, controversy, corroboration, debate, difference of opinion, disagreement, dispute, explanation, grounds, illumination, justification, proof, rational argument, reason, substantiation, support, understanding, verification. Evidence does not mean truth, as nothing is absolutely true in itself. Evidence means such things as a process of discussion and rational debate, sound and philosophical argument, which does not violate the informal logical fallacies (that is, misuses of language), understanding and clarification on the basis of the examination of theory and practice. Charlton states, "Statistical meta-analysis cannot provide medical decisions as it claims to do, nor should it be the basis of medical practice." It is "the seductive offer of precision without the need for understanding" [12]. Good and Hardin state, "Statistical procedures. . .should never be quoted as the sole basis for making a decision" [13]. Evidence is a broad, and in-depth analysis. As will be seen, EBM in many ways does not see evidence this way. Evidence does not stand alone as data or something, which can be observed. It needs to be clarified by theory, analysis and specific examples. There is no evidence-as-such, though EBM appears to suggest that there is and that it has it. Evidence (like consciousness and knowledge) is not something one has. It is something one does. Also, doing EBM involves reflection, reasoning, critiquing, questioning, reviewing, examining. It is a higher level than a mere summary of trials.

We need philosophical analysis to determine what evidence is. This is philosophical evidence-based medicine (PEBM). "Defining evidence, deciding what qualifies as evidence. . .are complicated matters with deep philosophical and huge consequences" [14]. Every time physicians oppose philosophy of medicine they make themselves more servants of an enculturated system and less scientific and less able to make rational decisions.

19.3 EBM as Statistics

"And as statistics only". . . Hopkins states, "A high percentage of inadequately designed or incompetently analyzed studies . . .are published in prestigious medical journals. Statistics used to determine inferences, decisions about effects of relationships in behavioral sciences is of questionable value and misleading as compared with other methods of behavioral analysis" [15]. Their results are often impractical or unusable in clinical practice. Analyses of how statistics deceives and misleads should be taken in account [16]. Greenhalgh says we need clinical narrative not EBM or statistical evidence alone. Reasonable EBM would consider this. It is a mistake to think that clinical observation is totally objective [17]. Facts are theory laden and interpreted. We think that the individual case is not just anecdotal but rather gives us some of the most reliable evidence available. It is the statistical evidence which must be most suspect. We are treating individuals, not statistical samples as a whole. Often few cases conform to a standard, uncomplicated disease or evidence-based guideline, 10% in the case of hypertension [18]. Two x-rays may appear identical, but in the different contexts one is interpreted as tuberculosis, the other as cancer.

Whereas clinical practice is seen as unscientific by the EBM model, the EBM model is seen as unscientific by the clinical model [19]. According to Montgomery, "Physicians. . .are not engaged in a quantifiable science, but in a rational, interpretative practice" [20]. "Medical knowledge. . .models rarely account for more than 30% of whatever outcome we are investigating" [21]. Medicine is not a mathematical or statistical science, but is rather based on reason, specific contexts of patient examination, narrative interpretations and clinical practice. To extend Montgomery's view we may say that medicine requires the rational critical approach of the philosophy of medicine. The author's conclusions are significant but her arguments could have been made stronger by the use of such an approach. Rationality is seen as being larger than science [22]. One could extend this view to assert that the philosophy of science/medicine is larger than science/medicine. We may say that adequate clinical experience is a form of philosophical inquiry.

Albert and coauthors state, "Physicians make probability claims all the time even when they are fully aware that no statistics currently bear on the claim in question" [23]. "The foundations of probability and statistics are still very unsettled. Logicians are not certain of what, if anything, we are measuring or how we should measure it when we assess chances and reasons from statistics" [24]. Results of such reasoning are often poor. "Probability causality" is self-contradictory [25].

Thompson shows that the literature of analytical chemistry, which uses statistics routinely, reveals a high proportion of inappropriate applications and conclusions [26].

Charlton states, "Statistical malpractice has an almost limitless potential for abuse" [27].

Statistics is also often misused especially in scientific experimentation, in medicine, psychology, and therapy. A Wall Street Journal reporter calls attention to the misuse of statistics and other numerical data by scientists and the media, some of which constitute fraud (especially omission of inconvenient data, deceptive surveys) and the imposition of ideology upon data (e.g., The Bell Curve) [28].

Statistics as used in EBM is not the same as statistical surveys. The latter are more scientific and useful. Evidence differs for each discipline. For the statistician statistical evidence is best, for the philosophical counselor it is sound holistic reasoning. What is to be the evidence for medicine? In popular EBM approaches the attempt is to reduce evidence to statistics. It treats all evidence as statistical and experimental evidence to the exclusion of reason in a broad and more adequate sense. It claims to present facts that it does not present. A new definition of evidence is needed in medicine. "Types of relevant evidence range far beyond that derived from randomized trials and meta-analyses" [29].

EBM is just a metaphor. It is an error to take it literally to assert: All medical research must be mathematical or statistical. If it is, it should be called, Statistics-based medicine, not evidence-based medicine. What is to count as evidence? Evidence for the philosopher is certainly not the same as evidence for the statistician, scientist, therapist or physician. It is in most cases opposite of evidence for the theologian. Evidence is not mere consensus or statistical analysis or trials.

19.4 EBM Often Investigates the Obvious and Trivial

Hand states, "Much statistical analysis and design is misdirected" [30]. Experiments are often done to "prove" the most obvious things not needing the millions spent for the research. Michelson accordingly states there are medical events on where it is "so obvious (to the casual observer) that no self-respecting study will ever be undertaken (let alone published) [31]. These are best dealt with by the judicious use of common sense." One of the most significant of these are the experiments to try to find out if working 48 h shifts and 100 h weeks is harmful to physicians and nurses. What kind of insensitivity and lack of understanding must such researchers have to think this? Perhaps they themselves have never stayed on active medical duty for 48 h (The research is presented in the section on physician stress and overwork).

Berghella, Baxter and Chauhan evaluated EBM for guidance for surgical decisions during cesarean delivery (CD). They used MEDLINE, PubMed, EMBASE and Cochrane searches with terms cesarean (section, delivery) pregnancy, plus technical terms, e.g., lateral tilt. Result was that the research was usually evaluated as being poor, not fair or not able to be assessed [32].

The following are examples of trivial or false EBM results. Leipzig presents the major EBM advances in 2004–2005 in the field of geriatrics [33].

They are given as: Vitamin D supplements reduce odds of elderly person falling by 22% (Presumably the EBM physician should then tell those who fell or have not yet fallen to take vitamin D. What would the EBM physician reply if the patient asked, "Would not vitamin C prevent falls just as well?"). Elderly exposure to sunlight over 15 min a day versus not regularly exposed to sunlight had 84% fewer hip fractures (This unspecific result is unreasonable and unlikely). Alzheimer's disease reduces life expectancy (This result is obvious). Stroke increases dementia risk (This result is obvious). Athletic and canvas shoes are associated with the lowest risk of a fall (This seems unlikely as it is the sole of the shoe, which is the most important, not the canvas top. It is, on the other hand, obvious if the shoes have high heels or are slippery. Should it be studied also whether people who have white shoelaces fall less?).

19.5 EBM Bears the Risk of Overgeneralization

Statistical generalizations are often empty and of little use as they must overlook all the most relevant differences. EBM searches for universals, which yield vague and metaphysical generalizations of little practical impact. Statistics do not give clinical results, but only statistical results. Experiments in EBM as in psychological experiments, are set up to show a correlation between several variables, e.g., between cancer and smoking. The variables are typically undefined or inadequately defined. They are often reduced to vague or simplistic terms, e.g., hate is defined as going away, love as coming toward. Quality of life is only defined in terms of physical movement.

The most concrete evidence one can have is personal evidence, contextual evidence and experienced evidence. Of course, one may err.

19.6 EBM Is Often Unintelligibly Complex

There is much EBM illiteracy. The statistics used in EBM are so complex they are unable to be understood or produced, by healthcare workers. A medical researcher today can hardly understand a research article or do a research project without a statistician. Guttman states, "Statistical errors [are] commonly found in peer-reviewed studies." "59% of papers in the *Journal of the American Academy of Dermatology* and *Archives of Dermatology* contain mistakes in methods" [34]. Previous studies had found errors in 45–95% of the published papers. The results and validity of the articles are therefore put in question. EBM reports are poorly understood by nearly all physicians. Physicians are not statisticians. Scientifically based clinical guidelines are therefore often ineffective in changing physician behavior. Thomas, Dayton and Peterson found that in spite of the widespread publication of clinical standards and practice guidelines, physicians have had difficulty understanding and applying them in the clinical setting [35]. They note that the guidelines given are not intended to replace human clinical judgment. National Guidelines Clearinghouse has clinical guidelines and protocols. If they are taken literally it eliminates reason in medicine. The physician will not have to think. But on the contrary, the clinician must ultimately evaluate and make the decisions [36]. Oliveri states that although the Cochrane Library is the best EBM source, it was the least consulted information source by physicians and had never been consulted by 49%. Only 18% always practiced EBM [37]. Compared to seldom or never users of the Cochrane Library, frequent users had significantly higher academic backgrounds. Most hospital doctors lacked knowledge of key methodological evidence-based medicine terms. Doctors are not scientists. And most doctors are still not capable of critically appraising an article in medical journals. They rather follow familiar patterns and rules. [38] Even with some expertise in EBM, in every trial report there are thousands of direct and indirect sources referred to to evaluate and check. The process becomes impossibly complex and physicians often do not have time to care for patients much less examine an infinite amount of questionable literature. These will go unevaluated. Few physicians are adequately trained to use medical databases such as Medline, Pubmed, etc. The process reduces to absurdity.

19.7 EBM Is Often Too Abstractionistic

Abstract terms are measured and correlated in EBM, which cannot be measured, e.g., experience, love, intelligence, nutrition, etc [39]. Vague causal relations, e.g., health versus illness, socio-economic class and cancer, educational level and coronary artery disease, etc. are measured without clear definitions of the concepts being investigated. The language in EBM typically commits the abstractionist fallacy. "Does eating more fiber reduce cancer?" This question consists of terms so abstract as to be unintelligible. In short, EBM is often using uncritical and fallacious language, as the following accounts will further illustrate. And then, the results will be unacceptable due to inadequate definition and uncritical language.

19.8 EBM as Appeal to Authority Fallacy

EBM sets its "gold standard" as the only standard. It is a captivation by a paradigm. Researchers use EBM studies as appeals to authority. If it is EBM it must be true. So critique is uttered as follows: "'Evidence-based medicine' is at best a meaningless substitute for 'medicine' and, at worst, a disguise for a new version of authoritarianism in medicine" [40]. Michelson in "Critique of (Im)Pure Reason: Evidence-Based

Medicine and Common Sense," maintains that EBM is "completely based on the proposition that 'truth' can be gleaned exclusively from statistical studies" [31]. Randomized controlled trial (RCT) is useful but cannot replace or have priority over other types of therapeutic evidence [41]. In any case, random is often not random. There is faulty generalization from a selected population. The current practices of statistical methods are not scientific [42]. In EBM some hold that the *only* valid evidence to be based on are randomized controlled trials (RCTs) while others hold it is meta-analysis [43]. Statistics is so central for Johann Gauss that he called deviations from the mean "errors," instead of just chance variations [44]. The exclusionary approach of EBM is expressed in the following: *Events that cannot be proven in a double-blind study are nonexistent* [2]. But just citing a trial is not evidence. One must oneself be able to and actually read and critically evaluate the statistical and other complexities of the trial and few healthcare workers have the expertise to do that. EBM cannot merely be trusted.

19.9 EBM and the Individual Case and Context

Tumors do not read textbooks.

"Expert knowledge cannot be reduced to a set of facts or propositions, but is linked to contexts of applicability" [45]. The aim of practice is not data collection, but complex problem-solving [46]. EBM and statistics deal with large numbers of people put into a group on some general principle. Statistics says nothing about the individual [47]. "Some statisticians have developed algebraic models [of diagnosis]; but since they have never seen the diagnostician at work, the models are hopelessly unrealistic" [48]. "These peculiarities [of diagnosis] are sufficient to render the methods of classical statistics almost useless" [48]. The risk of a certain operation may be 20% chance of survival, but in your case may actually be 95% when placed in a group having your more similar characteristics. Surgeons say that each operation is different, each has its surprises, and each is ultimately experimental. Every use of a medication is also experimental. It is argued that there is patient variation (e.g., the health risk of two people of the same age may be entirely different), complexity, imperfect taxonomy, and often too many variables to be able to use RCT (EBM) [49]. EBM holds that many cases must be looked at, but we by use of reason can conclude much by the in-depth examination of the individual and single case. Anecdotal evidence is important. Single case is important. Individual patient response is important. The expertise of the physician is built on years of observations of individual cases. It is partly, what is meant by being an expert. EBM examines many cases superficially and stereotypically. We may test a case by the theory and the theory by the case. General statements about groups yield only vague and general observations of little use in actual concrete practice. When we quantify we typically remove all of the qualities from the individual. What may be important in medicine is not just the average, or mean, but the extreme range of possibilities. Extreme ranges defeat statistics because they cannot then give a fixed cause or rule.

Lang and Hensrud suggest that EBM regarding medical research shows little result or progress. There are 140,000 cases of colorectal cancer, but only 25–30% of eligible patients receive colon cancer screening [50]. 50% of deaths are due to alcohol, drugs, tobacco, poor diet, lack of physical activity, risky sexual behavior, auto crashes, and stress [51]. Patients can do EBM also by computer. 78% of patients with lung cancer continue to smoke against physicians' orders. Non-adherence to health recommendations ranges between 20 and 90% [52]. In 12-week cardiac rehabilitation programs, 50% show reduced physical exercise levels within 3 months. In the medical setting, 50% of overweight men and 30% of overweight women were not trying to lose weight. If EBM were effective the above picture would be different [53].

19.10 Uncritical Use of EBM and Clinical Experience

Uncritical use of EBM undermines clinical experience, critical use supports it. EBM versus clinical experience is an either-or fallacy. One may use both methods, which is suggested in elaborate EBM definitions [4]. This is the difference between mere quantification versus qualitative research both of them dialectically interwoven. Thus, evidence-based medicine is best spoken of as mathematics-based medicine and statistics-based medicine on one hand, but on the other the application to the concrete situation remains as clinical challenge. "The probabilist's definition is purely mathematical, derived from arbitrary assumptions, and involves no allegiance to ontological or empirical questions" [54]. Clinical evidence is based on a vast variety of sources, which EBM, narrowly defined, may exclude in favor of ideal EBM gold standard principles. It is then a form of principlism meant to do away with the individual clinical physician's judgment. More critical approaches try to integrate both in decision-making [4]. If there is no trial, or conflicting trials, on a particular type of surgery should the surgeon not operate? EBM does not replace dialogue and debate, which is central to critical thinking. Manual driven decisions may not be preferred to experienced physicians. In mental health "practitioners are asked to adhere to a psychotherapy manual." Manualization is required for EB-therapy. The NIH wants to uniformize by means of manuals which often is not enough to adequately treat the individual patient [55]. Dialogue and debate are central to the philosophy of medicine. In the area of therapy what is to count as evidence has been unclear. Evidence in therapy has been "precious little. . .and it hasn't been any good." Only 10–25% of the medical decisions are based on high quality information. Only 15–20% of interventions did more good than harm [56]. In therapy involving dialogue and discussion as well as a deep understanding of personality and language-use, evidence is narrative.

Cook says that EBM is a reductionism, and that not all is reducible to published evidence [46]. Ethics is involved in clinical evidence and cannot be excluded. EBM does not take ethics into account. Clinical decisions involve ethics and values [57].

Practice is non-linear, complex, dynamic and in need of discussion. The aim of practice is not knowledge, but problem-solving.

It may be stressed that before, during and after EBM trials, critical thinking (speaking) and rational judgment must be used. EBM trials and journal reports should never be accepted without critical evaluation – critical not solely in regard to statistical evaluation. EBM is a tool, for not a criterion of good clinical practice. EBM cannot handle case example involving individual contexts and individual judgment. EBM does not involve critical philosophical thinking, a philosophy of medicine critical of EBM or statistics-based medicine. One great advantage of reasoning generally and case study in practice is that it is not amenable to statistical or mathematical analysis. Mathematics and statistics are essential tools but only for certain areas of quantitative analysis, and not fully sufficient for qualitative individual decision-making. "Practitioners…often find statistical hypotheses studies narrow, tedious, and too de-contextualized to be assimilated into their practice" [58].

19.11 EBM Often Excludes Relevant Causes and Variables

The "Null Hypothesis" is where there is no association between variables. P value is probability. If $p < 0.05$ null hypothesis is rejected. If $p > 0.05$ it has significance. Given enough people a shoe size can have statistical significance as being correlated with cancer. Statistical correlations cannot produce causes, but only statistical probabilities. According to Varney and Hagen, "A statistical association of an agent with a disease does not establish it as the cause of the disease" [59]. Also needed is the biological plausibility, consistency, specificity, strength of association, etc. The predictive value of a test may be unimpressive and the sample population should be similar in health to judge for the specific individual. It usually is not.

Association does not yield a cause [60]. Wang questions the use of statistical inference. Charlton states, "The seeking of algorithms for scientific decision-making is an offense best described as *statistical malpractice*" [61]. Science is concerned with causes, statistics with correlations [62]. It is an error to go from correlation to causation. Thoresen stated, "Statistically significant correlations or explained variances are not valid substitutes for evidence from experimentally-based studies" [63]. Correlation is more like poetic connotation. Causes require that all other possibilities be excluded (See discussion of cause in the Chapter 4). Statistical probabilities are not clinical probabilities. *Bayes' Theorem* suggests that in all but trivial cases there is uncertainty. We always have at least two explanations. Bayes' probability is what you believe it to be, anything at all. It is argument from a presumed model to potential data, inverse probability. Typical conclusions from EBM studies can only try to show that x may be correlated with y, not that they are correlated, and not that x causes y. The correlation is still simplistic, because it ignores context and specificity of compared items, so cannot succeed. Also, statistics does not yield physically required causality. There often needs to be shown that there is possible physiological causality as well [31]. The therapy classification system, DSM, has many different versions and means different things at different times so that a statistical analysis of the prevalence of, e.g., schizophrenia would be based on different things. We may have statistical significance, but clinical triviality.

One group is given an intervention or treatment and another not in order to determine if the treatment works. Most all other variables are ignored. Single substances are selected for single outcomes. Gershoff states, "It is nearly impossible to do controlled research on the effects of solitary substances" [64]. We can even suggest that if an experiment concludes that a single item is responsible for a single outcome, it is not true. There are only complex substances and treatments in complex combinations producing complex outcomes. The simplistic picture is that stimulus (s) produces response (y). Rather there are a series of stimuli, which produce a series of responses and these are all modified by the thinking of the individuals involved, even if they are mice. Further what we call a stimulus and response is arbitrary if one wishes to avoid causality and just deal with correlations. If we are to find causes we must consider all relevant variables and the variables must remain consistent over the experiment (See also discussion on causation in the Chapter 3). We may ask how a specific element or substance or treatment affects the whole body in various ways under various conditions in regard to health and disease. There are also time problem factors: during the trial there can be lifestyle changes, context changes, cognitive and emotional changes, bodily changes, etc. We must know the full context of the treatment and lack of treatment of each individual in order to assess causes or correlations. This is a complex clinical task and EBM trials do not cover all of such complexity. EBM cannot handle the complexity of judgment needed [65].

Practice is non-linear and too complex for EBM to suffice [46]. We cannot measure every possible variable. There are an unlimited number of things and combinations, which might be the cause(s). To pick one can easily result in the fallacy one cause taken instead of many. The major decision is to determine which factors to study.

For example, Soares argued that "Cigarette smoking affects uterine receptiveness" [66]. They contrasted a group smoking less than 10 cigarettes a day with a group smoking more than 10 cigarettes a day. Why should smoking 9 cigarettes a day be so much different than smoking 10 a day, or 8 rather than 11? Also there was no specification of the kind of cigarette, its length, its contents, whether it was a filter tip, whether the smoker deeply inhaled, how much of the cigarette was smoked, nor were nutritional and other significant controls investigated. Soares claimed that it did not matter because the results in general for the whole of each group were different.

The confounding of factors is a well-recognized and commonly committed error in EBM (Confounding fallacy). For example, cohort studies have shown that there is a dose-response relation of smoking as a risk factor for suicide. It may be the social and mental states, connected with smoking rather than smoking itself, which leads to suicide.

19.12 EBM Has Limited Self-Criticism

There are numerous ways in which EBM is self-critical. One is that editors have set research protocols and standards of research. Unfortunately they are so elaborate that few could adhere to them. If the protocols were adhered to the results, they would be guaranteed to be unacceptable for the reasons given in this chapter. Secondly, meta-analysis, and reviews of evidence are critiques of the studies on a particular topic. These studies typically find that the individual studies are too diverse to compare, and that many of the studies violate EBM methodologies. EBM groups people and conditions together which are very different. That is, it ignores differences. The trials themselves contain numerous differences from one another [67].

Meta-analysis claims to combine different studies as if they were a homogenous single study. They typically are rather vastly different studies combined [68]. Meta-analysis combines quite unlike studies illegitimately. For example, there may be different amounts of calcium intake or the hundreds of other factors, which may be present or absent. Meta-analysis is just a mechanistic calculation of an overall measure of effect and will often be biased. It is not a critical analysis. It is a summary of results. It depends on the orientation and biases of the researcher and so may be biased toward surgery or drugs or non-drug treatment, etc [69]. Goodman states that EBM meta-analysis reviews are problematic: often no conclusion could be arrived at, or too few studies have yet been done on the subject [70]. The third type of self-criticism is contained within each study itself. In the discussion and conclusion there can be many hundreds of qualifications regarding assumptions made, etc. in effect admitting that the experiment is inadequate. In the conclusion section of the reports it is typically stated that more or better studies are needed before any conclusions can be reached. None of the types of self-criticism involve rational philosophical argument, debate, interdisciplinary discussion or in-depth analysis.

19.13 EBM and Psychiatry

There is another sense in which EBM is a "gold standard." That is that hospitals and insurance organizations will often not support treatment or diagnosis unless it can be verified by EBM. EBM becomes a way of cutting costs. Cook holds that EBM is too closely tied to cost-saving strategies [46]. In terms of economics the trials themselves cost millions of dollars for clinically not useful, trivial, obvious, or faulty results. They are typically not cost effective. Goodman states that there are over 23.000 medical journals. Two million articles are published yearly. "Much of this output. . .is not worth the trouble" [71].

According to Nys and Nys, psychiatry is under pressure from an absolutistic and uncritical principle of autonomy and Managed Care economic systems, which force misdirected evidence-based, reductionist classificatory categories on psychiatric practice [72]. In terms of psychiatry, autonomy especially does not work in cases where patients are too ill to accept the treatment that they need (See Chapter 12 for a critique of that concept). One may be too ill to know that one needs help. According to Nys and Nys, economics determines and monitors psychiatric practice [73]. It may be noted that in the U.S. mental problems are not usually covered in individual health policies, and where covered have severe limitations of coverage. On a medical model mental disorders are reduced to somatic, physical disorders and regarded as not being genuine disorders and not covered or not made treatable [74]. This would practically eliminate all psychiatry and therapy. Typically, only when a condition cannot be treated medically does it become a task for the therapist. Here is also another failure of misunderstood "evidence-based medicine." The solution proposed by the authors is to appeal to "philosophy to keep psychiatry sane" [75]. Although Nys and Nys do not clearly present what is meant by philosophy and philosophy of medicine, we may suggest that it is a holistic approach involving the criticism of concepts and methods in medicine and therapy as is discussed and illustrated throughout this book.

19.14 EBM and Human Emotions

The most important theories and analyses of emotion are not to be found in EBM literature. The analysis of emotion is also especially weak in EBM literature. Cochrane Collaboration Library has little to nothing on the cognitive theory of emotion or anger. This is not the place to look for an analysis of the various emotions.

19.15 EBM and Ethics

In terms of EBM, ethics cannot be included as a basis of clinical decision-making. Cook agrees that ethics is outside of EBM [46]. Resnik gives examples of the inappropriate use of statistical methods and a plea for ensuring that ethics is included in experimental design and taught to scientists [76]. But ethics is involved in medical decision-making. Practice is non-linear, complex, and discussion is needed. Again, the aim of medical practice is not just knowledge, but to solve problems and make ethical decisions [46]. Thus according to Goodman, "All clinical decisions have ethical components" [77]. EBM is in this sense often insufficient for clinical diagnosis and decision-making. One must look to philosophical literature and the philosophy of medicine for an analysis of ethics. Less than 5% (1 in 20) probability being due to chance is statistically significant. This is only mathematical and may have no clinical significance [78]. One meaning of significance is ethical merit, value and worth. Statistical or mathematical significance is not ethical significance.

19.16 EBM Depersonalizes

The EBM statistical method dehumanizes the individual and patient. The patient is just a number or statistic to be manipulated mathematically. To represent a person by a number abstracts out all causal knowledge of people. The physician-patient relationship is eliminated. The physician does not actually try to treat the patient, but the ailment.

19.17 EBM Text Reviews

Instead of giving trials, meta-analysis, and reviews of evidence a few text case examples of criticisms of EBM from the literature will be presented.

19.17.1 Evidence-Based Spirituality

Wisneski and Anderson speak of psycho-neuroimmunology, which is supposedly about how our "psyches" (mind, emotions, and spiritual self) interact with the world and our body [79]. EBM is used here to support the supernatural. By "integrative medicine" is meant alternative medicine including religious practices and prayer.

Now we have here the use of EBM by a clinical professor of medicine based on supposed scientific trials and meta-analysis as well. If this is the outcome of EBM, then anything can be the outcome and what is shown is that one can give no credibility to EBM. On an entirely different and better-founded theory we may account for some of the outcome the authors speak of. Positive emotions such as hope and humor are created by our own cognitions and can positively affect the immune system and help healing in many cases. Opposite to the positive emotion based placebo is the "nocebo" which is a negative emotion, which can make the outcome or illness worse. This is an observation needing no EBM for evidence, but there is clinical evidence for it, and EBM "evidence" exists for it as well, for what it is worth. Healing can be altered by our thinking and emotions, but "subtle energy" has nothing to do with it. The authors would have done better to have developed a theory of emotions and given a critique of EBM.

19.17.2 EBM and Practical Medicine

On Peter von Wichert's account, "EBM is uncritical medicine." (translated from German) It ignores the patient and the pathological and physical basis of disease. People do not really know what EBM is but are "infected" with it "contagiously." "Complex studies are a highly artificial milieu and may in no way reflect the reality of the majority of patients" [80]. Reduction to statistics dehumanizes performance and is unscientific. He regards EBM as idealistic pseudo-science.

19.17.3 Evidence-Based Nursing

Nursing EBM may be illustrated by the popular texts of Ignatavicius and Workman *Medical-Surgical Nursing Care: Critical Thinking for Collaborative Care* [81]. They say that the collaborative approach is to include all healthcare practitioners. Then this would be a first step to including the philosophy of medicine. The authors claim to be presenting an EBM approach. The text uses the term "client" in place of "patient," thereby suggesting that healthcare is a business with a money-paying client. Their stress on holistic medicine is nevertheless wholly in a very positive direction. The nurse is asked to be an interdisciplinary, holistic educator and critical thinker. Critical thinking does not go very far without philosophical analysis which the authors indirectly admit when they state, "confusion remains about a precise definition of caring" [82]. (For an extensive critical philosophical analysis of caring, see Chapter 9) The authors also appeal to intuition, which is a pseudo-scientific notion with no clear definition [83]. Intuition is self-righteous mysticism. Basically, the volumes begin by supporting holism in medicine and a claim to support EBM, but throughout forget about critically challenging concepts and practice.

19.17.4 EBM and Logic

Jenicek and Hitchcock in their book on EBM deal with logic and critical thinking in medicine [84]. Medical schools do not require pre-medical courses in philosophy, logic or epistemology [85]. The book argues that the philosophy of medicine should be the basic perspective rather than EBM [86]. Michael O'Donnell is quoted as saying that EBM is "perpetuating other people's mistakes instead of your own" [87].

M Tonelli holds that EBM should use philosophy to go beyond the empirical evidence and investigate the complex variation of clinical judgment from one patient to another [88]. EBM is "educated guesses" [89]. Descriptions and definitions in medicine may be better regarded as being like "fuzzy logic," rather than on an either-or model which excludes shades of gray [90]. "EBM is a trend to reduce quality to quantity" [91].

19.17.5 EBM and Gender Medicine

Rogers attempted to determine if the principles and practice of EBM further women's health. She says that there is unnecessary surgery, ineffective and harmful treatments, which EBM can correct [92]. But it may be noted that this view was held independent of EBM literature. The latter was just used to support a prior view. Cochrane shows that in terms of medical indications restricted use of episiotomies are preferable over routine ones so that the reviews can be used to support the patient's judgments [93]. There should be a wider critique and discussion than is available in Cochran. Here is an appeal to authority. Rogers further notes that EBM ignores the social and political context and is biased against women. In research trials 85% of the participants are male [94]. Women research is basically only in the area of reproduction, not women's health.

Certainly, gender must be a factor to consideration in certain areas of medicine including better representation in management positions. It would however be irrational for women to themselves have a narrow feminist orientation, instead of a consequentialistic, fair one, which supports people in all situations and of both genders. Humanism, for example, is concerned to support the lives and well-being of all people. EBM is correctly seen as a narrow focus rather than involving larger context of women's lives.

19.18 EBM and Rational Medicine

Norcross, Beutler and Levant included in their book articles highly critical of EBM, showing a preference for qualitative over quantitative reasoning in medicine. There is nothing wrong with looking at experiments in medicine, psychiatry, alternative medicine, psychology, but they all require to be critically evaluated and no one can do that in advance of clinical and actual practice. The authors suggest that clinical experience as well as EBM is needed. This would not suffice, however. One would also need the criticism of the philosophy of medicine. The authors see EBM as an attempt to subvert qualitative reasoning by quantitative principles. "Neither human biology nor medicine escapes the fundamental rules of thought, reasoning, and decision making in general" [95]. Nor will a blind reliance on EBM protect an accused physician in the law courts. The authors propose a scientific method: identify the problem, clarify it, gather evidence, assess evidence, make overall judgment, etc. These procedures and terms are too vague to serve as a method. As with EBM, the notion of "evidence" is not clarified. One of the greatest tests showing the in/adequacy of EBM is what it has produced as dependable results. Inevitably the trial ends with the need to redo the trial and mega-analyses or summaries of trials show inadequacies and contradictions throughout. The "discussion" sections and "conclusion" sections of experiments are designed to release the authors from the shortcomings of the data and findings. They, however, do not do so. They are rather apologies for lack of success. The discussions and meta-arguments are examples of the philosophy of medicine and are not themselves evidence-based. It is these non-quantitative aspects of experiments, which are written in ordinary language, which are the beginnings of a philosophy of medicine [96]. EBM has not yet been able to determine if working 48 h shifts and around a 100 h a week is harmful – to the great exploitation of physicians by the economists (See section of book on *Enlightened Versus Normative Management. Ethics Versus Morals* and overwork of physicians).

P-value, probability value, or statistical significance only gives a clue that one might begin to look for a certain cause. "There is no way to somehow score the pros and cons of a causal relation. One must examine all criteria of causation one by one and judge" [97]. The authors advisedly request more specification and tighter definitions from EBM trials than ever given. We need to know which kind of smoking, how long, what were the other variables of one's life at the time. What is meant by cause here, e.g., direct or indirect or mere correlation? etc. What kind of lung cancer? EBM has generated a number of fallacies. The "aggregate bias fallacy" is fallacy of division, according to which what applies to the whole may not apply to the parts. A statistic finding applying to a large group may not apply to the individual, in fact the finding may be the reverse for the individual. Thus, the authors speak of a "statistic of one." The "naming fallacy" applies to schizophrenia, which should not refer to "split personality" or a material cause. It may rather be used to refer to a certain group of observed symptoms [98]. EBM fallacies are based on: inadequately stated hypotheses, conclusive ambiguities, that conclusions often fail to acknowledge negative findings, that conclusions use faulty reasoning, conclusions fail to consider alternative explanations. The last criticism is perhaps one of the most crucial. The authors add that EBM fallacies are due to: "An over-reliance on statistical findings at the expense of broader biological thinking" [99], lack of specification of context or individuals or types of practice/treatment, incomplete and poor arguments, "narrow and uncreative critical thinking," misleading use of statistics regarding confusion of risk with prognosis [99]. To be fair, one may say that philosophers themselves sometimes commit informal logical fallacies and present genuinely weak and irrelevant arguments. We wish here only to speak of the best philosophers and best of EBM. The authors' basic argument is that medical decision-making must go beyond EBM, epidemiology, clinical decision-making, and committing informal logical fallacies. It needs also educated, critical thinking which Pellegrino calls "rational medicine," but the traditional study is more properly called the philosophy of medicine. That critical thinking and reason and other terms have been used instead of the philosophy of medicine shows how little philosophy is known and how little it is taught. It needs also to be shown which part and schools of philosophy are most useful for medical practice. The authors themselves do, however, mention, that "domains of philosophy are increasingly present in all medical domains of evidence" [100].

19.19 EBM, Psychosomatics and Philosophy

Haldeman [101] gives the following account of back pain. He asks why we do not know the cause of back pain. The lifetime or long-term prevalence of back pain ranges between 60–90% regardless of technological advances. From 1971 to 1986, people disabled from back pain increased fourteen times faster than the population. "Pathology can exist without symptoms. . .symptoms can occur without pathology" [101]. "It is not possible to look at pathology and determine with any confidence the symptoms a patient may be suffering" [102]. Nor can one know from the pain what the physiological problem is. The incidence of back pain seems largely to depend on one's general health [103]. Obesity, lack of exercise, and smoking have been found to be causes of back pain. Exercise actually strengthens disks and the vertebrae [103]. These assertions suggest that back pain is largely determined by one's lifestyle, including stress. Most patients can spontaneously recover from even acute trauma to the spine [104]. Workers who like their jobs [one could extend this to having any positive emotions] are less likely to experience back pain. Negative psychosocial situations [and in our terms, negative emotions] are a good predictor of negative surgical outcomes [102]. Back pain then, does not seem to be reducible to the medical model of pathology. All of this argues for analysis of the philosophy of pain, and holistic philosophical medical treatment. One's lifestyle including one's goals and meaning of life have a direct bearing on the experience of pain and one's physiological conditions. We are physically molded by our philosophy.

Another example from the author's experience as an obstetrician is preterm labor. It is mostly ascribed to cervical infection. This is mostly the case, but the question is why the woman is inclined to infections and preterm labor. Is it a difficult relationship, fear of birth-giving, other psychosocial problems? This cannot easily be answered by a mechanical medical model, rather by a psychosomatic approach.

19.20 EBM and the Problem of the Placebo

The positive belief in EBM has a placebo effect.
Thank you for the placebos, doctor. I feel better already.

The EBM trials must be placebo controlled. A main reason to use placebos is for statistical purposes to show that the correlations themselves are significant and make a difference. Some drugs have often a 25% placebo effect so it is thought that any new medication should have better results than the 25% placebo effect. If the placebo effect alone is, as usually claimed, around 25%, one could recommend pills on that basis alone assuming they do no harm otherwise. For example, "In trials of therapies for depression it is common for a placebo to result in positive benefits in 25% or more of patients...and in treatments for hot flushes placebos may produce benefits in up to 50% of patients" [105].

The FDA has, since the late 1970s, required that all drugs be tested against placebos. But placebo is not defined and may include many different sorts of things. A placebo could be almost anything. Having no treatment and no pills can also be seen as healthful and as placebos. It is not clear what a placebo is. It is a vague abstraction. Thus, the alleged scientific status of EBM is undermined.

What is a placebo? Etymology is from Latin "I shall please." This suggests that one thinks in advance that something, such as a medicine will be good for him. Placebo is sometimes a mere ethical term, meaning good. It may thus serve to persuade, or induce an emotive attitude. "Take this. It will be good for you." Szawarski suggests that a placebo depends on an attitude, and is basically a "persuasive definition", a term used in the ethical theory of emotivism of Charles Stevenson [106]. An analysis of ethical terms and their connection to cognitive-emotive theory would have helped the author make the connection between ethics, persuasion and emotion. But certainly one can view placebo as an ethical term.

The placebo is typically defined as being inert, having no specific activity for the condition being treated. If so, by definition, it can have no placebo effect. It is false to define placebos as being not relevant to the condition being treated. This definition is based on the medical model that only physical things have an effect. Placebos are not inert, but have psychological and physical consequences. This is the definition of emotion of RET (See Chapter 7). There is no placebo or nocebo as such, but they are merely assessments causing bodily feelings. Placebos and nocebos are emotions. Thus, placebos become hope, comfort, etc, or as nocebo, fear, disappointment, sadness, etc. The term 'nocebo' became popular in the 1990s. A nocebo (Latin for "I will harm") is something that should be ineffective, but which causes symptoms of ill health. A nocebo effect is an ill effect caused by the suggestion or belief that something is harmful. The placebo is not one experience, but a changing experience as one reassesses and reevaluates the situation. In one sense, there is no such thing as a single placebo or nocebo. It is constantly changing. Instead of placebo-controlled experiments we have belief, and assessment controlled experiments where the latter are highly uncontrolled and variable.

In one sense, the placebo is not the same as non-treatment because there is still the psychological and emotional state of the patient whether given a placebo or not. Thompson found that if the physician says, "I do not know what ailment you have," or "Pills will not help," the improvement was 36–42%. If the physician says "You will be better soon," or "Pills will help," the improvement was 64% – dependent on the kind of illness of course, typically psychosomatic in the specific definition of psychosomatics [107]. We may note, however, that there is a divergence in effect and that the improvement favors the more positive evaluation of the physician. The placebo effect may be regarded as the psychosomatic effect. "Placebo effects are part of every treatment" [108]. But so are nocebo effects.

The physician him/herself is a placebo or nocebo. What a placebo is depends upon the meaning for us. Fifty-two percent of the colitis patients treated with placebo in 11 different trials reported feeling better – and 50% of the inflamed intestines actually looked better when assessed with a sigmoidoscope [109]. With a positive meaning it is a placebo, with negative meaning it is a nocebo [110]. Any treatment, diagnosis, surgery, medication, medical examination, etc. may be regarded as having a placebo effect. Angina pectoris responds to placebos [111]. "Inert surgery [in special cases] works nearly as well as real surgery" [112]. Moerman does not explain ethical meaning or assessments, but only says the perceived meaning of the situation determines if a treatment is a placebo or not. No theory of meaning is given. Writing and narratives are seen as healing methods [113]. For him the world is a metaphorical construction.

If placebo is *defined* as something that will cause a beneficial reaction in the patient, it is simply defined as being successful. Thus, a placebo can never, by definition, do harm. It becomes merely a value term. By thinking that something will be good for one it causes certain positive cognitions that may have physiologically beneficial effects. This is not surprising because it happens with the other positive emotions as well, such as love, caring, humor, hope, joy, enjoyment, aesthetic experiences. Jerome Groopman in *The Anatomy of Hope* notes that, a change of mind-set can alter neurochemistry both in a laboratory setting and in the clinic. He found relief himself from persistent back pain in the hope inspired by an empathetic fellow physician. Belief and expectation – the key elements of hope – can block pain by releasing the brain's [pain killing] endorphins and enkephalins, mimicking the effects of morphine [114].

There are as many types of placebo as there are positive emotions. All of these can have positive effects on the body. It is a paradigm of psychosomatic medicine. Thus, the notion of the placebo may be clarified by being analyzed as an emotion. We may speak of a "placebo emotion." An emotion is a positive evaluation, which causes a bodily feeling both of which together we call an emotion. Need a placebo always be positive or neutral? Cannot a placebo cause harm, make matters worse?

Negative assessments produce negative emotions that can have a harmful effect on the body. These may be called nocebos in place of placebos. If one thinks something will be harmful it may in fact do harm. None of this is very specific. There may be some benefit or other from the placebo and some harm or other from the nocebo. Some physical functions can be affected by cognition, others cannot. Worry, stress and depression may undermine the immune system, and positive emotions benefit it. Having a positive emotion may possibly not prevent a specific disease directly. Thus, we may think a placebo prevents cancer when it actually may just improve the immune system which delays or blocks cancer for a while. There is a double bind with placebo as a positive emotion because one would tend to have a negative emotion because of having a disease, but to get better one has to have a positive emotion. That is, one must accept a disease one cannot easily accept or it will make it worse. You have to like what you do not like.

The placebo may be falsely thought of as a mental thing as opposed to a physical thing. The placebo as spoken of in EBM commits the Mentalism Fallacy. There is no EBM evidence for mind or thought or any "mental state." (See full argument in the Chapter 18)

The popular view is that the left brain hemisphere controls: the verbal, conscious activity, the logical and linear, and asks how and why. The right hemisphere controls: the nonverbal, art, the intuitive, subconscious, holistic, and nonlinear. The claim, that this distinction is scientific, is unfounded and simplistic [114].

These terms are undefined and unscientific abstractionisms, e.g., "logical" seen as symbolic logic or as Aristotelian logic are unacceptable as they are actually irrational systems of thought (See Chapter 18). Also "intuition" is not a method of knowing, but a pseudo-concept like mystical thinking. Subconscious and conscious are in need of definition. Thus, the distinction between right and left parts of the brain was mis-described so that no differences could be made out. Additionally, the verbal-nonverbal difference was shown to be false and any speed function differences found were less than 50–60 milliseconds [115]. Both hemispheres, not just the left, were found to be able to handle linguistic skills. It could have been known beforehand because the verbal is connected with all of our thinking, perceptual and artistic abilities. Motoric speech function, not cognitive function, is however found mainly in the left hemisphere and the same thing is found in song birds [116]. We falsely think that the perceptual is totally separated from the linguistic, that art can just deal with nonlinguistic activity. Musical ability requires both parts of the brain, not just the right as was earlier claimed. Better or holistic thinking is not localized in left or right brain. The author likens the right/left brain mythologies to the pseudo science of phrenology, which localized certain traits as friendship to certain parts of the brain. Creativity is not localized in either hemisphere. The right/left brain mythology was applied to management with claims that the right hemisphere was better at management, advertising, marketing, etc. Evaluation was claimed to be left brain activity. $2000 half-day seminars were given promoting the left/right brain myths to managers [117]. Here is another example of the use of the experimental and EBM method, but with conceptual confusion. We must first know what thought, intuition, creativity, subconscious, etc. are before we can even look for them in neural activity. Hines concludes that in any case, even if we know an area were to relate to "friendship," it would not have any application to help us be better managers. There are ways to be a better manager and they do not involve neural mapping or electroencephalographic results. Thinking is not just alpha waves. Hines concludes that right/left brain findings and neuroscience have no relevance for management theory and practice [118].

Some deny there is a placebo effect at all. Brody thinks placebo cannot be defined because he thinks there must be literal and absolute definitions of words. He objects to calling them "inert" or "nonspecific," but then terms them "dummy" treatments [119]. He nevertheless does give a definition: "The placebo response is a change in the patient's health or bodily state that is attributable to the symbolic impact of medical treatment or the treatment setting" [119]. But symbol impact in non-medical settings can also have a placebo effect. It would have been better to say that a placebo is the positive assessments and emotions one has, and these do have a bodily effect. There is no placebo as such. This is consistent with Brody's recommendation that for the placebo to work the healthcare worker must provide positive reinforcement and care. The view presented that placebos work better with the acquiescent personality type, supports the view that placebos involve positive emotions. The basic conclusion can be extended beyond that of the author to the view that positive and holistic treatment of the patient can help healing by encouraging positive emotions. It is this, which is the placebo effect. A placebo is not a substance, but rather perceived care and support. Patients are said to need to feel in control, but this is another of many possible assessments, which can create positive emotions. One could conceive of someone who would develop negative emotions (and so a nocebo) as a result of having to have control. Brody refers to his view as a meaning model, but it is not just meaning alone which produces positive emotion. The problem is that he has no theory of emotions which could help him to clarify the nature of the placebo. On the other hand, the suggestion that patients write out narratives of their views is useful because it is an ordinary language approach to understanding what patients are actually thinking. However, the narrative would have to be analyzed in terms of philosophical analysis including ethics, emotions, and critical thinking.

Szawarski in his discussion of the placebo mentions and holds the Cartesian mind-body separation. As this view is rejected as being a mentalistic fallacy, and as most hold this view, the concept of placebo will be falsely based. One will falsely think of placebo as affecting the "mind," although no mind exists. Szawarski wrote, "What matters is a complex whole, composed of interrelated factors that work primarily on the patient's mind" [120]. He speaks of "the active role of the mind in the healing process" [121]. There is no mind. EBM based on the notion of mind is unscientific.

One assessment is that by believing one will become better one actually will. Here one may suggest that belief does not make something true. It may here not be the belief at all but the fact that it is a positive emotion. A belief may be held in a positive and negative sense. If it is held in a negative sense it may cause harm to one's body. The belief that one will become cured may or may not be physically beneficial. Belief in the sense of hypnosis, self-hypnosis, or suggestion may help one to do things one could not otherwise do, e.g., stop smoking. Thus, if a placebo is given in a trial to quit smoking a placebo may be effective. This interpretation of placebo would require an in-depth analysis of hypnosis that is not given in the usual EBM methods.

Placebos are typically thought to be objects or substances in themselves, like aspirin or sugar. A placebo is frequently a pill that seems to be a beneficial medication, but is not. It contains a harmless substance. What is tested here is unknown because it is not known what the subject's assessments are. Szawarski correctly points out that the total and detailed context must be specified in which each drug is taken, or treatment given. He comes close to presenting the use theory of meaning according to which the meaning of the drug/placebo is its use in a language-game. "What matters is what the drug looks like, its brand-name, who prescribes it, what are the properties of the patient, what kind of mental, social, economic situation he is in, etc" [122]. Consider the following assessments: "I hate medicine, but I agreed to take this, so I will." "It looks like a medicine and I think it will cure me." "It looks like medicine which might cure me." "I doubt that this will work." "You want me to take that pill. O.K." "This is without thinking about what it might do." "I do not agree to take a pill I know nothing about." "I know this pill might be a placebo or it might be the real medication to be tested. I am not sure which it is." "Thank you for the placebos, doctor. I feel better already." With assessment possibilities like these we cannot expect to have much gain or positive emotion from taking a placebo. Instead of a placebo we could have no treatment or therapy as opposed to actual full treatment or therapy. Giving the actual medication may also have diverse placebo effects. That is, the medication and treatment and just talking to the doctor may have placebo (or nocebo) effects. Just being in the doctor's office or using any technology may have a placebo or nocebo effect. Batavia speaks of the "*Hawthorne Effect*" as improved performance because one knows one is part of a study. In addition to the "placebo effect" one may speak of a "medication effect" or "treatment effect," or an unlimited number of other alleged effects. The distinction between a "real" pill or treatment and a "dummy" one dissolves: Doctor to patient: "You see, the placebo *is* a real treatment." Thus, both treatment and non-treatment can have placebo/nocebo effects. It only depends on the assessment of the patient at the time [123]. Szawarski states, "Anything and everything might be a placebo" [124]. Using the metaphorical technique of self-reflexivity, one could say that the positive belief in EBM has a placebo effect, and the positive belief in placebos can have a placebo effect.

In double blind EBM experiments neither experimenter nor subject knows which the placebo is. Now if the experimenter does not even know what placebos mean or are in the experiment and how many there really are it is experimental blindness meaning ironically that EBM trials are not just double but triple-blind. We never have just a placebo, but only a certain kind of placebo for a particular person for something. How anyone views a placebo determines what a placebo means. We need the specific assessments. Suppose we substitute for "placebo," "cognition". We would not know which cognitions were had. The same would be true of "belief." Which beliefs would they be? There is never a placebo, cognition, or belief as such, only certain ones and they need to be specified. On the other hand, much depends upon one's positive or negative assessment. A negative assessment can physiologically counteract the desired effect.

Because of the lack of definition and vagueness of the term, "placebo," many statements made in the literature are themselves problematic:

1. The EBM trials must be placebo controlled.
2. People will improve just by taking a placebo alone.
3. Placebo control makes trials more scientific.
4. "The placebo effect" as if there is only one effect, when there are many different effects.
5. "The placebo effect" is spoken of as if one knows what that is, when one does not.
6. There are "real" treatments or medications. But these may have a placebo effect.
7. There are "dummy" substitutes of treatments or medications (But these may have a "real" effect).

Gary Greenberg states that any given antidepressant has only about a 50% chance of working with any given person [125]. More than half of the 47 trials used by the Food and Drug Administration to approve the six leading antidepressants on the market, revealed the drugs failed to outperform sugar pills and in the trials that were successful, the advantage of drugs over placebo was slight. Billions of dollars may be spent on medicines which do not work and which have serious side effects. The problem with this statement is that it is unclear what placebo means. We are also not sure what is meant by saying that the drugs "work." If they only deal with the symptoms they do not work. Typically, between 35 and 45% of people given placebos supposedly improve. It is not clear here what is meant by "improve." If a candidate drug outperforms a placebo in two independent studies, and if it does so without negative side effects, the FDA will approve it for use. The author points out that the FDA does not consider the relative advantage that new drugs show over a "placebo." They only consider minimal statistical significance, regardless of whether it is 5 or 50 or 500% more effective than an inert pill. Some believe that the effectiveness of Prozac and similar drugs may be attributed almost entirely to the placebo effect. Again this statement has no meaning without definition of placebo. One can imagine a physician saying, "The pills do not work, but they have a great placebo effect."

Hróbjartsson and Gøtzsche found that clinical trials regarding 40 different clinical conditions, e.g., smoking, obesity, anxiety, show that they have little effect as treatments (except for continuous subjective outcomes and for pain treatment) [126]. The placebo is thought to be powerless. It supposedly only has a use in clinical trials as comparisons. It has been shown above that it does not have a scientific use in EBM because it is too poorly defined. Also, if the placebo is an emotion, it can have significant power to alter bodily functions. This view also contrasts with the above view that the alleged "real" treatments can have little effect.

The ethics of using placebos is raised. Of two groups is it ethical to give one beneficial treatment while the other has only a placebo? Trials are often stopped because of the adverse effects of the placebo group or the treated group. If a drug is truly beneficial is it ethical to deny it to the placebo group? Physicians as researchers must avoid using placebos when currently available drugs are known to save lives – such as in tests of drugs to prevent heart failure. A Declaration of Helsinki amendment calls for new drugs to be tested only against "the best current" treatment [127]. This would not be a placebo. For example in a vaccination trial a saline injection could be used but may not be ethically acceptable.

Is it ethical to deceive one group by telling them that they are being given a medicine or treatment when they are not? If they are being lied to they may claim medical harm. If they are told the truth that they are only taking a placebo it may not be effective. If they are told they may or may not be taking a placebo they may be apprehensive or feel that they are merely being experimented on without concern for their welfare. They may fear the risk of a new treatment or fear neglecting their health by non-treatment. This may not induce positive emotions. Also the term "placebo" may suggest pacification, like drugging patients to keep them quiet, or pretend to cure while merely treating the symptoms. Much of medicine and therapy treats the symptoms rather than the causes of illness.

The AMA states that placebo controls are needed for drug testing, but they must obtain informed consent from subjects or patients and must inform of the level of risk [128]. Each patient must also be made aware that they can terminate their participation in a study at any time. Protocols that involve conditions causing death or irreversible damage cannot ethically employ a placebo control if alternative treatment would prevent or slow the illness progression. Researchers should minimize the amount of time patients are given a placebo. Researchers must terminate the study because of either positive or negative results, thus protecting patients from remaining on placebo unnecessarily.

Miller and Brody, however, support placebo trials if the trials are needed and no harm is caused, which is in accordance with the 2002 World Medical Association restatement of the Declaration of Helsinki [127]. Placebos should not be used wherever an effective optimal treatment is already available. A good treatment can only be tested against what might be a better treatment, not a placebo. The authors however argue that voluntary-based trials should not be considered patient treatment and so placebos should be allowed. One might propose that it is better to compare two beneficial treatments rather than a placebo. A consequentialist pragmatic view (pragmatism) would consider the details of the total holistic and humanistic context in determining if substances, treatments or placebos should or should not be used. One could imagine a treatment being given by the wrong person at the wrong time. Certainly some therapy does not depend upon reality-based narratives, but only upon the client accepting the narrative, placebo-like, that is arrived at which is found to be useful. This ought to be no surprise as every model in science is merely a perspective or paradigm that constitutes how we view the world, not a description of some absolute reality.

Biller-Andorno asks if the use of the placebo effect in clinical medicine is an ethical blunder or ethical imperative? [129] The placebo trial should supposedly not be used when there are better treatments available. Its use is said to be deceitful and paternalistic without scientific rationale. The author opposes the use of deceit in the use of a placebo [130]. On the other hand, for depression patients 30% responded favorably to placebos, and for pain, placebos under certain conditions had the same result as treatment. Some drugs have often a 25% placebo effect. It is pointed out that the term placebo is seldom defined and if defined is not done so adequately, e.g., as the symbolic effect of the healer [131]. It is an unscientific term even when used in EBM trials. If placebo is defined as symbolic or sham procedures that produce positive physical and/or psychological changes, [131, 132] then we may add that any positive emotion would be a placebo. The author suggests the phrase "placebo enhanced therapy." The placebo may be a conditioned response. It may often be the case that we advise our friends in a therapeutic way without anyone suggesting that therapy is being done. Is not modeling behavior another placebo like influence? So now the question becomes what are the various influences, which may contribute to the therapeutic effect. The author's suggestion to ask the patient's permission to use a placebo may defeat the point of the placebo. Also there may well be cases where it is justified, e.g., where the patient may be lying or malingering or where one cannot tell if the symptom is psychological or physical, as with back pain, and a non-invasive test is needed. In therapy, "paradoxical intention" is used, advising the patient to do what the patient fears most. The patient may well have to be challenged in such ways. The power of suggestion is used in everyday persuasiveness, education, poetry and aesthetics. The author's point is wholly supported insomuch that if one uses placebo in the therapeutic setting it should be well defined and be used for the primary purpose of the care of the patient. And certainly one cannot evaluate the ethics of using placebos until the concept is well defined. However, in the experimental setting of research this would be too restrictive [128]. It may also be noted that, in effect, every use of even FDA approved drugs is to a large extent experimental. This is why they have adverse effect drug reporting agencies in various countries. In addition, as medicine and therapy often treat only the symptoms and not the proximal and distant causes, they are performing something like placebo medicine, not adequate and holistic medicine.

It is interesting to note that placebo groups often report side effects. "Well doctor, the placebo you gave me worked well, but I am suffering a lot of side effects." This can be explained by means of the cognitive theory of emotions. On the view that the placebo is a complex of emotions and cognitions we can see how the placebo can have side effects. Negative emotions produce negative bodily reactions.

Conclusion. Because the placebo may be thought of an emotion, on the cognitive theory of emotions, and because positive emotions produce positive physical states and negative emotions produce negative ones, we may specify what kinds of placebos are necessary to create the best emotions. It would seem that cognitive therapy in the form of rational critical thinking (speaking), philosophical counseling, may be thought of as effective placebo enhanced therapy. Philosophy may be thought of as a form of therapy [133]. The best and most effective assessments which can produce a placebo effect with the most hope of success are those which are based on rationality, trust and knowledge which induce positive emotions, rather than those based on irrationality, superstition, supernaturalism, rituals, rites, deceptions, or mechanical or undefined objects. Our emotions, lifestyle, and ability to think philosophically, ethically, humanistically and critically have an effect on our bodily state and health. They can genuinely improve the immune system and provide harmony and structure for our physical health. Our physical health is not isolated and separate from our thinking and emotions.

19.21 "Philosophy of Medicine" – Based Medicine Instead of Only "Evidence" – Based Medicine

EBM presents itself as reactionary, reductionary and exclusionary in many ways. It often reduces and excludes reasoning, discussion, critical thinking (speaking) and argumentation to a narrow largely statistical base and thus creates an unacceptable scientific method for clinical practice. By insufficiently focusing on critical thinking (speaking) and reasoning it does not allow physicians and clinicians to make their own rational judgments. While claiming to stress the importance of examining the medical literature it actually excludes and restricts most of the relevant critical literature with an authoritarian bias in favor of its own narrow method. It treats medicine as if it is separate from all other rational inquiry and disciplines. Cook states that not all decisions and evidence are reducible to EBM [46]. On the contrary, medicine is not isolated from the knowledge to be found in other disciplines. EBM is not "the whole of medicine" as some seem to think. One may think of EBM as being like sending the patient to the wrong specialist. One has a kidney problem and is sent to a statistician, the heart attack victim is recommended to a mathematician, one in need of an organ to an economist. "EBM is far from being the only player on the field" [134]. In this book it is argued that EBM should also include the broadest philosophical and higher level ethical and critical sources in order to be able to critically examine its own shortcomings. The result of trying to isolate medicine with its own EBM "gold standard" has led, as has been shown, to just what we imagine it would lead to: reduction of complex contextual clinical practice to only statistical findings which in addition often are problematically produced and not further questioned.

By contrast, the philosophy of medicine is not exclusionary and reductive, but includes the most comprehensive and most detailed, in-depth perspectives. One can do a philosophical critique of EBM, but not an EBM critique of philosophy. The Philosophy of Medicine cannot be reduced to EBM. The philosophy of medicine and philosophy of science must be more comprehensive and careful and adequate than either medicine or science because philosophy is the examination of concepts and methods in the various disciplines. Philosophers cannot merely use medical and scientific terms without first analyzing and carefully defining such terms. It must be made clear that EBM is neither critical thinking (speaking) nor an analysis of concepts and methods. EBM meta-analysis is really a summary, not a higher level of reflective criticism. There is supposedly self-criticism of EBM trials by experts. Which kind of experts? Do they also have a background in critical thinking (speaking), ethics, emotion and philosophy? The same is true of clinical "experts." This book illustrates such clarifications. This chapter illustrates the clarification of the method, EBM, though it does not claim to be the final word on the analysis. It is an attempt to begin to put the concept of EBM into question, to engage in a critical dialogue.

"The framework of clinical medicine is always incomplete or faulty" [135]. One reason for this is that we have limited knowledge in medicine. Medical classification systems are often inconsistent and arbitrary [136]. Diagnoses are often haphazard, determined by which illness happens to be going around, or on the basis of too few symptoms or superficial without sufficient checking of knowledge or research. Genuine causes are often not pursued to any extent [137]. "Establishing a diagnosis prior to any treatment is often a clinical luxury" [138]. Often one only knows the diagnosis by the response to the treatment [139]. The diagnostician is regarded as a detective or even as a "gambler." The kinds of probabilities are often controversial and unscientific, e.g., Bayes' Theorem versus regression equations. Probabilistic approaches are not realistic to follow [140]. One might add that there are no fixed rules for good reasoning although there are logical fallacies one must be acquainted with in order to avoid them. One must be trained in critical and philosophical thinking, ethics and emotions in order to make sound decisions. This is the study of the philosophy of medicine. The authors think the "clinicians remain the best diagnostic instrument", [141] and it is the case that the physician's reason must be relied on rather than set principles or statistics, but the physician must also be trained to use research results adequately and critically evaluated. But the individual physician might also be a problem and often is, indeed. EBM aims at making individual physician`s decision-making intersubjectively comparable, transparent, "reproducable". The "state of the art" (a term to be defined itself again and again) should be considered in contextual individual decision-making and in addition, the practicing physician should assume that the treatment is wrong, it at least needs correction during the course of the disease unless it can be proven otherwise. Broad spectrum drugs are used because the cause cannot be determined. Diagnoses are given without sufficient basis. This is also true of psychotherapy. These authors are misled in saying that the physician should give the most likely diagnosis [142]. Rather, no, or specified as hypothesis, diagnosis should be given if one does not know it. For sound decisions, one needs critical thinking (speaking), education, clinical experience, sufficient medical training and information, access to the relevant literature (Evidence-Based Medicine), reference books and software, test results, constant patient observation and interaction with the patient and patient reports, expert opinion, knowledge of similar cases, etc. In sum, the reason and decision of the working physician must be relied on and cannot in the end be replaced by mathematics, statistics, fixed rules or principles or on fixed ethical systems only helped by it to a critically examined degree.

Summary: EBM attempts "objectivity" or what is held for it in its own inherent reasoning. EBM aims at ruling out subjective inadequate and ignorant decision-making of the individual clinician trying to provide basic "general" standards.

EBM involves basically problems such as: false causes, insufficient data, incorrect data, undefined data, conceptual confusion, illogic, mistakes due to lack of knowledge of ethics and emotion, discussion and communication problems, false or excessive unnecessary assumptions, failure to note changing and contextual factors, unknown effectiveness of treatment, uncertain diagnosis, treatment, individual response. What is seen is that to be a scientific method it must involve our full and comprehensive critical reasoning, which is found in philosophy and especially in the philosophy of medicine for the problems involved. Clinical experience and critical philosophical reasoning remains a base of EBM application which still is in need of critical evaluation and creative management of the individual physician. EBM should include the philosophy of medicine and science.

References

1. Wittgenstein, L. 1958. *Philosophical investigations*, 3rd edn, 232e. New York: Macmillan.
2. Biesalski, H., and Grimm, P. 2005. *Pocket atlas of nutrition*. Junkermann, S. tr, 2nd edn, 210. Stuttgart: Thieme.
3. Straus, S., Richardson, W., Glasziou, P., and Haynes, R. 2005. *Evidence-based-medicine*, 3rd edn. New York: Elsevier.
4. Sackett, D.L., Rosenberg, W.M., Muir Gray, J.A., Haynes, R.B., and Richardson, W.S. 1996. Evidence based medicine: What it is and what it isn't. *BMJ* 312:71–72.
5. Charlton, B. 1996. Statistical malpractice. *Journal of the Royal College of Physicians of London* 30:112–114; esp. 113.
6. wwww.nhlbi.nih.gov/about/framingham/, www.channing.harvard.edu/nhs/index.html, www.ucl.ac.uk/whitehallII/Cohortprofile.pdf.
7. Wang, C. 1993. *Sense and nonsense of statistical inference: Controversy, misuse and subtlety*, 26. Dekker, New York: Marcel.
8. Kant, I. 1785/1968/1996. *Grundlegung zur Metaphysik der Sitten*, ed. Weischedel, W., vol 5 and 6, Frankfurt am Main, suhrkamp TB Wissenschaft: De mundi sensibilis atque intelligi bilis forma et principiis Kant, I. 1785/1998. *Groundwork of the metaphysics of morals*, ed. Gregor, M. New York: Cambridge University Press; Kant, I. 1788/1997. *Critique of practical reason*. Gregor, M. tr, New York: Cambridge University Press.
9. Kuhn, T.S. 1962/1970. *The structure of scientific revolutions*. Chicago: University of Chicago Press.
10. Black, D. 1998. The limits of evidence. *Journal of the Royal College of Physicians of London* 32:23–26; esp. 23.
11. Wang, C. 1993. *Sense and nonsense of statistical inference: Controversy, misuse and subtlety*, 180. New York: Marcel Dekker.
12. Charlton, B. 1996. Statistical malpractice. *Journal of the Royal College of Physicians of London* 30:112–114; esp.113.
13. Good, P., and Hardin, J. 2003. *Common errors in statistics and how to avoid them*, 3. New York: Hoboken.
14. Norcross, J., Beutler, L., and Levant, R. eds 2006. *Evidence-based practice in medical health: Debate and dialogue on the fundamental questions*, 7. Washington, DC: American Psychological Association.
15. Hopkins, B. 1998. A critique of the usefulness of inferential statistics in applied behavior analysis. *The Behavior Analyst* 21:125–137.
16. see Bailar, J. 1986. Science, statistics and deception. *Annals of Internal Medicine* 104:45–61.
17. Greenhalgh, T. 1999. Narrative based medicine: Narrative based medicine in an evidence based world. *BMJ* 318:323–325; esp. 323.

18. Greenhalgh, T. 1999. Narrative based medicine: Narrative based medicine in an evidence based world. *BMJ* 318:323–325; esp. 324.
19. Montgomery, K. 2006. *How doctors think: Clinical judgment and the practice of medicine*, 14. Oxford: Oxford University Press.
20. Montgomery, K. 2006. *How doctors think: Clinical judgment and the practice of medicine*, 6. Oxford: Oxford University Press.
21. Katerndahl, D. 2005. Is your practice really so predictable? Nonlinearity principles in family medicine. *Journal of Family Practice* 54:970–977; esp. 971.
22. Montgomery, K. 2006. *How doctors think: Clinical judgment and the practice of medicine*, 44. Oxford: Oxford University Press.
23. Albert, D., Munson, R., and Resnik, M. 1988. *Reasoning in medicine: An introduction of clinical inference*, 65. Baltimore, MD: John Hopkins University.
24. Albert, D., Munson, R., and Resnik, M. 1988. *Reasoning in medicine: An introduction of clinical inference*, 69. Baltimore, MD: John Hopkins University.
25. Wang, C. 1993. *Sense and nonsense of statistical inference*, 40*: Controversy, misuse and subtlety*. New York: Marcel Dekker.
26. Thompson, M. 1994. Perspective. Statistics: The curse of the analytical class. *Analyst* 10:119, 127n.
27. Charlton, B. 1996. Statistical malpractice. *Journal of the Royal College of Physicians of London* 30:112–114; esp.113.
28. Crossen, C. 1994. *Tainted truth: The manipulation of fact in America*. New York: Simon and Schuster.
29. Black, D. 1998. The limits of evidence. *Journal of the Royal College of Physicians of London* 32:23–26; esp. 23–24.
30. Hand, D. 1994. Deconstructing statistical questions. *Journal of the Royal Statistical Society, A., Part 3*, 157:317–356; esp. 317.
31. Michelson, J. 2004. Critique of (im)pure reason: Evidence-based medicine and common sense. *Journal of Evaluation in Clinical Practice* 10:156–157.
32. Berghella, V., Baxter, J., and Chauhan, P. 2005. Evidence-based surgery for cesarean delivery. American *Journal of Obstetrics and Gynecology* 193:1607–1617; esp. 1608–1609.
33. Leipzig, R. 2005. An evidence-based clinical update 2004–2005. *American Geriatrics Society Meeting* 1–27.
34. Guttman, C. 2005. Dermatology. *Times* 26:16.
35. Thomas, K., Dayton, C., and Peterson, M. 1999. Evaluation of Internet-Based Clinical Decision Support Systems. *Journal of Medical Internet Research* 1: e6 (URL: http://www.jmir.org/1999/2/e6/).
36. Goodman, K. 2003. *Ethics and evidence-based medicine*, 98. Cambridge: Cambridge University Press.
37. Oliveri, R. 2004. Hospital doctors` self-related skills in and use of evidence-based medicine – a questionnaire survey. *Journal of Evaluation in Clinical Practice* 10:219–226.
38. McAlister, F.A., Graham, I., Karr, G.W., and Laupacis, A. 1999. *Journal of General Internal Medicine* 14:236–243 .
39. Batavia, M. 2001. *Clinical research for health professionals*, 76. Boston: Butterworth-Heinermann.
40. Shahar, E. 1997. A Popperian perspective of the term "Evidence-Based Medicine". *Journal of Evaluation in Clinical Practice* 3:109–116; esp.109.
41. Black, D. 1998. The limits of evidence. *Journal of the Royal College of Physicians of London* 32:23–26; esp. 24.
42. Wang, C. 1993. *Sense and nonsense of statistical inference: Controversy, misuse and subtlety*, 33. New York: Marcel Dekker.
43. Worrall, J. 2002. What is evidence-based-medicine? *Philosophy of Medicine* 69:316–330.
44. Batavia, M. 2001. *Clinical research for health professionals*, 64. Boston: Butterworth-Heinermann.

45. Reed, G. 2006. Clinical experience. In *Evidence-based practice in medical health: Debate and dialogue on the fundamental questions*, eds. Norcross, J., Beutler, L., and Levant, R., 13–23; esp. 22, 23 Washington, DC: American Psychological Association.
46. Cook, M. 2004. Evidence-based medicine and experience-based practice – clash or consensus? *Medicine and Law* 23:735–743.
47. Batavia, M. 2001. *Clinical research for health professionals*, 77. Boston: Butterworth-Heinermann.
48. Murphy, E. 1997. *The logic of medicine*, 2nd edn, 7. Baltimore, MD: John Hopkins University Press.
49. Black, D. 1998. The limits of evidence. *Journal of the Royal College of Physicians of London* 32:23–26.
50. Lang, R., and Hensrud, D. Eds. 2004. *Clinical preventative medicine*, 23. Chicago, IL: AMA.
51. Lang, R., and Hensrud, D. Eds. 2004. *Clinical preventative medicine*, 25. Chicago, IL: AMA.
52. Lang, R., and Hensrud, D. Eds. 2004. *Clinical preventative medicine*, 60. Chicago, IL: AMA.
53. Lang, R., and Hensrud, D. Eds. 2004. *Clinical preventative medicine*, 60 ff. Chicago, IL: AMA.
54. Murphy, E. 1997. *The logic of medicine*, 2nd edn, 81. Baltimore, MD: John Hopkins University Press.
55. Reed, G. 2006. Clinical experience. In *Evidence-based practice in medical health: Debate and dialogue on the fundamental questions*, eds. Norcross, J., Beutler, L., and Levant, R., 9 Washington, DC: American Psychological Association.
56. Goodman, K. 2003. *Ethics and evidence-based medicine*, 3. Cambridge: Cambridge University Press.
57. Hope, T., Savulescu, J., and Hendrick, J. 2003. *Medical ethics and law: The core curriculum*. London:: Churchill Livingstone Elsevier. Preface.
58. Stiles, W. 2006. "Case Studies." In *Evidence-based practice in medical health: Debate and dialogue on the fundamental questions*. eds. Norcross, J., Beutler, L., and Levant, R., 57–73:63 Washington, DC: American Psychological Association.
59. Varney, H. in: Lang, R., and Hensrud, D. eds. 2004. *Clinical preventative medicine*, Ch. 2. Chicago, IL: AMA.
60. Batavia, M. 2001. *Clinical research for health professionals*, 43. Boston: Butterworth-Heinermann.
61. Wang, C. 1993. *Sense and nonsense of statistical inference: Controversy, misuse and subtlety*, 112. New York: Marcel Dekker.
62. cf. Freedman, B. 1999. From association to causation. *Statistical Science* 14:243–258.
63. Thoresen, C. ed. 1999. Spirituality and health: Is there a relationship? *Journal of Health Psychology* 4:291–300.
64. Gershoff, S. 1996. *Tufts university guide to total nutrition*, 65–66. New York: Harper Perennial.
65. Stiles, W. 2006. "Case Studies." In *Evidence-based practice in medical health: Debate and dialogue on the fundamental questions*, eds. Norcross, J., Beutler, L., and Levant, R., 57–73 esp. 61 Washington, DC: American Psychological Association.
66. Soares, S. 2006. Cigarette smoking affects uterine receptiveness. *Human Reproduction* 21(Suppl 1):i46.
67. Black, D. 1998. The limits of evidence. *Journal of the Royal College of Physicians of London* 32:23–26.
68. Webb, G. 2006. *Dietary supplements and functional foods*, 30. Oxford: Blackwell.
69. cf. Reed, G. 2006. Clinical experience. In *Evidence-based practice in medical health: Debate and dialogue on the fundamental questions*, eds. Norcross, J., Beutler, L., and Levant, R., 19 Washington, DC: American Psychological Association.
70. Goodman, K. 2003. *Ethics and evidence-based medicine*, 26. Cambridge: Cambridge University Press.
71. Goodman, K. 2003. *Ethics and evidence-based medicine*, 6. Cambridge: Cambridge University Press.

72. Nys, T., and Nys, M. 2006. Psychiatry under pressure: Reflections on psychiatry`s drift towards a reductionist biomedical conception of mental illness. *Medicine, Health Care and Philosophy* 9:107–115.
73. Nys, T., and Nys, M. 2006. Psychiatry under pressure: Reflections on psychiatry's drift towards a reductionist biomedical conception of mental illness. *Medicine, Health Care and Philosophy* 9:107–115; esp. 107–109.
74. Nys, T., and Nys, M. 2006. Psychiatry under pressure: Reflections on psychiatry`s drift towards a reductionist biomedical conception of mental illness. *Medicine, Health Care and Philosophy* 9:107–115; esp. 110 .
75. Nys, T., and Nys, M. 2006. Psychiatry under pressure: Reflections on psychiatry's drift towards a reductionist biomedical conception of mental illness. *Medicine, Health Care and Philosophy* 9:107–115; esp. 115.
76. Resnik, D. 2000. Statistics, ethics and research: An agenda for education and reform. *Accountability in Research* 8:163–188.
77. Goodman, K. 2003. *Ethics and evidence-based medicine*, 34. Cambridge: Cambridge University Press.
78. Webb, G. 2006. *Dietary supplements and functional foods*, 31. Oxford: Blackwell.
79. Wisneski, L., and Anderson, L. 2005. *The scientific basis of integrative medicine*. Boca Raton, FL: CRC Press.
80. von Wichert, P. 2005. Evidenzbasierte Medizin (EbM). Deutsches Ärzteblatt 102: C1242–C1243; esp. C1242. See von Wichert's end bibliography of critiques of EBM..
81. Ignatavicius, D., and Workman, M.L. 2002. *Medical-surgical nursing care: Critical thinking for collaborative care*, 4th edn, 2 vols. Philadelphia, PA: WB Saunders.
82. Ignatavicius, D., and Workman, M.L. 2002. *Medical-surgical nursing care: Critical thinking for collaborative care*, 4th edn, 2 vols, 10. Philadelphia, PA: WB Saunders.
83. Ignatavicius, D., and Workman, M.L. 2002. *Medical-surgical nursing care: Critical thinking for collaborative care*, 4th edn, 2 vols, 11. Philadelphia, PA: WB Saunders.
84. Jenicek, M., and Hitchcock, D. 2005. *Evidence-based practice: Logic and critical thinking in medicine.* Chicago, IL: American Medical Association Press.
85. Jenicek, M., Hitchcock, D. 2005. *Evidence-based practice: Logic and critical thinking in medicine*, XV. Chicago, IL: American Medical Association Press.
86. Jenicek, M., and Hitchcock, D. 2005. *Evidence-based practice: Logic and critical thinking in medicine*, 19. Chicago, IL: American Medical Association Press.
87. Jenicek, M., and Hitchcock, D. 2005. *Evidence-based practice: Logic and critical thinking in medicine*, XX. Chicago, IL: American Medical Association Press.
88. Tonelli, M. 1998. The philosophical limits of EBM. *Academic Medicine* 73:1234–1240.
89. Jenicek, M., and Hitchcock, D. 2005. *Evidence-based practice: Logic and critical thinking in medicine*, 81. Chicago, IL: American Medical Association Press.
90. Jenicek, M., and Hitchcock, D. 2005. *Evidence-based practice: Logic and critical thinking in medicine*, 84 ff. Chicago, IL: American Medical Association Press.
91. Jenicek, M., Hitchcock, D. 2005. *Evidence-based practice: Logic and critical thinking in medicine*, 90. Chicago, IL: American Medical Association Press.
92. Rogers, W. 2004. Evidence-based medicine and women: Do the principles and practice of EBM further women`s health? *Bioethics* 18:50–71; esp. 53 .
93. Rogers, W. 2004. Evidence-based medicine and women: Do the principles and practice of EBM further women`s health? *Bioethics* 18:50–71; esp. 54.
94. Rogers, W. 2004. Evidence-based medicine and women: Do the principles and practice of EBM further women`s health? *Bioethics* 18:50–71; esp. 60 .
95. Norcross, J., Beutler, L., and Levant, R. eds. 2006. *Evidence-based practice in medical health: Debate and dialogue on the fundamental questions*, 90. Washington, DC: American Psychological Association.
96. Norcross, J., Beutler, L., and Levant, R. eds. 2006. *Evidence-based practice in medical health: Debate and dialogue on the fundamental questions*, 162 presents a similar critique. Washington, DC: American Psychological Association...

97. Norcross, J., Beutler, L., and Levant, R. eds 2006. *Evidence-based practice in medical health: Debate and dialogue on the fundamental questions*, 161. Washington, DC: American Psychological Association.
98. Other fallacies of EBM are expanded on: Norcross, J., Beutler, L., and Levant, R. eds 2006. *Evidence-based practice in medical health: Debate and dialogue on the fundamental questions*, 167–255. Washington, DC: American Psychological Association.
99. Norcross, J., Beutler, L., and Levant, R. eds 2006. *Evidence-based practice in medical health: Debate and dialogue on the fundamental questions*, 173. Washington, DC: American Psychological Association.
100. Norcross, J., Beutler, L., Levant, R. eds. 2006. *Evidence-based practice in medical health: Debate and dialogue on the fundamental questions*, 255. Washington, DC: American Psychological Association.
101. Haldeman, S. 1990. Failure of the pathology model to predict back pain. *Spine* 15:718–724; esp. 719.
102. Haldeman, S. 1990. Failure of the pathology model to predict back pain. *Spine* 15:718–724; esp. 722.
103. Haldeman, S. 1990. Failure of the pathology model to predict back pain. *Spine* 15:718–724; esp. 721.
104. Haldeman, S. 1990. Failure of the pathology model to predict back pain. *Spine* 15:718–724; esp. 720.
105. Webb, G. 2006. *Dietary supplements and functional foods*, 27. Oxford: Blackwell.
106. Szawarski, Z. 2004. The concept of placebo. *Journal of Science and Energineering Ethics* 10:57–64.
107. Thompson, W. 2005. *The placebo effect and health: Combining science and compassionate care. Prometheus books*, 54. New York: Amherst.
108. Thompson, W. 2005. *The placebo effect and health: Combining science and compassionate care. Prometheus books*, 15. New York: Amherst.
109. Talbot, M. 2000. *The placebo prescription*. New York Times Magazine, January 9, 2000.
110. Moerman, D. 2002. *Meaning, medicine and the placebo effect*, 14. Cambridge: Cambridge University Press.
111. Moerman, D. 2002. *Meaning, medicine and the placebo effect*, 57. Cambridge: Cambridge University Press.
112. Moerman, D. 2002. *Meaning, medicine and the placebo effect*, 66. Cambridge: Cambridge University Press.
113. Moerman, D. 2002. *Meaning, medicine and the placebo effect*, 99. Cambridge: Cambridge University Press.
114. Groopman, J. 2004. *The anatomy of hope: How patients prevail in the face of illness*. New York: Random House.
115. Hines, T. 1987. Left brain/right brain mythology and implications for management and training. *Academy of Management Review* 12:600–606; esp. 600.
116. Hines, T. 1987. Left brain/right brain mythology and implications for management and training. *Academy of Management Review* 12:600–606; esp. 601.
117. Hines, T. 1987. Left brain/right brain mythology and implications for management and training. *Academy of Management Review* 12:600–606; esp. 603.
118. Hines, T. 1987. Left brain/right brain mythology and implications for management and training. *Academy of Management Review* 12:600–606; esp. 605.
119. Brody, H. 2000. The placebo response: Recent research and implications for family medicine. *Journal of Family Practice* 49:649–654; esp. 650.
120. Szawarski, Z. 2004. The concept of placebo. *Journal of Science and Energineering Ethics* 10:57–64; esp. 61.
121. Szawarski, Z. 2004. The concept of placebo. *Journal of Science and Energineering Ethics* 10:57–64; esp. 62.

122. Szawarski, Z. 2004. The concept of placebo. *Journal of Science and Energineering Ethics* 10:57–64; esp. 59.
123. Batavia, M. 2001. *Clinical research for health professionals*, 108. Boston: Butterworth-Heinemann.
124. Szawarski, Z. 2004. The concept of placebo. *Journal of Science and Energineering Ethics* 10:57–64; esp. 60.
125. Greenberg, G. 2003. *Is it Prozac? Or placebo?* Mother Jones Nov/Dec 2003.
126. Hróbjartsson, A., and Gøtzsche, P. 2001. Is the placebo powerless? An analysis of clinical trials comparing placebo with no treatment. *NEJM* 344:1594–1602.
127. Declaration of Helsinki, Clinical trials: http://www.wma.net/en/30publications/10policies/b3/index.html 53rd WMA General Assembly, Washington 2002 (Note of Clarification on paragraph 29 added); Miller, F., and Brody, H. 2002. What makes placebo-controlled trials unethical? *American Journal of Bioethics* 2:3–9 [World Medical Association 2000. Declaration of Helsinki: Ethical principles for medical research involving human subjects. *JAMA* 284:3043–5. World Medical Association 2001. Note of clarification on paragraph 29 of the WMA Declaration of Helsinki. Geneva: World Medical Association. Available from: http://www.wma.net/e/ home.html].
128. American Medical Association Principles of Medical Ethics. (E-2.075) The Use of Placebo Controls in Clinical Trials. (Issued June 1997 based on the report "Ethical Use of Placebo Controls in Clinical Trials," adopted June 1996.
129. Biller-Andorno, N. 2004. The use of the placebo effect in clinical medicine – ethical blunder or ethical imperative? *Science and Engineering* 10:43–50.
130. Biller-Andorno, N. 2004. The use of the placebo effect in clinical medicine – ethical blunder or ethical imperative? *Science and Engineering* 10:43–50; esp. 47 .
131. Biller-Andorno, N. 2004. The use of the placebo effect in clinical medicine – ethical blunder or ethical imperative? *Science and Engineering* 10:43–50; esp. 46.
132. NIH study definition. NIH Guide: The Placebo Effect in Clinical practice, grants.nih.gov/grsnts/guide/rfa-files/RFA-AT-02-001.html (accessed Sept., 2003).
133. Shibles, W. 1974. *Emotion: The method of philosophical therapy*. Whitewater, WI: Language Press.
134. Black, D. 1998. The limits of evidence. *Journal of the Royal College of Physicians of London* 32:23–26; esp. 25.
135. Albert, D., Munson, R., and Resnik, M. 1988. *Reasoning in medicine: An introduction of clinical inference*, 180. Baltimore, MD: John Hopkins University.
136. Albert, D., Munson, R., and Resnik, M. 1988. *Reasoning in medicine: An introduction of clinical inference*, 171. Baltimore, MD: John Hopkins University.
137. Albert, D., Munson, R., and Resnik, M. 1988. *Reasoning in medicine: An introduction of clinical inference*, 183. Baltimore, MD: John Hopkins University.
138. Albert, D., Munson, R., and Resnik, M. 1988. *Reasoning in medicine: An introduction of clinical inference*, 184. Baltimore, MD: John Hopkins University.
139. Albert, D., Munson, R., and Resnik, M. 1988. *Reasoning in medicine: An introduction of clinical inference*, 185. Baltimore, MD: John Hopkins University.
140. Albert, D., Munson, R., and Resnik, M. 1988. *Reasoning in medicine: An introduction of clinical inference*, 188. Baltimore, MD: John Hopkins University.
141. Albert, D., Munson, R., and Resnik, M. 1988. *Reasoning in medicine: An introduction of clinical inference*, 189. Baltimore, MD: John Hopkins University.
142. Albert, D., Munson, R., and Resnik, M. 1988. *Reasoning in medicine: An introduction of clinical inference*, 197. Baltimore, MD: John Hopkins University.

Chapter 20
Lying in Medicine

Omnis homo mendax. (All people lie)

Lies are imaginary gardens with real people in them [2].

Pretty much all of the honest truth telling there is in the world, is done by children. (Supreme Court Justice, Oliver Wendell Holmes)

Abstract A theory of lying is presented. A lie is to believe one thing and to express another. The liar must be aware that the belief and statement are different. On this new definition: A lie is not the same as making a false statement. A lie is not the same as not telling the objective truth. A lie in itself is neither good nor bad, but just a contradiction between belief and statement. There are consequences of lying: We gain faulty information on which to base decisions. We fail to understand what or how the liar really thinks and feels which is especially important in medicine. Communication is undermined as well as relationships, which are based on communication. Trust is undermined. Lying promotes more lying and encourages others to lie. A lie (or truth) may benefit all in the short run, but not in the long run, or vice versa. We may not realize or be able to know in advance the harmful consequences that even the smallest lie may have.

Keywords Definitions of lying · new theory of lying · self-lie · consequences of lying · justifiable lies · logic of flattery · beneficial lying · mental reservation · hypocrisy · truth

20.1 Introduction

We are commonly told by those around us that we should never tell a lie, and along these lines Kant is believed to have held that we should never in any case tell a lie regardless of the consequences [1, 2]. But other philosophers such as Jeremy Bentham believe that the consequences should determine whether or not we should tell a lie. What if by lying we could save the lives of many people and harm no one? Should we do so? Sissela Bok wrote, "Most doctors, in a number of surveys stated that they do not, as a rule inform patients that they have an illness such as cancer"

B. Maier, W.A. Shibles, *The Philosophy and Practice of Medicine and Bioethics*, International Library of Ethics, Law, and the New Medicine 47,
DOI 10.1007/978-90-481-8867-3_20, © Springer Science+Business Media B.V. 2011

[3]. This may in many countries, such as in the U.S. or in Europe, be a violation of the physician's code of ethics. And it is referred to 30 years ago. Similarly, it was pointed out that euphemisms are regularly used and not such words which tell the whole truth because the doctor (paternalistically or just caring) thinks the whole truth to be unbearable for the patient, at least at the very moment of diagnosis and at least not all at once.

Before we can answer such questions we must find out what a lie is. Various forms of lying are: hypocrisy, dogma, insecurity, euphemism, libel, slander, perjury and defenses against perjury, feign, fake, pretend, fraud, impostor, false arguments, rumor, hearsay, fib, "white" lie, palter, mendacious, false promises, psychological terms such as: antisocial personality, confabulation (giving fake accounts of events), Machiavellianism (lying to achieve one's goals), malingering (faking illness for excuse to avoid work), Münchhausen Syndrome (telling tall tales, usually involving faking an illness so as to be constantly hospitalized), mythomania (creating myths or exaggerating), pathological liar, etc.

Also the use of medical jargon can be a way of not telling the truth.

20.2 Definitions of Lying

The typical dictionary definition and popular view is that lying is an untrue statement made with intent to deceive.

1. Must a lie always be a false statement? If a physician believes and states that a patient has cancer, but it turns out to be false, it is not a lie. If the physician does not believe that a patient needs medication, yet states that she does, and it turns out that she does, it is a true statement, but nevertheless a lie. Thus, a lie does not depend on the objective truth or falsity of a statement. A lie is not merely an objectively false statement.
2. Must a lie deceive in order to be a lie? We may lie because forced to by policy or a superior, yet hope that no one is thereby deceived. One may lie as a joke or as an irony. Similarly, a statement is true whether or not it has harmful consequences. Being a lie does not depend on whether or not one intends to deceive anyone, nor does it require that they be deceived. One may just lie out of habit. We may thus reject the common definition of lying. Fainzanz gives a one-sided definition of lying: that a lie must be produced to make another person believe it [4].

20.3 A New Theory of Lying

A lie is to believe one thing and to express another. The liar must be aware that the belief and statement are different. On this new definition:

A lie is not the same as making a false statement.
A lie is not the same as not telling the objective truth.

We can subjectively lie, yet tell what is popularly thought to be the truth. We sometimes tell people what they want to believe (cf. placebo effect, or to induce hope which can be a self-fulfilling prophecy). Taking Saw Palmetto may have a 30% placebo effect for benign prostatic hyperplasia (BPH). To say it will help correct the problem would only be true if the patient thought so, thus one cannot predict that it will help. Should one tell the patient that a drug has mainly a placebo effect thereby possibly precluding it having the beneficial effect? Also, a lie should not be confused with mere disagreement.

A lie in itself is neither good nor bad, but just a contradiction between belief and statement (cf. tact). It is like the wind, pleasant in summer, freezing in winter. But it is a lie (or wind) nevertheless.

It is important to note that, being subjective, truth may differ for each individual. Medical indications may be read differently from the data available.

We cannot lie about things we do not know or only have beliefs about. It is also a lie to say we believe, when we actually know, or that we know when we only believe something to be true, e.g., "I believe you have cancer," when we actually know it is present. A patient may state that s/he has an ailment, which s/he does not have, yet not being lying. Religious people have beliefs lacking any factual basis, yet may not be lying. They just do not know the objective truth.

One need not be blamed for merely lying, except if the truth of communication is presupposed. We may be praised for a lifesaving lie and blamed for a harmful lie. We may contradict ourselves without lying if we are not aware of the contradiction or if we accept contradictions. Normative and moral statements and actions often are highly contradictory.

20.4 Self-Lie

The final belief is to believe in a fiction, which you know to be a fiction.
Wallace Stevens
The most common lie is the one, one tells to oneself. Nietzsche
Nothing but saying makes it so. Shakespeare, Hamlet

Self-lie refers to our lying to ourselves. This is like keeping secrets from ourselves, or accusing ourselves of something we know we did not do. To lie to one-self is supposedly to believe what you do not believe. How can one do that? You believe in a deity, but you know you lack all evidence for this belief. On the new definition, we can lie to ourselves by telling ourselves other than what we believe. In this way we are not required to believe what we do not believe, we just say other than what we believe to ourselves. It is a pointless thing to do. What we do in self-lie is hold contradictory beliefs. Physicians and scientists often believe in the scientific method and hold contradictory supernatural beliefs side by side. One may also try to lie to oneself (e.g., rationalize) to create or justify a questionable, but advantageous, belief. This is a form of self- indoctrination. One, for example, reads only one-sided or dogmatic books, e.g., anti-abortion literature. Some synonyms of self-deception are: insincerity, self-persuasion, self-hypnosis, logical fallacies such as circularity, assuming what you are trying to prove, sublimation, repression, censorship, intuition, etc. In short, self-deception is faulty thinking gained by suppression of reason. What is important to know is how we are lying (being inconsistent) to ourselves, or deceiving ourselves.

"The worst enemy you can encounter will always be you, yourself; you lie in wait for yourself in caves and woods" [5]. If we lie to ourselves by holding beliefs without evidence, e.g., weekday scientists who are supernatural on Sunday, we disqualify ourselves from being reliable health care workers regarding others. Nietzsche stated, "'Faith' means not wanting to know what is true" [6]. Shakespeare put it this way:

> This above all: to your own self be true,
> And it must follow, as night the day,
> You cannot then be false to any man. (Hamlet)

20.5 Consequences of Lying

A lie may cause harm and/or benefit to one-self and/or others. That is a lie may: harm one or another or both, harm or benefit no one, prevent harm to one and/or both, prevent benefit to one and/or both, benefit one or another or both, hurt no one, but benefit someone, etc.

Thus, lies can be harmful, helpful or both. It would be strange for someone to tell a lie which harms both oneself as well as others and benefits no one. On the other hand, it would be strange not to tell a lie which benefits everyone and harms no one.

Should we always tell the truth? It could also be perverse to tell the truth, which seriously hurts everyone. If by telling a lie we can save lives and hurt no one, should we not tell the lie? In each case one must carefully consider the consequences. Also, always telling harmful truth can do much harm and be insulting. Some consequences of lying are the following:

1. We gain faulty information on which to base decisions (Bluffing is often allowed in employer-employee negotiations). "A great deal of lying and deception...is openly condoned or encouraged by both business and labor" [7]. "Many forms of bluffing in labor negotiations are legal" [8]. This has the problem that the employee-employer demands may be unrealistic, the institution goes bankrupt, and the employees lose their jobs. It would be better to be honest about the financial picture and actual situation.
2. We fail to understand what or how the liar really thinks and feels which is especially important in medicine.
3. Communication is undermined as well as relationships, which are based on communication.
4. Trust is undermined.
5. Lying promotes more lying and encourages others to lie.

6. A lie (or truth) may benefit all in the short run, but not in the long run, or vice versa.
7. We may not realize or be able to know in advance the harmful consequences that even the smallest lie may have.

Questions arise as to whether or not the physician should tell a patient with a disease the truth if by so doing it will thereby shock the patient and endanger the quality of his life. Should one lie to save the lives of many people if no one is thereby harmed? We may answer as follows. It depends on a full and thorough analysis of the consequences, including the above consequences as well. It is best to first consult with others involved and to make a joint decision, e.g., with other physicians and those involved with written records being kept (See also the Chapter 17).

Because lying undermines communication, general fairness, etc. it is best never to lie unless it can be fully inter-subjectively justified. It is not necessary to even tell small lies such as having someone tell a caller that one is not in when one is, or saying one is busy, when one is not. But if one caused the death of numerous people because one would not tell a lie, which hurt no one, one could not be proud of one's fatal honesty. "They are all dead, but at least I did not lie." There are also cases imaginable, in which by not lying, or even by telling the truth, one is letting others die.

The position here, however, is not that lying is justifiable in general. It is that one has good reason to be more honest than anyone else even in the smallest detail (white lies), but that there are cases where the consequences fully justify lying.

Most people including those who say that lying is not justifiable already typically lie in thousands of small or large ways. They lie for social reasons, to be polite, sign letters with "yours truly," swear by oath to tell the whole truth in the courtroom though they do not and never can know the whole truth, make fixed promises which are actually conditional, etc. Medical advertising and health politics typically present untruths. The Green party is active in generally blocking advertising and replacing it by objective information about each product or service.

The question is not "Do we lie?" but "Are our lies justifiable?" "Justifiable" is an open context term and has to be exposed to critical ethical examination. It is here argued that all small as well as large lies and life-lies are not justifiable unless careful examination of short and long-range consequences can prove otherwise. Questions may be raised as to whether or not one should lie to criminals, the insane, small children, the irrational, etc. If the courts and rules are unfair to healthcare workers, as they often are, to tell the truth to irrational officers or bad managers is discouraged. That is, if the truth is twisted and made into lies and if the consequences are anti-humanistic, or cause harm to the patient, it is reasonable to withhold the truth or even lie. Irrational and unfair people may not be trusted with the truth. In each case, the full and adequate consequences must be considered before one can decide. The issue must be decided by reason, not merely by blind rule. We ethically cannot simply claim that lying is never justifiable regardless of consequences. This indicates another shortcoming of principlism in bioethics.

One of the problems in the clinic is that patients lie about their condition, e.g., about smoking, drinking, drug taking, taking prescribed medicine, sexual activity, etc. The physician and patient have ethical codes (e.g., AMA) according to which both have to be honest with one another. Roger Higgs wrote that up until the 1980 American Medical Association (AMA) Principles of Ethics, doctors used truth as medicine and often did not disclose to patients their ailment or its severity thinking that, for example, it would lead to loss of hope, worsen their condition, or be too disappointing to cope with [9]. In 2000 the AMA principles state that the physician shall deal honestly and openly with patients and colleagues and also that patients are responsible to be truthful. Full disclosure is expected of the physician. Unfortunately, honesty and truthfulness are not defined. Also, it is to be doubted if this is actually carried out in practice on the part of the physician or the patient. In general, however, we would support such rules. There may be cases, however, where the truth or the way in which the truth is told would cause such harm that it cannot be told in a certain situation and at a certain time, or only step by step. Nuyen argues that it is not always good to tell the truth [10] and that it is acceptable for a doctor to lie to a patient about the patient's critical condition when the patient does not have long to live anyway [11]. WHO Patients Rights declaration states the patient has a right to the truth unless it could be seriously damaging to him or her. The French medical ethics code makes a similar provision only for the benefit of the health and education of the patient. The policy of America giving foreign aid only if people are not told about abortion and contraception is a large-scale lie. The Catholic Church and their anti-abortion centers are accused to also withhold and distort the medical facts, e.g. about defects in fetuses and resulting disabilities in children to be and abortion alternatives [12]. It often requires a critical evaluation of what the state of the patient is and what they can make of the information given them. This view, however, would require further evaluation in terms of holistic consequences. For example, the patients may wish to put their affairs and estate documents in order. In sum, absolutistic, fixed rules cannot override a rational consideration of specific circumstances and consequences.

To whom "belongs" the truth? If it concerns the patient and it will be up to him to deal with the consequences learning or not learning about it, the "truth" belongs to him. He has a right to be told.

Physicians not telling the patients the whole truth and errors were not reported for correction. Henry states, "The ethical principle that speaks to truth telling and understanding is respect for patient autonomy" [13]. The autonomy principle comes from Beauchamp and Childress and is a form of principlism [14]. It is what Henry means by an "ethical framework." It is not an ethical framework at all. The author does admit that there are problems with the principle [15] (See Chapter 12).

There are concepts like "mental reservation". "If a feverish patient, for example, asks what his [or her] temperature is, the doctor is advised to answer: 'Your temperature is normal today,' while making the mental reservation that it is normal for someone in the patient's precise physical condition" [16]. "Mental reservation" means you can lie to medical and other people within a certain preconditioned framework. It is a confused concept in terms of communication.

20.6 Logic of Flattery: Beneficial Lying

I lie with her and she with me.
And in our faults by lies we flattered be [17].

We may in flattery tell a beneficial lie which harms no one in order to reveal a subjective truth. Flattery can be to tell the truth by lying. Our supposed faults can be seen as lies and fictions, and so be of no concern. We are all disabled in the sense that each one has abilities others do not have. A blind person can develop senses and abilities the sighted do not have. Beauty is not objective, but only depends on our normative culture and evaluation, which in turn depends upon our ability and willingness to appreciate. On the critical, ethical and philosophical level, one creates an aesthetic, which goes beyond the mere everyday societal norms. For a positive attitude, it is we who must keep the aesthetic alive. Flattery creates and keeps friendship, and positive emotions alive. The physician may say after an internal exam, "You are as beautiful inside as you are outside." The physician can have such an honoring attitude regardless of the prevailing standards of beauty. By flattery, an aesthetic attitude and an ideal are created. In genuine flattery an objective lie is told to reveal a subjective truth. "False flattery" is when one lies both about the facts and about one's subjective emotions. Flattery may also be objectively true on the view of most people's standards, but subjectively false in terms of more enlightened knowledge. One may praise a patient's progress, while realizing that much more could have been done. Either the objective or subjective truth, or both may be desired. Knowing this, allows the nurse or physician to respond appropriately. Flattery as subjective appreciation is as important in medicine as it is in love relationships.

Frankl encouraged seeing people not just as what they actually are, but what they at best could be and appreciating them already that way [18].

20.7 Hypocrisy

Actors are the only honest hypocrites [19].
One may smile, and smile, and be a villain [20].

Hypocrisy is a type of contradiction. Thus, like contradiction, we may have 1. hypocrisy in definitions, 2. inconsistency hypocrisy, e.g. in regard to our beliefs, 3. experiential hypocrisy, e.g., we perform or observe contradictory behaviors. People claim to be rational, but have never engaged in critical thinking. Our beliefs are often indoctrinated rather than based on evidence, and so they can be inconsistent and hypocritical. It is hypocrisy to be religious without first being acquainted with the arguments for and against religion. We claim to be loving and kind, yet selfishness, militancy, anger, and revenge appear to be predominant characteristics.

Hypocrisy is pretending or professing to have beliefs, feelings or virtues one does not actually, or in practice, have. It is to be what one is not. It often involves selfishness while attempting to appear altruistic [21].

"Person" (persona) originally meant "mask." Persona refers to fictional characters. Thus, it involves pretense and insincerity. Humor uncovers the masks. Actors must pretend, yet be sincere. If they can fake that they have arrived. In some ways, everyone is a hypocrite. One uses cosmetics, or pretends at times to be what one is not. We play roles. The young pretend to be old; the old, young; the large, thin; the thin, large; the ugly beautiful, the militant peaceful; the ignorant intelligent, etc. In not unusual cases, heads of institutions or departments pretend to support the advancement of the employees, but actually work behind the scenes to undermine them. The beliefs and behavior of people in any town are typically contradictory and inconsistent. We often try to appear to know more than we do. People oppose euthanasia, but support war. They oppose abortion for others, although they are for their own autonomous choices. They at the same time support bombing which causes the deaths of the unborn and children. That is, abortion is strongly opposed, but it is thought to be all right to bomb pregnant mothers.

It may be suggested, then, that hypocrisy involves conscious pretense, lie, inconsistency, contradiction, disharmony between one's belief and one's action. The hypocrisy may also be an inconsistency between thought, word, or action: You do not do what you say or believe, or think other than you say or do, act other than you think or say, etc. If these conflicts are not consciously produced, one cannot speak of genuine hypocrisy. It may be due to ignorance, or indoctrination. It is ironical that one may not be aware of one's hypocrisy. We may be tempted not to materialize our beliefs, or not have the ability to put our beliefs into practice. Often people hold the view that conviction is worthless unless converted into action. This may not always be true. Most people do not realize how inconsistent their behavior is. Basically, unconscious or accidental hypocrisy is due to lack of critical thinking, failure to inquire. If one does not know about ethics, one cannot even be ethical or unethical. Ethics is above the heads of those who have not found out what ethics is about. One may speak of "complacent hypocrisy" which means that one cannot claim to be ethical if one does not inquire. In this sense, one is a hypocrite if one does not inquire. John Dewey had similarly stated that ethics rests on inquiring [22]. Hypocrisy also depends on the viewpoint. It may be that I do not think I am being inconsistent, but someone else does. They think I am a hypocrite, but I do not. Pure hypocrisy is when one is consciously and deliberately inconsistent or contradictory in behavior, thought or statement. This may be a positive, good thing if I am trying to reform an unfair system. We may infiltrate a perverse organization as an undercover agent would. Hypocrisy as described above has no place in health care or any other areas of an enlightened society and should be exposed for the purpose of improving ourselves as well as our society.

References

1. See Shibles, W. 1985. *Lying: A critical analysis*. Whitewater, WI: Language Press; Shibles, W. 1988. A revision of the definition of lying as an untruth told with intent to deceive. *Argumentation* 2:95–113 (Lying book reviewed in *Revue Philosophique* March 1987:123; *Revue de Metaphysique et de Morale* (Paris) vol. 2 1987. *American Rationalist* 1986. 30:103 Also see forms of lying in Shibles, W. 2002. Shibles eBook on humor at http://facstaff.uww.edu/shiblesw/ or search: Shibles humor.

2. Shibles, W. 2002. *Lügen und Lügen lassen: Eine kritische Analyse des Lügens*. Maier, B. tr, Mainz: Lermann Verlag.
3. Bok, S. 1978. *Lying. Moral choice in public and private life*, 228–229. New York: Pantheon; 1989/1999 *Vintage books*, New York; Random House; Bok, S. 1983. *On the ethics of concealment and revelation*, esp. ch III, 5 Secrecy and self-deception. Vintage Books: New York.
4. Fainzanz, S. 2002. Lying, secrecy and power within the doctor-patient relationship. *Anthropology and Medicine* 9:117–133; esp.119.
5. Nietzsche, F. 1977. *A Nietzsche Reader*. Hollingdale, R.J. tr. New York: Penguin; *Thus Spoke Zarathustra* I. 17: "The Way of the Creator." 90.
6. Nietzsche, F. 1977. *A Nietzsche Reader, Antichrist* 55. Hollingdale RJ tr. New York: Penguin; Nietzsche, F. 1960. "On Truth and Lie in an Extra-Moral Sense." 3 vols, Schlechta, K. ed, München.
7. Carson T, Wokutch, R., and Murrman, K. 1982. Bluffing in labor relations: Legal and ethical issues. *Journal of Business Ethics* 1:13–22; esp.20.
8. Carson T, Wokutch, R., and Murrman, K. 1982. Bluffing in labor relations: Legal and ethical issues. *Journal of Business Ethics* 1:13–22; esp.13.
9. Higgs, R. 1994. Truth-telling, lying and the doctor-patient relationship. In *Principles of health care ethics*. eds. Gillon, R., and Lloyd, A., 499–509; esp.502 New York: John Wiley.
10. Nuyen, A. 1999. Lying and deceiving: Moral choice in public and private life. *International Journal of Applied Philosophy* 13:69–79 esp. 78.
11. Nuyen, A. 1999. Lying and deceiving: Moral choice in public and private life. *International Journal of Applied Philosophy* 13:69–79 esp. 76.
12. Fainzanz, S. 2002. Lying, secrecy and power within the doctor-patient relationship. *Anthropology and Medicine* 9:117–133.
13. Henry, L. 2005. Disclosure of medical errors: Ethical considerations for the development of a facility policy and organizational culture change. *Policy, Politics and Nursing Practice* 6:127–134 esp.130.
14. Beauchamp, T., and Childress, J. 1994. *Principles of biomedical ethics*, 4th edn. Oxford: Oxford University Press.
15. Henry, L. 2005. Disclosure of medical errors: Ethical considerations for the development of a facility policy and organizational culture change. *Policy, Politics and Nursing Practice* 6:127–34 esp.131.
16. Bok, S. 1978. *Lying. Moral choice in public and private life*, 37. New York: Pantheon; 1989/1999. *Vintage books*. New York: Random House.
17. Shakespeare, W. 1964. Shakespeare's Sormets. Sormet 138. Ed. Rowse A. L. London: Macmillan.
18. Frankl, V. 1946. *Man's search for meaning*. London: Rider; Frankl, V. 1959. *From death camp to existentialism: A psychiatrist's path to a new therapy*. Lasch, I. tr, Boston: Beacon Press.
19. Hazlitt, W. 1931. *On the fear of death. Essays*. ed. Gray, C., 302–313 New York: Macmillan.
20. Shakespeare, W. 2006. Hamlet i.5.107. Edn Thompson, A., and Taylor N. London: Alden Shakespeare.
21. cf. Shibles, W. 1992. Altruism versus egoism: A pseudo-problem: A cognitive-emotive analysis. *International Journal of Applied Philosophy* 7:21–29.
22. Dewey, J., and Tufts, J. 1908/1932. *Ethics*. New York: Holt.

Chapter 21
The Rhetoric of Death and Dying

Abstract Medicine aims at preventing death, yet it is not clear how we may understand what death is. "Death" is a word too familiar to us to be understood. The "Metaphorical Method" helps us to make death extraordinary. The rhetoric about death is a contribution to philosophy of medicine and to narrative medicine, as how to deal and to communicate death against the incommunicable and to avoid being silenced and isolated as a dying patient and his relatives. It is to enable healthcare-workers to develop an attitude open to communication with dying patients and their significant others. Because thought is mainly language use, careful attention should be paid to the metaphors and models the dying patient lives by. The criteria to bring about an appropriate death are: 1.conflict reduction, 2.proper understanding of the patient in terms of the image he has of oneself, 3.restoration of important social relationships, and 4.satisfaction of his wishes as much as possible. Also those who are losing their loved ones have to be taken care of. An adequate theory of emotion helps to cope with grief and bereavement.

Keywords Death · dying · Metaphorical Method · poetic metaphor · death as language-game · rhetoric of death · emotion of grief and bereavement · death denial · humanistic view of death · death – medical profession

21.1 Definitions of Death

In medicine there is the main goal of preventing death, yet it is not clear how we may understand what death is. The following shows the various ways in which death can and cannot be understood. Death is often said to be: the end of life, destruction of the human organism, extinction. The nature of death and its meaning for us is limited by our language and knowledge. It is therefore a philosophical and interdisciplinary problem as to what the nature of death is. "Death" is a word commonly used as any other word and so is very familiar. An examination of the problems associated with death shows, however, that it is far from familiar. There is a question as to whether it can be understood at all. The understanding of it depends upon the very limits of knowledge itself. This is dealt with especially in the philosophy of death. It will be useful, then, to consider the term initially as an unfamiliar or foreign term.

B. Maier, W.A. Shibles, *The Philosophy and Practice of Medicine and Bioethics*, International Library of Ethics, Law, and the New Medicine 47, DOI 10.1007/978-90-481-8867-3_21, © Springer Science+Business Media B.V. 2011

21.1.1 General Definitions

The numerous philosophical and interdisciplinary definitions of "death" are presented throughout this chapter. How we are to conceive of a person depends upon our own purposes, as well as on our knowledge of human behavior. An analysis of philosophical psychology involves a clarification of the nature of "person," "self," "action," "consciousness," "thought," etc. The view we have of elementary "particles," electromagnetic "forces," and physical matter (as opposed to animate organisms) will determine whether anything really dies at all. In summary, when one speaks of death one must know the specific context and paradigm which is involved. Death does not refer to a single context or an example, but has a great many meanings. It is not a single instantaneous process. And it seems not to be a process at all. "She is dead," is therefore an equivocal statement [1].

21.1.2 Medical-Psychological Definitions of Death

The five criteria of death established in 1968 by the Council for the International Organization of Medical Science are: (1) loss of response to environment, (2) cessation of reflexes and muscle tone, (3) cessation of spontaneous respiration, (4) sudden fall in blood pressure, (5) a flat EEG. Hadassah Gillon, however, found a brain injured patient who satisfied these criteria yet lived [2]. He proposes an added test be oxygen consumption in the brain. There may be oxygen consumption even with a flat EEG (electro-encephalogram). Distinctions may also be made between the following: (1) Cellu-vegetative death. (2) Organ death. (3) Organismal death, e.g., heart and lungs stop. (4) Psychic death. Cerebral brain cells are permanently destroyed in about 5 minutes after breathing ceases. This produces a permanently vegetative state. The body may be kept "alive" by heart, lung, and kidney machines. Relevant to abortion, it is not clear when psychic life begins. Electroencephalograph waves appear only after the embryo is about 43 days old. Self-consciousness begins only after several years. 5) Vegetative death. The cells of the vegetative controlling part of the brain die after five to 8 minutes. The medical definitions of death are constantly changing.

21.1.3 Death and Abortion

One issue concerns whether or not the embryo can intelligibly be defined as a person or human being. To ask if the embryo is in itself a real "person" or "human being" is to ask a misleading question. It assumes that "person" or "human being" name entities in themselves and that by certain procedures we can discover if the names fit. These terms are rather naming-fallacies. We may regard anything as a person or human being if we wish. We do so to suit our purposes. But it makes no sense to say that an embryo must descriptively be defined as a person. To apply "person" to such new uses or situations is to personify. To apply "human" to new uses or situations is to humanize and anthropomorphize. Thus, on this view, if an embryo is not in itself a person or human being, one need feel no guilt or grief in this regard in aborting.

21.2 Death: The Literature

21.2.1 The Poet's View

Emily Dickinson spent her life exploring death by means of poetry. Having grown up in a climate of religious dogma and Puritanism, she explored religion in an attempt to believe in it, tried to experience conversion, but found she could not honestly do so. She did not accept Christ, rejected religious orthodoxy, and stopped attending church. Dickinson was not a theoretical philosopher and did not create a systematic philosophy. Her exploration was more subtle, more concrete, more by suggestion. Metaphorical juxtaposition broke down old meanings creating new more diverse and more suggestive ones. It allowed her to cope with death by means of poetic inquiry, ongoing inquiry and so escape from narrow dogma in a delicate, indirect and subtle way. Phrases and parts of poems cited are from *The Poems of Emily Dickinson* [3].

"Mortality is fatal." (Poem 3. L 41) A perfectly circular or redundant statement showing by its repetition that death is solid, final, dangerous. Though fatally serious the surprising circular juxtaposition of the obvious creates humor. Humor provides a release from problems and tension and can in some cases give insight.

"Death, but our rapt attention/ To immortality." (Poem 7, L 15) If we take the view that death does not name an entity, death is our present experience of thinking of it or of unending life. And our very concern with death is what death is. It is what we imagine it to be – our "rapt attention."

"We lose – because we win/ – Gamblers – recollecting which/ Toss their dice again." (Poem 21, L 1–3) It makes no sense to win unless it makes sense to lose. To lose is then to affirm winning or the concept of winning. Similarly it makes no sense to live unless it makes sense to die. To die is then to affirm living or the concept of living. We then toss our dice again in the risk, in the game of life and death. Death gives meaning to life and vice versa.

"Ah Little Rose – how easy/ For such as thee to die!" (Poem 35, L 11–12) The price of our consciousness is our thinking that we will die.

"We who saw the launching/ Never sailed the Bay!" (Poem 43, L 12–13) We are aware of the death of others, but neither have evidence of nor know what our own death will be like.

"Dust is the only Secret – / Death, the only One/ You cannot find out all about/ In his 'native town.'" (Poem 153, L 1–4) Though in a small town one knows the people and all about its happenings, though one's native town or one-self should be best known to one, as regards one's death in the "native town" of the cemetery we find that "native" phenomenon we know nothing of. We know little of the atomic nature of our bodies and the common dust, which it supposedly becomes.

"It's such a little thing to weep – / So short a thing to sigh – / And yet – by Trades – the size of *these/* We men and women die!" (Poem 189, L 1–4) We treat the dying and death by means of trades, like any other trades, but they are not. Trades of the living are used to represent trades of dying; such things as sighs and weeping are much too lively and too insignificant to do justice to challenging sighless death.

"After great pain, a formal feeling comes – / The Nerves sit ceremonious, like Tombs." (Poem, 341, L 1–2) This line, perhaps one of Dickinson's finest, says much more than can be said in any other way. Pain is an experience, formal is not. "Formal feeling" seems a contradiction. After pain, one returns to normal (one sense of formal here). After life one returns to normal, death. Formal is both normal and static. The line is almost a casual explanation or law according to which death is explained as being as desirable as the normal condition, which often follows pain. An especially pleasurable feeling comes when pain is overcome. Here, after life and pain comes a release and with it a static formal state. Death is covertly made to seem desirable and the line therefore seems to give hope.

"I tie my hat – I crease my Shawl – /. . . We cannot put Ourself away." (Poem 443, L 1, 13) We think we can put ourselves away as we can fold our clothes and put them neatly on a shelf. One ordinarily speaks as if one can witness one's own funeral. We imagine and have a picture of ourselves as lying in a coffin with people standing around. To imagine such things is to know more than we have evidence for. It is like putting oneself away. Theologians in this respect often speak as if they can perform their own autopsies. Also, putting someone away implies that one is mad. One cannot put one's consciousness away as one can put a mad person away. One can, however, go mad.

"I heard a Fly buzz – when I died – ." (Poem 465, 1) "uncertain stumbling Buzz – ." (L 13) The buzzing fly is an impressive image connoting the weakening of the senses when organized sounds and speech become mere noise and buzzing; the contrast of buzzing with the "stillness" of death; the contrast of a quite common and ordinary unimportant thing, such as a fly, with death. It is a metaphorical contrast or "sinking" of an abstract, theoretical notion of death, down to the level of our everyday life experiences. In this last connotation the fly becomes important. Flies often are associated with decomposing flesh. The image of the buzzing fly shows even more. It gives one a commonplace experience of death or dying. A fly is death.

"We do not play on Graves – / Because there isn't Room – "/ "And People come –"/ "And put a flower on it – / And hang their faces so – / We're fearing that their Hearts will drop – / And crush our pretty play – ." (Poem 467, 1–10) Innocent, joyful play of children is contrasted with the negative, destructive, and serious emotions of adults toward death. Perhaps emotions such as grief are unnecessary and adults could attain once more the joyful innocence of children. There perhaps should be room for such play. Life should be stressed, not negative emotions, at death. She wrote, "Endow the living – with the Tears – /You squander on the Dead." (Poem 521, L 1–2)

"I had no time to Hate – / Because/ The Grave would hinder Me/ And Life was not so/ Ample I/ Could finish – Enmity – ." (Poem 478, 1–6) Facing death makes one examine one's life more carefully. The metaphorical deviation or conceit is to think that one cannot hate enough, whereas hating is usually thought undesirable. The point is made by reducing the situation of hating to a hyperbole and to absurdity. She says, "Death reorganizes estimate."

"If I should die/And you should live – / And time sh'd gurgle on – " / .... "'Tis sweet to know..." ... "That Commerce will continue – ." (Poem 54, 1–3, 11, 13) This "sinks" the seriousness of death to the level of everyday active business and trading of wares. It produces humor and irony by the great contrast between active business, colloquial "gurgling" time, and "inactive, timeless" death. That it is sweet to know commerce will continue is covertly suggestive that one is immortal, because it suggests that one will be around to know that commerce will continue. Her statement is a parody of this common view.

"As subtle as tomorrow/ That never came,/ A warrant, a conviction,/ Yet but a name." (Poem 1713, L 1–4) We think of death in terms of models, which are subtle, persuasive and captivating: tomorrow, beyond, possible, eternal, warrant, conviction, etc. These sound as though they describe or explain. Yet upon analysis they are seen to be naming-fallacies, mere names.

Dylan Thomas is a poet, and so it is natural that he should turn his inquiry to metaphors about death. Metaphors are aesthetic and allow for the release from difficult or insurmountable problems; render wonder and paradox; dissolve traditional categories and limitations, thereby promoting insight and inquiry; unite diverse elements such as: opposites, e.g. life-death, serious and the absurd; poetry, love, language, death, all of which may inextricably involve one another. For Thomas, then, we live and die our metaphors.

Dylan's father, like Dylan, did not believe in religion. He faced death honestly and explored it by means of poetic insight. It was always more than he knew, more than anyone knows, more than poetry could tell yet poetry could tell more than most. A metaphor tells us more than we can literally say in prose. His poetry could tell him more than he could know. The following examples are from his *Collected Poems* [4].

What follows also serves as an insightful scientific hypothesis: "A process in the eye forwarns/ The bones of blindness; and the womb/ Drives in a death as life leaks out" [5]. "A darkness in the weather of the eye/ Is half its light." It appears there is a biological *memento mori* (remember death) in the bones, a biological death instinct. Because death is built-in, to give birth to a child is to give death to a child. We are ignorant of such processes, but Thomas projects them as hypotheses. Metaphors may be regarded as hypotheses for exploration. "A darkness in the weather of the eye" may be "half its light" because we know so little about perception, yet it is our very ignorance, which may save us. If what we see death to be is true, the picture is quite dim, but our ignorance about much of the perceptual process gives us some hope, some "light." It is that of which we are now ignorant which gives us hope. These are clarifications and hypotheses to be explored. Also, it requires darkness to see, and it requires death to live. Perception involves the model of an internal and a separate external world. But a clear distinction between inner (or pseudo-psychological) mental events and external events has never been clearly made out.

Just as the distinction between inner and outer breaks down to some extent, so also does the distinction between life and death. An absolute difference between such terms is apparent not real: "Light breaks where no sun shines;/ Where no sea runs, the waters of the heart/ Push in their tides" [6]. We cannot think of death as nothing at all, an absolute negative because "nothing" and "negative" refer to conscious positive configurations of events within our experience. There are no absolute negative statements as such. Negative statements describe positive events. "There is no life," refers to a certain positive relationship of objects. The extravagant metaphors break down our more narrow logics and reveal that there is yet mystery – that "death is nothingness" will not do: "When logics die,/ The secret of the soil grows through my eye,/ And blood jumps in the sun" [7]. In other poems also the notion of negation and nothingness is explored poetically: "The world's turning wood. . .will. . .undie" [8], "undead water" [9], "What's never known is safest in this life" [10].

"Do not go gentle into that good night,/ Old age should burn and rave at close of day;/ Rage, rage against the dying of the light" [11]. This poem contains reverse ritual, that is, whereas religion and funeral ceremonies stress ritualistic repetition in an attempt to comatose one into belief in miracle, repetition is used here to stress honest inquiry and praise of life and nature. The typical Christian view is that one should accept death, not inquire into it or try to avoid it; all is in God's hands. Thomas opposes the dishonesty of this and he opposes a mere quiet giving-in to death as if death is good, as if death is a mere goodbye.

Time is explored in an attempt to comprehend death. If all time is transcended by non-temporal or aesthetic experience of the moment or of miracle, then death is overcome. This adds up to a paradox of a fixed time or moments versus an endless eternity. Time, humans as one with nature, and paradox are explored in: "Time sings through the intricate dead snowdrop. Listen" [12].

We live and die the demons and hells we ourselves create. We live our views of death. If our language is vague we hypnotize and psychologize ourselves into belief by vagueness and delusions as with the words: eternity, timeless, god, afterlife, other world. "Were vagueness enough and the sweet lies plenty,/ The hollow words could bear all suffering/ And cure me of ills" [13]. And all this Thomas put in one short line: "Death is all metaphors" [14].

21.2.2 Wittgenstein on Death

21.2.2.1 Death as a Language-Game [15]

What mainly characterizes recent philosophy is the central concern with and analysis of language. We note, however, that there are ordinary-language strains in numerous previous philosophers. Our notions of death then must be related to and reduced to paradigms of what we know of what is present before us in the here and now.

Wittgenstein presents the following ordinary-language views: "At death the world does not alter, but comes to an end" [16].

Thus, it is difficult to imagine one's own death, to imagine the world and experience coming to an end. Absolute nothingness or an absolute end seems not to be within our experience. Of another's death we can say his world alters. We are conscious of it altering. But we cannot say of our death that it alters, because "alter" implies "alters for my consciousness," and there supposedly is no consciousness. I can alter only from the viewpoint of another. But at death my consciousness of the viewpoint of another comes to an end also. What is meant by "end" here must remain in question. The end of life is not necessarily like the end of a road or end of a trip. It may be more like the end of knowing or more carefully, an unknown sort of end. It is the end of that which determines that there are ends in the first place. It is a paradoxical ending.

"Death is not an event in life: we do not live to experience death. If we take eternity to mean not infinite temporal duration but timelessness, then eternal life belongs to those who live in the present. Our life has no end in just the way in which our visual field has no limits" [17].

That death is not an event in life and so cannot be experienced reminds us of Wittgenstein`s statements: "What we cannot speak about we must pass over in silence" [18], and "The limits of my language mean the limits of my world" [19].

We cannot talk about or perhaps even name it for it is not clear what we are naming. We have experienced being unconscious or asleep and we think death must be like that. We think this in our conscious life. We do not think this out of consciousness. In a sense, death is only something within life.

His statement "We do not live to experience death," is circular. It reduces to: We do not experience non-experience. It would seem that the one thing we certainly cannot experience is non-experience. But then we cannot phrase this statement meaningfully because "non-experience" has no use, no meaning for us. We want to talk of absolute non-experience where we only experience qualitative and relative degrees of experience.

"Our life has no end in just the way in which our visual field has no limits," may suggest that the notion of limit or end (of life) is not perceivable. We never experience the boundary of vision or of life. One might say in this regard that concerning death there is much to draw on and little to draw on. Even to speak of "within" and "outside" our experience presupposes a picture or model which we have. This yields a paradox or puzzle. We are creating metaphors for death.

The following are from Wittgenstein's *Lectures on Religious Belief* [20].

"Suppose you say: 'I have the idea of myself being a chair after death?'" [21].

Why is it any less absurd to say I will survive my death? What more evidence is there for the survival of a soul than for one turning into a chair after death? One could imagine a dog-lover wanting to be a dog or a gardener wanting to be soil or a lily. Would things work out better for the gardener?

In speaking of beliefs, Wittgenstein asks how they may be compared with one another. Does it involve grounds or a state of mind? He suggests that the appeal will be to various paradigms or pictures. What is offered for a belief, e.g., for an afterlife, may be what we are willing to stake on the belief, that there is "retribution," or the idea that "This will be punished." Again, our view of an afterlife as well as of death depends on our paradigms, models, and metaphors.

"Are you clear when you'd say you had ceased to exist?" [21]. "I ceased to exist at all" has no use or intelligible meaning. We often talk about "life after death" as if we were talking about ourselves in the past, as if we could witness our own funeral. "We are all here using the word 'death,' which is a public instrument, which has a whole technique [of usage]" [22]. When we talk of death we are not talking of an unknown or the other side of life, but of techniques or uses of our language in a concrete situation of our life. Since there is no death on this view, there is no living and so this becomes simply, the meaning of a word is its use in a language-game. "Death" is not a private or mental term, something I alone picture, but a term which we have learned in the context of an inter-subjective language. The use of this word is an inter-subjective use.

"'My idea of death is the separation of the soul from the body' – if we know what to do with the words" [22]. That is, these words have no use. They do not matter. They solve no problem, answer no question and do not serve as a description, nor are they usually used for such purposes as to greet people with. However, they may be used to greet someone with, to put someone at ease. Consider as a greeting, "You will die." "The soul separates from the body" is not like "The cover separates from the book," or "The oil separates from water." "If he connects [Here a black scribbled line is drawn] with death, and this was his idea, this might be interesting psychologically" [23].

In trying to conceive of death we make analogies to pictures, or objects, or things associated with what we imagine death and dying to be. To describe death we use such models or metaphors as darkness, absence, un-fulfillment, abyss, etc. Our models of death are within our experience and so constitute their object, rather than merely illustrate it. They say something about us, something psychological. "The whole weight may be in the picture" [24]. Death may be seen as black chaos, scattered burnt-out feelings, used-up thoughts, a scribble, one's electrons though scattered continuing to circle their microscopic orbs.

In presenting a Wittgensteinian analysis of death, R Liveritte points out that to be ill is not to be dying [25]. To say someone is dying is a guess. Also a person cannot die, because it is not something one can do or not do. If it were something I could do, I could supposedly refuse to do it. I can only know that I am ill, not that I am dying. Thus, I cannot experience "pangs of death." Liveritte states, "My death is not an event in my life that I can avoid or not avoid" [26]. We can thus become misled by confusing pictures.

"'The separation of soul from body'... If one says this, I won't know yet what consequences one will draw. I don't know what one opposes this to" [23]. Another way of showing that "the soul separates from the body," has no use is to see that nothing follows from such a view, and that one might hold that the reverse is the case without consequence either. Nothing would seem to verify, falsify, or count against having a soul and the claim that the soul separates from the body. It does not serve well as a regulative idea or useful fiction.

"All I wished to characterize was the conventions one wished to draw. If I wished to say anything more I was merely being philosophically arrogant" [24]. Death talk reduces to concrete experiences, events, uses, or techniques in specific language-games we play. Our notions of death are determined by our analogies and metaphors, which cannot be taken literally or as the only possible ones. To do so would be to misunderstand the use of our language and to misuse our language as is often done in metaphysics. To understand what death is, we should look and see how and in what contexts death language is used. Its meaning will only be to see its use. In this sense, to know what death is is to know how to speak our language. We can know nothing about death, because "know" does not apply.

The analysis of death as a language game given here presupposes the view that language has epistemological primacy. We can play language-games with death, but error arises when we misuse language or imagine that we can go outside of language into an extra-linguistic, real world, or mentalistic world of "thinking" in "concepts" to render death. Death is not there, but within the language. Critical "thinking" is critical speaking. Rather, we may *say* that whatever is known is *said* or *written* about "death." There are as many meanings of death/ "death" as there are language-uses.

Misuse of language creates faulty views about dying. Most of what we "know" about death may be false. There are no ultimate explanations although we may play a language-game of explaining, describing, believing, knowing, and so forth. It is only to play one language-game or another. The question of death raises the question of the limits of knowledge, that is, of language. What I call the "Metaphorical Method," the rhetoric of death, may be used to explore these language-uses. We can thereby expand the possibilities of language to its limits. In our everyday life there is nothing extraordinary about death. The "Metaphorical Method" helps us to make death extraordinary (See Chapter 1).

21.2.2.2 The Question: What is Death?

Does death exist? First tell us what death means, then we will see. The question about death is otherwise a faulty question with no possible answer. Do we mean death or Death with a capital "D," or the word "death" in quotes? Do we mean death outside of language, somehow beyond language? What is it to have knowledge about death? Can we *know* anything *outside* of language? Does knowledge extend that far? That depends upon what knowledge is – and isn't.

21.2.2.3 Death is not a Thought or Concept

It is still curious to note that people, and perhaps most philosophers among them, still hold that they have such things as thoughts, ideas, cognitions, intentions, minds, imaginations, inner emotional entities. They do not. All such alleged are pseudo-psychological entities and commit the fallacy of mentalism. Although mentalism is in the literature, it is not taken as seriously as it should be. Thus, also, knowing and knowledge are not mental states. Thinking and knowing are not like digestion.

We can only metaphorically say that we digest what we read, or find it indigestible. If thinking and knowing are mentalistic fallacies, then what sense can we make of them? Is there another operational definition by which we can account for what we speak of when we refer to knowing and thinking? What must be assumed to have epistemological primacy is not ideas or Descartes' "I think," nor Sartre's or Husserl's *cogito*, but rather language. When we say, "She thinks well," we do not refer to a physical process. To think is to say or write something. We have the wrong picture here. We think that we just have non-linguistic ideas and then, as if by magic, put them into words – that we ex-press them, press them out, like pressing grapes into wine. But ideas are not grapes, and we do not pour them into words. We do not know what we think until we "see", hear what we say. Our words "reason" and "concept" are created by language. Geiger wrote, "Language is everywhere primary; the concept arises from the word. . .. Language created reason; before language man was devoid of reason" [27]. "We do not first have concepts, then language." Yet people still think they associate ideas together, like atoms. It was called mental chemistry. Such "ideas" are not on the right track. Wittgenstein noted, "When I think in language, there aren't 'meanings' going through my mind in addition to the verbal expressions: The language is itself the vehicle of thought" [28]. That is, language is the vehicle and the vehicles are empty.

21.2.2.4 The Epistemological Primacy of Language

When starting points are looked by means of which to account for knowledge it is language, which has epistemological primacy. Thought presupposes language, but language does not presuppose thought. Language only presupposes language, itself. Thus, we are in what I call a "linguo-centric" epistemological predicament in regard to our understanding of death. This means that we can only play language-games with death. Whatever is known or said about death is *said* about death (See Chapter 18).

Statements about death should then be bracketed, kept within their boundaries, put in quotation marks. The scientific method is usually said to ultimately rest on observation and empirical sensations. It does not. It rests on and presupposes language-use. If there is no language, there is no sensation, no observation, no science, and no other symbolic systems. There is no theory of perception. We wish to penetrate the bubble of language, to burst it, and if we do we become "speechless." Death is the destruction of language. It is said that after death we will know everything, or that you cannot know death until you die. What does "know" mean here? To get out of language is not like getting out of the room. We make the strange seem too familiar. We keep trying to escape the limits of language, or trying to take it where it cannot go.

We want to say, "Look here, inside the sphere is language and outside the sphere is death." But there is no outside. The outside is within the sphere of language. It is nothing one can visualize – a one-sided sphere. Our paradigms or analogies for the nature of death are all models within our experience. Even to speak of "within" and "outside" our experience presupposes a picture or model that we have. This yields a paradox or puzzle. We are creating verbal metaphors for death. The genre, Black Literature, is spoken of as the death of literary forms. But destruction remains within language. We did not get out to destroy it. If we claim to get out we have not done so. Our explanations and descriptions are absolutely fascinating to us, for whom alone they seem to make sense.

21.2.2.5 The Death of Mentalistic Meaning

In terms of theories of meaning, people typically think that words just stand for or symbolize ideas, that the words themselves are just vehicles of thought used to communicate ideas from one person to another. Language is not important, only ideas are. We think that we would still think as now do even if we had never learned a language, but we are not in a position to do so as we already have language, and so we would not know how we would think without one. In fact we would not think at all. Would we have experience, for example, without the word "experience"? Not in the first person or "I" case, because we do not have the word "experience," and we would not have any other word for experience either. It is not just a semantic problem. For the same reason, people and animals cannot see, do things or perceive, without language. We, with language, can only say that perhaps they have some sort of "animal language" of their own. Yet people personify and anthropomorphize to think that animals and fetuses are aware of their death. Whenever we think we have found something outside of language it comes back in and can be rendered by the question, "Can you see", know or have x without ever having learned a language, where you substitute a word for x. The answer is always, "No." I have argued elsewhere that meaning may rather be seen as non-mentalistic associations we make with language [29].

21.2.2.6 There is no Non-linguistic Knowledge of Death

We cannot have knowledge of death without language. There is no non-linguistic knowledge. Knowledge is a complex form of language-uses. Knowledge does not rest on alleged forms of knowing which supposedly transcend language, for example, intuition, special visions, revelation, direct knowledge, knowledge by acquaintance, "reine Gestaltwahrnehmung", etc. This should not be surprising because theories in science do not describe the world. They constitute it. Nothing, not even the word "nothing," is outside of language. Nagel thus makes a mistake in saying, "Death is nothing and final" [30]. Even language is constructed by and about language [31]. Death exists only within language-uses. Without language there is no death. Because knowledge is confined to language-use all we can know about death is what the limits of language allow us to know. For Nielsen "'What is unsayable is unsayable,' is a significant tautology" [32]. Wittgenstein held that, "*The limits of my language* mean the limits of my world" [33]. Thus, we can only explore death and dying through the language of philosophy, poetry, medicine, etc. Wittgenstein said, *Death is not an event in life* [34]. However, when he adds: *We do not live to experience death*, his statement too becomes a contextually living statement. There are no limits to such language exploration. We have only a rhetoric of death. As we have seen above, for the poet Dylan Thomas, "Death is all metaphors."

21.2.2.7 Language-Games Again and Again

Because there are criticisms of language-games, some clarification is needed. By language-games I mean language-uses and all we can do with language. This does not mean that we just use language for a purpose. The meaning is identical with the use. It is within the full language situation. We use the very tool we investigate. There is not a separation between the language, self and object, not a mind-body, self-object separation [35]. "I-see-the-dead," has dashes between the words to indicate that we cannot separate I from the seeing, or seeing from what is seen, or the self from the seeing of the object. Each is involved in the other. There is no seeing without an object and no object without the seeing. The self is inseparable from seeing and objects as well. That is, the language-game includes and constitutes everything involved in the context. Language-use does not mean the use of language in a separate world of selves, objects and ideas. Language-use is the world. It contains everything already. These language-uses also constitute the all-inclusive living context. They include the pragmatics of language. The dash between language and use suggests their inseparability. In Wittgenstein's words, "What has to be accepted, the given, is so – one could say – *forms of life*" [36]. We can only participate in language-uses, not stand outside of them. Thus, no language-use has priority over any other. To explain death, does not have priority over imagining or rationalizing a death. Explaining is just one among many language-games we can play, but we must play one or another. There are no ultimate explanations. Our language-uses are part of a living language so that whatever we say about death is restricted to a living context. Goethe said, "To die, the grave. I do not understand these words."

21.2.2.8 What Death is Not

Now we also know what death is not. It is not outside of language and it is not misuses of language. Error arises when we use words in the wrong language-games. It is not circularities, tautologies, abstractionisms, undefined and unintelligible terms like soul and eternity, faith, dogma, spirit, etc. What goes for evidence is constrained and ascertained by rational narrative and discourse, but not by consensus which is an "appeal to the majority" fallacy. A description of death is not to be voted in or out. What goes for evidence is created in dialogue. There is a grammar of evidence [37]. As it is careful or careless the word "death" will be also. Outside of the abstractionist fallacy, one fallacy has been found to be the most pervasive: that we can somehow go outside of or transcend language. Plato used a lot of words to disclaim the validity of language as knowledge. Meta-physics is beyond science, super-natural is beyond nature or knowledge. The meaning of beyond knowledge or the supernatural is ignorance. Religions claim belief in the supernatural. Emily Dickinson observed, "To believe what you do not believe does not exhilarate" [38]. Religion does not tell about death because the supernatural cannot talk.

21.2.2.9 Imagery and Sensation

Two further candidates for the category of the non-linguistic are imagery and sensation [39]. On Norwood Hanson's view our theories and knowledge become part of our seeing. "The knowledge is there in the seeing and not an adjunct of it" [40]. There are infinite ways in which things can be seen. But once again, if there is no language, there is no sensation or imaging in the first person case. We never just see. Imagery is a synonym of metaphor. Pain is psychogenic. Death, the man with a scythe, is referred to verbally as "the man with the scythe." We project the image that death is cold and dark, but it is the living who are cold and dark. These are living images. Death is not like an unheated room.

21.2.2.10 Can We Imagine Death?

Of course, we do not have an imagination any more than we have a mind. Thus we cannot get around language by speaking of such a mentalistic thing. In any case, it seems that people cannot imagine death or at least they are not very good at it. If one's cat dies it is a tragedy; if a billion people die overseas it is a statistic. Madeleine Albright, former United States Secretary of State, said about the 500,000 deaths of Iraqi children the U.S. caused in Iraq: "We think the price is worth it" [41]. The media took virtually no notice of the remark. Also, an American pilot observed, "A day without bombing is like a day without the sun." Another exclaimed, "And to think they pay me for this." For the hunter there is the thrill of the kill. Imagining death requires a sensitivity, which we do not seem to have. In *Alice in Wonderland*, "The Queen had only one way of settling difficulties, great or small. 'Off with his head,' she said without even looking around" [42].

We speak as if we can survive death – can get in and out of life as we get in and out of a car. Suppose we say as some do, "I believe I will live forever." We do have great power to believe such things. We could reply, "What interesting beliefs you have. What hope in the face of past experience?" Or we could say, "Yes, life is good," for that seems to be what could be meant. The claim treats something outside of our language and experience as if it were inside it. We ask, "After you die will you still love me?" We speak about death as if we can witness our own funeral. Emily Dickinson wrote, "If I should die...'Tis sweet to know ...that Commerce will continue – that gentlemen so sprightly conduct the pleasing scene" [43]. It is as if I/we did not die. We have cemeteries with a view. Flew argues that to say, "I can witness my own funeral" has no literal but only a pragmatic use, for example, to refer to the consideration of one's life as a whole. He argues that to literally speak of surviving death is simply self-contradictory [44]. How could one have missed knowing this? Would it help one's understanding of the contradiction when rephrasing it as, "The dead are alive"? But they would think that that is true as well. The usual meaning of "death" is that one does not live, does not survive. Flew notes that such afterlife statements have "headline value" and are "shockers" [45]. They make us wonder what the story is so we keep looking for something deeper because we cannot believe it is as nonsensical as it seems. The absurdity is again seen if we substitute: "We will speak when we cannot speak." But now we may think that we have before us pure poetry. Dylan Thomas wrote, "Where no sea runs, the waters of the heart push in their tides" [4]. To say we survive death is one way to destroy our rationality. Once that is given up others can command our obedience to do or believe anything. You can imagine your own funeral after death, but it just shows what living imaginative language can do. We can also imagine ourselves turning into a butterfly. What would "I" now mean?

21.2.2.11 Illustrations

Someone says, "I" will survive, but "I" is an equivocation. Compare, "The corpse survived."

One form the rhetoric of death takes is illustrated by black humor literature. The comic is combined with the tragic with no possible resolution. Death is seen to be a contradiction, beyond our grasp, whereby the only thing left is for us to note the paradox and laugh. Humor is the acceptance of a mistake or deviation which then produces bodily feelings. We have to accept perceived tragedies in the world in order to survive emotionally. It is essential, for example, in emergency rooms. If we can laugh at the negative, the tragic, betrayal or death, we can thereby accept them. Similarly, "blatant vice humor" means acceptance of the vice. The rhetoric of black humor is to juxtapose humor and death and dying in various ways in order to better describe the human situation and to explore its possibilities. A paradox is a funny place to go when you die.

"'My idea of death is the separation of the soul from the body' – if we know what to do with these words" [46]. But "the soul separates from the body" seems like our common expression "give up the ghost." This sort of separation of soul from body is not the sort of thing one can learn. It is only a verbal technique or use and cannot serve as a description. It has no descriptive use or function. It may have a consoling function, however.

"If he connects [a picture of scribbled lines] with death, and this was his idea, this might be interesting psychologically" [46]. "The whole *weight* may be in the picture" [47]. In trying to conceive of death we make analogies to pictures or objects or things associated with what we imagine death to be. To describe death we use such metaphors as darkness, absence, loneliness, abyss, etc. Our models of death are within our experience and so constitute their object rather than merely illustrate it. They say something about us.

21.3 Dying

The meaning of dying depends upon the definition of death. It often refers to the pain and suffering resulting from disease or bodily damage, which is expected to lead to death.

21.3.1 The Cognitive-Emotive Theory

In order to properly understand the function of emotion in relation to death and dying one must first be clear about what an emotion is (For a full analysis see Chapter 7).

Brief summary of the cognitive theory of emotion:

a. Emotion is not just a bodily feeling. Instead say, "I think-feel emotion."
b. Emotion is cognition, which causes feeling.
c. The emotive cognition is typically a value assessment.
d. Emotion can be changed, by changing the cognition. We cause our own emotions.
f. Emotion is not innate or unalterable.
g. We cannot have exactly the same emotion twice, because both cognition and feeling change.
h. Negative emotion such as grief is due to faulty assessments such as failure to accept reality, failure to understand that we can only do what is within our power, misuse of value terms, such as thinking that something is bad in itself.
i. Emotion is not the sort of (mentalistic) thing that can be "released." We are wrongly told to "release" our grief, guilt and bereavement.
j. Emotion, as such, is not a cause of behavior. Only cognition-causes-feeling can be a cause.
k. Because a judgment or statement is cognition plus feelings, any statement or judgment may be regarded as an emotion.
l. There are meta-emotions such as emotion about emotion, or fear of fear.

Some of the implications of the above theory of emotion regarding death and dying follow. On the cognitive-emotive approach, it has been found that negative emotions are largely a result of confusions in reasoning. Such confusions lead to all kinds of physical and psychological disorders.

a. The rational-emotive theory helps one to face death more rationally.
b. It does not seem to be the case that the present typical funeral ceremony properly deals with human emotions.
c. Our knowledge or ignorance of death and dying will determine the feelings we have when confronted with them. Thus, the more informed we are about them the better we will be able to cope with them.
d. Once one understands how guilt works it is seen that guilt experienced at the death of someone close can be avoided. Instead of "feeling" guilty because, for example, one was not considerate of the deceased before death, one understands that guilt does not help oneself or the dead. In place of guilt one can learn to be more attentive to people in the future. Guilt may be due to not having done those medical things necessary to prevent the death. Guilt is still not appropriate. In place of it one can offer one's organs in case of accidental death (e.g. by filling out the back of one's license), contribute to medical research and encourage research into the understanding of death itself. In this regard, it is problematic that people, though well-intended, leave large sums of money for the funeral ceremony. In this regard, it would be appropriate if a percentage of estates were to go toward medical and death research.
e. Analysis of anxiety, dread, and emotions supposedly uniquely experienced regarding death, may now be analyzed [48].

f. Because emotions are not the sorts of things, like steam, which can be "released," the widespread belief that grief, bereavement and other supposedly pent-up emotions must be "released" by the funeral ceremony or in other ways is rejected. The funeral director is not a grief therapist and the typical funeral ceremony worldwide is said to be an irrational, enculturated activity. Sometimes one even could call it "barbaric." It may be pointed out, however, that the funeral director is obliged to give the kind of service people in a particular culture ask for – whether or not it is beneficial or rational.

   "Releasing emotions" does not get at the cause of the emotion. Emotion is assessment which guides feeling. If the assessment is unrealistic, faulty, or confused, then negative emotions will most likely be induced. The client's cure will come from the clarification of beliefs, not from "release of energy," "release of emotions," or other fictive or magical "psychic forces." If the funeral ceremony is for the purposes of "releasing emotions" it is misguided. The funeral ceremony, on the cognitive-emotive theory, would rather be for the purpose of clarifying one's assessments and emotions in a rational, healthful way.

g. Rational assessment in regard to dying depends upon our knowledge of death. We see death in terms of our metaphors for death, e.g., sleep, nothingness, journey to heaven, etc. If we do not think carefully about death we may fear it as an unknown or create superstitions about it. That is, in order to create better emotions concerning death we must become clearer about what death is.

We fear death without a clear knowledge of what it is we fear. It is an emotion without a clear object. Thus, this emotion is sometimes called "anxiety." It may also involve depression, dread, etc. An analysis of "anxiety" shows it not to be an internal state. Such would be an extremely misleading and unhelpful concept [48]. According to the cognitive-emotive analysis of emotions, anxiety is a descriptive assessment, which guides feeling, but in this case one's assessment cannot be too clear because people know very little about death. This means that in attempt to cope with anxiety concerning death one must find out more about death. It forces us to inquire into death. The prevailing views concerning death are superstitious or philosophically inadequate. There is death-denial rather than open, honest inquiry into and clarification of the concept of death and dying. Thus, to improve our emotions regarding death and dying we must improve rational assessments by inquiring into, and clarifying our concepts.

21.3.2 The Cognitive-Emotive Theory of Grief and Bereavement

Grief and bereavement as such do not improve the understanding of death. They are based on faulty assessments such as failing to accept the fact that people die, that is, death-denial. People often grieve for most of their lives. If we were to ask someone close to us how long and how much they want us to grieve for them after they die, they would often say, "Not at all." In contrast to that, one might hear the reply,

"At least a year," or "Follow the guidelines in a book on etiquette." Rather, open inquiry in advance, assessment at the time, and communication are essential for avoiding destructive emotions. It is important for the bereaved to communicate their views about dying and death or to communicate to a sympathetic, careful listener. Communication is perhaps the most important form of therapy. We may note that we often do not grieve when an elderly person dies. This suggests that perhaps grief may be arbitrary. Also, we do not grieve for people who are distant from us.

Assessment when one has the feeling or emotion of grief and bereavement is necessary because it is at this time that one must challenge one's previous patterns of thinking. In the case of shock it is too late to reassess at its initial stage. Thus, advance assessment is needed. In sum, in place of grief or bereavement or irrational funeral ceremonies or rituals, or at least in addition to them, the bereaved need to know how to properly deal with their emotions, obtain the desired information about death, and obtain sound advice about how to psychologically and economically proceed with their lives. As Mitford, Harmer, Shibles, and others have pointed out, funeral merchandising and ceremonies can often interfere with these goals [49].

DSM IV states concerning pathological grief, "After the loss of a loved one, the symptoms persist for longer than 2 months or are characterized by functional impairment, morbid preoccupation with worthlessness, suicidal ideation, psychotic symptoms, or psychomotor retardation" [50].

DSM excludes grief from the category of mental disorder. It does so because grief is a "culturally sanctioned response."

Grief behavior usually is culturally imposed. In one culture it is thought to be harmful for children to attend funerals, in another the child plays games at funerals and enjoys it with no harmful aftereffects.

On the cognitive-emotive theory the following would be recommended:

(a) The adoption of the view of emotions as assessment, which guides feelings.
(b) The employment of this view of emotions to effect changes in present funeral practices. This would make them more rational practices and bring more dignity to one as an intelligent human being.
(c) The employment of this theory of emotions in treating the bereaved as well as the dying patient. This would involve open, honest, discussion with the dying patient. Because thought is mainly language use, careful attention should be paid to the language used and captivating metaphors of those affected.
(d) Because emotions are descriptive, empirical assessments they force us to promote and engage in further inquiry into death, aging, dying.
(e) Negative emotions should be regarded as diseases or forms of sickness which can and should be dealt with.

On the view presented above, without a full scale clarification of our ideas concerning emotion, death and dying we become slaves of our emotions. We live and die our metaphors. We die a thousand deaths before our own, live in fear of fictions, and erect monuments to our lack of inquiry.

21.4 Philosophy of Religion

21.4.1 Views in Theology

Religion involves language. It would appear, however, to wish to go beyond it. Can one get outside of one's language – beyond language? Contemporary theologians now attempt to make obscure religious terms meaningful by reducing them to concrete experiences, which are intelligible. To take "eternal life" literally rather than as a language use, or to take it as metaphorical is to create a myth as well as to ignore what the person who uses the term is really experiencing. The terms "beyond" and "transcend" are especially relevant here. They lead to false analogies.

Ian Ramsey wrote, "Too often have people talked as if the way to solve theological problems was by great familiarity with God, when what was needed was a patient and thorough examination of the language being used about him" [51]. Religious language such as "God," eternal life, etc. seem to name entities, the devout want them to name. But we never find what they are supposed to name. Why? The answer given is that they name performances and activities experienced at the moment, have a meaning only in the linguistic and situational context of the moment, and only this use of such words is their meaning. To think they name things is a naming-fallacy as unfounded as thinking that ideas, mind or time name objects in themselves. "Eternal," means on this view, only to point to such things as loneliness, despair, wishes, or separation. For Ramsey, "Hell" is not a place, for example, with tables and chairs, not a place at all. DM MacKinnon wrote, "One cannot if one is honest, ignore the extent to which metaphysical arguments, like those concerning immortality, have gained plausibility from a refusal to attend to the logic of our language" [52]. Ramsey's approach to religion is very close to that of ordinary-language philosophers. The meaning of a term is its use in a language-game (i.e., as part of a language context and a specific situational context). To desire God may then mean only to wish one were more secure and less fragile. Ramsey further sees that metaphor, especially, characterizes religion. Death may then be examined by the exploration of metaphors.

People think that "eternal life" names a state or entity or place somewhere. We have to see what eternal could possibly mean here. It seems to mean something like, "I do not want to die." As to the notion of infinite time or the eternal, this is a vague abstraction. If time is merely change and there is no time as such at all, then eternity has no meaning in terms of time itself. That is, if there is no time as such, there is no eternal time either. To think so is a naming-fallacy. Rather we can simply look and see how the words "eternal life" are used in a specific, actual, living situation. How they are used will then constitute their meaning.

21.4.2 Old Testament

In general, the Old Testament (OT) regards death as an unknown. Very little is said to make it intelligible. The same is true of the *New Testament* although it adds the doctrine of resurrection. People are seen to become one with nature. Job 17:26 "They shall lie down alike in dust, and the worms shall cover them" [53].

Eccl. 9:4 "To him that is joined to all the living there is hope" [54].

Immortality is often thought of as only the survival of the community or one's ancestors. The emphasis is on relations in a religious society not on another life. It is not clear what death has to do with one's moral behavior. One's obedience and devoutness is judged. The judgment takes place at resurrection: Eccl. 7:1 "The day of death is better than the day of one's birth" [55]. In the Middle Ages and Renaissance this became the Christian slogan: *mors melior vita* (death is better than life). Accordingly, universal suicide was sometimes preached and practiced.

Eccl. 9:10 "Whatsoever thy hand findeth to do, do it with thy might, for there is no work, nor device, nor knowledge, nor wisdom, in the grave, whither thou goest" [56]. Death is final without divine enlightenment. Therefore live.

Ezek. 18:4 "The soul that sinneth, it shall die" [57]. "Soul" means whole person. One who chooses to oppose society creates for oneself a social death, which leads to a physical death. One dies also in that no one will remember positively such a person. Death may be thought of in the context of human social injustices and situations, rather than as descriptions of places. Judaism, for example, which bases its beliefs on the OT has no dogmatic belief about an afterlife.

21.4.3 New Testament

It is usually thought that the New Testament (NT) gives the only answer to death and immortality, but there is no analysis or clarification of the concept of death at all. Its concern is mainly to induce people to live according to the social laws of a religious community and the Ten Commandments. Talk of God (eternal life, heaven, hell, etc.) is metaphors for human, living interrelationships. Some writers give metaphysical interpretations of soul, etc., but such views are not made intelligible to us, nor is the view of a new resurrected body intelligible, outside of the explanation of its being a product of self-hypnosis, belief, or faith. In the philosophy of religion it is typically found, even by St. Thomas Aquinas, that there are no rational arguments for the existence of a God or many gods. On this view, the Christian, for example, has neither the answer to the nature of death nor is "salvation" intelligible. The Christian does not look at the arguments in the philosophy of religion or critically inquire into the concept of death, and the Bible says virtually nothing about either.

With the New Testament (NT) comes an increasing belief in resurrection of the individual instead of just a communal resurrection of Israel. Rom. 6:5 "For if we have been planted together in the likeness of his death, we shall be also in the likeness of his resurrection" [58]. The NT is symbolic, mythic and metaphorical, the stress is on reform rather than on eternal damnation. The Protestant stress is on life and experience here and now. OT is a statement about life, not death. It is, also, argued against the OT that there is no contradiction in the view that there is a God and no survival or afterlife, even if God were somehow made into an intelligible concept.

There is good reason why members of the various religions have not contributed to an understanding or clarification of the notion of death. In the first case, the Bible is in the mythical not the scientific tradition. Secondly, religious statements are held dogmatically or on faith. It is thought that whatever the Bible says about death is the truth and so there is no need for further inquiry. Religion in this way held back inquiry and investigation into the concept of death. The church condemned physicians and medical research, holding that the only cause of disease is sin and the only cure is to be religious. Thus, there has been little interest in and almost no research done on, the inquiry into the nature of aging and death. Religion has supposedly answered the question a priori. Thus, we today do not know about the causes of aging and serious and honest inquiry into the concept of death has hardly begun.

The religious emphasis on immortality may also be seen as death-denial. It is put in the form that one must die in order to live religiously, that is, it is not immortality so much as resurrection, but death nevertheless is denied. We become victorious over death. But the victory comes by leading a religious life. It is perhaps more a victory over life than a victory over death. Also, if one thinks one will live forever there is little incentive to support medicine or medical research.

On the other hand, religion does not give us a critical philosophy of life or ethics either. The person who is dead is the person who is not yet aware of honest, open, inquiry into oneself and one's environment. What survives in religion is religion itself. Humans are superstitious. Therefore, even to make honest inquiry into death acceptable it must be made religion. To be understood by the populace this inquiry must then have its rituals, myths, poetry, and music. We light a candle for inquiry.

One religious view is that at the moment of death the soul leaves the body and a beatific vision or ecstasy is experienced. The soul is often thought of as being rooted throughout the body and thus pain is experienced when the soul leaves and its roots are pulled out. But the ecstasy experienced is not, as the dogma would have it, that we at death are pure soul and can see God. Physicians have shown that rather the anesthetic action of carbon dioxide on the central nervous system as well as the effect of toxic substances produce such ecstasy.

21.5 Humanism

Contemporary humanism is based on and defined by *The Humanist Manifesto I and II* and *A Secular-Humanist Declaration* [59]. Although there is perhaps no rigid humanistic view of death, the general view has been presented by Corliss Lamont [60]. It involves the following beliefs: (1) Immortality is an illusion. Belief in immortality is a dishonest belief for which there is no evidence and is a mere superstition arising out of our desire to live forever. All supernaturalism, personal immortality, and conscious survival are rejected. Humanists also reject divine purpose. (2) Belief in immortality or an afterlife is harmful. Christianity has made humans disregard this life in favor of the next, thereby causing great suffering and inhumanity to humans and nature. By accepting this life as the only one, humanists think that humans will make it a better world to live in, instead of mere preparation for an illusory next life. Lamont says, "It is best not only to disbelieve in immortality, but to *believe in mortality*" [60]. (3) Death is a natural event and is in accordance with our scientific views of it. It is a part of biological evolution, and life comes to an end. The material ingredients of the body, alone survive. (4) Our present funeral practices are inadequate to the extent that they are based on superstition, religious or otherwise. If there is no future life we need not be concerned about the dead. Some hold that funeral practices should be based on an informed and intelligent philosophy of death. Any ritual performed should have a rational basis in one's relationship to others and to nature. It should be a humane ritual rather than based on religious illusions of an afterlife, eternal punishment, hell, etc.

21.6 The Rhetoric of Death Using the Metaphorical Method

Death is all metaphors (Dylan Thomas)

21.6.1 Introduction to the Metaphorical Method

The central concern with the analysis of language is as true for existentialism as well as analytic philosophy. Wittgenstein's ordinary-language approach, perhaps the most significant one in this century, will be mainly discussed here. We note, however, that there are ordinary-language strains in numerous previous philosophers. For Marcus Aurelius the ordinary-language philosophy element comes in his concern with the purpose of the present action performed, or language actually spoken and his stress on that which is in front of our eyes, but which we often fail to see due to our vague theorizing or over-imagining. In one important sense we know all about death. It is something like our observing a leaf falling from a tree. On this point, Aurelius in his *Meditations* speaks of a "disposition, which gladly accepts all that happens, as necessary, as usual" [61] and of a method "Use plain discourse" [62]. "Be intent only on that which you are now doing and on the instrument by which you are doing it." "Everything which happens happens justly" [63]. "He who has seen present things has seen all" [64]. "Everything which happens is as familiar and well known as the rose in spring and the fruit in summer; for such is disease, and death, calumny, and treachery, and whatever else delights fools or vexes them" [65].

Our notions of death would accordingly be related to and reduced to paradigms of what we know of what is present before us in the here and now. For example, we can only know and speak of death in terms of our actual use of "plain discourse." The reduction to the common and everyday is one interpretation of what is meant by Aurelius' stress on living according to nature. "Plain discourse" does not mean everyday language, but critical, rational, philosophical clarification (See Chapter 18 for a full analysis).

The analysis given here presupposes the view that language has epistemological primacy, much as does Wittgenstein's *Philosophical Investigations*. Also, Heidegger states, "We – humans – are a conversation, the being of which is found in language" [66]. Thought presupposes language, but language does not presuppose thought. Language presupposes language. Thus, we are in a linguo-centric epistemological predicament in regard to our understanding of death. This means that we can play language-games with death, but that error arises when we unwittingly mix them up, use words in the wrong language-games, or imagine that we can go outside of language – and so outside of knowledge speak of death in itself. Rather, we may say that whatever is known or said about death is *said* about death. The question of death raises the question of the limits of knowledge – i.e., of language. There are no ultimate explanations although we may play a language-game of explaining, of metaphysics or theology. It is only to play one language-game or another. The problem of death involves the realization that most of what we "know" may be false.

Such statements about death should then be bracketed, kept within their boundaries, put in quotation marks. It is in this sense that we may interpret Plato's statement in the *Apology* that no one knows what death is. "To be afraid of death is only another form of thinking that one is wise when one is not; it is to think that one knows what one does not know. No one knows with regard to death..." [67]. As mentioned above, Wittgenstein said that death is not an event in life when he adds: "We do not live to experience death" [68]. This statement too becomes a contextually living statement. With the rhetoric of death we explore these language-games: juxtapose, reduce to absurdity, create new insight, etc. and in these ways expand the possibilities of language to its limits.

Because there are as many meanings of "death" as there are language-games, it is easy to equivocate. We speak of "death" (cosmic) as if it were merely the living pain of dying. We fear death as we fear things in life, think death is darkness because we fear darkness. Our fears of death are often in this way ordinary fears of life experiences, rather than fear of a cosmic death. For the average person there is nothing extraordinary about death. The *Metaphorical Method* presented here helps us to make death extraordinary. We also speak of death for terrestrial ethical reasons – whereas from the cosmic perspective such ethics may be irrelevant [69]. Death talk reduces to concrete experiences, events, uses, or techniques in specific language-games we play [70]. Our utterances about death are determined by our analogies and metaphors, which cannot be taken literally or as the only possible ones. To do so would be to misunderstand the use of our language and to misuse our language as is often done in dogmatic metaphysics. There is no literal description of death. Death is not a fact. Metaphor has meaning, which cannot be reduced to the literal [71]. To understand what death is we should look and see how and in what contexts death language is used. Its meaning will only be its use. The metaphorical method employs the different types of metaphor in order to analyze or create an assertion, theory, or method, as will be shown. Some examples of kinds or uses of metaphor are: visual metaphors, (as illustrated by the use of the visual *I Ching* hexagrams to be discussed), tension metaphors, combinations of opposites or oxymora, reciprocal metaphors, reversals, far-fetched metaphors, satire, personification, poetic metaphor, insight metaphor, juxtaposition or parataxis, metonymy, connotations, associations, expansion of metaphors, concise metaphors, similes, comparisons, analogies, seeing-as, as-if or heuristic metaphors, paradoxical metaphors.

The metaphorical method of inquiry involves constructing sentences combining the term to be clarified with an entirely different term. The form is "A is B," for example, "Life (or death) is a journey," "Death: that undiscovered country." Also, we may construct alternate or deviant relations or verbs as in "AR" or "ARB" where "R", is some relation. E.g.: (AR) "Death continues." (ARB): For Hegel the thesis *dissolves into* the antithesis, which *grows* into a synthesis. For example, death is a continuance of life growing into a synthesis in the cosmos. Some other relations to explore are: A causes B, A of B, A in B, and so on. We make the familiar seem strange, e.g., Death in life, and life in death. "What is life?"

In everyday life there is nothing extraordinary about death. The Metaphorical Method helps us to make death extraordinary. Metaphor deviates from usual associations and allows us to arrive at insights we would not otherwise have achieved. The following rhetorical techniques constituting the Metaphorical Method may be used for the exploration and criticism of any term or statement, including that of death [72].

21.6.2 Rhetorical Techniques for the Exploration of the Concept

The following rhetorical techniques may be used for the exploration of any concept.

21.6.2.1 Abstraction (Fallacy)

Suppose we say, "She lived, she died." This is not obvious. Death is a vague abstraction. The problem is that it is too familiar. "Die," "dead," and "death" are dead metaphors. "Death" itself is a dead metaphor.

When we speak of death it is like trying to go beyond language. Is death, then, like ZpˆØ¿†∞$^{\Omega}$? And now we see that we cannot "say" this either – that it is also not the break-up of language. Death is in one sense only a living experience.

Suppose one says, "There was a time when we did not exist and we do not worry about it; so why should it bother us that a time will come when we shall no longer exist?" There is a false analogy here because "time" is a subjective and living narration only meaningful for the living. Do not ask, "What is time to a stone?" There are also diverse theories of time. Additionally, "cease to be" is a living phrase, and so not applicable to death. We cannot speak of the "end of life" as we speak of, for example, the "end of the day." The term "life" is problematic as well [73]. Nagel said, "Death is nothing and final" [74]. Absolute nothingness or an absolute end seems not to be within our experience. Of another's death we can say his world alters. We are conscious of it altering. But we cannot say of our death that it alters, because "alter" implies "alters for my consciousness," and there supposedly is no consciousness. I can alter only from the viewpoint of another. But at death my consciousness of the viewpoint of another comes to an end also. What is meant by "end" here must remain in question. The end of life is not necessarily like the end of a trip.

It may be more like the end of knowing or an unknown end – an oxymoron. It is the end of that which determines that there are ends in the first place. It is a paradoxical ending.

Suppose, say a football lands in a cemetery. Now, we want to say there are dead people there, that perhaps somehow the dead are being disturbed. There is a profound contrast and connotation between a dead person and a ball used in living play. Similarly, there is only a grammatical and usage difference between "understand death," in quotes and without quotes. By not using quotes we have not gotten at death itself. "I will still be me after death"; "I will not be me after death." Both statements are problematic. There is no self (or many selves) as such to die, or we can equivocate with "I." We seem to think that some morning we will wake up dead. Our language is stretched: "I once was a person, lived in 1999, etc." We use words but seem not to understand them. We think that we will "return," or come back as something else.

21.6.2.2 All-Statements or None-Statements (Also "Always" and "Never" Statements)

At death we will know everything. The deity is all knowing. All is relative. We can't know anything. Nothing is more true than anything else (relativism). No one has answers, only questions. Everything is absurd (Existentialism). Nothing is true (Buddhism). All language is misleading (Plato). One may also see if the statements made violate any of the other informal logical fallacies.

21.6.2.3 Allegory (cf., Symbol)

A metaphor is expanded to generate meanings on several different levels. It may expose the subject or character of death, e.g., The wafer stands for the body – "Isn't that cannibalism?" "We read off God in nature." – "Then, would you read this pencil?" "God is every smile." – "Then what do we need God for if we have the smile?" Similarly, one may develop an "archetype," that is, a metaphor repeated so often that it seems to be a universal idea or truth.

21.6.2.4 Ambiguity

We expose death by the use of ambiguous terms. "The poles are kissing as they cross" [75]. "He went to the kitchen and then suddenly died." This is a faulty parallel. One does not die like one goes to the kitchen. One may, however, "die" as part of our living experience in the same sense that one falls down. To collapse is not the same as dying. It is the wrong paradigm (cf. *category-mistake*). "Death" equivocates between naming a special transcendent state, and referring to a common experience such as falling down.

21.6.2.5 Analogy and Simile (cf., Category-Mistake, Simile)

Analogy and Simile show parallels and mistaken parallels. We think we "know" of death from an analogy to our living experience of other people and animals around us, which we say have "died." But that is still a living experience. A conscious experience cannot give the experience of not being conscious. "Sleep is a little death" is not a parallel because we wake up. Death is that which can never be told about because it is the breakdown of language. When we speak of death it is like trying to go beyond language.

"No one has come back from the dead to report about it." If someone had come back from death with a, so to speak, first hand report it would still be a living report outside the context of death. I am alive and listen to the report. Perhaps I take some notes, or record the report. Death is a linguistic experience of the living. In a sense death is human, a linguistic personification. Language and consciousness are life. The concept of death breaks them down. Death is the number of questions we wish to ask. William Hazlitt wrote, "There was a time when we were not: This gives us no concern – why then should it trouble us that a time will come when we shall cease to be?" [76]. The origin of this statement is most likely from the Stoics. There is a false analogy here because "time" is a subjective and living concept only meaningful for the living. Additionally, "cease to be" is a living concept, and so not applicable to death. We cannot speak of the "end of life" as we speak of, for example, the "end of the night."

21.6.2.6 Behavioral Metaphor

"But now if I say 'The bread and wine change into the body and blood of Christ,' what can I mean? If you test the bread and wine before and after the communion service, it appears to be exactly the same; there isn't any test which would show any change in it at all" [77]. Exclamations, particles and interjections are thought to be behavioral expressions of emotion having little genuine meaning. It has been shown, rather, that they are rich in meaning [78]. Suppose one goes to a funeral and after the long event returns home, sits down with a beer and says, "Well." "Well" here means a great deal. To find out something about it we must know the context, intonation, etc., but what it means will be metaphorically richer than its paraphrase.

21.6.2.7 Category-Mistakes (cf., Context Deviation)

Terms of one situation are used to apply to another. The following examples suggest that what we usually say about death involves category-mistakes. This is because we speak of death, which is in language and experience as if it were outside of them: (a) "Afterlife" (b) "Life is short" (c) "He is dying." One cannot be "dying" if one does not know what death is. (d) Whether we say we can or cannot have knowledge about death we remain well within our living linguistic experience. (e) Suppose, say, loud music comes from a radio in a cemetery. Now, we want to say there are dead people there, who need to be left quiet. (f) I can remember my "dead" grandfather, but not my dead grandfather. Similarly, we can "understand death," but cannot understand death. Both statements are problematic. (g) There is no self (or many selves) as such to die, or we can equivocate with "I." (h) "To die and go we know not where, to lie in cold obstruction" [79]. But "cold" is not to the point, nor is "go." Nor does "recycled" fit. (i) Suppose we say as some do, "I believe I will live forever." We do have great power to believe such things. This can mean something like, "I am now writing this sentence," or "See, the sky is clouding over." (j) When it comes to time words, death calls time into question. "But you will stay in the ground for such a long time." This is like, "The bus will be very late ". For similar reasons we cannot speak of either knowing or being ignorant of death. On this view, we can neither imagine nor not imagine death. We cannot say about death, "I don't know about it," or "It is a question."

21.6.2.8 Grammatical Term Metaphors

Death may be seen as a grammatical mark: brackets, the asterisk, comma for a coma, death as a period or question mark, a footnote to life, the funeral as an exclamatory mark. Wittgenstein similarly wrote about, "A whole cloud of philosophy condensed into a drop of grammar" [80]. Death! With or as an exclamatory mark reduces death to an emotive term. This is part of the grammar of death [78].

21.6.2.9 Circularity

This is the metaphor of combining identical things or assuming what is to be proven. For example, whatever happens at all happens as it should [81]. Thus, death is supposedly not a problem. We may accordingly analyze the synonyms of death, e.g., "end of life, decease, end, demise, pass away, expiration, termination," etc. We may then rewrite our question about death in various possible circular formulations, such as, "Death is the end of life," "If you believe you will be saved, then you will be saved" inasmuch as you believe it. "'What is unsayable is unsayable,' is a significant tautology" [82].

Also, we may construct and explore meta-language and self-applicable statements such as fear of fear, death of death. We may say, "So far in my life I have been eternal." "Eternal" like death is a concept limited to our lived experience – as is also this very statement. In this sense we are eternal – the eternal present. This is also a circular statement in the sense, like saying that I experience the present now.

One may say that in everyday life one does not try to understand death, one only speaks of the things in one's life when one does so, such as going away or inheriting. In this sense, we are like animals, which do not know that they will "die." So death for them always begs the question (assumes what it is supposed to prove). Thus, death is not a significant problem for people, which requires research. Reverence for life is rare. Religions can be cavalier regarding death because as long as they keep on speaking of death there is none – one "lives forever" as long as one can say so. There is also circularity in the statement that there is no meaning in life except that we give to it. In compari son, one could say, there is no meaning in *death* except that we give to it.

Epicurus says, "Death means nothing to us, because that which has been broken down into atoms has no sensation and that which has no sensation is no concern of ours" [83]. This assertion presupposes all knowledge, which we do not have. It also treats something outside of our language and experience as if it were inside.

21.6.2.10 Connotation (cf., Free Association)

The now empty bed, in which you made love so many times with the partner, who died, can be more shocking than seeing the dead person. We associate death with black (or with white in parts of Asia). Death is seen as darkness, cold, a skull – things feared only among the living. A skull is not death and in one sense has nothing to do with death. We are shocked by a drowned child's teddy bear, which has been washed to shore after a boat sinks.

21.6.2.11 Context Deviation (cf., Category-Mistake)

With context-mistake a statement is taken literally, the false seems to be true. We may purposely use the terms to be analyzed, in a wrong context. Tension is created. This may give insight. Example: Can you hear time pass? If we do not know what time is how can life or death be eternal? "If I said, 'Smith always answers when you speak to him,'... you know what I mean.....But now suppose I say, 'God answers prayer'.....You ought to begin to wonder what I mean by 'answer.'...And now it's not clear that I really mean *anything* by the word 'answer'" [84].

Examples: Do you die fast or slow? You can't define death, only a word. Where is your death? Compare: Where is your mind? How long did the enjoyment of life/death last? Does death last forever? Say a last "goodbye" when you go to sleep. Death: Can you hear the silence?

21.6.2.12 Contradiction Humor (cf., Poetic Metaphor)

Types of contradiction are: (a) opposites and analytic contradictions: verbal, relatively objective contradictions in definition (definitional contradictions). (b) incongruity contradiction: connotative contradiction mainly between secondary meanings, actions, and perceptions. (c) synthetic contradictions: based on experience between statements, actions, perceptions.

(a) Opposites and Analytic Contradictions:
    John Austin wrote, "In general, it will pay us to take nothing for granted or as obvious about negations and opposites. It does not pay to assume that a word must have an opposite, or one opposite" [85]. The abstract meanings of "opposite" and "contradiction" must be given concrete meanings. "Opposite" has the metaphorical meaning of "facing each other: in a boxing ring," or "on different sides of a color chart." "Opposite" is a synonym of "unlike." "Contradictory" has the added meaning that the unlike terms cannot ordinarily both be true at the same time in the same sense. "Life after death," is contradictory. We do not understand such a statement.
    When on one level a statement is contradictory, but on another level makes sense, there is a humorous contrast developed such that there is unity in difference or truth in apparent falsity. This mechanism is mentioned in Beardsley's

"logical absurdity" theory according to which, because the first level of meaning of a statement does not make sense, it must for clarity be taken on a second level [86]. We may also think of death as humor in these terms as well:

After death we will be happy.
The best life is the afterlife.
All death is rape [87].
Be two people at once (treat dead as if still alive).
Dead life.
Good death.
Happy death.
I died.
Incongruous congruity.
Learn the obvious (about death).
Life in death.
Life: an important triviality.
Life after death.
Meaning in meaninglessness.
Order is disorder (Dadaism).
Possible impossibilities.
Practical impracticality (hope).
Real unreal.
The real is surreal (Dadaism).
Say it nonverbally.
Solve a problem by escape.
The unborn want to live.
Truth in falsity.
Ugliness in beauty.
Understand the inexplicable.
Unintelligible intelligibility.
Useless passion (Sartre).
Vague clarity.

Examples of faulty thinking: Knowledge of the unknown. The natural is supernatural; the physical is metaphysical; after you die, you will know everything. To deny thought itself to achieve salvation is to deny oneself to save oneself.

(b) Incongruity Contradiction: Connotation
Examples: Gentle death. Stony people. Loving death. Convict on the way to the gallows in winter asked for a scarf to prevent his neck from getting cold. If you believe there are fairies in catsup, you are insane; if you believe there are ghosts and angels, you are religious. Each one has a right to his or her own irrational belief.

(c) Synthetic Contradiction: Contradiction Based on Experience
Examples:

A bad death.
A perfect death.
Are you dead?
Can you doubt doubt?
Casket with an adjustable bed.
Cemetery with a view.
Death with dignity.
He was even late to his own funeral.
After death, I will not be me.
I will still be me after death.
Kill for self-defense.
Life is absurd.
After you die will you still love me?
Speed limit by cemetery is 10 mph.
Unborn people.
We can never know reality in itself.
We live to die (Teleological fallacy).
We only kill to preserve peace.
What was the world like before it was created?

21.6.2.13 Defense Mechanisms

Death denial and religion have been thought of by some as defense mechanisms against death and all of one's fears. The irony is that defense mechanisms do not defend, but distort reality, for example, rationalization and wish fulfillment.

a. Denial. "My religion guarantees that if I behave in a certain way, I will never die."
b. Rationalization (Compare "wishful thinking."). "If you believe, then you will be saved."
c. Repression (Not face an issue. Censorship). Do not acquaint yourself with the arguments in the philosophy of death or philosophy of religion which then could undermine your beliefs.
d. Sublimation. A metaphorical substitution of one's true desires in the false belief that it will guarantee one eternal life.
e. Symbolization. God is a glass of water in the middle of a desert. This cross will protect me from all evil. . .common, normatively held beliefs.
f. Wish fulfillment. "I was lonely, poor, tired, tripped over my own feet and couldn't do anything right. Then I found God."

21.6.2.14 Deviation

a. Deviation from the familiar. We believe in an afterlife, we just don't believe there is a god.
b. Deviation from the ideal. Isn't it too quiet in heaven?
c. Deviation from language. We die within life. What about an after-life then?
d. Deviation from the practical. Live forever.
e. Deviation from tradition. There is no death as such.
f. Deviation from the usual. In Tibet, the Buddhist belief is that a hair from the scalp must be removed in order to allow the soul to escape.

21.6.2.15 Euphemism

The word "death" is avoided wherever possible, and there is general death denial. This leads to depression and shock. "Depart" or "go away" are used in place of "death," corpses are "loved ones," the corpse is put in a "slumber room," a "death certificate" is a "vital statistics form," embalming fluid is called "lifelike" fluid. There are cemeteries with a view, caskets lined with velvet with adjustable beds, and similar practices, including religious ones, which present the belief that people do not really die.

21.6.2.16 Free Association (cf., Connotation)

Death is a wild garden.

21.6.2.17 Hopelessness

We are temporarily alive.

21.6.2.18 Irony

"Born to die" is cosmic irony. "The aspects of things that are most important for us are hidden because of their simplicity and familiarity". One is unable to notice something – because it is always before one's eyes" [88]. So, e.g. that we die.

21.6.2.19 Juxtaposition

We may use the *I Ching* [89] whereby one question is asked, but a number of replies can be given. One could by juxtaposition read all of the hexagrams as a reply to this one question, although this is not part of the usual *I Ching* method. The question asked was: "Please give me insight into the nature of death." First reply received was:

a. "Difficult Beginnings": "The birth of every new venture begins in some confusion because we are entering the realm of the unknown. . .nothing less than complete chaos, it ultimately presages a time of order and efficiency.
b. Inexperience: We can handle much, but not death. We do not know what to do about it. We need a questioning attitude. Unrealistic fantasies and obsessions [religion] may consume us. Humor is generated from this model because it is stated that we may have "beginners luck" regarding death.
c. Conduct: Disorder and chaos of death cannot touch you if you behave with dignity, reevaluate values and keep an open mind. Our powers are not adequate to cope with death. Thus avoid expectation and demands. Success comes to those who can weather this storm while maintaining their principles (cf., defense mechanisms, religious fictions). Things are struggling to take form in the self as well – an identity crisis. Accept these changes in your self without combating them. . .. Allow fate to manipulate external events.
d. Stagnation. No growth, nothing can be accomplished, only misunderstandings. Withdraw. This is the ultimate breakdown of language. The situation is blocked in regard to the object of your inquiry. Only by reorganizing your priorities will you transcend it." It is something that we cannot deny or oppose, but must bear whatever it is.
e. Repair (Decay). We bring about our own demise by lack of critical thinking in our lives, not paying attention to our nutrition, lack of adequate support for research into medicine, lack of research into the understanding of the expression of death, in short, lack of critical thinking and inquiry. The same question is re-asked and thus by juxtaposition given a second or third reply.

21.6.2.20 Metaphor and Metaphorization

This is a central technique of gaining insight. It takes limitless forms. Because we cannot have a full understanding of death, we may use metaphor to explore its various possible meanings.

a. We may construct sentences combining the term to be clarified with an entirely different term. The form is "A is B," for example, "Life (or death) is a journey,"

"Death: that undiscovered country," "Death: dinner for none," etc. We fear that "death is dark and cold."

b. Also we may construct alternate or deviant relations or verbs as in "AR" or "ARB" where "R" is some relation. Examples: (AR) "Death eats up our lives." Some other relations to explore are: A causes B, A of B, A in B; e.g., "The reader is to get *inside* the subject," A on B, A or B, A with B, A and B, A to B, and so on. That is, replace A, R, and/or B with the unlike terms or unusual relation so as to give insight into the terms in question. Examples: "At death we will go into the blue sky." "Will you meet me there?" Death is when life is expected in return. We make the familiar seem strange, e.g., "What is life?" or "How do we know we die?" "Man *is* the microcosm: I am my world" [90]. "Death is the limit of all metaphors." "Death is the destruction of language." Shakespeare wrote, "Cowards die many times deaths before their death" [91].

We commonly hear the view that one expects to be with one's god or spouse after death. This may be regarded as a negative effect of family. Some believe their cat will be there also.

21.6.2.21 Metonymy

Substitute an attribute or association of a thing for the thing, e.g., "Went away" for "died." "Kicked the bucket."

21.6.2.22 Personification

We erroneously treat things as humans or humans as things. "The fertilized egg is a person." "Death is the evening of life." "Death is a grim reaper." We give inanimate things gender. In German *Der Tod* ("death") is masculine. The state of death is often personified as an organism. We ask, "Will we be lonely after death?"

21.6.2.23 Poetic Metaphor

Poetry creates connections in language, which give insight, for example, "Death is measured with a broken watch." The following poem also exemplifies *contradiction* or *oxymoron*.

JOIE
DE MORT
Commissioned to
kill with a
misfired education
uniformed thoughts
lined up in a row
skewered by false
belief

they target the heart
bomb hungers and
winged symptoms,
let me count the
ways,
build torpedoes for
fish
rifles for lovers
are generous with
ammunition
and embrace the
world with their
arms.
They know the
beauty of the hunt
under a silver-plated
sky,
provoke the sunset
with negotiated fire
escalated to the
mirrors for the stars
and in their sights
see faces in their
shoes
and salute their dead.
Who kills
can never live to die,
must self-destruct
before it is too late
to win a death.
Warren Shibles

21.6.2.24 Probability

We may assess the degree of certainty of the various views presented. E.g., Absolute certainty (dogma); warranted hypothesis which can be changed in view of future evidence; merely 50% possibility. Is it a belief or view, which will not be changed even in the light of future evidence? It is not possible or probable that there are gods, or spirits, or that the world was created by them, because probability does not apply to undefined concepts or the supernatural – or to a horse race with Pegasus or the fictional Black Beauty running in it. The use of "certainty" is circumscribed by contexts. The following statements are thus problematic: "Death: A chance at the unknown." "Eternal punishment," "I know that I will go to a better place after I am dead."

21.6.2.25 (Faulty) Questions and Riddles

It was thought that one might unravel the mystery or riddle of the universe – as if a magic word would be the key to it all. Riddles involve questions. "Faulty" question means that there are vague words, contradictions, meaningless words used, improper grammatical mistakes made, metaphors, and any mistake or deviation involved. One way of solving some riddles, and mystical, philosophical, or scientific questions, is to show that the question does not make sense. No question will be completely answered in every way, and it is a mistake to assume that it will be. There is no such thing as *the* answer to a question. Questions about cause are especially misleading. What causes death? If we do nothing to inquire into it, there is a sense in which we cause our own deaths. Ironically we in this way kill ourselves (See also the Chapter 11). What does everything really mean? Regarding existence Wittgenstein asked, "How extraordinary that anything should exist?" [92].

In *Alice in Wonderland,* Alice drinks from a bottle and shrinks to a height of ten inches. She then wonders if she will snuff out altogether, like the flame of a candle. She asks, "I wonder what I should look like then?" [93]. If we do not know what "death" means then questions such as "Does death exist?" do not make sense. So the solution to some puzzles or answers to some questions is to dissolve the question.

21.6.2.26 Reduction to Absurdity

If a fertilized egg is a potential person, so also is a sperm. After death we will all have better penmanship. If I am a good musician in this life, I will become a piano in the next. "I have an afterlife," is not like "Bicycles have a refurbished afterlife."

21.6.2.27 Rejuvenate Metaphors

"Cemetery" is a dead metaphor meaning, "sleeping place."

21.6.2.28 Reversal Humor

Examples: We live in a timeless world. He believes strongly, but in no particular thing. There is no self. We live death. We die life. I believe in eternal death. I'll die forever. In a sense, we cause our own death.

21.6.2.29 Substitution

Shakespeare renders sleep as "the death of each day's life, sore labor's bath, balm of hurt minds, great nature's second course, chief nourisher of life's feast" [94].

21.6.2.30 Tension Metaphors

Juxtapose two things so as to create tension or paradox, e.g., Life is death. Such metaphors create a mystery, paradox. "Is my body me, or am I my body?" "To die is not to die at all."Consciousness is what it is and is not what it is" [95]. "Where no sea runs, the waters of the heart push in their tides" [6]. "Since there is no gate, let me tell you how to pass through it." (Zen saying)

21.6.2.31 Uselessness

Music piped into caskets. Burial rituals.

21.6.2.32 Therapeutic Metaphor

Metaphor may be used to avoid the literal, escape from narrow or oppressive categories, avoid taboo or unacceptable language, provide release, and give indirect ways of saying something. Euphemism is substitution of an agreeable word for one we wish to avoid uttering, e.g., "pass away" for "die." Schizophrenics and others are able to speak metaphorically about things they cannot face directly or more literally. By means of metaphor one is able to distance oneself from an object, person, or situation. Therapy also involves showing that what one takes literally is really metaphorical.

21.7 Death and Medical Profession

The rhetoric about death is a contribution to philosophy of medicine and to narrative medicine, as how to deal and to communicate death against the incommunicable. It is to avoid being silenced and isolated as a dying patient and his relatives. It is to enable healthcare-workers to honestly and critically inquire into death and dying, to develop an attitude open to communication with dying patients and those who experience their loved ones die and also to help them to cope with the encounter of death in their profession as well as in their lives.

Dying patients must be allowed and helped to communicate and clarify their thinking about death and they may be helped by use of a more sound view of emotions. Because thought is mainly language use, careful attention should be paid to the language and captivating metaphors of the patient. We have to be careful listeners. Communication is especially important in grief therapy, and treatment of the dying patient. In fearing death we are fearing loss of what we conceive to be our self. This forces a new view upon oneself. To understand the dying patient we must find out as much as we can about him, his views upon life, his environment, his significant other(s). We must find out about the metaphors and models he lives by [96]. Once, this is known the criteria for the therapy (accompany, being there) to bring about an appropriate death are: 1. conflict reduction, 2. proper understanding of the patient in terms of the image he has of oneself, 3. restoration of important social relationships, 4. satisfaction of his wishes as much as possible [97].

Also those who loose a loved one have to be taken care of. Seeing that a loved one dies suggests that one too will die. The dead person "belongs" to us as a part of ourselves. Death cannot be thought of because our use of language gains its meaning from contexts involving life. But death has no context because it is the end of contexts. To embrace those fallen from these contexts is the task of those taking care of the bereaved. This means to be attentive listeners, to be there for them.

21.8 Final Personal Remarks

At the end of this chapter I realize that what I have written here is also language-use, rhetoric of death. It is appropriate then that I may close with these two simple words: Death, indeed!

Warren died July 17th 2007 working on this book until July 11th and thus giving all the insights he had. I continued to work on the book for almost more than 2 years and finished it as his legacy. . . .. in order to cope with his death and to be more adequate in my profession as a physician, and finally to openly and honestly inquire into my own death and be prepared.

Death I Understand

What is more than death
but only falling asleep?
What is left over?
Life.
Do not feel sorry
for the dead,
but for the living
who cause death.
Betrayal I understand.
Kindness is the surprise.
Death I understand.
Life is the surprise.
Warren Shibles

"What begins as a study of death ends up, to some extent, as a study of life and love" [98].

References

1. For additional definitions of death see Walton, D. 1979. *On defining death*. Montreal: McGill-Queens University Press. For bibliographics on death and dying, see Fulton, R. 1977. *Death, grief and bereavement: A bibliography 1845–1975*. New York: Arno Press; Guthmann, R., and Womack, S. 1978. *Death, dying and grief: A bibliography*. Nebraska: Pied; Miller, A. 1977. *Death: A bibliographical guide*. Metuchen, NJ: Scarecrow Press; Pacholski, R. 1988. Death themes in visual arts: Resources and research opportunities for death educators. *Death Studies* 10:59–74; Poteet, G. 1976. *Death and dying: A bibliography*. Troy, New York: Whitston.
2. AMA Council on Ethical and Judicial Affairs Report B-A91, Decisions near the end of life. 1992. *JAMA* 267:2229–2233; Gillon, H. 1969. Defining death anew. Science News.
3. Dickinson, E. 1955/1958/1979. *The poems of Emily Dickinson*, ed. Johnson, T., 3 vols; Cambridge, MA: Harvard University Press. *Complete poems of Emily Dickinson*, ed. Johnson, T. London: Faber.
4. Thomas, D. 1953/1957. *Collected poems of Dylan Thomas*. New York: A New Directions Book.
5. Thomas, D. 1953/1957. *Collected poems of Dylan Thomas*, 6. New York: A New Directions Book.
6. Thomas, D. 1953/1957. *Collected poems of Dylan Thomas*, 29. New York: A New Directions Book.

7. Thomas, D. 1953/1957. *Collected poems of Dylan Thomas*, 30. New York: A New Directions Book.
8. Thomas, D. 1953/1957. *Collected poems of Dylan Thomas*. New York: A New Directions Book xvi.
9. Thomas, D. 1953/1957. *Collected poems of Dylan Thomas*, 43. New York: A New Directions Book.
10. Thomas, D. 1953/1957. *Collected poems of Dylan Thomas*, 59. New York: A New Directions Book.
11. Thomas, D. 1953/1957. *Collected poems of Dylan Thomas*, 128. New York: A New Directions Book.
12. Thomas, D. 1953/1957. *Collected poems of Dylan Thomas*, 134. New York: A New Directions Book.
13. Thomas, D. 1953/1957. *Collected poems of Dylan Thomas*, 56. New York: A New Directions Book.
14. Thomas, D. 1953/1957. *Collected poems of Dylan Thomas*, 80. New York: A New Directions Book.
15. For an analysis of death see Shibles, W. 1974. *Death: An interdisciplinary analysis*. Whitewater, WI: Language Press. Especially relevant is the chapter "Wittgenstein": 69–80.
16. Wittgenstein, L. 1961/2001. *Tractatus logico philosophicus*. Trs. Pears, D., and McGuiness, B., 6.431 London: Routledge.
17. Wittgenstein, L. 1961/2001. *Tractatus logico philosophicus*. Trs. Pears, D., and McGuiness, B., 6.4311 London: Routledge.
18. Wittgenstein, L. 1961/2001. *Tractatus logico philosophicus*. Trs. Pears, D., and McGuiness, B., 7 London: Routledge.
19. Wittgenstein, L. 1961/2001. *Tractatus logico philosophicus*. Trs. Pears, D., and McGuiness, B., 5.6 London: Routledge.
20. Wittgenstein, L. 1966. Lectures on religious belief. In *Lectures and conversations on aesthetics, psychology and religious belief*. ed. Barett, C., 52–72. Oxford: Blackwell.
21. Wittgenstein, L. 1966. Lectures on religious belief. In *Lectures and conversations on aesthetics, psychology and religious belief*. ed. Barett, C., 52–72, esp. 65. Oxford: Blackwell.
22. Wittgenstein, L. 1966. Lectures on religious belief. In *Lectures and conversations on aesthetics, psychology and religious belief*. ed. Barett, C., 52–72, esp. 68–69. Oxford: Blackwell.
23. Wittgenstein, L. 1966. Lectures on religious belief. In *Lectures and conversations on aesthetics, psychology and religious belief*. ed. Barett, C., 52–72, esp. 69. Oxford: Blackwell.
24. Wittgenstein, L. 1966. Lectures on religious belief. In *Lectures and conversations on aesthetics, psychology and religious belief*. ed. Barett, C., 52–72, esp. 72. Oxford: Blackwell.
25. Liveritte, R. 1971. Some thoughts on death anxiety. *ETC* 28:21–37.
26. Liveritte, R. 1971. Some thoughts on death anxiety. *ETC* 28:21–37 esp. 29.
27. Geiger, L. 1869/1878. *Der ursprung der sprache (The origin of language)* 2nd edn. Stuttgart: 141.
28. Wittgenstein, L. 1958. *Philosophical investigations*, 3rd edn, 329. New York: Macmillan.
29. Shibles, W. 1995. *Emotion in aesthetics*. Dordrecht, Netherlands: Kluwer Academic Publishers.
30. Nagel, E. 1986. *The view from nowhere*, 224. New York: Oxford University Press.
31. Wittgenstein, L. 1964. *Philosophische bemerkungen* (Philosophical investigations), 54. Oxford: Blackwell; "*Ich kann mit der Sprache nicht aus der Sprache heraus.*" ("I cannot by means of language get out of language.").
32. Nielsen, K. 1970. In defense of atheism. In *Perspectives in education, religion and the arts*. eds. Kiefer, H., and Munitz, M., 143 Albany, NY: SUNY.
33. Wittgenstein, L. 1961/2001. *Tractatus logico philosophicus*. Trs. Pears, D., and McGuiness, B., 5.6 London: Routledge.
34. Wittgenstein, L. 1961/2001. *Tractatus logico philosophicus*. Trs. Pears, D., and McGuiness, B., 6.4311. London: Routledge.

35. See also Ryle's view that sensations are not neat (isolated). Ryle, G. 1949/1984. *Concept of mind*, Ch. 7. Oxford: Orig. Hutchinson, London/Barnes and Noble.
36. Wittgenstein, L. 1958. *Philosophical investigations*, 3rd edn, 226. New York: Macmillan. Note also 19: "To imagine a language means to imagine a form of life." .
37. Wittgenstein, L. 1958. *Philosophical investigations*, 3rd edn, 373. New York: Macmillan; "Grammar tells us what kind of object anything is.".
38. Dickinson, E. 1955/1958/1979. *The poems of Emily Dickinson*, ed. Johnson, T., 3 vols. Cambridge, MA: Harvard University Press. *Complete poems of Emily Dickinson* ed. Johnson, T., III, 1741 London: Faber.
39. See Shibles, W. 1974. Imagery and visual perception. In *Emotion: The method of philosophical therapy*: 445–446.
40. Hanson, N. 1958/1961. *Patterns of discovery*, 22, esp.1–30. Cambridge: Cambridge University Press.
41. Stated on 5/12/96 in an interview on CBS's Sixty Minutes with Leslie Stahl. http://en.wikipedia.org/wiki/Madeline_Albright. "Media uncurious about Iraq policy's effects." Fairness and Accuracy in Reporting. Nov.–Dec. 2001 (April 18, 2006) http://www.fair.org/index.php?page=1084.
42. Carroll, L. 1960. *Alice's adventures in wonderland and through the looking glass*, 81. New York: New American Library.
43. Dickinson, E. 1955/1958/1979. *The poems of Emily Dickinson*, ed. Johnson, T., 3 vols. Cambridge, MA: Harvard University Press. *Complete poems of Emily Dickinson*,. ed. Johnson, T., I, 54 London: Faber.
44. Flew, A. 1955–1956. Can a man witness his own funeral? *Hibbert Journal* 54:242–252; esp.243.
45. Flew, A. 1955–1956. Can a man witness his own funeral? *Hibbert Journal* 54:242–252; esp.244.
46. Wittgenstein, L. 1966. Lectures on religious belief. In *Lectures and conversations on aesthetics, psychology and religious belief.* ed. Barett, C., 52–72 esp. 69. Oxford: Blackwell.
47. Wittgenstein, L. 1966. Lectures on religious belief. In *Lectures and conversations on aesthetics, psychology and religious belief.* ed. Barett, C., 52–72 esp.72. Oxford: Blackwell.
48. See Shibles, W. 1994. Anxiety: A pseudo-concept. *International Journal of Applied Philosophy* 9:43–52.
49. Mitford, J. 1998. *The American way of death revisited.* New York: Knopf; Harmer, R. 1963. *The high cost of dying*. New York: Crowell-Collier; Shibles, W. 1974. *Death: An interdisciplinary analysis*. Whitewater, WI: Language Press.
50. *DSM IV Diagnostic and statistical manual of mental disorders* 1994, 4th edn. Washington: 336.
51. Ramsey, I. 1967. *Talk of God. Royal Institute of Philosophy Lectures*, vol II. London: Macmillan. Preface.
52. MacKinnon, D.M. 1964. Death. In *New essays in philosophical theology*. eds. Flew, A., and MacIntyre, A., 262 New York: Macmillan.
53. The Holy Bible, New International Version, 2010 Grand Rapids, MI: Zonderbon. Job.17:26.
54. The Holy Bible, New International Version, 2010 Grand Rapids, MI: Zonderbon. Eccl.9:4.
55. The Holy Bible, New International Version, 2010 Grand Rapids, MI: Zonderbon. Eccl.7:1.
56. The Holy Bible, New International Version, 2010 Grand Rapids, MI: Zonderbon. Eccl.9:10.
57. The Holy Bible, New International Version, 2010 Grand Rapids, MI: Zonderbon. Ezek.9:4.
58. The Holy Bible, New International Version, 2010 Grand Rapids, MI: Zonderbon. Rom.6:5.
59. Kurtz, P. 1973/1984. *The humanist manifesto, I. and II and a secular-humanist declaration, 12th printing*. New York: Prometheus Books.
60. Lamont, C. 1965/orig. 1935. *The illusion of immortality*. New York: Ungar.
61. Aurelius Marcus 1964. *Meditations*. Long, G. tr, 41. New York: Doubleday/Penguin.

62. Aurelius Marcus 1964. *Meditations*. Long, G. tr, 90. New York: Doubleday/Penguin.
63. Aurelius Marcus 1964. *Meditations*. Long, G. tr, 36. New York: Doubleday/Penguin.
64. Aurelius Marcus 1964. *Meditations*. Long, G. tr, 65. New York: Doubleday/Penguin.
65. Aurelius Marcus 1964. *Meditations*. Long, G. tr, 42. New York: Doubleday/Penguin.
66. Heidegger, M. 1968. *Existence and being*, 3rd edn., 277. London: Vision; also Chicago, IL: Regenery.
67. Wittgenstein, L. 1961. *Notebooks 1914–1916*. ed. Von Wright, G., Anscombe, G. tr and ed., 15, 29ab. New York: Harper.
68. Wittgenstein, L. 1961/2001. *Tractatus logico-philosophicus*. Pears, D., and McGuiness, B. trs, 6.4311. London: Routledge; for discussion of the term "life" cf. Maier, B. 2005. Unsere Empfindsamkeit für und unsere Sorge um das Leben von Menschen auf dem Hintergrund moderner Biotechnologien. In: "Leben nach Maß" – zwischen Machbarkeit und Unantastbarkeit. Biotechnologie im Licht des Denkens von Albert Schweitzer . Schüz, G. ed., 115–142. *Beiträge zur Albert – Schweitzer – Forschung*, vol 10, Frankfurt am Main: Peter Lang Verlag. www.bundeskanzleramt.at/bioethik/Detail: http://www.bka.gv.at/Docs/2006/4/3/BMaier_Schüz.pdf.
69. For a full analysis of death, see Shibles, W. 1974. *Death: An interdisciplinary analysis*. Whitewater, WI: Language Press; Shibles, W. 1999. I Ching as a metaphorical method of insight. *Journal of Chinese Philosophy* 26:343–376.
70. We are all here using the word 'death,' which is a public instrument, which has a whole technique [of usage]." Wittgenstein, L. 1961. *Notebooks 1914–1916*. ed. von Wright, G.; Anscombe, G. tr. and ed., 68–69. New York: Harper.
71. Shibles, W. 1971. *An analysis of metaphor*. The Hague: Mouton; Shibles, W. 1999. I Ching as a metaphorical method of insight. *Journal of Chinese Philosophy* 26:343–376.
72. See also Shibles, W. 1971. Pictures of reason. In *Philosophical pictures*. Whitewater, WI: Language Press; Shibles, W. 1971. *An analysis of metaphor*. The Hague: Mouton; Shibles, W. 1999. I Ching as a metaphorical method of insight. *Journal of Chinese Philosophy* 26:343–376; Shibles, W. 1973. Die metaphorische Methode. Merz, K. tr, *Deutsche Vierteljahresschrift für Literaturwissenschaft und Geistesgeschichte* 48:1–9; Shibles, W. 1995. *Emotion in aesthetics*. Dordrecht, Netherlands: Kluwer Academic Publishers.
73. Maier, B. 2005. Unsere Empfindsamkeit für und unsere Sorge um das Leben von Menschen auf dem Hintergrund moderner Biotechnologien. In: "Leben nach Maß" – zwischen Machbarkeit und Unantastbarkeit. Biotechnologie im Licht des Denkens von Albert Schweitzer. Schüz, G. ed. *Beiträge zur Albert – Schweitzer – Forschung*, vol 10, , 115–142. Frankfurt am Main: Peter Lang Verlag. www.bundeskanzleramt.at/bioethik/Detail: http://www.bka.gv.at/Docs/2006/4/3/BMaier_Schüz.pdf.
74. Nagel 1986. *The view from nowhere*, 224. New York: Oxford University Press.
75. Thomas, D. 1953/1957. *Collected poems of Dylan Thomas*. New York: A New Directions Book.
76. Hazlitt, W. 1931. *On the fear of death. Essays*. ed. Gray, C., 302–313 New York: Macmillan.
77. Wilson, J. 1968. *Philosophy*, 17. London: Heinemann.
78. Shibles, W. 1989. An analysis of German emotive particles and interjections. *Papiere zu Linguistik* 40:71–81; Shibles, W. 1989. Some remarks on particles and interjections in English and German. *Zeitschrift für Anglistik und Amerikanistik* 37:241–245.
79. Shakespeare, W., *Measure for Measure* iii.l.119.
80. Wittgenstein, L. 1958. *Philosophical investigations*, 3rd ed., 222 New York: Macmillan.
81. Aurelius Marcus. 1964. *Meditations*. Tr. Long, G., 41 New York: Doubleday/Penguin.
82. Nielsen, K. 1970. In defense of atheism. In *Perspectives in education, religion and the arts*. eds. Kiefer, H., and Munitz, M., 143. Albany, NY: SUNY.
83. Epicurus 1963. *The philosophy of Epicurus*. ed. Strodach, G., 178–196: esp.196. Esp. Letter to Menoeceus: Northwestern University Press.
84. Wilson, J. 1968. *Philosophy*, 16. London: Heinemann.

85. Austin, J. 1964. Plea for excuses. In *Essays in philosophical psychology*. ed. Gustafson, D, 1–29; esp.17. New York: Doubleday.
86. Beardsley, M. 1962. The metaphorical twist. *Philosophy and Phenomenological Research* 22:293–307.
87. Dworkin, A. 1983. *Right-wing women: The politics of domesticated females*. New York: Putnam's Sons.
88. Wittgenstein, L. 1958. *Philosophical investigations*, 3rd edn, 129. New York: Macmillan.
89. Shibles, W. 1999. I Ching as a metaphorical method of insight. *Journal of Chinese Philosophy* 26:343–376.
90. Wittgenstein, L. 1961. *Notebooks 1914–1916*. ed. Von Wright, G.; Anscombe, G. tr. and ed., 84. New York: Harper.
91. Shakespeare, W. Julius Caesar. Act II, Scene II.
92. Wittgenstein, L. 1958/1965. *The blue and brown books*, 3–12. New York: Harper.
93. Carroll, L. 1960. *Alice's adventures in wonderland and through the looking glass*, 22. New York: New American Library.
94. Shakespeare, W., Macbeth II,ii,37.
95. Sartre, J.P. 1956. *Being and nothingness*. Tr. Barnes, H., 531–553 ("My death") New York: Philosophical Library.
96. Lakoff, G., and Johnson, M. 1980. *Metaphors we live by*. Chicago and London: University of Chicago Press.
97. Huseboe, S., and Klaschik, E. 2003. *Palliativmedizin*, 3rd edn. Berlin, New York, Tokio: Springer.
98. Shibles, W. 1974. *Death: An interdisciplinary analysis*. Whitewater, WI: Language Press. Last sentence of the preface.

Index

A
Abortion, 498
  argument, 345
Abstract definition, 18
Abstraction fallacy, 519
Abstractionism, 30
Abstractionist fallacy, 342, 460
Accidents, 362
Addictions, 189
Ad hominem fallacy, 30
Adult stem cells (AS), 337
  research, 346
Age-adjusted death-rate by cancer, 363
Aging, age, 365
AIDS, 365
Allowing death, 262
All-statements, 520
Altered nuclear transfer, 346
Altruism, 240
Altruism versus egoism, 230
Altruistic emotion, 236
Alzheimer's disease, 365
Analogy, 520
Analysis of prevention, 356
Anger, 146–148
Another "You", 336
Anti-inquiry, 87
Apathy, 142
Arguments from the philosophy of religion, 128
Assessment, 146
Assisted suicide, 272
Association, 45, 463
Autonomy, 68, 88, 281–295, 390–391

B
Bad doctor, 37
Beauty, 308–309
Beliefs, 335, 344
Beneficial lying, 493
Bioethical Principlism, 396
Bioethics, 82
Bioethics Consultation, 386
Biological life, 269
Biotheology, 129
Blame, 39, 54, 217, 246
Bodily feeling, 139, 143
Body, 299, 305–306
  language, 312
  parts, 310
Bonding, 203
Burnout, 189

C
Caesarean section, 92
Cancer prevention, 362
Captivation by a model, 25
Cardiovascular disease (CVD), 372
Care, 201, 206
  as emotion, 205
  theories, 202
Caring, 206
Case control study, 455
Case method of clinical ethics, 401, 403
Casuistry, 401–402
Categorical Imperative (Kant), 77, 411
Category mistakes, 521
Causation, 45
Cause(s), 18, 45–46, 463
  of cancer, 364
Cesarean section, 164
Circularity, 522
Circular statement, 16
Claim of certainty, 24
Clinical experience, 462
Clinical practice, 463
Cloning, 337
Cognition, 138–139

B. Maier, W.A. Shibles, *The Philosophy and Practice of Medicine and Bioethics*, International Library of Ethics, Law, and the New Medicine 47,
DOI 10.1007/978-90-481-8867-3, © Springer Science+Business Media B.V. 2011

Cognitive caring, 212
Cognitive emotive theory of bereavement, 512
Cognitive emotive theory of caring, 211
Cognitive-emotive theory (CTE), 138, 140, 154
Cognitive emotive theory of grief, 512
Cohort study, 454
A Common Faith (J Dewey), 119
Communication, 163
Complementary and Alternative Medicine (CAM), 113
Concept of the body, 300
Concept of continuity of care, 192
Confounding factors, 464
Conscience, 79
Consensus, 27, 384, 399
Consent, 288
Consequences of lying, 490–491
Consequentialism, 123
Consistency principle, 77
Context deviation, 523
Contradiction humor, 523–524
Coronary heart disease (CHD), 151, 372
Correlation, 51, 463
Created for research embryos, 346
Criteria for causation, 46
Criteria of death, 498
Critical choices, 286
Criticism of principlism, 396
Criticisms of the principle of autonomy, 282
Criticisms of principlism, 398–401
Culture, 65

D
Data-driven studies, 164
Death, 368
  denial, 516
  as language-game, 502–505
  and medical profession, 530
  of mentalistic meaning, 507
  personal remarks, 531
  requirement, 319
Decision-making, 165
Declaration organ donation policy, 374
Defense mechanisms, 525
Defensive medicine, 40
Definition(s), 15
  of altruism, 227–228
  of body, 300
  of care, 213
  of death, 497
  of EBM, 454
  of egoism, 227–228
  of embryo, 340
  of ethics, 336
  of lying, 488
  of morals, 336
  of philosophical counseling, 410
  of philosophy, 300, 412
Degrees of cause, 53
Deontic logic, 435
Derivation, 338, 344, 442
Descriptive statement, 73, 232
Deviation, 525
Dewey's Pragmatism, 408
Diabetes, 368
Dignity, 391–393
Direct killing, 262
Disease, 357, 368
Distal cause, 47
Dogma, 26, 31
Double-blind EBM experiment, 475
Double-blind randomized placebo controlled trial, 454
Drugs, 369–371
DSM-IV, 47, 152
Duty, 76
Dying, 510

E
EBM, 478
EBM prayer, 130–131
EBM as statistics, 457
Economic incentive approach for organ donation, 326
Education, 362
Effect(s), 48
  of overwork, 187
  of sleep deprivation, 194
Egoism, 240
Egoistic emotion, 236
Either-or fallacy, 31
Embarrassment, 95
Embryonic development, 338, 341
Embryonic germ cells, 337
Embryonic stem cells (ES), 337
  research, 349
Emotion, 138–141, 144, 166, 415–416, 510
  of bereavement, 513
  of grief, 513
  of interest, 155
  literacy, 166
Emotional abuse, 153
Emotional causation, 53
Emotional harm, 153
Emotional intelligence, 141
Empathy, 219–221

Enculturation, 66
Enlightened pacifism, 255
Enterprise liability, 38
Epidemiology, 45
Epistemological primacy of language, 430, 506
Equality, 68, 261, 327
Equivocation, 17
Errors, 167, 373
Ethical caring, 205
Ethical examination, 343
Ethical management, 162
Ethical methods, 407
Ethical statement, 73
Ethical terms, 74
Ethics, 73, 80–81, 123
  of the body, 313
  committees, 121, 384
  consultation, 407
  counselor, 387
  literacy, 168
  of using placebo, 477–478
Ethics counseling (EC), 386
Euphemism, 525
European Bioethics, 388
European Biolaw, 388
Euthanasia, 271–272
Evidence, 51
  definition, 456
Evidence-based medicine (EBM), 453
Evidence-based nursing, 467
Evidence-based spirituality, 467
"Excentricity of man", 312
Existential Theory approach, 411
Exploration of the concept of death, 519

F
Face, 309–310
Facts, 455
Fallacy of abstraction, 435–436
Family, 86, 325
  values, 85
  approval for organ donation, 324
Federal funding for embryonic stem cell research, 348
Feeling, 138, 212
Female circumcision, 81, 100
First person caring, 214
Formal logic, 431–437, 439–442, 444
Formal logical fallacy, 435
Form of life, 394, 428, 508

G
Gender, 308
  medicine, 468
General definitions of death, 498
Genes, 367
Golden standard, 455
Good physician, 137
Good Samaritan law, 265
Guilt, 511

H
Hand-washing, 369
Hawthorne Effect, 475
Health, 357–358, 406
Healthcare worker, 11
  patient relationship, 12
Healthful lifestyle, 361
Health insurance, 253–254, 376
Healthy philosophical lifestyle, 189
Helper's syndrome, 221
Hippocratic Oath, 115, 254–257
HIV, 365
Holistic philosophy of medicine, 405
Hope, 150
Hospital on strike, 181
Humaine medicine, 350
Human being, 342
Humanism, 70, 120, 122–123, 169, 222, 241, 409
Humanistic care, 205
Humanistic view of death, 516
Human papillomavirus, HPV, 366
Humor, 9, 150
Hunger, 355, 357
Hypocrisy, 493–494
Hypothesis, 50
Hysterectomy, 373

I
Ideal autonomy, 285
Ignorance, 30
Illustrations of death, 510
Imagery of death, 509
Incomplete information, 286
Indirect killing, 262
Individual case, 461
Indoctrination, 55
Informal logic, 436
Informal logical fallacies, 29, 404
Insensitivity to killing, 95–96
Institutional structures, 403
Integrity, 393–395
Intellectual illiteracy, 371
Intrinsic, 75
Intuitionism, 32
Irony, 526
Irrational forms of caring, 208–211

Irrational medical thinking, 24
Irresponsible lifestyle, 360
Is-ought fallacy, 90
IVF left over embryos, 346

J
Job satisfaction, 407
Justifiable lies, 491
Juxtaposition, 526

K
Kant, 76
Karoshi, 184
To kill, 96
Killing, 245, 249, 259, 261, 263

L
Lack of exercise, 366
Language, 429
  game, 394, 427, 508, 518
  use, 139
Language-game of causation, 55
Laws, 78
Legal causality, 55
Leib, 311–312
Let die, 245–248, 262
Letting die, 263–264, 270
Lie, 488
Life, 268–269, 303, 342
  expectancy, 356
Linguo-centric predicament, 430, 443
Literal definition, 16, 20
Literalist fallacy, 58
Live, 268
Logic of flattery, 493
Longevity, 367
Lottery, 325

M
Majority rule, 260
Malpractice insurance, 40
Management, 161
Mandatory organ donation declaration, 329–331
Many question fallacy, 33
Mathematics, 431–433
McNaughton rule, 82
Meaning(s), 438, 444
  of life, 304
  of cause, 47
Medical language, 8
Medical risk, 260
Medicine, 67
Medicotheology, 129
Mental disorders, 374–375
Mentalism, 153, 434
Mentalistic fallacy, 139, 151, 505
Mental reservation, 492
Metaemotion, 217
Meta – metaphor, 5
Metaphor, 2, 443–444
  about death, 499–501
  to myth fallacy, 6
Metaphorical definition, 20
Metaphorical method, 1, 517–519
Metaphorical models of cause, 49
Metaphorization, 526
Metaphysical causes, 117
Methods of philosophical counseling, 410
Military medical service, 92
Military medicine, 94
Mind, 139, 305
Misdiagnosis, 38
Misleading metaphors, 345
Mismanagement, 174
Mistakes, 167, 373
Misuse of ethical terms, 74, 249
Misuse of language, 16
Models of management, 169
Moral caring, 205
Moralization of embryos, 341
Moral(s), 80
  management, 162
  statement, 232
  status of an embryo, 341
Mother-infant bonding, 204
Murder, 262
Mutuality of caring, 218

N
Narrative, 8, 393
  ethics, 393–394
  method, 411
  theory, 306
National Institute of Health (NIH), 113
Naturalistic ethics, 202, 408
Naturalistic ethics of the body, 302
Naturalistic fallacy, 71, 90
Naturalistic theory of ethics, 69, 260
Negative emotions, 142, 137, 270
Negative emotions as philosophical language fallacies, 147
Negligent causality, 58
New Testament, 515
New theory of lying, 488

Nocebo
  definition, 471
  as emotion, 471
Noddings
  care theory, 202
  caring, 203
Noncausal events, 59
None-statements, 520
Non-ethical, 68
No non-linguistic knowledge of death, 507
Nosocomial infections, 369
Number of work hours, 183

O
Obedience, 26
Objectivist Egoism (A. Rand), 233–235
Old Testament, 514–515
On-call, 185
Open communication, 26
Ordinary language approach, 72, 428
Ordinary language basis of altruism, 235
Ordinary language basis of egoism, 235
Ordinary language case method, 404
Oregon Healthcare Prioritizing, 250
Organ donation failure, 374
Organ replacement waiting list, 317
Other, 231
Outcomes, 170
Overgeneralization, 459
Oversimplification of causation, 60
Overtime, 186
Overwork, 38, 170, 182, 185, 188–190, 192–193
Oxymoron, 6

P
Pacifism, 254
Pain, 143–144, 308
Paradigms of the body, 302
Passionate rational emotion, 142
Paternalism, 287
Patient, 11
  code of ethics, 291
  duties, 293
  error, 38
  responsibility, 291
The Patient's Hippocratic Oath, 218
Perception, 23, 301
Perceptual metaphor, 3
Person, 300
Personal caring, 210, 214
Personality, 309
Personal life, 269
Personal lifestyle problems, 378
Personhood, 343
Personification, 5, 527
Person as a soul, 125
The philosophical body, 312
Philosophical counseling, 274, 406, 412
Philosophical critique of EBM, 479
Philosophical EBM, 456
Philosophical practice, 404, 409, 412
Philosophy, 67, 383
  of EBM, 454
  education, 414
  and Ethics of Personality, 223
  of language fallacies, 148
  of medicine, 1, 28, 302
  of medicine-based medicine, 479
  of medicine counseling, 413
  practitioner, 415
  of religion, 115, 128, 514
  of science, 385
Phronesis, 46
Physicians on strike, 173, 179
Pity, 149
Placebo, 472
  definition, 471
  as emotion, 471, 478
Placebo/nocebo effect, 471, 473
Pluripotent cells, 338
Poetic metaphor, 527–528
Poet's view of death, 499–502
Positivist law, 396
Poverty, 356, 376
Power, 171
Pragmatic case method, 403
Pragmatic definition of cause, 60
Pragmatic Theory of Ethics, 168
Pragmatism, 442
Prayer, 130
  as medical treatment, 129
Precautionary principle, 340
Preference utilitarianism, 262
Preferential treatment, 250
Pregnant woman, 311
Presumed consent, 321
Presumed organ donation, 322–323, 331
Presumed refusal to organ donation, 320
Preventable mistakes, 35–36
Prevention, 252, 357–359
  of cancer, 363
Primacy of ordinary language, 442
Principle(s), 389, 396–397
  of autonomy, 281
  of non-contradiction, 77
Probability, 463, 528

Propositions, 438–439
Proximal cause, 47
Psychiatric patient autonomy, 291
Psychiatry, 465
Psychosomatics, 143, 470
Publicity, 171

Q
Qualified manager, 162
Quality, 172
 of life criteria, 127
Quantitative cause, 60
Quantity, 172

R
Rational altruism, 241
Rational bonding, 204
Rational caring, 215, 222
Rational egoism, 241
Rational Emotive and Behavioral Theory (REBT), 211
Rational emotive theory of emotion, 511
Rationalization, 33
Rational medicine, 469–470
Reason(s), 54
 for mistakes, 36
Reciprocity of caring, 205
Reductionism, 25
Relativism, 412
Release theory of caring, 216
Release theory of emotions, 145
Religion, 73, 112, 119–120, 122
Religious, 119
Religious beliefs, 111
Reproduction, 311
Requirements for good management, 163
Research donors, 348
Respect, 267, 288
Response, 52
Responsibility, 290
Resurrection, 515
Reverence for life, 267
Rhetoric of death, 517
Riddles, 529
Right, 75
Risk–benefit balance, 340
Risky behavior, 375

S
Sacred, sanctity, 127
Salary schedules, 173
Salary transparency, 173
Salvage value, 93
Samaritan help, 264
Sanctity-of-life doctrine, 125–126
Scarce medical resources, 376
Science, 408
Scientific method, 301
Seeing-as, 3
Self, 11, 231, 266, 307, 428–429
Self-cause, 61
Self-caused illness, 251
Self-determination, 290
Selfishness, 233–234, 239
Self as language construct, 305
Self-lie, 489
Self-pity, 149
Self, selves, 306
Self-talk, 435
Sensitivity, 97
Sex, 308
Sexuality, 311
Sexually transmitted disease, 366
Sick leave, 188
Significance, 51
Situational ethics, 403
Sleep deprivation, 190–191
Social death, 515
Social responsibility, 395
Sociological theory of law, 396
Solidarity, 395
Somatic cell nuclear transfer (SCNT), 347
 embryo, 338
Staffing, 175
Starvation, 355, 376–377
Statistical decision-making, 27
Statistical evidence, 458
Statistical malpractice, 458
Stem cells, 337
Steps of EBM research, 453
Stimulus, 52
Stoics, 142
Story, 306
Strike of physicians, 191
Structural life style problems, 378
Suicide, 175, 270–273
Supernaturalism, 112, 117
Syllogism, 441
Symbolic logic, 434
Sympathy, 238–239
Synonyms of cause, 48

T
Teleological fallacy, 62
Temporal factors in causality, 49
Therapeutic cloning, 337, 345
Therapeutic metaphor, 11

Thinking, 429, 435
Thought, 139
Total quality management, 217
Totipotent cells, 338
Toxins, 370
Transplantation, 116
  medicine, 310
  waiting list, 318
Triage, 248
"Trolley example", 257, 261
"Trolley example" equal, 258
"True", 16
Truth, 440
Tsunami disaster, 98
Types of causality, 50
Types of definition, 18
Types of metaphor, 4
Types of understanding, 177

U
Uncaring, 215
Uncritical use of EBM, 462
Understanding in management, 176
Unified Field Theory of Science, 436
The Uniform Determination of Death Act, 319
Unique emotions, 145
Universalization, 79
Universals, 78
Universe of discourse, 19
Unneeded operations, 373
The un-philosophical body, 307
Use in context definition, 21
Use of language, 428
Utilitarian altruism, 233
Utilitarian formula, 233, 260
Utilitarianism, 74, 259, 266

V
Value, 127
  contradictions, 83
  of human life, 126
  of life, 303
  terms, 237
Vulnerability, 395

W
Wants, 70
Will, 76
Withholding treatment, 247
Word field, 19
Word field of altruism, 229
Word field of autonomy, 282
Word field of caring, 207
Word field of egoism, 229
Work, 182–183

Y
You-also fallacy, 34

Breinigsville, PA USA
14 December 2010
251335BV00012B/31/P

9 789048 188666